AF566925

JOURNAL OF CHROMATOGRAPHY LIBRARY – volume 20

electron capture

theory and practice in chromatography

JOURNAL OF CHROMATOGRAPHY LIBRARY

Volume 1 Chromatography of Antibiotics
by G.H. Wagman and M.J. Weinstein

Volume 2 Extraction Chromatography
edited by T. Braun and G. Ghersini

Volume 3 Liquid Column Chromatography. A Survey of Modern Techniques and Applications
edited by Z. Deyl, K. Macek and J. Janák

Volume 4 Detectors in Gas Chromatography
by J. Ševčík

Volume 5 Instrumental Liquid Chromatography. A Practical Manual on High-Performance Liquid Chromatographic Methods
by N.A. Parris

Volume 6 Isotachophoresis. Theory, Instrumentation and Applications
by F.M. Everaerts, J.L. Beckers and Th.P.E.M. Verheggen

Volume 7 Chemical Derivatization in Liquid Chromatography
by J.F. Lawrence and R.W. Frei

Volume 8 Chromatography of Steroids
by E. Heftmann

Volume 9 HPTLC — High Performance Thin-Layer Chromatography
edited by A. Zlatkis and R.E. Kaiser

Volume 10 Gas Chromatography of Polymers
by V.G. Berezkin, V.R. Alishoyev and I.B. Nemirovskaya

Volume 11 Liquid Chromatography Detectors
by R.P.W. Scott

Volume 12 Affinity Chromatography
by J. Turková

Volume 13 Instrumentation for High- Performance Liquid Chromatography
edited by J.F.K. Huber

Volume 14 Radiochromatography. The Chromatography and Electrophoresis of Radiolabelled Compounds
by T.R. Roberts

Volume 15 Antibiotics. Isolation, Separation and Purification
edited by M.J. Weinstein and G.H. Wagman

Volume 16 Porous Silica. Its Properties and Use as Support in Column Liquid Chromatography
by K.K. Unger

Volume 17 75 Years of Chromatography — A Historical Dialogue
edited by L.S. Ettre and A. Zlatkis

Volume 18 Electrophoresis. A Survey of Techniques and Applications. Part A: Techniques
edited by Z. Deyl

Volume 19 Chemical Derivatization in Gas Chromatography
by J. Drozd

Volume 20 Electron Capture. Theory and Practice in Chromatography
edited by A. Zlatkis and C.F. Poole

JOURNAL OF CHROMATOGRAPHY LIBRARY - volume 20

electron capture

theory and practice in chromatography

edited by

A. Zlatkis

Chemistry Department, University of Houston, Houston, TX 77004

C.F. Poole

Chemistry Department, Wayne State University, Detroit, MI 48202

ELSEVIER SCIENTIFIC PUBLISHING COMPANY
Amsterdam – Oxford – New York 1981

ELSEVIER SCIENTIFIC PUBLISHING COMPANY
1, Molenwerf, 1014 AG Amsterdam
P.O. Box 211, 1000 AE Amsterdam, The Netherlands

Distributors for the United States and Canada:

ELSEVIER/NORTH-HOLLAND INC.
52, Vanderbilt Avenue
New York, NY 10164

ISBN 0-444-41954-3 (Vol. 20)
ISBN 0-444-41616-1 (Series)

© Elsevier Scientific Publishing Company, 1981
All rights reserved. No part of this publication may be reproduced, stored in a retrieval system or transmitted in any form or by any means, electronic, mechanical, photocopying, recording or otherwise, without the prior written permission of the publisher, Elsevier Scientific Publishing Company, 1000 AH Amsterdam, The Netherlands

Printed in The Netherlands

Contents

List of contributors

U.A.Th. Brinkman, Department of Chemistry, Free University, De Boelelaan 1083, Amsterdam, The Netherlands

F. Bruner, Istituto di Scienze Chimiche, Università di Urbino, Piazza Rinascimento 6, 61029 Urbino, Italy

D.I. Carroll, Institute for Lipid Research, Baylor College of Medicine, Houston, TX 77030, U.S.A.

E.C.M. Chen, School of Science and Technology, University of Houston at Clear Lake City, Houston, TX 77058, U.S.A.

W.P. Cochrane, Laboratory Services Division, Food Production and Inspection Branch, Agriculture Canada, Ottawa, Ontario, Canada

I. Dzidic, Institute for Lipid Research, Baylor College of Medicine, Houston, TX 77030, U.S.A.

F.C. Fehsenfeld, Aeronomy Laboratory, Environmental Research Laboratories, National Oceanic and Atmospheric Administration, Boulder, CO 80302, U.S.A.

P.D. Goldan, Aeronomy Laboratory, Environmental Research Laboratories, National Oceanic and Atmospheric Administration, Boulder, CO 80302, U.S.A.

E.P. Grimsrud, Department of Chemistry, Montana State University, Bozeman, MT 59717, U.S.A.

E.C. Horning, Institute for Lipid Research, Baylor College of Medicine, Houston, TX 77030, U.S.A.

R.E. Kaiser, Institute of Chromatography, Bad Dürkheim, G.F.R.

F.W. Karasek, Chemistry Department, University of Waterloo, Waterloo, Ontario, Canada.

J.E. Lovelock, Department of Engineering and Cybernetics, University of Reading, Berkshire, Great Britain

R.B. Maybury, Laboratory Services Division, Food Production and Inspection Branch, Agriculture Canada, Ottawa, Ontario, Canada

M.P. Phillips, Department of Chemistry, Cooperative Institute for Research in Environmental Sciences, University of Colorado, Boulder, CO 80309, U.S.A.

C.F. Poole, Department of Chemistry, Wayne State University, Detroit, MI 48202, U.S.A.

R.I. Rieder, Institute of Chromatography, Bad Dürkheim, G.F.R.

R.E. Sievers, Department of Chemistry, Cooperative Institute for Research in Environmental Sciences, University of Colorado, Boulder, CO 80309, U.S.A.

P.G. Simmonds, Organic Geochemistry Unit, School of Chemistry, University of Bristol, Bristol, Great Britain

G.E. Spangler, Bendix EPID, Baltimore, MD 21204, U.S.A.

R.N. Stillwell, Institute for Lipid Research, Baylor College of Medicine, Houston, TX 77030, U.S.A.

J. Vessman, Department of Analytical Chemistry, AB Hässle, Mölndal, Sweden

W.E. Wentworth, Department of Chemistry, University of Houston Central Campus, Houston, TX 77004, U.S.A.

A. Zlatkis, Department of Chemistry, University of Houston, Houston, TX 77004, U.S.A.

Preface

Since its introduction by Lovelock, about 20 years ago, the electron-capture detector (ECD) has had an enormous impact in trace analysis by gas chromatography. Nowhere is this more apparent than in environmental and biomedical studies, where the detector was often foremost in forging the link between minute concentrations of biologically active substances and their physiological responses. Although new detectors have appeared since the introduction of the ECD, none has succeeded in replacing it in many of its original applications. Its use remains dominant in such areas as the study of the concentration and fate of pesticides in the environment, the impact of fluorocarbon aerosol propellants on stratospheric ozone concentrations, the health risk from toxic chemicals in the workplace and the vicinity of disposal sites, the therapeutic monitoring and metabolic fate of drugs in man and for the determination of biologically active molecules such as hormones in physiological fluids.

Although deceptively simple in design and operation, the very high sensitivity of the ECD brings with it a certain number of problems. Most analysts have mixed feelings concerning its general use; it is either a "trusted friend" or "fiendish enemy". This is often a judgment fashioned by time, as confidence comes with experience. However, it is not necessary that experience be obtained at the expense of time and labor. This book was designed specifically to provide, in one source, the most comprehensive coverage of the theory, practice, and major application areas of the ECD in chromatography. To make this work possible, the editors consider themselves very fortunate in being able to draw upon the combined talent and experience of an internationally recognized team of distinguished authors.

As by right and merit, we leave the introductory remarks to the inventor and chief practitioner of electron-capture detection chromatography. We hope this personal odyssey and historical framework will set the scene for the remaining chapters. The theory of the detection mechanism and the construction and operating principles of the detector are reviewed in the next two chapters. As not all compounds are chromatographically stable or have a significant detector response, a method is required to bring these compounds within the compass of the detector. Derivatization techniques are described for both organic and inorganic compounds in two companion chapters. An additional chapter highlights natural electrophores with structures not immediately associated by many analysts with a large detector response. The ECD can also be used as an ion–molecule reactor by doping the carrier gas with a collision gas of high cross-section for thermal electrons. Sometimes called selective electron-capture sensitization, this new technique offers a great deal of promise for the sensitive detection of molecules with only a low ECD response. Chapters on atmospheric pressure ionization mass spectrometry and plasma chromatography are included as these techniques, as well as being sensitive analytical procedures in their own right, provide useful information of the chemical structure of the products generated in the electron-capture process. The remaining chapters of the book deal with the calibration of the ECD, its use as a detector for liquid chromatography and a description of its major application areas in environmental and biomedical studies.

It is hoped that all analysts, whatever their level of experience, will find something of value in this book.

Houston ALBERT ZLATKIS
Detroit C.F. POOLE

Chapter 1

The electron-capture detector – A personal odyssey

JAMES E. LOVELOCK

When I first came to Houston in 1958 and felt the touch of its warm humid air on my skin it seemed like a place where interesting things might happen. On that visit I travelled light carrying only a briefcase holding my pyjamas, a toothbrush and two small items of high technology: a gas chromatographic column of unusually high resolving power and an ultra-sensitive vapor detector for use with it. I was met at the airport by Al Zlatkis and taken directly to his laboratory at the University of Houston. There we set it up and in no time were running samples of products from the Houston Petrochemical Industry; mixtures called "platformates" and "reformates" which I had never heard spoken of before. They gave the most glorious and unbelievably excellent chromatograms resolving components hitherto only suspected to be present. It was not high class pure science but it surely was one of those days when everything goes right technically and life as a scientist seems to be very worthwhile.

That visit to Houston set the pattern of what was to become for me, not just an odyssey but a way of life. Ever since I seem to have been travelling somewhere with a small and simple apparatus seeking to measure something but usually finding something else which was more interesting instead. The great simplicity of its construction combined with an exquisite sensitivity has made the electron-capture detector a perfect companion for this kind of travel. This introductory chapter concerns the origins of this device, and about my travels around the world with it and also of the arguments its use has engendered.

It all began in 1948 when I was working for the British Medical Research Council on the problem of the common cold. In those days precious little was known about the science of this subject but as often is the way in the states of scientific ignorance the man in the street was quite sure he knew all that there was about the common cold. To him it was quite simple, you caught colds in the winter by getting cold hence the name. Now the Medical Research Council was a government institution and therefore not entirely unaffected by public opinion and political pressure. We thought that it might be wise therefore to consider the possibility that colds were in fact caught through getting cold. My job was to determine the extent of chilling objectively so that it could be compared with the clinical data on the frequency of colds in different indoor climates. The three factors important in chilling are temperature, humidity and air movement. The first two are easy enough to measure but the air movements in a closed room, draughts as the English call them, were so slight as to be undetectable by the simple anemometers then available.

In those days of long ago it was customary to build rather than to buy instruments. Indeed scientists were expected to invent and in such circumstances I soon found myself with two novel anemometers. The first was an ultrasonic device which exploited the change in wavelength of sound due to air motion. It worked very well but was still too insensitive to detect the one foot per minute air motion we needed to measure. The second method I tried was an ion drift anemometer. Positive ions move in air at a speed of 1 cm/s in a field of 1 V/cm. The slow drift of these ions was easily perturbed by winds of the speed we needed to measure. It was great fun to make such an anemometer and to find that it worked even better than expected [1].

When I say "make", I mean it literally. Everything from the electronic equipment, which in those days used fire bottles, to the sensing head had to be made by hand. Even the radioactive source needed to ionise the air was made by scraping the dial paint from some gauges taken from the flight deck of World War II military aircraft. These provided a rich harvest of radium or mesothorium. I made the sources by ashing this paint and then resuspending the ash in a thin lacquer which was painted onto the anemometer ion source. Fig. 1.1 illustrates one of these ionization anemometers.

It worked well and I was even able to take it on an expedition to the Arctic in winter time. The only snag was that its response was perturbed by cigarette smoke, indeed it was as sensitive to this as is a reformed smoker. To discover the source of this drawback I tried exposing it to a number of different gases and smokes and found that in addition to

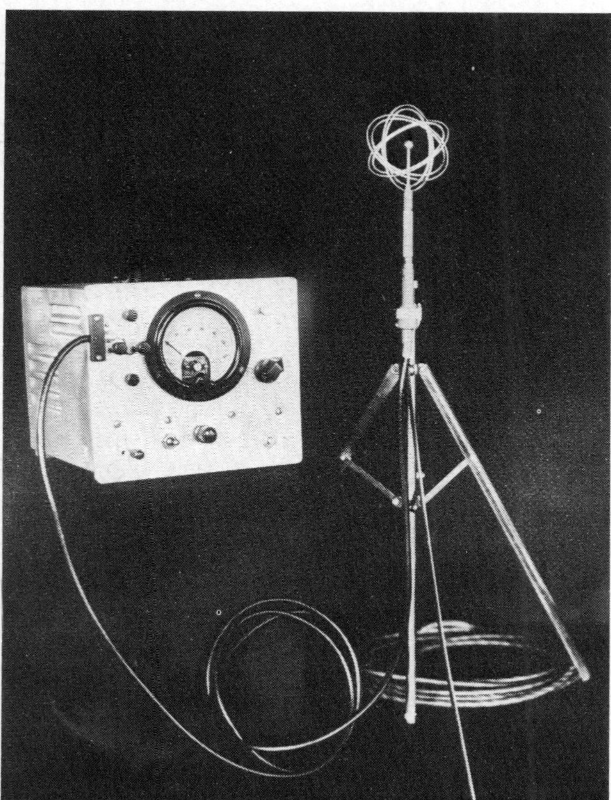

Fig. 1.1. Ionization anemometer.

smoke, halocarbon gases also disturbed its function. Looking back I realise that the key to invention is need. We did not at that time need a device to detect low levels of chlorofluorocarbons and so the electron-capture detector was in a sense prematurely invented.

In 1951 having by then discovered little more about that enigmatic disease the common cold, other than it was not caught by chilling, I was moved back to our parent institute in London. My new task was to work on the preservation of life in the frozen state. It was splendid, spectacular and performing circus sort of science. It culminated in 1953 with the successful freezing and reanimation of a golden hamster [2]. This would seem to have little to do with gas chromatography or electron capture but in the strange and serendipitous ways of science it so happened that a key factor in determining the resistance of a living cell to freezing was the composition of the lipids of the cell membrane. Cells whose membrane lipids were highly saturated were more easily damaged by freezing than were those whose membrane lipids were rich in polyunsaturated fatty acids. Before gas chromatography the analysis of fatty acids was a slow and laborious business requiring samples much larger than the few milligrams we then had available. By great good fortune my laboratory was only a few yards away from that of Martin and James where they were using their newly invented gas chromatograph to analyse fatty acid methyl esters. It was not long before a collaboration was established that not only answered the fatty acid problem but also led to the development of a range of sensitive ionization detectors, which Archer Martin knew were needed to exploit the full potentialities of gas chromatography.

Perhaps I should say that the National Institute for Medical Research in London in those days was a very unusual Laboratory. There were very few barriers to prevent free collaboration between scientists of all disciplines. Such freedom no doubt accounted for the phenomenal productivity it then had.

The first ionization detector I made was modelled on Boer's design of the ionisation cross section detector. It was in effect a gas thickness gauge. The denser the gas the greater the number of ions and consequently the larger the flow of current. This detector works best with light carrier gases such as helium or hydrogen; indeed it is very similar to the thermal conductivity detector in its performance. Helium was very expensive in Europe and hydrogen was unacceptable for use in a high-temperature apparatus expected to run overnight unattended. I was obliged therefore to use nitrogen as the carrier gas as did Martin and James with their gas density balance. It was easy enough to confirm Boer's performance figures but these were miserably insensitive when compared with those of Martin's gas density balance. The first ionisation detector did not seem to be promising. Sometimes when confronted with a failed experiment or an unsatisfactory device it is better to cut ones losses and pass on to something different. In this instance, however, I remembered the success of the ion anemometer and of how its sensitivity was very dependent upon the applied potential I thought it at least worth trying a few experiments to see if different ranges of applied potential would improve for the better the performance of the ionisation cross section detector.

It was easiest to try low potentials first. I soon found that if the detector was polarised with less than 30 V the ion current in pure nitrogen became a little less but that when other substances were present very much less. Tony James had supplied me with a test mixture of fatty acid methyl esters, from methyl propionate to methyl caproate. With the detector operated at its recommended potential namely, 100 to 300 V, 1 mg of this mixture gave four small peaks. When I tried it with only 10 V and the recorder connection reversed so as to reveal negative peaks positively, the 1-mg sample gave what seemed to be a never ceasing range of offscale peaks. I thought that the search was over and that now we had a truly sensitive detector. I asked Tony James and Archer Martin to come and try it, which they did, bringing with them an allegedly pure sample of methyl caproate. I shall never forget the look of amazement on Tony James' face as peak after peak was drawn from a small sample of this substance and none of them with retention time of methyl caproate or of any other fatty acid ester. We now know that what was seen were traces of electron absorbing impurities in the sample but at that time it seemed to be a useless and wholly anomalous device.

In spite of this disappointment I continued to play with it whenever there was time and by trying compounds chosen at random from the lab shelves I discovered a certain sort of sense in its behaviour. It seemed to respond to polar compounds like ketones and alcohols but not to hydrocarbons and ethers. But when I tried a mixture of compounds made up in the nonpolar solvent carbon tetrachloride the ion current fell to zero and there remained, resisting all attempts to restore normal operation. Looking back I realise how unwise it was to connect the column to the detector using a silicone rubber seal which became an almost permanent source of the vapour of that intensely electron absorbing substance carbon tetrachloride.

For the ordinary gas chromatograph we clearly needed something more sensitive than the original ionisation detector but less temperamental than the electron-capture detector. I wondered if other ionisation processes might be exploited to detect the vapors of organic compounds in a nitrogen carrier gas.

This line of thought led by a happy accident to the discovery of the "argon" detector [3]. This was a device which exploited the ionizing collisions between rare gas metastable atoms and vapor molecules for detection. For a brief few years this was the principal sensitive detector used in gas chromatography. It was, of course, replaced by the even better flame ionisation detector. During the spring of 1959 I made a brief visit to New York to read a paper on the "argon" detector and there met Sandy Lipsky. He invited me to visit his laboratory at Yale for the coming academic year. My first thought was to decline this invitation for we had had as a family spent a rather difficult year in Boston in 1955 as a young post doctoral research fellow. Sandy Lipsky soon convinced me that things at Yale would be different and indeed they were. Kindness, hospitality and an environment conducive to creative work all were freely provided. At Yale, relieved of the need to spend most of my time on biochemical problems I was able to develop the argon detector especially for use with capillary columns.

The other main line of work in New Haven was the development of the electron-capture detector. We chose to develop it as a qualitative device, especially as one which could be used to distinguish functional groups among the vast range of compounds resolved by capillary columns from natural products [4]. At that time the mass spectrometer was too expensive and too insensitive for this need. Another reason for applying this detector to qualitative analysis was that we rather doubted its utility in quantitative analysis. The versions polarised by a small fixed potential were erratic in performance and sometimes even gave signals which were entirely false and irreproducible.

During this time at Yale, the key to the cure of the electron-capture detector's bad habits came from an encounter with Dr McAffee of Bell Telephone Laboratories. He had developed a pulse method for observing electron attachment in a drift tube. From this idea it occurred to me that most of the difficulties with the detector could be resolved if the electrons were collected by brief pulses of high potential rather than by letting them drift under the influence of a weak electrical field. The high-potential pulses overcame the all too frequent contact potentials and space charges which unpredictably either enhanced or opposed electron collection.

When I returned to London in 1959 I was able to spend more time in developing the electron-capture detector further and also on other ionisation detectors such as the photoionisation and the triode argon detector. So far as I am aware the first application of the electron-capture detector to environmental problems was by a group at the Shell Laboratories in Kent: Goodwin et al. [16], although this work was either simultaneous with or closely followed by that of Watts and Klein [17] of the U.S. FDA. In spite of the difficulties of using the primitive detectors then available these capable professionals successfully and accurately applied the method to the important environmental problem of pesticide residue analysis. The demonstration by their work and that which followed it of the ubiquitous distribution of pesticides throughout the global environment did much to fuel the environmental revolution which followed. They lent veracity to the otherwise unprovable statements of that remarkable book by Rachel Carson, Silent Spring. One thing about those early days of that revolution not often sufficiently mentioned is I think the positive role of those scientists of the multinational chemical industry and the government agencies.

As the experience in the analysis of different molecular species by electron capture became established an odd and interesting association between electron capture and biological activity grew ever more apparent. The great bulk of electron absorbing substances fell into two categories, those which were important components of the biological system of energy transport or those which were highly toxic, especially to this system or as carcinogens [5].

It was tempting to speculate that the free electron might be a fundamental particle of biology as well as of chemistry and physics. It was a challenging coincidence that each alternate acid of the Krebs cycle was one of the very few organic compounds which reacted vigorously with free electrons. These included pyruvate, oxaloacetate, fumarate, ketoglutarate and *cis*-aconitate. Not only these but also the coenzymes ubiquinone and nicotinamide adenine dinucleotide were also electron attaching as were compounds such as the thyroid hormones which are able to mediate in this system. Finally so were the iodo- and nitrophenols which are toxic and act by uncoupling oxidative phosphorylation. It is still unclear as to whether this association is real or coincidental but there is no doubt that a remarkably high proportion of electron absorbers are biologically active and it is this which has made it so important a device in environmental science.

Whenever nowadays I come across a chemical substance which is strongly electron absorbing I tend to regard it cautiously. I well recall it being argued that the apparent association between carcinogenesis and electron capture was illusory since so many of the halocarbons were not carcinogenic. Vinyl chloride and trichloroethylene were both quoted as examples of substances safe enough for use as anesthetics in surgery. Now of course we know them to be carcinogenic. After I had written this paragraph I again experienced that feeling of déja vu on reading in Chemical and Engineering News that certain of the phthalate esters were suspect as a health hazard. These ubiquitous plasticisers have long been a nuisance to analysts as an electron absorbing contaminant. Do they also have a more sinister role?

In 1961 I came to Houston again at the invitation of Evan Horning to participate in the work of his newly established Lipid Research Laboratory at Baylor College of Medicine. This provided at last an opportunity to spend full time in discovering how the electron-capture detector really worked. In this I was fortunate to have nearby both Dr. Zlatkis who by then had become a close friend and colleague and Drs. Wentworth and Chen of the University of Houston. This fruitful collaboration led to the first plausible kinetic model of the electron-capture detector [6]. Unlike so much theoretical modelling, the work by Wentworth and his colleagues led to practical improvements in the use of the detector, not least of which was its application to the direct analysis of atmospheric nitrous oxide by using the detector at a temperature of 300°C or higher [7]. Measurements of nitrous oxide and other atmospheric gases brings me to that part of my story which concerns expeditions to remote places with a home-made portable electron-capture gas chromatograph to discover trace quantities of interesting compounds.

In 1966 we purchased a holiday cottage in far western Ireland and during the first summer there I noticed that whenever the wind came from a certain quarter the air was very hazy. Fig. 1.2 is a view across Bantry Bay on a clear day; Fig. 1.3 on a day when the wind was from the East. The haze looked and smelt like smog. When I returned to England I asked colleagues who were meteorologists or atmospheric chemists if it was possible for smog to travel the 1000 kilometres from Europe to Ireland. They were sceptical and suggested instead that what I had seen was more probably a natural exhalation of the Irish Bogs. I was not convinced by their plausible arguments and next summer,

Fig. 1.2. Bantry Bay on a clear day.

Fig. 1.3. Bantry Bay when wind was from the east.

much to the annoyance of my wife and family, I took with me on holiday a portable gas chromatograph. My objective was the measurement of the atmospheric concentration of fluorocarbon 11 (trichlorofluoromethane) before, during and after one of the outbreaks of hazy air. I chose to look for fluorocarbon 11 (F11) rather than a typical smog chemical such as ozone or peroxyacetyl nitrate because the fluorocarbons alone are unequivocally man made. The other substance I knew would be regarded by my doubting colleagues as possible emissions from natural sources.

The measurements soon established the presence of a small steady background concentration of F11 in the clear air, about 50 parts per trillion by volume. When the haze appeared the concentration rapidly rose to the region of 150 parts per trillion and there remained until the wind changed direction and clear air returned. The measurements proved that the haze was indeed man made smog and later observations revealed a close correlation between the concentrations of F11, ozone and other smog chemicals. While not denying the interest and importance of regional air pollution which has now become a topic in its own right, I found that my own interest was now stirred mostly to wonder what was the source of the small background of F11 in the clear air. Was it pollution blown right across the Atlantic from America, or could it be that these very inert and stable compounds were accumulating in the atmosphere on a global scale?

To try to answer these questions I set sail aboard the Research Ship Shackleton in November 1971 bound for the southern hemisphere. I measured the concentration of F11 and other halocarbons and also that of the natural sulphur carrier dimethyl sulphide. Measurements were made several times daily in both air and the sea water. The voyage lasted nearly six months and the measurements in Antarctica and on the return to England were made by my colleagues Bob Maggs and Roger Wade. Fig. 1.4 illustrates the global distribution of F11 we found. These observations clearly indicated an accumulation of F11 in the atmosphere. They formed the basis upon which Molina and Rowland

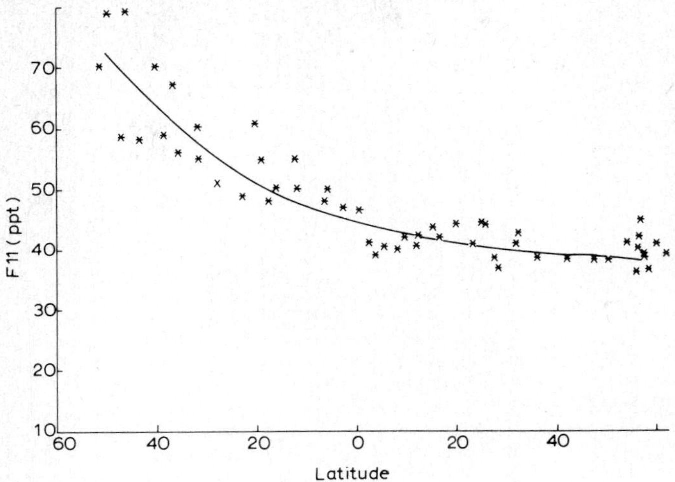

Fig. 1.4. Global distribution of F11.

were able to build their famous theory of the chlorine catalysed depletion of stratospheric ozone.

The holiday cottage in Ireland where it all began has now become a monitoring station, one of a global network dedicated to the task of observing the concentration and growth rate of the halocarbons in the global environment. Soon after this voyage was completed the peaceful environment of Ireland was perturbed by political problems. It was not long before the electron-capture detector proved to be sensitive in social as well as in environmental problems. The need to detect explosives soon revealed its capacity sensitively to detect the vapors of nitrate esters and nitro compounds. I was glad, because of my close friendships and associations in Ireland not to be personally involved in this new application.

I was fortunate to have another sea voyage in the offing. This time aboard the German Ship Meteor, which sailed from Hamburg to Santo Domingo in the Caribbean and whose journey was through the Bermuda Triangle and across the Sargasso Sea. The stated intention of my part in this expedition was to gather further information about the global distribution of the halocarbons. This was done but much more interesting was the discovery that the air pollutant peroxyacetyl nitrate (PAN) increased steadily in abundance the further we sailed away into the remote regions of the Atlantic. There was no doubt that it was a natural product since its abundance closely followed that of the sunshine rising by day and vanishing by night.

Fig. 1.5 illustrates the growth and decline of PAN concentration during the daylight hours on a clear and cloudy day over the Sargasso Sea. Fig. 1.6 is a view of the German Research Ship Meteor taken from a rubber boat mid Atlantic. It was necessary to make measurements a long way from the ship for it was a veritable floating city and consequently a prime source of atmospheric pollution. I did sometimes wonder if my complaints about the excessive use of aerosol dispensers by the sailors would give them cause to leave me behind.

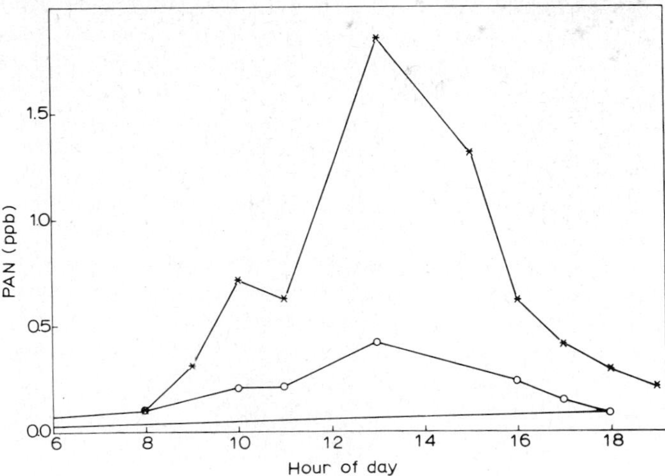

Fig. 1.5. Concentration of peroxyacetyl nitrate over the Sargasso Sea.

Fig. 1.6. Research ship "Meteor" in mid-Atlantic.

This is not the place to discuss the theory of the electron-capture detector since it is covered in later chapters, but it is an opportunity to mention a few interesting theoretical aspects of the device which rarely otherwise are revealed. Thus I find it helpful to think of the detector as a small reaction vessel holding a dilute suspension of the reagent chemical, gaseous free electrons. Then it is interesting that the free electron in thermal equilibrium with a gas at ambient temperature behaves as if it were a very large particle indeed, larger even than most of the molecules it encounters. The apparent obesity of the electron is reflected by the large De Broglie wavelength it possesses (7 nm at room temperature). This large size, or cross section, is a way of expressing the great rapidity of electron reactions and accounts for the sensitivity of the detector. Lastly, the chemical reaction between electrons and molecules is second order and many of the problems of analysis arise from this fact. If the electron-capture detector was insensitive and the number of molecules present was vastly greater than the number of electrons the device would be splendidly linear and predictable in its response. Unfortunately with the compounds it detects sensitively, which are those the analyst seeks to measure, the numbers of molecules in the detector are comparable with the number of electrons. In such circumstances as text book physical chemistry would tell you, the response of the detector to varying sample size is unlikely to be either linear or easily predictable.

There is a saying that "there are no bad instruments only bad analysts" which implies that however unusual the relationship between sample size and signal, careful calibration can always get results. But calibration with a gas at a concentration of a few parts per trillion is easier said than done. Intercalibration exercises by R. Rasmussen and by the U.S. Bureau of Standards have both shown that even among professional laboratories the scatter results reported for the stable and easily analysed fluorocarbons can be more than 100%.

My personal solution to this problem is in keeping with the spirit of the odyssey. I moved my home and laboratory to a remote country region about 1 kilometer from the nearest other dwelling and close to the Atlantic Ocean. Here I converted a barn into a 50 cubic meter exponential dilution chamber. For the fluorocarbons at least the analysis by electron-capture detector is tamed and absolute accuracies of 5% and a precision of 0.5% have been achieved.

The analyst is probably not too perturbed by the assertions of academic scientists that theory proves that the detector does not work. I well recall the fierce attack on the method and its measurements by Stockdale and his colleagues in 1964 [8]. Their paper in Nature was a cry of sheer exasperation straight from the heart. It so well expressed the justifiable annoyance of professional chemical physicists who saw their neat and orderly pastures of work on ion molecule reactions trodden over by what seemed to be a mob of clumsy peasants. It is only fair to add that within a year the same authors confirmed the validity of measurements using the electron-capture detector [9].

Present day controversy seems to centre on the role of positive ions in the detector. It was a pleasure to read Eric Grimsrud's papers on this aspect in which theory and practical observation are so well balanced [10]. On the matter of positive ions no one who has used the thermionic ion source detector introduced by Jim Sullivan can fail to have noticed how differently an electron-capture detector functions when there are no positive ions within it. Lack of sensitivity used not to be a complaint levelled at the electron-capture detector but once the possibilities of using electron attaching compounds as tags or tracers was realised, then that ultimate of molecular detection was the new destination.

We are still far from taking a grab sample of air or water and finding in it one molecule of tracer. The best we can now do is to detect between 10^5 and 10^6 molecules of tracer in a cubic centimeter of air. The use of certain fluorinated hydrocarbons as tracers is already becoming established as a method of following air masses across regions covering hundreds or even thousands of miles [11]. It is obviously applicable to other tracing needs such as the movement of water masses in the oceans. The gain in detectivity which enabled me to go from detection at the parts per 10^{12} level towards a part per 10^{14} was achieved by signal processing using gas switching techniques.

Every year that has passed in this odyssey I have expected to find this simple device that anyone could make, superseded by some impressive flight of high technology. Instead it seems to be expanding its use into new territories as the elegant experiments of Dr. Sievers, Dr. Fehsenfeld and their colleagues have shown [12]. So far neither laser infrared methods nor the natural developments of electron capture itself, such as those pioneered by Dr. Karasek in the plasma chromatograph [13] or by Dr. Horning in negative ion mass spectrometry [14] have decisively beaten it in the detectivity olympics.

The odyssey continues. I still travel with instruments incorporating electron-capture detectors but now in a quest to discover if the Earth and the biosphere form a single system, Gaia, with the powerful capacity actively to determine its and our environment [15].

REFERENCES

1 J.E. Lovelock and E.M. Wasilewska, *J. Sci. Instruments*, 26 (1949) 367.
2 A.U. Smith, J.E. Lovelock and A.S. Parkes, *Nature (London)*, 173 (1954) 1136.
3 J.E. Lovelock, *J. Chromatogr.*, 1 (1958) 35.
4 J.E. Lovelock and S.R. Lipsky, *J. Amer. Chem. Soc.*, 82 (1960) 431.
5 J.E. Lovelock, *Nature (London)*, 189 (1961) 720.
6 W.E. Wentworth, E.C.M. Chen and J.E. Lovelock, *J. Phys. Chem.*, 70 (1966) 445.
7 W.E. Wentworth and R.R. Freeman, *J. Chromatogr.*, 79 (1973) 322.
8 J.A. Stockdale, G.S. Hurst and L.G. Christophorou, *Nature (London)*, 202 (1964) 459.
9 J.A. Stockdale, G.S. Hurst and L.G. Christophorou, *Nature (London)*, 203 (1964) 1270.
10 E.P. Grimsrud and D.A. Miller, *Anal. Chem.*, 50 (1978) 1141.
11 J.E. Lovelock, *Nature (London)*, 203 (1971) 379.
12 M.P. Phillips, R.E. Sievers, P.D. Goldan, W.C. Kuster and F.C. Fehsenfeld, *Anal. Chem.*, 51 (1979) 1819.
13 M.J. Cohen and F.W. Karasek, *J. Chromatogr. Sci.*, 8 (1970) 330.
14 E.C. Horning, M.G. Horning, D.I. Carroll, I.D. Dzidic and R.N. Stillwell, *Anal. Chem.*, 45 (1973) 936.
15 J.E. Lovelock, *Gaia – A New Look at Life on Earth*, Oxford University Press, Oxford, 1979.
16 E.S. Goodwin, R. Goulden and J.G. Reynolds, *Gas Chromatography with Electron Capture Ionisation Detection for Rapid Identification of Pesticide Residues in Crops, 18th Int. Congr. on Pure and Applied Chemistry, Montreal, August 1961.*
17 J.D. Watts and A.K. Klein, *Determination of Chlorinated Pesticide Residues by Electron Capture Gas Chromatography, 75th Annual Meeting of the Association of Official Agricultural Chemists, Washington, DC, October 1961.*

Chapter 2

The design and operation of the electron-capture detector

COLIN F. POOLE and ALBERT ZLATKIS

CONTENTS

2.1. INTRODUCTION

The electron-capture detector is a member of the family of ionization detectors which include the cross-section detector, the argon detector, the electron mobility detector and the flame ionization detector [1]. Their success rests on the fact that at normal temperatures and pressures, gases behave as perfect insulators. Consequently, the presence of very few charged species can be readily observed by application of an electric field. In the

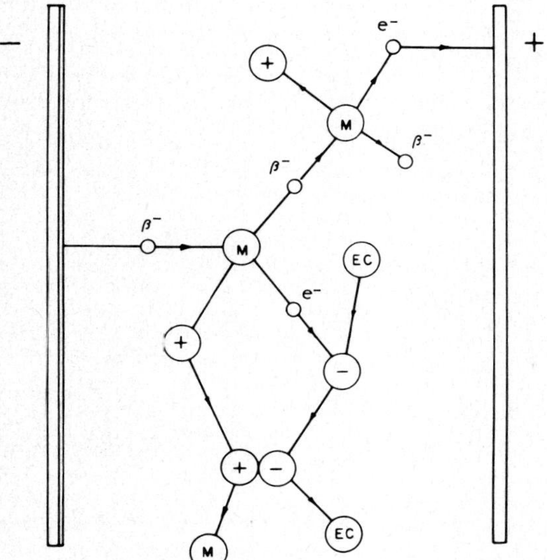

Fig. 2.1. Processes occurring in an electron-capture detector.

electron-capture detector a radioactive isotope is used as a source of β-electrons which bombard the carrier gas resulting in the formation of a plasma of positive ions, radicals, and thermal electrons by a series of elastic and inelastic collisions. The application of a potential difference to the electron-capture cell allows the collection of the thermal electrons, which constitutes the detector standing current. The electron-capturing compound introduced in the carrier gas stream reacts with the thermal electrons to produce negative ions of larger mass. The rate of recombination between negative and positive ions is manyfold faster than between electrons and positive ions. The decrease in detector current due to removal of thermal electrons by recombination in the presence of electron-capturing compounds forms the quantitative basis of the detector operation. Pictorially, the processes occurring in the electron-capture cell can be represented as shown in Fig. 2.1. The operation of the electron-capture detector for gas chromatography has been reviewed previously [2–8].

2.2. ELECTRON-CAPTURE DETECTOR CELL DESIGN

The electron-capture detector has appeared in three basic designs, the parallel plate [1,9,10], coaxial cylinder [11], and the asymmetric or displaced coaxial cylinder [12–17] models. Essentially, the detector consists of an ion chamber containing a source of β-electrons and two electrodes. It is the position and shape of the electrodes which give the detector its name. The parallel plate detector (Fig. 2.2) was convenient for the early experimentalist as its very simple geometry and construction enabled the relative position of the radioactive foil and the collector electrode to be varied without the need of constructing another detector. If tritium is used as the source, there is little to choose in terms of performance for similar source areas between the parallel plate and the coaxial detector. The lower specific activity of the ^{63}Ni source compared to tritium requires the use of a large foil area which is more easily accommodated in the coaxial design (Fig. 2.3). In this way, the overall detector volume is minimized and the source area maximized. The use of ^{63}Ni as a source allowed the use of much higher temperatures than could be used with tritium. This influenced the final design of the detector, in that

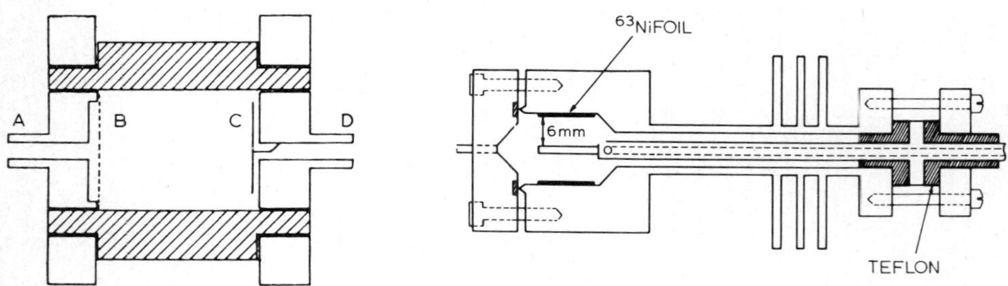

Fig. 2.2. Parallel plate electron-capture detector. A, carrier gas inlet and anode; B, diffuser; C, source of ionizing radiation; D, carrier gas outlet and cathode.

Fig. 2.3. Coaxial high temperature ^{63}Ni electron-capture detector.

insulating materials capable of withstanding high detector temperatures could not be found. This was solved by insulating the electrodes at a position remote from the heated detector chamber. In Fig. 2.3, the detector body is formed from stainless steel (this is also the cathode) and contains an axial anode insulated with Teflon and with the insulation protected from excessive heat by cooling fins. In the asymmetric configuration (Figs. 2.4 and 2.5) the cylindrical cathode, which may also serve as the detector body, is separated

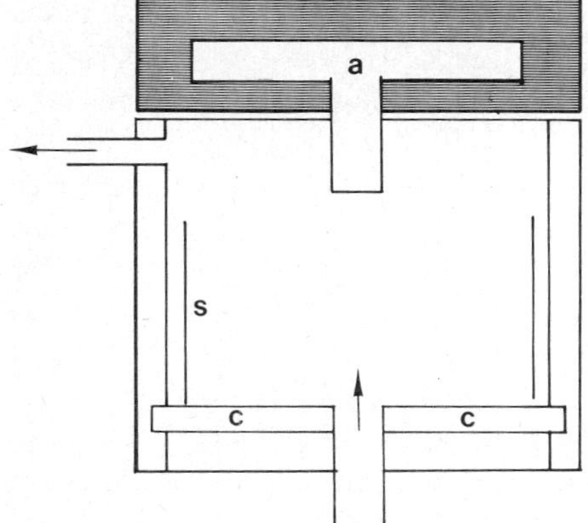

Fig. 2.4. Asymmetric electron capture detector, also called the pin-cup electron-capture detector. a, anode; c, cathode and s, source.

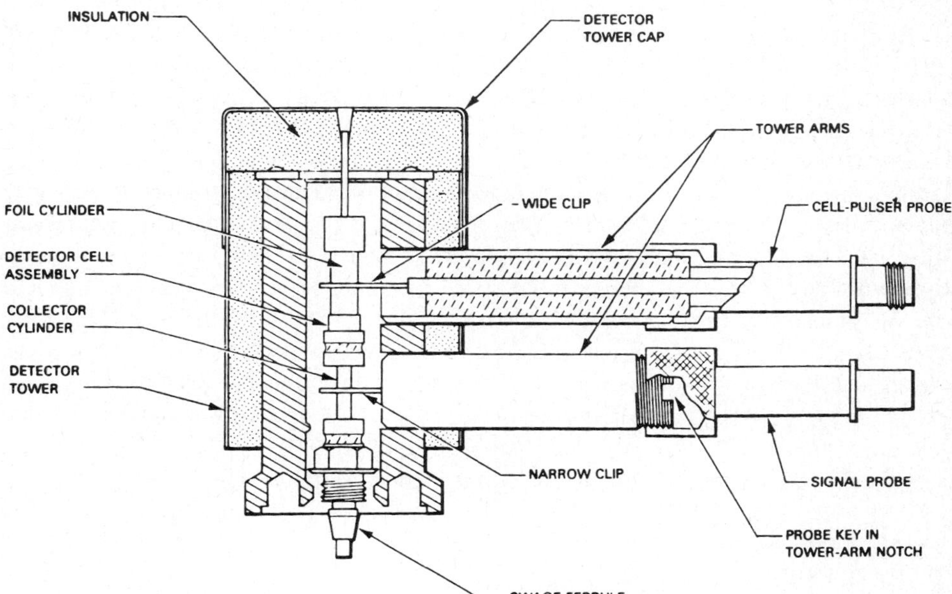

Fig. 2.5. Asymmetric electron-capture detector, also called the displaced coaxial cylinder electron-capture detector.

by a glass, ceramic, or Teflon insulator from a small anode. If the electrode spacing is arranged to maximize their separation, then the applied electric field is longitudinally asymmetric and minimizes the effect of positive space charge by concentrating the field in the vicinity of the anode while making the field near the cathode less intense. This detector design is really only useful in the d.c. mode of operation [12–16]. If, on the other hand, the same cell design criteria are maintained but the detector cell miniaturized so that the inter-electrode spacing is minimized, then operation in the pulse-modulated mode is possible [17]. By locating the anode entirely upstream of the ionized gas volume exposure to long-range β-particles is minimal, and the direction of gas flow minimizes diffusion and convection of electrons to the collector electrode. However, the free electrons are sufficiently mobile that modest pulse voltages (e.g., 50 V) are adequate to cause the electrons to move against the gas flow and to be collected during the time the pulse is on.

In the design of a practical detector, shape is not too important but dimensions are. The detector volume should be as small as possible and less than the volume of the carrier gas in which any component elutes, otherwise column efficiency will be degraded. For the higher energy ^{63}Ni source the overall detector volume is about 2–4 ml. Such a large detector volume is incompatible with capillary columns normally operated with flow-rates less than 10 ml/min, unless make-up gas is added to the column effluent [18–20]. The electron-capture detector is a concentration sensitive detector and the use of make-up gas to preserve column efficiency serves to dilute the eluting vapors and reduce sensitivity. At best, the addition of make-up gas to the column effluent is a compromise in which neither column efficiency nor detector sensitivity can reach its true optimum value.

The lower β-particle energy and the commercial availability of tritium sources of higher specific activity favor its use in the construction of low volume detectors. Devaux and Guiochon [18] described the construction of low volume detectors using, for the most part, plane parallel electrode configurations. This design restricted the source area which resulted in a low detector standing current and therefore sensitivity. Fenimore et al. [21] designed a moderately low volume detector employing coaxial geometry and a scandium tritide source. The coaxial geometry allowed a substantial increase in source area and hence standing current over that realized with the plane parallel configuration detectors. This detector could be used at elevated temperatures and was applied to capillary column gas chromatography without resorting to the use of make-up gas. A similar detector design with very minor modifications was described by Pellizzari [22]. Brechbühler et al. [23] have described a microvolume electron-capture detector which was constructed from a polyimide resin and glass lined stainless steel tubing with a dead volume of 140 μl. The displaced coaxial detector described previously has a dead volume of approximately 300 μl and can be used with capillary columns [17]. For optimum performance, make-up gas should be used with this detector.

2.3. RADIATION SOURCES

In its present form, the electron-capture detector utilizes a source of ionizing radiation with a metallic support, for it is convenient to use the source as one of the chamber electrodes. The ideal source would produce a small number of ion pairs per disintegration in order for the fluctuations in ion current and hence noise level to be minimal. At the same time, the total ion pair formation should be large so that the resulting electron current during the passage of an electron-capturing substance can be measured conveniently without introducing other sources of noise. An α-emitter, which would produce 10^5 ion pairs per cm of travel, would generate high detector noise. A very high activity γ-emitter could fill the above requirements because the ion pair yield per cm through a gas is extremely low. The radioactive hazard associated with such a primary radiation source, however, makes its use impractical. The best compromise among these demands appears to be found in isotopes emitting low energy β-particles (minimum number of ion pairs per particle) at relatively high specific activities (maximum total ion pair formation) without the potential hazard presented by γ-emitters. A number of radiation sources meeting the above requirements have been evaluated [24]. Their physical properties are summarized in Table 2.1. In commercial terms only the pure β-emitters tritium and ^{63}Ni are important, although conceivably several of the entries in Table 2.1 could also be used if the technical difficulties in preparing suitable foils could be overcome and a price or safety advantage demonstrated for them. Of the two sources, tritium and ^{63}Ni, tritium would be preferred due to its lower energy β-emanation (0.018 MeV) compared to ^{63}Ni (0.067 MeV) and the fact that foils with a higher specific activity are more readily prepared and are less expensive. This has important implications for the design of low volume detectors which were discussed previously. The principal advantage of ^{63}Ni sources is their high temperature operation stability (to 400°C) compared to Ti^3H_2 (225°C) and Sc^3H_3 (325°C). At high temperatures, source contamination from the chromatographic system is minimized and an enhancement of sensitivity with many compounds which undergo dissociative electron attachment can be achieved.

Although not currently available in commercial detectors, two new sources have been described in the literature which may become more widely used in the near future.

TABLE 2.1

GENERAL CHARACTERISTICS OF POSSIBLE RADIATION SOURCES

Source	Particle	Energy (MeV)	γ (MeV)	Half-life (years)
^{3}H	beta	0.018	0	12.5
^{99}Tc	beta	0.292	0	$5 \cdot 10^5$
^{226}Ra	alpha	4.795	0.64	1622
		4.611	0.420	
		4.21	0.260	
			0.188	
^{241}Am	alpha	5.476	0.06	457
		5.433	0.04	
			0.10	
^{90}Sr	beta	0.55	0	28
^{63}Ni	beta	0.067	0	85
^{85}Kr	beta	0.67	0.25	10.8
^{147}Pm	beta	0.23	0.121	2.6

Operating the electron-capture detector in the constant-current pulse-modulated mode can give rise to a non-linear response for low concentrations of strongly electron-absorbing compounds. To avoid this problem, a source giving low noise with operating currents less than those observed with β-emitters was needed. Such a source is the Auger electron emitter ^{55}Fe [25]. Auger electrons are formed in an extra-nuclear re-adjustment which follows radioactive decay by orbital electron capture and for ^{55}Fe yields a multiplicity of lines with energies between 5.387 and 5.640 keV. This energy is considerably lower than the β-particles emitted from tritium and ^{63}Ni and should produce a more favorable value for the number of ions generated by each ionizing particle. It was shown that the noise level for a ^{63}Ni source was approximately the square root of ten greater than the ^{55}Fe source at specific activities giving matched saturation current levels.

An electron-capture detector has been constructed which uses photoionization to produce thermal electrons, thus eliminating the conventional use of a radioactive source for this purpose [26]. Thermal electrons are generated by radiating the detector cell with photons from a light source to ionize a volatile compound added continuously to the carrier gas. Most of the carrier gases used in gas chromatography are not ionized by the radiation employed and for this reason an excess of an easily ionizable substance is continuously bled into the detector. The ionization of this compound is the source of thermal electrons for the electron-capture process. Two systems were investigated, an argon resonance lamp with ethylene in the detector cavity and a Lyman hydrogen resonance lamp with triethylamine. The latter gave the best results and for five test compounds the sensitivity of the photoionization electron-capture detector was 10 to 150 times less than that of a tritium electron-capture detector.

Considerable attention has been given to the potential problem of occupational radiation exposure due to the operation of electron-capture detectors in gas chromatography laboratories. Two potential sources of exposure can be identified. Those laboratories which build or disassemble their own detectors are obviously at most risk. These laboratories should be registered for handling radioactive materials and conform to the safety guidelines developed for such practices. The alternative source of exposure is the continuous low level emanations from detector leakage which may have an accumulative radiological effect on laboratory workers. The kinds of radiological disorders which may arise from handling radioactive materials used in gas chromatography detectors have been summarized by Taylor [27].

Kahn and Goldberg [28] determined that operation of a tritium electron-capture detector resulted in a low-level contamination of the workplace which exceeds the safety limit set by the atomic energy commission (U.S.A.). The extent of the health hazard from these sources is dependent on the source activity, detector operating temperature and inversely proportional to the size and air-exchange rate of the room. They recommended venting the detector effluent into a fume hood to minimize occupational exposure.

In detailed studies of the effect of detector operating variables on the loss rate of the radiative isotope from the source in electron-capture detectors the following general comments have been noted [15,29]:

(1) A new foil when first installed should be heated for at least 12 h to reach a reasonably stable emanation rate.

(2) The equilibrated emanation can vary over a range of 20-fold due to batch to batch variations in new sources.

(3) Average ^{3}H emanations after 12 h at 230°C for new Ti^3H_2 sources were 10 μCi/min with a low of 1 μCi/min and a high of 16 μCi/min.

(4) For Sc^3H_3 at 335°C, the equilibrated emanation rate was 5 μCi/min.

(5) Tritium emanations are not significantly different for nitrogen, helium, argon, and argon with 5% methane as carrier gas.

(6) If hydrogen is used as the carrier gas, then the emanation rate from Ti^3H_2 sources was 4–10 times greater than nitrogen carrier gas values at 225°C.

(7) The use of a Sc^3H_3 source at 325°C and hydrogen as the carrier gas resulted in the loss of 95% of tritium activity within 16 h.

(8) ^{63}Ni sources are stable to about 400°C. Above this temperature, heating for extended periods resulted in a loss of activity. This loss is not due to evaporation of the radioactive isotope from the foil but results from the diffusion of the isotope into the foil backing support. The beta particles lack the necessary energy to escape the metal support and thus the specific activity and therefore the detector standing current is diminished.

2.4. METHODS OF MEASURING DETECTOR CURRENT

The free electron concentration in the detector cell can be measured in the continuous mode by application of a d.c. voltage or in the discontinuous mode by applying a pulse of short width, long period and of sufficient amplitude to collect all the available electrons. A variation of the latter mode of operation is the constant-current pulse-modulated detector. This due to its commercial importance will be described in some detail later.

The d.c. mode of operation has several disadvantages from both the practical and theoretical point of view. The application of an applied field to an electron-capture cell results in an increase in detector current until a saturation plateau is reached at which point all the electrons produced by the ionization source are collected (Fig. 2.6). The introduction of an electron-capturing compound into the cell causes a decrease in cell

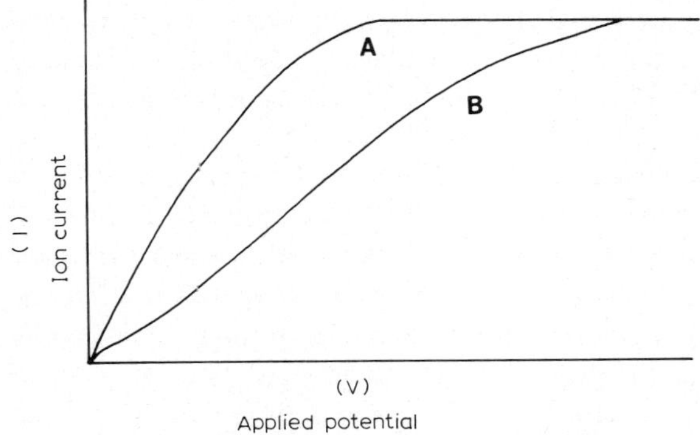

Fig. 2.6. Relationship between current and applied potential in a d.c. electron-capture detector. A in the presence of pure carrier gas and B with carrier gas containing a trace of a strongly electron-capturing compound.

current (i.e., a reduction in the concentration of free electrons) at all potentials until at a sufficiently high applied field both electrons and negative ions are collected simultaneously and no effect due to electron capture is noted. If the applied field is increased to still higher values, eventually ionization of the carrier gas itself occurs and a sharp breakaway in the detector current voltage curve results. The voltage range of interest for the operation of the electron-capture detector is the range up to and including the onset of the saturation plateau region. Knowing the chromatographic conditions to be employed, the optimum conditions of potential required for detector operation can be obtained from a plot of detector current against voltage. The optimum applied voltage occurs at the knee of the current–voltage curve, approximately 85% of the detector saturation current. Exceptions may occur with compounds capturing by a dissociative mechanism [30].

In the d.c. mode, maximum detector sensitivity occurs at low applied voltages. This is also the region in which the detector may behave in an anomalous manner due to space charge effects, contact potential effects, and non-electron-capture ionization processes. Space charge effects are a result of the different mobility of electrons and positive ions in an applied field. The electrons are readily collected at the anode, while the slow moving positive ions drift to the cathode resulting in a new accumulation of positive ions in the region of the cathode. This cloud of positive ions sets up a potential in opposition to the applied voltage as well as moderating the energy and flux of the ionizing radiation. The introduction of a weakly electron-capturing vapor (e.g., column bleed, etc.) can drastically change the composition of the space charge cloud with a concomitant change in electron concentration. At high field potentials, the space charge cloud is compressed around the cathode which enhances its effect on the electron concentration. Also at high field potentials, the secondary electrons formed by collision between carrier gas molecules and β-particles have insufficient time to reach thermal equilibrium and are rapidly collected at the anode. This reduces their life-time in the cell and makes them less available for reaction with organic vapors. Furthermore, the negative molecular ions formed by reaction between the organic vapors and the secondary electrons may also be collected at the anode before they have chance to undergo collision with a positive ion, thus producing an erroneous value for the detector current. When operating in the d.c. mode, detectors with asymmetric geometry are less affected by space charge phenomena. The response of a detector with a small or remote anode is relatively independent of applied potential since the field is weak in the vicinity of the cathode and the separation of positive and negative ions cannot easily occur.

A substance eluting from the gas chromatographic column may temporarily adsorb onto one of the electrode surfaces which in turn can give rise to a contact potential. Such potentials are usually of the order of a few volts and may either enhance or oppose the potential applied to the cell [10]. Contact potentials are less of a problem when the detector is operated at high temperature as eluting vapors are less likely to condense on an electrode, or at high applied voltages, when the contact potential is small compared to the applied voltage. In a weak field, a contact potential opposing the applied field gives an erroneously large response on the recorder with peaks often showing severe tailing. If the contact potential enhanced the applied field, then the detector shows a diminished response and the front edge of the peak may drift below the baseline before the baseline is re-established. When the contact potential opposes the applied field, there is always the possibility of a false response for a compound which generates a potential at the electrode surface but is not itself electron capturing.

Many of the problems encountered with the constant voltage d.c. electron-capture detector can be overcome by the use of the pulse sampling technique. The applied potential is characterized by being of sufficient magnitude and width to collect all the thermal electrons and of long period so that the detector operates essentially in the "no field" mode. Thus between pulses there is sufficient time for the concentration of thermal electrons to be replenished by the ionizing β-radiation and for their energy to reach thermal equilibrium. In pulse sampled electron-capture detection, the pulse variables which must be considered in achieving optimum operation are the amplitude (V_a), width (t_w), and interval (t_p). The amplitude and width should be of sufficient magnitude to afford complete collection of electrons as indicated by a plateau in plots of detector current vs. either of these variables (Figs. 2.7 and 2.8). The conditions for maximum collection should not, however, be exceeded to any great extent in order to preserve the advantages of field-free operation. The choice of optimum pulse interval requires careful consideration. Here a compromise is to be reached between the difficulty of measuring very small

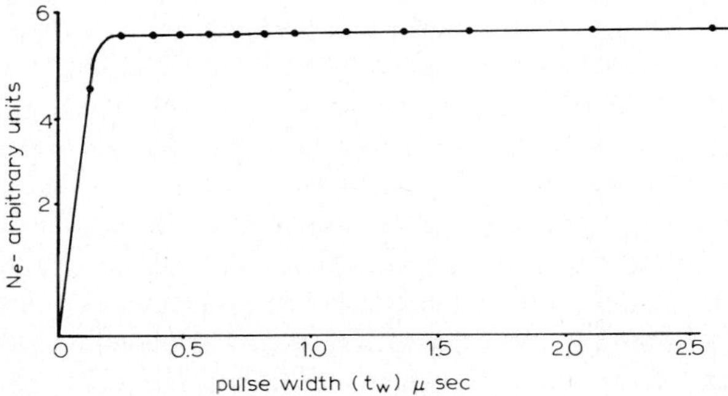

Fig. 2.7. Electron concentration vs. pulse width for a pulse sampled electron-capture detector.

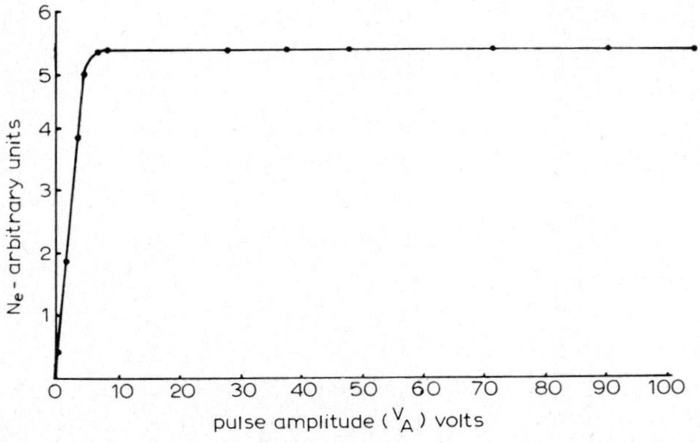

Fig. 2.8. Electron concentration vs. pulse amplitude for a pulse sampled electron-capture detector.

currents if long pulse intervals are used and the need for a certain minimum interval to allow both "no field" operation and sufficient time for the concentration of thermal electrons to reach a maximum concentration. The relationship between the electron concentration and pulse period (t_p) is shown in Fig. 2.9. The optimum value of t_p is in excess of 1000 μs, but this may not always be so readily usable practically (because of the very small detector currents involved) and lower values may be preferred. An often overlooked advantage of the pulsed sampling technique is the ability to control conveniently detector sensitivity by changing the period between pulses.

As stated above, the sensitivity of the detector varies with the pulse space and usually increases with larger time intervals between pulses as this maximizes the time available for recombination of positive and negative ions, up to the limit where recombination of positive ions and electrons becomes important. As the detector is in the "no field" mode for most of the operation, the secondary electrons have plenty of time to reach thermal equilibrium. In the "no field" mode, the electrons do not drift out of the plasma so that negative ion formation occurs in the region where positive ions are also present and where recombination can most efficiently take place. Also the duration of the brief pulse is insufficient to collect the slower moving negative molecular ions. Likewise, space charge separation occurs to a much smaller extent.

A variation of the pulse sampling method is the pulse-modulated constant-current electron-capture detector [17,31]. Instead of observing the cell current at a constant pulse frequency, in the new method of operation, the cell current is fixed with respect to a reference value and the frequency of the pulse is modulated so that the difference between the cell current (I_{cell}) and the reference current (I_{ref}) is zero throughout the chromatographic separation. Since pulse frequency is the variable quantity in this method of operation the output signal is a voltage proportional to that frequency. A block diagram of the pulse-modulated electron-capture detector circuit is shown in Fig. 2.10. The circuit in fact forms a closed-loop electronic feedback network in which the cell current (I_{cell}) is combined with an external reference current (I_{ref}) such that the difference ($I_{ref} - I_{cell}$) is the input to an electrometer; the electrometer output feeds into the pulse-generating network such that the pulse frequency is determined by the magnitude of the

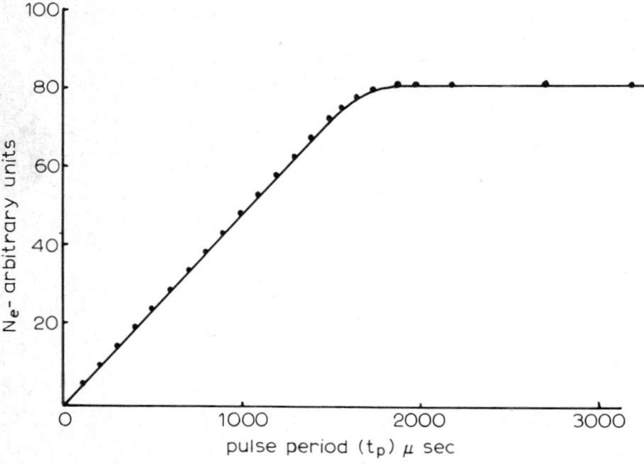

Fig. 2.9. Electron concentration vs. pulse period for a pulse sampled electron-capture detector.

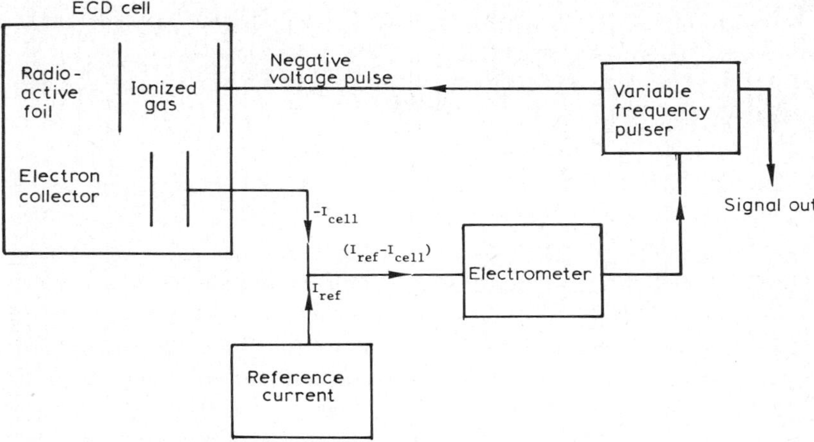

Fig. 2.10. Block diagram of the electronic components of a constant-current pulse-modulated electron-capture detector.

electrometer output; and the frequency of voltage pulses in turn determines the magnitude of I_{cell} (with the proviso that $I_{cell} = I_{ref}$). The two principal advantages of this method of operation are that the response of the detector against concentration has a greater linear range than the pulse sampling method and secondly, the detector operation is less disturbed by traces of interferents entering the detector cell.

It has been shown that the average current I_{cell} is related to the concentration of free electrons in the cell $[e]$ and the frequency f of the applied field by

$$I_{cell} \propto [e]\, f$$

When only carrier gas is in the cell $[e]$ is very large and f is low. This pulse frequency with pure carrier gas is called the base frequency, f_0, of the system. When an electronegative sample enters the electron-capture detector, electron capture occurs and some of the electrons $[e]$ are removed as negative ions. This decrease in $[e]$ is matched by an increase in f in order to keep I_{cell} constant. The output signal of the detector is the frequency difference $(f_A - f_0)$, where f_A is the frequency corresponding to a sample concentration [A] of electron-capturing material within the electron-capture cell. The function $(f_A - f_0)$ is linearly related to the concentration [A] over about four orders of magnitude for many compounds. As a precautionary note, it is now well established that certain strongly electron-capturing compounds may produce S-shaped calibration curves at low concentrations with this method of operation [32].

2.5. LINEARIZATION OF THE ELECTRON-CAPTURE DETECTOR RESPONSE

In the pulse sampling mode, the linear range of the electron-capture detector response against concentration extends over about two orders of magnitude. One method of increasing the linear operating range, namely the constant-current pulse-modulated detector, was described above. An alternative approach using the pulse sampling mode with pulses having a long period and signal analog conversion has been described [33,34]. The output from an electron-capture detector in the pulse sampling mode can be represented by the equation [35]

$$\frac{(I_b - I_e)}{I_e} = Ka$$

where I_b = detector standing current, I_e = current observed during the passage of an electron-capturing compound, K = electron-capture coefficient, and a = sample concentration.

Pictorially, the various measurements are shown in Fig. 2.11. This equation is true for long pulse periods as under such conditions equilibration of the electron concentration will occur and the conditions required for the steady state solution of the kinetic model used to derive the above equation are valid. The time required for equilibration will depend on the source, detector geometry, and the carrier gas used. A circuit diagram for performing an analog conversion of the detector signal is shown in Fig. 2.12. The cell current is amplified and converted to a voltage corresponding to the current I_e as an electron-capturing compound migrates through the cell. This voltage is fed simultaneously to a logarithmic divider network and to the inverting input of a subtractor amplifier. A voltage corresponding to I_b (standing current) supplied by a variable source was fed to the noninverting input, and the output, $I_b - I_e$, of this amplifier was also fed to the logarithmic divider together with a scale factor voltage. The I_b voltage was set to give a zero value of $I_b - I_e$ in the absence of sample elution, and the scale factor voltage was chosen such that a given preset ratio of $I_b - I_e$ to I_e could be calibrated with the recorder scale. The linear range obtained by the ratio method will normally exceed four orders of magnitude in the range of 0–90% saturation of the detector standing current. A definite departure from linearity was observed at higher values of detector saturation due probably to the error introduced by computing $(I_b - I_e)/I_e$ when I_e is small compared to I_b.

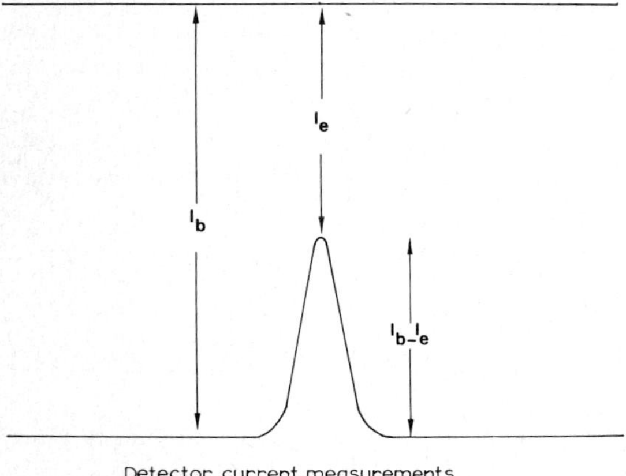

Fig. 2.11. Detector current measurements.

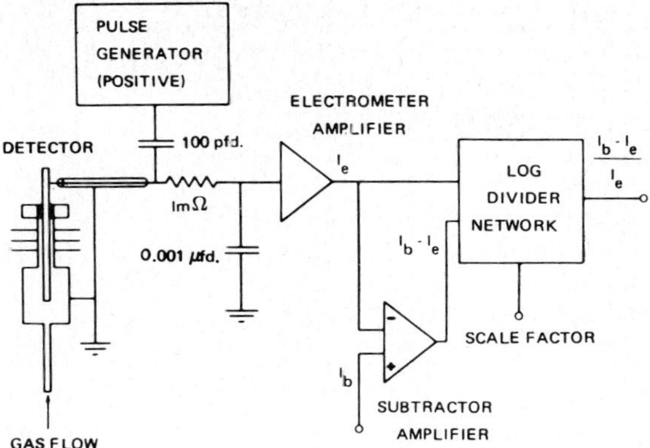

Fig. 2.12. Linear electron-capture system.

2.6. MISCELLANEOUS DESIGN CRITERIA

The response of the electron-capture detector is temperature dependent, as would be predicted from the mechanism of the electron attachment process [36]. This has two effects on the detector design. First of all, provision must be made for varying the temperature of the detector cell over its complete operating range, the upper limit of which is set by the thermal stability of the radiation source. Secondly, for good reproducibility of measurements, the detector cell should be well thermostated and capable of holding a given temperature to better than $\pm 0.2^{\circ}C$.

The choice of carrier gas for use with the electron-capture detector is limited to hydrogen, the noble gases and nitrogen. Hydrogen is rarely used as its escape into the laboratory constitutes an explosion hazard, and at high temperatures with tritium detectors, the carrier gas will strip the tritium from the foil [15]. Pure argon and helium are unsuitable as they readily form metastable ions which can transfer their energy of excitation by collision with solute vapors resulting in undesirable ionization effects (Penning reaction). The addition of 5–10% of methane to argon removes the metastable ions as fast as they are formed by deactivating collisions and also serves to reduce and maintain the secondary electron energy at a constant thermal level. Argon containing methane and oxygen-free nitrogen are the most commonly used carrier gases. The electron drift velocities in argon–methane are much higher than in nitrogen and as a consequence not all detector designs are equally suited to the use of both gases. Some commercial detectors which specify the use of argon–methane as the carrier gas will show degraded performance if nitrogen is used. Scrupulous cleaning of the carrier gas is required to remove water and oxygen, otherwise the detector may respond in an anomalous manner. Such responses are made use of in selective electron-capture sensitization methods which are described elsewhere in this book (Chapters 4 and 5).

REFERENCES

1 J.E. Lovelock, *Anal. Chem.*, 33 (1961) 162.
2 W.A. Aue and S. Kapila, *J. Chromatogr. Sci.*, 11 (1973) 255.
3 D.J. David, *Gas Chromatography Detectors*, Wiley-Interscience, New York, 1974, p. 76.
4 E.D. Pellizzari, *J. Chromatogr.*, 98 (1974) 323.
5 J. Sevčík, *Detectors in Gas Chromatography*, Elsevier, Amsterdam, 1976, p. 72.
6 C.F. Poole, *Laboratory Practice (London)*, 25 (1976) 309.
7 E.R. Adlard, *CRC Crit. Rev. Anal. Chem.*, 5 (1975) 13.
8 A. Zlatkis and D.C. Fenimore, *Reviews in Anal. Chem.*, 2 (1975) 217.
9 J.E. Lovelock in C.L.A. Harbourn (Editor), *Gas Chromatography 1968*, Institute of Petroleum, London, 1969, p. 95.
10 J.E. Lovelock, *Anal. Chem.*, 35 (1963) 474.
11 P.G. Simmonds, D.C. Fenimore, B.C. Pettitt, J.E. Lovelock and A. Zlatkis, *Anal. Chem.*, 39 (1967) 1428.
12 M. Scolnick, *J. Chromatogr. Sci.*, 7 (1969) 300.
13 W.L. Yauger, L.M. Addison and R.K. Stevens, *J. Ass. Offic. Anal. Chem.*, 49 (1966) 1053.
14 J. Lasa, T. Owsiak and D. Kostewicz, *J. Chromatogr.*, 44 (1969) 46.
15 C.H. Hartmann, *Anal. Chem.*, 45 (1973) 733.
16 J.F. Uthe and J. Solomon, *J. Chromatogr.*, 95 (1974) 169.
17 P.L. Patterson, *J. Chromatogr.*, 134 (1977) 25.
18 P. Devaux and G. Guiochon, *Chromatographia*, 2 (1969) 151.
19 K. Grob, *Chromatographia*, 8 (1975) 423.
20 L. Rejthar and K. Tesařík, *J. Chromatogr.*, 131 (1977) 404.
21 D.C. Fenimore, P.R. Loy and A. Zlatkis, *Anal. Chem.*, 43 (1971) 1972.
22 E.D. Pellizzari, *J. Chromatogr.*, 92 (1974) 299.
23 B. Brechbühler, L. Gay and H. Jaeger, *Chromatographia*, 10 (1977) 478.
24 G.R. Shoemake, D.C. Fenimore and A. Zlatkis, *J. Gas Chromatogr.*, 3 (1965) 285.
25 D.J. Dwight, E.A. Lorch and J.E. Lovelock, *J. Chromatogr.*, 116 (1976) 257.
26 W.E. Wentworth, A. Tishbee, C.F. Batten and A. Zlatkis, *J. Chromatogr.*, 112 (1975) 229.
27 M.P. Taylor, *J. Chromatogr.*, 9 (1962) 28.
28 L. Kahn and M.C. Goldberg, *J. Gas Chromatogr.*, 3 (1965) 287.
29 G.R. Shoemake, J.E. Lovelock and A. Zlatkis, *J. Chromatogr.*, 12 (1963) 314.
30 R.A. Landowne, *Anal. Chem.*, 42 (1970) 1468.
31 R.J. Maggs, P.L. Joynes, A.J. Davies and J.E. Lovelock, *Anal. Chem.*, 43 (1971) 1966.
32 J.J. Sullivan and C.A. Burgett, *Chromatographia*, 8 (1975) 176.
33 D.C. Fenimore, A. Zlatkis and W.E. Wentworth, *Anal. Chem.*, 40 (1968) 1594.
34 D.C. Fenimore and C.M. Davis, *J. Chromatogr. Sci.*, 8 (1970) 519.
35 W.E. Wentworth, E. Chen and J.E. Lovelock, *J. Phys. Chem.*, 70 (1966) 445.
36 C.F. Poole, *J. Chromatogr.*, 118 (1976) 280.

Chapter 3

Theory of electron capture

W.E. WENTWORTH and E.C.M. CHEN

CONTENTS

3.1. INTRODUCTION

3.1.1. Theory of the electron-capture detector

Many analytical techniques have been developed from physical and chemical principles which were well understood from a theoretical perspective. This was not the case for the electron-capture detector (ECD). Although gas phase negative ions and electrons were discovered in the late nineteenth century, very little fundamental work was attempted on the reactions of gaseous thermal electrons with organic molecules. This created a unique opportunity for developing a theoretical model for the ECD which could lead to the determination of important molecular data. The development has followed basically the scientific method whereby one would propose a model based on the observations, develop new experiments to test the model, and improve the model based upon the results of the new data. This procedure has been carried out over the past twenty years in our laboratory for a specific version of the ECD. The first complete description of the model was presented by Wentworth, Chen and Lovelock [1]. The model is dynamic and is being continually reviewed as new data are obtained in our laboratory and in others. The historical development of this model, sometimes called the "classical" model, will be presented in the next section. In the third section of this chapter, the model as it stands today will be described. This will be followed by a comparison of the model with experimental data obtained with other techniques since the early development of the theory and finally a section on correlations and predictions will be presented.

The model which has been developed in our laboratory applies to a plane parallel plate or a concentric electrode ECD with a tritium source where the electrons are collected by a pulsed voltage at a fixed frequency. An inert carrier gas, usually argon + 10% methane, is utilized which reduces the electron energy to a thermal distribution and quenches metastable species formed by the radiation. The response is primarily due to the initial ion–molecule reactions and subsequent reactions are considered to merely alter the nature of the negative ions and not release electrons. One exception to this is detachment from the first formed negative ion. Positive ion–molecule reactions other than neutralization have not been considered. Radical reactions involving electron attachment are considered negligible. A number of modifications to this basic system will be discussed in this introduction for comparison with the "classical model".

3.1.2. Modifications of the electron-capture detector

Because of the extreme sensitivity and selectivity of the ECD, it has been used extensively in analytical chemistry so that many operational variations have been developed to improve the practical use of the detector. Some of these are strictly physical, others strictly chemical, while some affect both the physical and chemical processes. The physical alterations include different geometries, electron sources, and means of electron collection, while the chemical changes involve altering the reactions by doping the carrier gas with reactive species. A different aspect of the chemistry of the ECD has been observed by monitoring the terminal ions or by measuring and identifying the neutral reaction products in the effluent.

There are two alternative methods of linearizing the response in the ECD: the application of a variable d.c. voltage and the measurement of the frequency of the applied pulse required to maintain a constant current, the "constant current pulsed mode". The original ECD described by Lovelock and Lipsky [2] used a d.c. voltage and because of the simplicity of the power supply, many commercial units still use a d.c. voltage in spite of the experimental problems which have been encountered. The constant current mode of collection was introduced by Maggs, Joynes, Davies and Lovelock [3] as a method for obtaining a linearization function which could be easily automated.

Many different radioactive sources have been proposed and used for the ECD but the most common are ^{3}H and ^{63}Ni [4,5]. Several non-radioactive sources based upon discharges [6], thermionic emission [7] and photoionization [8] have been proposed and demonstrated, but their use is limited. Lovelock and co-workers have described two alternative modes of operating the ECD, the coulometric mode where electrons react quantitatively with the test molecule [9] and electron-capture spectroscopy where the energy of the electrons is modified by the application of a radio frequency field and the response is measured as a function of the electron energy [10].

The terminal ionic products from the ECD have been monitored by using the ECD as an ion source for a mass spectrometer. This has been termed atmospheric pressure ionization (API) mass spectrometry. A ^{63}Ni source is generally used in this application [11–13]. The ions can sometimes be characterized by their drift times. The neutral products of the ECD have been investigated by passing the effluent of the ECD through a second ECD in series. In some cases a gas chromatographic column was placed between the two detectors [14].

Recently, a new mode of ECD operation has been introduced which involves doping the carrier gas with oxygen or nitrous oxide to enhance the response to certain compounds. The reactions proceed by way of a molecular oxygen anion or an atomic oxygen anion. In the case of nitrous oxide, the carrier gas must be nitrogen. This technique is called Selective Electron Capture Sensitization (SECS) [15,16] or could be called negative ion capture.

3.1.3. Theory of the modifications

The theory of many of the techniques described above can be considered as simple modifications of the "classical" theory, but others must be treated in a different manner because different phenomena are being investigated. In this section the current status of the theoretical treatment of these variations will be presented.

The d.c. electron-capture detector has been investigated in a fundamental manner by Aue and co-workers [17,18]. This investigation was prompted by the observation of high responses for some compounds in a d.c. system operated at elevated pressures. It was postulated that the response in the d.c. mode, apart from the initial electron-capture reactions, is due to space charge effects which promote recombination. The basic electron-capture reactions are presumably consistent with the model described in this chapter. The effect of the d.c. system upon the energy of the electrons was calculated by Śliwka and co-workers [19,20]. These calculations showed that the average energy of the electrons in a d.c. system is higher than in a pulsed system as anticipated earlier by Lovelock.

The theoretical treatment of the coulometric mode has presented some difficulties since in some cases each molecule removes more than one electron. The d.c. model of Aue and Kapila [17] attributed this to space charge effects while Siegel andMcKeown [21] and Rosiek et al. [19] consider the phenomenon to be a chemical one in which the products of the electron-capture reaction can also capture electrons. Broś and Page [23] have also investigated the coulometric mode theoretically and have concluded that the conditions for coulometric response are contradictory to those necessary for the optimum operation in the constant current mode.

Maggs et al. [3] proposed that the reciprocal of the frequency necessary to maintain a constant current is proportional to the electron concentration. The exact conditions necessary for this proportionality have not yet been defined although a considerable amount of data has been collected in the constant current mode. Patterson [24] has obtained some basic data on a special geometry using both modes of electron collection. However, to truly define the differences between the two modes, a comprehensive study of both modes must be done simultaneously. In our laboratory, preliminary investigations of the constant current mode have been carried out using a computer solution to the kinetic model but the results have not been verified experimentally.

Grimsrud and Kim [13] and Siegel and McKeown [21] have investigated the ^{63}Ni source in an API unit from a theoretical point of view. Ayala et al. [25] have presented an experimental comparison of the ECD using a tritium source and a nickel source. The nickel betas have an increased range and that may have an effect on the average energy of the electrons due to the higher energy of the nickel betas. This is not definite and is still being investigated. Grimsrud has proposed a modified model of the ECD based upon his studies of a nickel source in an API. The model envisions an annular region for the reaction zone in which positive ions react with negative species. The dimensions of this region are modified by "space charge" driven migration [26,27]. Siegel and McKeown [21] considered a "field free" ECD where no electrons are collected and concluded that the electrons and the positive ion concentrations are equal and fill the cell. The application of the voltage will certainly collect the electrons and modify the concentrations.

The study of the terminal ions and neutrals obtained from the ECD are not conclusive. This is partially due to the fact that the API does not obtain a representative sample of the ions in the ECD. Grimsrud has addressed this problem and has obtained a partial solution by sampling the mass spectrometer in a field free condition but with the pulsed mode of operation. If a true representation of the ionic species in the ECD can be obtained, then the API can be used to obtain fundamental information on the primary and secondary reactions in the ECD. The theory of the neutrals formed in the ECD reaction has not been examined.

Since the SECS technique is relatively new, little work has been done in the theoretical area. The primary reactions have been identified but fundamental studies to establish the mechanisms have not been carried out. One exception is the enhancement of the CO_2 response. Chen et al. [29] included the reaction of O_2^- with CO_2 into the kinetic model and successfully matched the temperature data obtained by Simmonds [30] to this model. In summary, the theoretical studies of the modifications are limited. The variable voltage d.c. system, the constant current mode, and SECS could offer great promise for the future. Some of these operational variations of the ECD are discussed in more detail in other chapters of this text.

3.2. THE HISTORICAL DEVELOPMENT OF THE THEORY OF ELECTRON CAPTURE

3.2.1. The original system

The ECD originally described by Lovelock and Lipsky [2] was different from the modern version in almost every possible way. The carrier gas was helium, nitrogen, hydrogen or possibly argon. A pin and cup geometry, a radium source, and a d.c. voltage were utilized. The major purpose of the detector was the qualitative identification of molecules which attach electrons. This was done by noting the voltage required to change the decrease in current due to electron capture to an increase due to ionization. In spite of these differences, two important reactions were identified as:

$$AB + e^- \rightleftharpoons AB^- + \text{energy} \tag{3.1}$$

$$AB + e^- \rightarrow A + B^- \tag{3.2}$$

It was also recognized that recombination reactions played a major role in the observed response. The term "electron affinity spectroscopy" was attached to the technique even though the term electron affinity is generally defined as the energy of reaction 3.1 at zero Kelvin. This nomenclature created some confusion in the early development of the theory. There were no attempts to assess the quantitative aspects of the detector but it was observed that the detector was extremely sensitive towards certain compounds and that the response did not vary linearly with concentration of the test material except at very low concentrations. It was suggested that a pulsed system would increase the sensitivity of the detector.

3.2.2. The pulsed electron-capture detector

In 1961, Lovelock [31–33] described the use of a pulsed voltage to collect thermal electrons generated by beta particles from a tritium foil in a plane parallel plate geometry. The carrier gas was argon with a "cooling and quenching" additive such as hydrogen, methane, or carbon dioxide. Using this system under fixed conditions of voltage, frequency, and flow-rate, the relative responses of about 430 compounds were determined.

In order to place quantitative values on the responses, it was necessary to define a relationship between the electron current and the concentration of the molecules. By analogy to light absorption, a "Beer's Law" relationship was proposed. The limit of this form as concentration of capturing substance becomes small is consistent with a relationship developed later in this chapter (eqn. 3.31)

$$\lim_{a=0} [-\log([e^-]/b)] = \frac{b-[e^-]}{[e^-]} = K_0 a x \tag{3.3}$$

where $[e^-]$ is the concentration of the electrons in the presence of the test species AB, b is the electron concentration in the absence of AB, K_0 is the electron absorption coefficient, a is the concentration of the test species AB, and x is a dimensional factor related to the distance between the electrodes. The electron absorption coefficient is characteristic of the compound and was observed to span over seven orders of magnitude. In order to eliminate instrumental variations, the coefficient was measured relative to that of chlorobenzene which was assigned a value of unity. Unfortunately, this quantity was termed the "relative electron affinity" instead of the relative affinity for electrons. The compound chosen as a reference was unfortunate since it undergoes dissociative capture and has a large temperature dependence.

The physical significance of the pulsed mode was recognized by Lovelock who stated "The density of the free electrons in the chamber was measured by applying to the anode, every 10 μs, a rectangular pulse of 50 V in amplitude and 0.5 μs in duration. The voltage and duration were chosen to just collect all of the electrons set free in the chamber or in other words, just capable of withdrawing a saturation current from the chamber. Such a pulse was, however, much too short to collect any of the relatively immobile negative molecular ions" [31–33]. This description formed the basis of the theory of the ECD operated in the pulse sampling mode.

The chemical aspects of the electron-capture detector were also discussed in these articles. The differences between the non-dissociative capture, reaction 3.1, and the dissociative capture, reaction 3.2, processes were defined in a qualitative manner. Molecules with extended pi systems captured electrons non-dissociatively while those with highly electronegative groups such as the halogens reacted in a dissociative manner. These generalizations have been modified in the ensuing years.

The significance of the recombination reactions was again noted. The importance of the relative rates of recombination of electrons and negative ions was specifically emphasized. In ref. 32, it was stated that "the rate of recombination between positive and negative molecular ions is between 10^5 to 10^8 times greater than between free electrons and positive ions". This has been interpreted by some authors to mean that the recombination rate constants are different by a factor of 10^5 to 10^8. This is not necessarily true since the negative ion concentration need not be equal to the electron concentration. The negative ion concentration should be larger than the electron concentration since the negative ions are not pulsed out of the reaction zone. Consequently, the larger negative ion recombination rate can be attributed, at least in part, to the greater negative ion concentration. However, it is questionable whether this difference is as large as 10^5 to 10^8. It appears that the rate constant for negative ion recombination is about an order of magnitude greater than electron recombination for these ions [1].

3.2.3. Response of the electron-capture detector and molecular electron affinities

In 1962, Wentworth and Becker [34] and Becker and Wentworth [35] postulated that the ECD response for compounds which do not dissociate upon electron attachment is directly related to the equilibrium constant for reaction 3.1 and hence is related to the electron affinity of the molecule via the statistical mechanical expression for the equilibrium constant. As a consequence, the Beer's Law relationship proposed by Lovelock was not valid for the concentration dependence of the ECD response. Fortunately, the data obtained by Lovelock were measured at low (less than 10%) electron capture and the functional relationships reduce to identical forms. From the equilibrium postulate, the electron-capture coefficient should be defined as:

$$\frac{b - [e^-]}{[e^-]} = Ka \tag{3.4}$$

where the symbols have the same meaning as before.

Because there were no accurate estimates of molecular electron affinities available in the literature at the time, this "potential" method for determining molecular electron affinities could not be tested directly. However, the comparison of the relative values of the electron affinities of a series of aromatic hydrocarbons with other properties known to be related to the electron affinities strongly supported the postulate. These data included half wave reduction potentials, ionization potentials, charge transfer complex absorption spectra, and theoretical Hückel calculations [35,36].

More significantly, this postulate suggested two types of experiments which could be used to verify the model. These were the concentration dependence, as suggested by eqn. 3.4, and the temperature dependence suggested by the statistical mechanical expression for the equilibrium constant.

3.2.4. Development of the kinetic model

On the basis of the encouraging results just described, a study of the thermal electron attachment to both dissociative and non-dissociative molecules was begun. These studies resulted in a detailed kinetic model for the chemical and physical processes occurring within the ECD for non-dissociative molecules which provided a direct relationship between the electron-capture coefficient and the rate constant for attachment, detachment, and recombination reactions. This differed from the earlier equilibrium constant at low temperatures, because the recombination rate constant predominates over the detachment rate constant as the mode of negative ion loss. The high temperature variation of the electron-capture coefficient yields data on the electron affinity of the molecule while the low temperature data yield information on the attachment rate and the activation energy for attachment. Some molecules only exhibit one temperature region.

The model led to a family of second order differential equations which were solved in an approximate manner by considering that the positive ion concentration in the reaction zone was greater than the electron concentration and hence remained essentially constant. This also inherently took into consideration the build up of various species during the repetition of the pulses [1].

The model was extended to dissociative compounds and the solutions related the electron-capture coefficients to the rate constants for attachment, detachment, recombination, and dissociation. In certain cases, the temperature dependence gives the activation energy for dissociation [37]. The dissociative process can be a simple bimolecular reaction in which the molecular ion dissociates rapidly (within a vibration) or it can involve a stable molecular ion which is formed prior to dissociation. A detailed molecular description of the dissociative attachment processes has been presented in terms of molecular potential energy curves for the neutral and the negative ion [38–40]. All of the electron attachment processes have been classified in terms of four mechanisms described by the nature of the potential energy curves [41]. The classification depends upon the depth of the negative ion potential energy curves and the crossing point with the potential energy curve for the neutral molecule.

Śliwka et al. [22] explicitly considered the effect of the pulsed mode on the concentration of the electrons during the field free time and during the application of the pulse. In addition, the effect of the repeated application of the pulse was considered. However, they did not consider a variation in the positive species. On the basis of recent experimental data, the positive ion concentration changes as the sample elutes. Thus the second order equations must be solved. Wentworth and Chen [42] included the positive species as reactants in the kinetic model and solved the family of equations numerically. At the end of each pulse, the electron concentration was set to zero and a constant fraction of the positive species was removed. When these boundary conditions are utilized, then the expression for the electron-capture coefficient is equal to a constant times the capture coefficient obtained from the earlier solution. Thus the predicted temperature variations and the energy parameters obtained therefrom are still valid. The absolute values of some of the rate constants will however be modified. The solution also predicts a change in the positive species as observed experimentally.

From this description of the development of the theory of the ECD, it can be seen that many of the original ideas are still incorporated in the kinetic model and that the basic operation has been one of refinement and adjustment rather than replacement. Also the new data obtained from the techniques described in Section 3.1 will eventually lead to a more accurate model of the ECD.

In a recent publication, Connor has reconsidered the theoretical model for the electron-capture detector [43]. His solution to the kinetic model is based upon the assumption that the positive ion concentration is equal to the electron concentration. His argument is based partially on the work of Siegel and McKeown [21] who have concluded that the electron and positive ion concentrations must be equal under field free conditions. Infinitely long pulse intervals would be equivalent to a field free condition and under these conditions, very likely the electron and positive concentrations are equal. However, at practical pulse intervals of 1–5 ms, this fact is uncertain. Experimental studies will have to be performed to clarify this point. Fortunately, the solution to the kinetic equations gives the same concentration dependence providing the concentration is reasonably low (<50% capture).

3.3. MODELS AND MECHANISMS

3.3.1. Description of the kinetic model

The electron-capture detector can be envisioned as a chemical reactor in which thermal electrons are reactants. The application of the pulse removes all of the free electrons and thus initiates a reaction cycle with the time of the reaction equal to the time between the pulses. The concentrations of the positive and negative ions and the electrons will reach a steady state which will be different in the presence and in the absence of the capturing species. The magnitudes of these will be governed by the rate constants for the chemical reactions and the respective loss mechanisms for the species present in the ECD. Thus the kinetic model must specify the pertinent chemical reactions, the various loss modes, and the rate constants and expressions for each of the species. The model can then be described by a system of differential equations which must be solved for the pertinent boundary conditions indicated by the loss mechanisms. Since the only species which is monitored in the ECD is the free electron concentration, only the initial reactions and any electron releasing reactions need to be specified in the model.

The first step in setting up the model was the determination of the effects of the various pulse parameters upon the overall processes. These variations were anticipated by Lovelock [31] and were verified by Wentworth et al. [1]. Typical plots of the number of electrons collected per pulse, N_e, vs. t_w, t_p, and V (pulse width, pulse time, and voltage) are shown in Fig. 3.1. In general, plateau values were obtained and the values of the pulse parameters were chosen in the plateau region. These plots can also be used to establish that an ECD is operating properly. If plateau values are not achieved, then something is wrong with the experimental set up. In addition, the plot of N_e vs. t_p can indicate the presence of a contaminant in the carrier gas or bleeding from the column.

The data for N_e vs. t_w can be used to infer dimensions for the reaction zone. Measurements of N_e vs. t_p at various electrode spacings carried out by Lovelock can be used to estimate the depth of the reaction zone [44]. Finally, Lovelock and Watson [45] presented an oscillographic trace of the arrival of an electron pulse at the anode which placed an upper limit on the depth of the reaction zone. All of these experiments indicate that the depth of the reaction zone is 2–3 mm with a tritium source at room temperature.

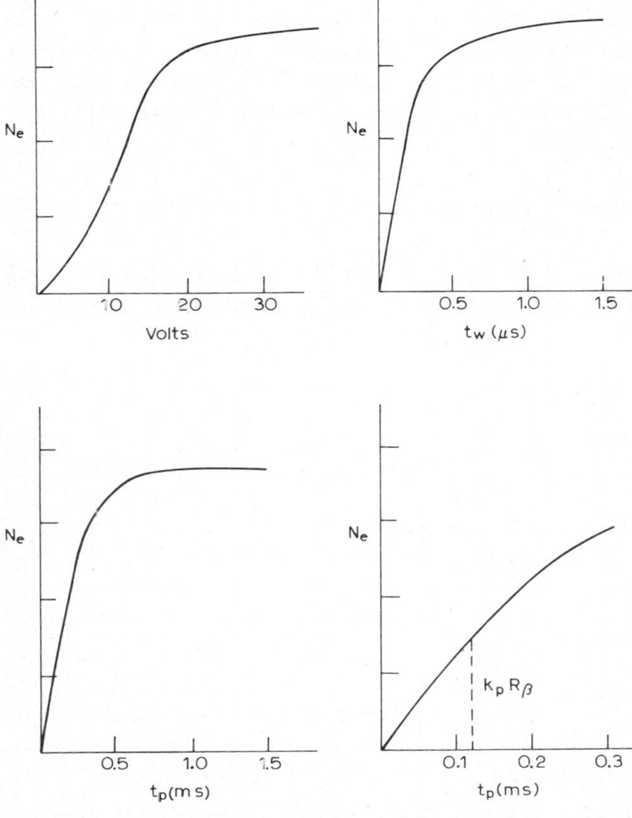

Fig. 3.1. The number of electrons collected per pulse (N_e) versus voltage (V), pulse width (t_w) and pulse period (t_p).

The variation of N_e with temperature, pressure, and flow-rate at specified values of the pulse parameters has been measured [46–48]. With a tritium source and a "clean" system, the variations are slight and can be explained by the normal gas laws. In the event of contamination, or oxygen intrusion, the N_e values will vary considerably so that the degree of variation can be used to indicate the quality of the experimental apparatus. In general, a flow-rate of 150 ml/min and a pressure slightly above atmospheric is utilized.

The electrons are generally assumed to have a thermal distribution. This has been verified by altering the energy and measuring the capture coefficient as a function of energy. Examples of this are given in refs. 1 and 10. Thermal electron attachment rate constants for reference molecules such as SF_6 and C_7F_{14} [49] determined with the ECD are in good agreement with values determined in other experiments where the electron energy is known to be thermal. This suggests a thermal distribution of the ECD. This is discussed further in Section 3.4.

The physical loss mechanisms for the positive ions have not been clearly defined. However, it is clear that loss through ventilation and diffusion is small compared to the kinetic time framework considered in the reaction model. When a field is applied to collect the electrons in a pulsed manner, there will be a temporary build up of the positive species until the rate of loss of the positive species is equal to the rate of collection of the electrons. Then, a fraction of the positive ions will be collected when the pulse is applied which is the positive ion loss assumed in the numerical solution to the coupled differential equations. The physical loss of negative ions is small since they would have to be collected at the opposite electrode.

3.3.2. Chemical reactions in the electron-capture detector

The reaction sequence in the kinetic model is summarized in the following, where AB represents any polyatomic molecule capable of capturing or attaching an electron:

$$\beta + (\mathrm{Ar} + 10\%\ \mathrm{CH_4}) \xrightarrow{k_\mathrm{p} R_\beta} \mathrm{e}^- + \oplus + \mathrm{R} + \beta^* \tag{3.5}$$

$$\mathrm{e}^- + \oplus \xrightarrow{k'_\mathrm{D}} \text{neutral species} \tag{3.6}$$

$$\mathrm{e}^- + \mathrm{AB} \xrightarrow{k_{12}} \mathrm{A} + \mathrm{B}^- \tag{3.7}$$

$$\mathrm{e}^- + \mathrm{AB} \xrightarrow{k_1} \mathrm{AB}^- \tag{3.8}$$

$$\mathrm{AB}^- \xrightarrow{k_{-1}} \mathrm{AB} + \mathrm{e}^- \tag{3.9}$$

$$\mathrm{AB}^- \xrightarrow{k_2} \mathrm{A} + \mathrm{B}^- \tag{3.10}$$

$$\mathrm{AB}^- + \oplus \xrightarrow{k'_{\mathrm{N1}}} \text{neutral species} \tag{3.11}$$

$$\mathrm{B}^- + \oplus \xrightarrow{k'_{\mathrm{N2}}} \text{neutral species} \tag{3.12}$$

where β^* designates the β particle with reduced energy as a result of the ion pair formation. Generally β contains sufficient energy to cause subsequent formation of ion pairs. Each ion pair formation requires ca. 40 eV and a single β particle with energy in the keV or MeV range can form numerous ion pairs in the carrier gas. The rate constant k_pR_β is used to represent the overall rate of ion pair formation while the other k's are standard kinetic rate constants.

In this kinetic scheme several reaction steps may involve collision with a neutral species to either add or remove the necessary energy for the reaction to take place. This would increase the order of the reaction step. However, at high pressures (ca. 1 atm) these neutral species will be at comparatively high concentrations, compared to the capturing species AB or the ionic species, and will remain constant. For this reason they are not shown explicitly in the kinetic mechanism.

If the electron attachment process forms AB^-, which does not dissociate according to reaction step 3.9, then it is impossible to differentiate kinetically between reaction steps 3.7 and 3.8 in terms of electron capture. For this reason in the subsequent rate expressions we have left out reaction step 3.7, and Mechanism β, as we define it later, will inherently describe also electron attachment via reaction 3.7. Generally reactions 3.7 and 3.8 differ in that reaction 3.7 frequently can have a significant activation energy whereas for reaction 3.8 the activation energy is generally small.

The rate expressions describing the change in concentration of the various species between pulses for reactions 3.5, 3.6 and 3.8–3.12 are given by:

$$\frac{d[b]}{dt} = \frac{d[\oplus_0]}{dt} = k_pR_\beta - k'_D[\oplus_0][b] \tag{3.13}$$

$$\frac{d[e^-]}{dt} = k_pR_\beta - k'_D[\oplus][e^-] - k_1[AB][e^-] + k_{-1}[AB^-] \tag{3.14}$$

$$\frac{d[\oplus]}{dt} = k_pR_\beta - k'_D[\oplus][e^-] - k'_{N_1}[\oplus][AB^-] - k'_{N_2}[\oplus][B^-] \tag{3.15}$$

$$\frac{d[AB^-]}{dt} = k_1[AB][e^-] - k_{-1}[AB^-] - k'_{N_1}[\oplus][AB^-] - k_2[AB^-] \tag{3.16}$$

$$\frac{d[B^-]}{dt} = k_2[AB^-] - k'_{N_2}[\oplus][B^-] \tag{3.17}$$

where $[b]$ = concentration of electrons when no capturing species is present, $[\oplus_0]$ = concentration of positive species in the absence of capturing species, and the remaining ion concentrations $[e^-]$, $[\oplus]$, $[AB^-]$, $[B^-]$ are those when the capturing species AB is present.

3.3.3. Numerical solution

In the previous analysis of this kinetic model [1] the positive ion concentration was assumed to build up to a constant value as a result of electron withdrawal by the applied negative pulsed potential. Furthermore, it was assumed that the positive ion concentration did not change significantly when the electron-capturing species is present. This simplified the mathematical analysis since the differential equations could be solved explicitly for the electron concentration. The effect of the pulsed potential was inherently accounted for in the build up of the positive ion concentration.

However, in a recent study [26] it was shown that the positive ion concentration decreases at the same time that the electron concentration decreases. For this reason a rigorous solution of the differential equations must consider the positive ion concentration as a variable. In addition the mode of removing the positive ions must be specified and the solution of the differential equations repeated through numerous pulsed cycles until a steady state is reached. This procedure must be carried out for each specified pulse period, t_P, and concentration. For a complete set of concentrations at various pulse periods these calculations can be time consuming. This is especially true for the higher concentrations where the differential equations show greater changes in rates and the numerical solution requires smaller time increments in the integration.

The differential equations 3.13–3.17 were integrated numerically assuming the concentration of AB remains constant, which is a reasonably good assumption under normal electron-capture operation. The procedure for the numerical integration is described in ref. 42. At the completion of the integration over the time of the designated pulse period, t_P, the electrons were removed as anticipated by the application of a negative potential commonly practiced in pulsed electron-capture operation. A small fraction (f) of the positive ion concentration was also removed at this time. This mode of positive ion removal simulates the collection of a fraction of the positive ions at the cathode where the pulsed negative potential is applied. This mode of positive ion removal would apply most likely to the parallel plate tritium ECD where the positive ions are in close proximity to the cathode. Other processes for positive ion removal could also be important, especially at long pulse intervals. Most certainly for the ^{63}Ni detector other modes of positive ion loss must be considered since the positive ions are distributed throughout the cell, far removed from the cathode.

This process was repeated as in the operation of an electron-capture detector until a steady state of electron concentration at the duration of the pulse period was attained. The criterion for steady state was a change in electron concentration of less than one part in 10,000. In practice, extrapolation techniques were used so that steady state could be reached more rapidly, thus decreasing the computational time. The details can be found in ref. 42.

3.3.4. Kinetic mechanisms

In an earlier review paper [39] the electron attachment mechanisms were classified into four mechanisms based on the nature of the potential energy curves of the negative ion in relationship to that of the neutral species. In a later paper [50] three temperature regions were defined in terms of the relative magnitude of the rate constants and identification with the previously defined four mechanisms was made. We will follow this same classification where we have replaced k_L with $k'_{N1}[\oplus]$.

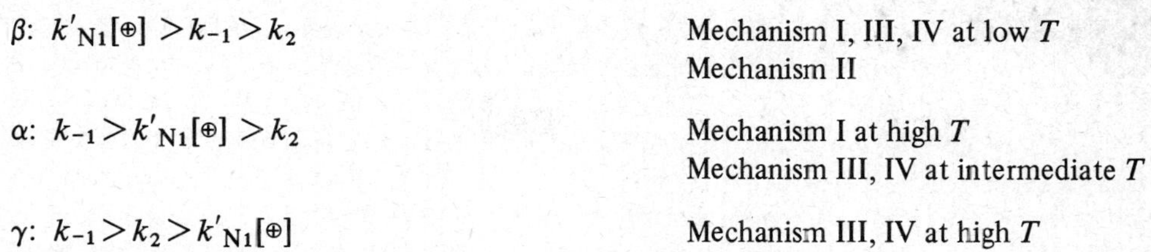

β: $k'_{N1}[\oplus] > k_{-1} > k_2$ — Mechanism I, III, IV at low T; Mechanism II

α: $k_{-1} > k'_{N1}[\oplus] > k_2$ — Mechanism I at high T; Mechanism III, IV at intermediate T

γ: $k_{-1} > k_2 > k'_{N1}[\oplus]$ — Mechanism III, IV at high T

The expressions for the capture coefficients at steady state, K_∞, have been expressed in terms of k'_D and k'_N since we are now assuming that the positive ion concentration can change. In principle a fourth kinetic region could occur where $k_2 > k_{-1} > k'_{N1}[\oplus]$ and this would give K_∞ identical to Mechanism β. However, there have been no published data supporting this fourth mechanism so we will not consider it at this time.

The kinetic regions and the potential energy curves are illustrated in Fig. 3.2. The kinetic regions can be identified generally from a graph of ln $KT^{3/2}$ vs. $1/T$. If the graph has a negative or zero slope with an intercept in the range 28–35, then the data are in the beta region (some of the smaller molecules such as oxygen have an intercept as low as 22). If the graph has a large negative slope with an intercept greater than about 28, then it is a gamma region and if the graph has a positive slope with an intercept in the region 10–18, then it is in the alpha region. A high negative slope can be either a beta or a gamma region and can only be differentiated on the basis of the formation of an intermediate stable negative ion prior to dissociation.

The four mechanisms based on the potential energy curves involve (I) the formation of

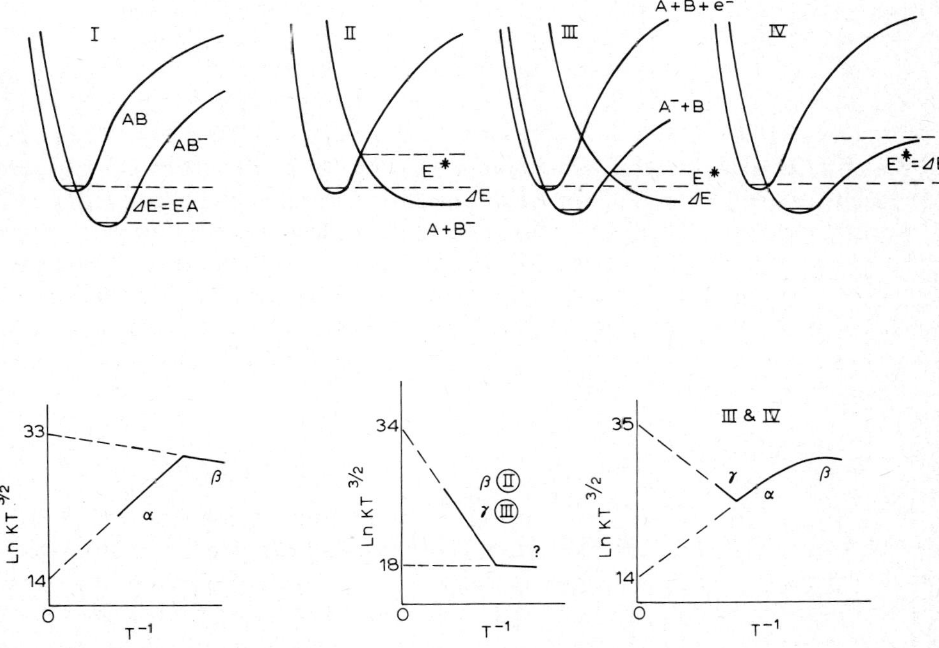

Fig. 3.2. Electron-capture mechanisms and temperature dependence.

a non-dissociative molecular ion, (II) a single bimolecular electron attachment step leading immediately to dissociation via a dissociative potential energy curve, (III) a two-step dissociative process where the formation of a molecular negative ion is followed by a dissociative step which proceeds via a dissociative potential energy curve, and (IV) a two-step dissociative process which first involves the formation of a molecular negative ion followed by dissociation by means of the same potential energy curve. Mechanisms I and III involve a negative ion curve with a large dissociation energy but with Mechanism III this curve is crossed by a dissociative curve. In Mechanism II, the negative ion curve has a small or zero dissociation energy while for Mechanism IV, the dissociation energy of the negative ion is thermally accessible.

3.3.5. Electron concentration dependence on pulse period

The electron concentration in the electron-capture detector, when no capturing species is present, is found by solving the differential equations given in eqn. 3.13. The initial slope of the electron concentration versus pulse period is given by $k_{\mathrm{p}}R_{\beta}$ which in turn is dependent on the activity of the radioactive source. The value of $k_{\mathrm{p}}R_{\beta}$ does not affect the slopes of the electron concentration versus t_{p} curves, but only the overall magnitude. In our calculations we selected a value of $4 \cdot 10^{-10}$ mole/l·s which is typical of a tritium source of 150 mCi.

In our initial solution of eqn. 3.13 we examined the electron concentration as a function of pulse period for different k'_{D} and f values. It was noted that the ratio of electron to positive ion concentration remained constant for any curve and in fact the ratio was simply the fraction (f) of positive ions removed at the end of the pulse period

$$f = \frac{b}{[\oplus_0]} \tag{3.18}$$

This is understandable since the removal of electrons by pulsing at the end of any pulse period must equal the removal of positive ions other than by recombination. Since $[\oplus_0]$ is given by b/f in eqn. 3.13, the steady state expression at $t_{\mathrm{p}} = \infty$ is given by

$$\frac{\mathrm{d}b}{\mathrm{d}t} = 0 = k_{\mathrm{p}}R_{\beta} - \frac{k'_{\mathrm{D}}}{f}(b^{\infty})^2$$

The initial slope at $t_{\mathrm{p}} = 0$ ($b = 0$) (see Fig. 3.1) is

$$\left.\frac{\mathrm{d}b}{\mathrm{d}t}\right|_{t_{\mathrm{p}}=0} = k_{\mathrm{p}}R_{\beta}$$

The constant $k_{\mathrm{p}}R_{\beta}$ can be evaluated from the initial slope and k'_{D}/f from

$$\frac{k'_{\mathrm{D}}}{f} = \frac{\text{initial slope}}{(b^{\infty})^2}$$

In order to illustrate the agreement between the numerical solution to eqn. 3.13 and

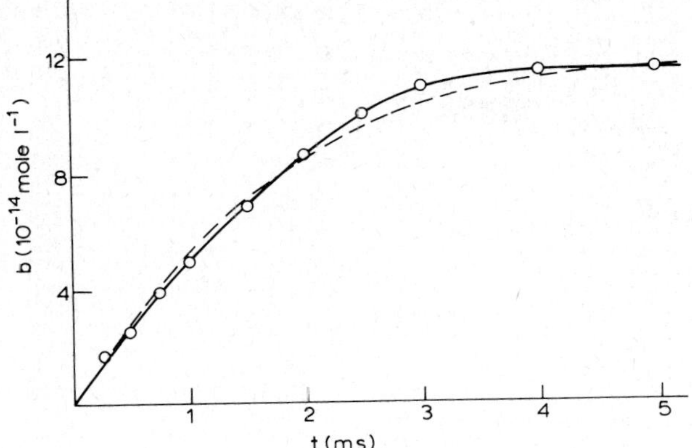

Fig. 3.3. Number of electrons collected per pulse versus pulse period (t_p): experimental (ref. 51) ——; calculated – – –.

experiment, we show in Fig. 3.3 the experimental graph of Van de Wiel and Tommassen [51] for ionization of argon + 10% methane. The calculated curve agrees well with the experimental curve except in the region of 3000 μs. The experimental curve tends to plateau sooner than the calculated curve and this can only occur by removal of more positive ions at shorter pulse intervals, allowing the electron concentration to increase. Our calculations remove more positive ions at shorter intervals compared to removal by other mechanisms, such as diffusion. Consequently, we cannot account for the discrepancy in the two curves. If one assumes the positive ion concentration remains constant, independent of pulse interval, this results in an even greater discrepancy with the experimental curve.

It should be noted, in Fig. 3.3, that no maximum occurs in the calculated curve, regardless of the choice of constants. This is in agreement with Van de Wiel and Tommassen [51], who have shown that the maximum in this type of graph arises from contamination of the detector with O_2.

3.3.6. Concentration dependence at steady state

In order to arrive at the concentration dependence predicted by the kinetic model, the differential rate expressions in eqns. 3.13–3.17 were solved numerically over the pulse period and this was repeated until a constant concentration of ionic species was observed. The calculations were made for different concentrations of the capturing species and pulse intervals for the three mechanisms described previously. The major results of these calculations are reported in ref. 42 but for general purposes, the concentration dependence at steady state or long pulse intervals is desired and can be obtained by setting eqns. 3.13, 3.14, and 3.16 to zero.

$$0 = k_p R_\beta - k'_D [\oplus_0] b \tag{3.20}$$

$$0 = k_p R_\beta - k'_D [\oplus] [e^-] - k_1 [AB] [e^-] + k_{-1} [AB^-] \tag{3.21}$$

$$0 = k_1 [AB] [e^-] - k_{-1} [AB^-] - k'_{N1} [\oplus] [AB^-] - k_2 [AB^-] \tag{3.22}$$

Solving eqn. 3.22 for $[AB^-]$ and eqn. 3.20 for $k_p R_\beta$, substitution into eqn. 3.21 reduces to

$$\frac{b}{[e^-]} = \frac{[\oplus]}{[\oplus_0]} + \frac{k_1(k_2 + k'_{N1}[\oplus])}{k'_D[\oplus_0](k_{-1} + k_2 + k'_{N1}[\oplus])}[AB] \tag{3.23}$$

Obviously if $[\oplus] = [\oplus_0]$ then the expression reduces to that derived previously [50]. However, it has been our observation from the numerical solutions that regardless of the concentration of AB, pulse period, or mechanism, at steady state the ratio of concentration of electrons to positive ions at steady state remains approximately constant, equal to the fraction (f) of positive ions removed at the end of each pulse period.

$$\frac{[e^-]}{[\oplus]} = f \tag{3.24}$$

As with eqn. 3.18, this result can be rationalized on the basis of the removal of equal amounts of electrons and positive ions, thus requiring a higher concentration of positive ions to offset the fact that only a fraction (f) is removed. In any event, eqns. 3.18 and 3.24 can be justified on an empirical basis and substitution into eqn. 3.23 reduces to:

$$\frac{b^2 - [e^-]^2}{b[e^-]} = \frac{k_1(k_2 + k'_{N1}[\oplus])}{k'[\oplus_0](k_{-1} + k_2 + k'_{N1}[\oplus])}[AB] \tag{3.25}$$

This equation can be reduced to simpler expressions for each of the mechanisms by using eqn. 3.24 and noting the $[\oplus_0]$ = constant so that $k'_D[\oplus_0] = k_D$

Mechanism β: $k'_{N1}[\oplus]) > k_{-1} > k_2$

$$\frac{b^2 - [e^-]^2}{b[e^-]} = K_\infty[AB] = \frac{k_1}{k_D}a \tag{3.26}$$

Mechanism α: $k_{-1} > k'_N[\oplus] > k_2$

$$\frac{b^2 - [e^-]^2}{[e^-]^2} = K_\infty[AB] = \frac{k_1 k'_{N1}}{k_{-1}k_D}a \tag{3.27}$$

Mechanism γ: $k_{-1} > k_2 > k'_{N1}[\oplus]$

$$\frac{b^2 - [e^-]^2}{b[e^-]} = K_\infty[AB] = \frac{k_1 k_2}{k_{-1}k_D}a \tag{3.28}$$

Obviously from eqns. 3.27 and 3.28 we expect the function $(b^2 - [e^-]^2)/b[e^-]$ to be linear with concentration at long pulse intervals for Mechanisms β and γ. Mechanism β is probably the most prevalent since it encompasses dissociative as well as non-dissociative attachment. For Mechanism α we expect the function $(b^2 - [e^-]^2)/[e^-]^2$ to be linear with concentration at long pulse intervals.

It is interesting to note that both of these functions reduce to $(b - [e^-])/[e^-]$ at low capture where $[e^-]$ approaches b:

$$\frac{b^2 - [e^-]^2}{b[e^-]} = \frac{b - [e^-]}{[e^-]} \cdot \frac{b + [e^-]}{b} \approx \frac{b - [e^-]}{[e^-]} \cdot 2 \tag{3.29}$$

$$\frac{b^2 - [e^-]^2}{[e^-]^2} = \frac{b - [e^-]}{[e^-]} \cdot \frac{b + [e^-]}{[e^-]} \approx \frac{b - [e^-]}{[e^-]} \cdot 2 \tag{3.30}$$

The factor of 2 in each case would be incorporated into K.

However, at concentrations where you get a high percent capture the functions in eqns. 3.26, 3.27 and 3.28 will differ considerably from $(b - [e^-])/[e^-]$. At high capture the two functions should show deviations in the opposite directions. For Mechanisms β and γ, as given in eqns. 3.26 and 3.28, a graph of $(b - [e^-])/[e^-]$ versus concentration should show positive deviations from linearity since $(b + [e^-])/b$ decreases from 2 down to a lower limit of 1. However, for Mechanism α as given by eqn. 3.27, a graph of $(b - [e^-])/[e^-]$ should show negative deviations from linearity since $(b + [e^-])/[e^-]$ increases from 2 to unlimited values as $[e^-]$ decreases to zero.

3.3.7. Temperature dependence

If the ECD data are taken at low (10–20%) electron capture, then the response function reduces to

$$\lim_{a \to 0} \frac{b - [e^-]}{[e^-]} = K_\infty a \tag{3.31}$$

where K_∞ is given by eqns. 3.26, 3.27, and 3.28 for the various mechanisms. This is very important since most of the earlier ECD data were reduced according to the functional form of eqn. 3.31 and were taken at low capture. With this experimental restriction, the previous electron-capture coefficient, which we will designate as K', is equal to $K_\infty/2$ so that the temperature dependence will be exactly as before. For the alpha mechanism,

$$K' = \frac{1}{2} \frac{k_1}{k_{-1}} \frac{k'_{N1}}{k'_D} = \frac{1}{2} \frac{k'_{N1}}{k'_D} \frac{A}{T^{3/2}} e^{EA/RT} \tag{3.32}$$

$$\ln K'T^{3/2} = \ln \frac{k'_{N1}}{2k'_D} + \ln A + \frac{EA}{RT} \tag{3.33}$$

For the beta mechanism

$$K' = \frac{1}{2} \frac{k_1}{k_D} = \frac{A_1 e^{-\frac{E_1^*}{RT}}}{k_D} \tag{3.34}$$

so that the rate constant for thermal electron attachment and the activation energy (either dissociative or non-dissociative) can be determined.

For the gamma region

$$K' = \frac{1}{2}\frac{k_1 k_2}{k_{-1} k_D} = \frac{1}{2}\frac{A_1 A_2}{A_{-1}}\frac{1}{k_D} \cdot e^{-(E_1^* + E_2^* - E_{-1}^*)/RT} \tag{3.35}$$

The combination of the activation energies can be determined from the temperature dependence of the ECD response. Thus the results shown in Fig. 3.2 are still the temperature dependencies to be expected from this solution at steady state.

3.4. COMPARISON OF THEORY WITH EXPERIMENTAL RESULTS

3.4.1. Introduction

The ideal test of the theory would be the calculation of the concentration and temperature dependence of the electron-capture coefficient from values of the rate constants, electron affinities and/or activation energies determined by some other established experimental procedure. Unfortunately, during the development of the model and even now, very little such data are available. At present, the only compound with all of the pertinent rate constants available is oxygen.

Therefore it was necessary to obtain confirmation of the model by comparing the quantities obtained from the ECD with the comparable quantities obtained by other methods. In the case of the non-dissociative compounds, this consists of the molecular electron affinity and the rate constants for thermal electron attachment. Presently there are about a dozen molecular electron affinities which have been determined by both the ECD data and some other technique. Approximate rate constants obtained from the ECD and some other technique are available for a like number of compounds.

In the case of the dissociative compounds, there are two electron affinities of radicals which can be compared and there are three rate constants for thermal electron attachment which have been accurately measured by the ECD method and some other method. However, in general the dissociative mechanism must be compared with literature data by the use of semi-empirical potential energy curves for the neutral molecule and the negative ions. The experimental quantities used for comparison include the dissociative attachment cross section, vertical and adiabatic electron affinities, activation energies for dissociative processes have also been correlated to the exothermicity of the reaction for aliphatic halides.

The kinetic model has also been verified by correlation of the molecular parameters obtained from the kinetic model with other theoretical and experimental parameters such as discussed earlier. These studies are still quite important because of the limited amount of data available for direct comparison.

To summarize, there do not appear to be any serious discrepancies between molecular parameters derived from the kinetic model with ECD data and those derived from other experimental techniques. In fact the correlation is so good that we propose the use of molecular parameters from other techniques to predict electron-capture response for compounds that have not been used in analyses. This will be discussed in Section 3.5.

3.4.2. Electron affinities and electron attachment rate constants

At present, there are about a hundred molecular electron affinities that have been reported in the literature and about a half of these have been determined with the ECD. A distinction is made here between stable molecules and radicals and unstable free radicals. The electron affinities of NO [52], O_2 [51,52], CS_2 [53], COS [53], N_2O [54], NO_2 [55], SF_6 [56], CH_3NO_2 [52], $C_6H_5NO_2$ [41], biacetyl [57], 1,4-naphthoquinone [58], and the acetate radical [59] have been determined with both the ECD and an independent experimental technique. In a recent paper Chen and Wentworth [53] have favorably compared the results obtained for these compounds with the ECD and the results with other techniques. The conclusions reached in that article will be summarized in this section. First a discussion of the other experimental techniques used to measure electron affinities will be presented.

Two absolute methods for determining molecular electron affinities have been developed. The term "absolute" means that the experimental measurements yield an estimate of the electron affinity with only the use of fundamental constants; that is, the method does not have to be calibrated. These techniques are the measurement of the energy change of reaction 3.1 from the temperature dependence of the equilibrium constant, and the measurement of the threshold energy of the reaction:

$$AB^- + h\nu \rightleftharpoons AB + e^- \tag{3.36}$$

The first method is a standard equilibrium procedure which is applicable to any reaction while the second is a photodetachment process which depends upon the determination of the threshold.

There are two general thermochemical procedures which have been used extensively for the determination of molecular electron affinities: the magnetron surface ionization method [60] and the ECD method, also called the pulse sampling method for the study of thermal electron attachment (PSTEA). In the magnetron technique, the equilibrium constant is determined by measuring the free electron and the negative ion concentrations formed in the presence of a known concentration of sample. The separation of the ions and the electrons is accomplished by a magnetic field. For non-dissociative molecules, the electron affinity is determined from the temperature dependence of the equilibrium constant. This procedure has been generally applied to molecules with electron affinities greater than 1.0 eV and is hence complementary to the ECD method. The equilibrium constant can also be determined by measuring the rate constant for the forward and the reverse reactions. The electron affinities of O_2 [61] and NO [62] have been determined in this manner using a drift tube and the flowing afterglow, respectively.

The experiments involving photons have been carried out in two ways: the determination of the threshold for the production of the electrons, photodetachment (PD), and the measurement of the energy of the detached electrons, photoelectron spectroscopy (PES). The PES technique has given the most precise measurements of the electron affinities of O_2 [63] and NO [64] while the PD procedure has been applied to NO_2 [65,66]. Thus only these three molecules have electron affinities which have been determined by more than one absolute method.

The electron affinities of a number of molecules studied in the ECD have been determined by techniques yielding relative electron affinities. If a charge transfer reaction from one negative ion to a neutral takes place rapidly, then it is assumed that the reaction is exothermic and that the electron affinity of the second species is greater than that of the first. In this manner, the relative electron affinities can be obtained. If the species are in thermal equilibrium, this is called thermal charge transfer (TCT). If the negative ion is

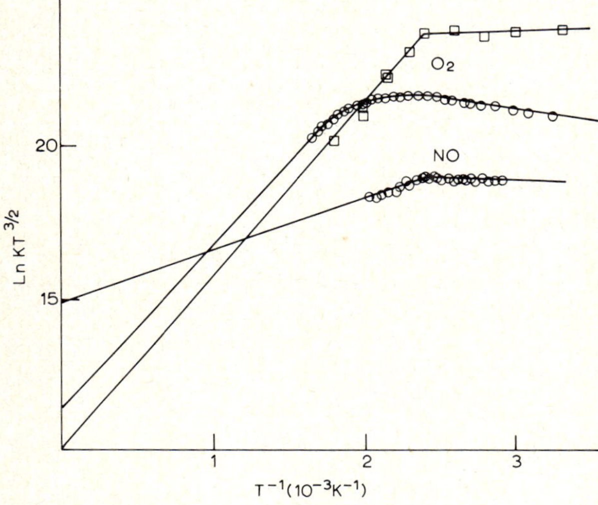

Fig. 3.4. ln $KT^{3/2}$ versus $1/T$ for NO and O_2.

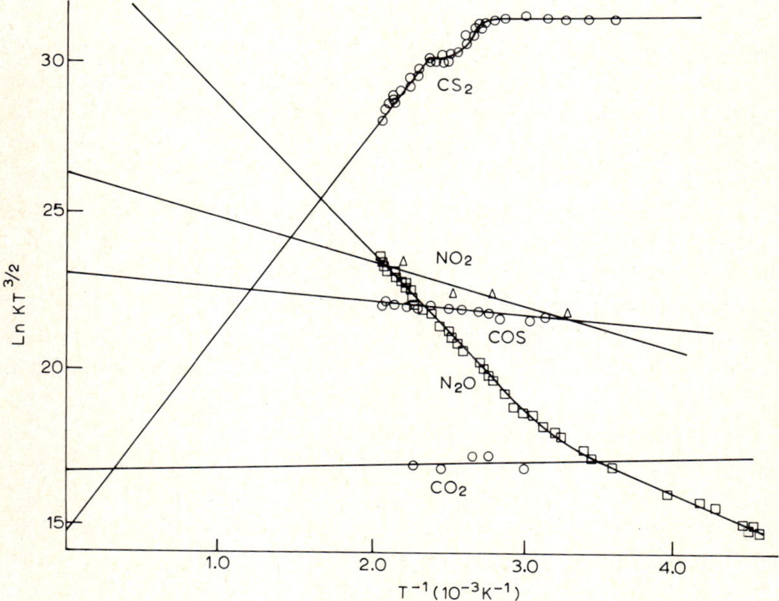

Fig. 3.5. ln $KT^{3/2}$ versus $1/T$ for CO_2, COS, NO_2, N_2O and CS_2.

given additional energy as in a beam, then the procedure is termed endothermic charge transfer (ECT) and the threshold for the reaction must be determined and included in the calculation. Relative electron affinities can also be obtained when ion pairs are formed in a mass spectrometric experiment. If the impacting species is an alkali metal beam (AMB) then the measured threshold and the known alkali metal ionization potential are used to obtain the electron affinity. Ion pair formation can also be initiated by an electron beam (EIP) in which case the threshold of positive ionization and the threshold for ion pair formation must be measured. The determination of electron affinities from polarographic half wave reduction potentials in aprotic solvents yields very good relative values. Absolute values can be obtained by an empirical relationship of gas phase electron affinities with the solution values [67]. However, this evidence will only be used for one compound in this section and will be discussed in more detail in Section 3.5.

Some molecular electron affinities have also been estimated from lifetimes of negative ions, the cross section for attachment and the vibrational properties of the neutral and the negative ion by means of a theoretical relationship (THEOR). This procedure has been applied to biacetyl, SF_6, nitrobenzene and 1,4-naphthoquinone [68].

The temperature dependence of the electron-capture coefficients are shown for diatomic molecules in Fig. 3.4, for triatomic molecules in Fig. 3.5, and for polyatomic

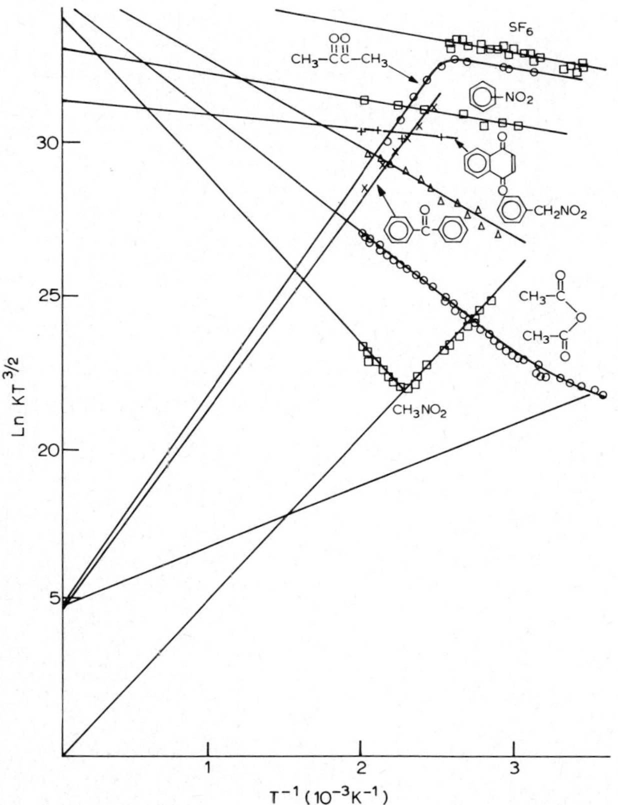

Fig. 3.6. ln $KT^{3/2}$ versus $1/T$ for acetic anhydride, nitromethane, benzophenone, α-nitrotoluene, naphthoquinone, nitrobenzene, biacetyl, and sulfur hexafluoride.

molecules in Fig. 3.6. The compounds with both alpha and beta regions are easy to identify from the slopes and the intercepts as, for example, for oxygen and biacetyl. The gamma region for acetic anhydride, nitromethane, and α-nitrotoluene, is also easily identified. The potential energy curves are more difficult to establish. A single beta region could signify any of the four mechanisms; a single alpha region, Mechanism I or IV; and a single gamma region, Mechanism III or IV. A combination of an alpha and a beta region implies Mechanism I or IV; and a combination of a beta region with a gamma region, or all three regions implies Mechanism III or IV. The exact mechanism can only be established with auxiliary data. However, the presence of an alpha region always leads to the determination of the electron affinity of the molecule; the presence of a gamma region always leads to either a precise value of the electron affinity of the radical or a limit to the electron affinity of the radical in Mechanism IV or III, respectively; and a beta region will always lead to an estimate of the rate constant for thermal electron attachment and the activation energy for the attachment reaction.

The electron affinities obtained from the data in Figs. 3.4–3.6 are listed in Table 3.1, column 1, while the absolute electron affinities from the literature are given in column 2 and the other literature values are listed in column 3. The literature values are identified

TABLE 3.1

ELECTRON AFFINITIES OF MOLECULES AND RADICALS (ECD AND LITERATURE VALUES)

For abbreviations, see text. Method and reference in parentheses.

Molecule	EA(ECD) (eV)	EA(Lit., absolute) (eV)	EA(Lit., other methods) (eV)
NO	0.1 ± 0.1 [52]	0.028 (Th, [62])	$\geqslant 0.09$ (ECT, [69])
		0.024 ± 0.01 −0.08 (PES, [63])	$\geqslant 0.015$ (ECT, [70])
			$> 0.1 \pm 0.1$ (AMB, [71])
			$\leqslant 0.46$ (AMB, [72])
O_2	0.5 ± 0.1 [51]	0.43 ± 0.01 (Th, [61])	0.45 ± 0.1 (ECT, [73])
			$\geqslant 0.48$ (ECT, [69])
			0.5 ± 0.1 (ECT, [74])
			$\geqslant 0.46 \pm 0.05$ (AMB, [75])
NO_2	2.11 ± 0.2 [57]	3.10 ± 0.05 (PD, [66])	$\geqslant 2.32$ (ECT, [76])
	$> 1.8 \pm 0.2$ [57]	2.36 ± 0.10 (PD, [65])	$\leqslant 2.50$ (ECT, [76])
			$\geqslant 3.6$ (TCT, [77])
			$\geqslant 1.8$ (TCT, [78])
			$> 2.5 \pm 0.1$ (ECT, [74])
			2.04 (ECT, [69])
			2.1 (ECT, [79])
			2.28 ± 0.1 (ECT, [70])
			2.5 ± 0.1 (AMB, [71])
			2.50 ± 0.05 (AMB, [80])
CS_2	0.58 ± 0.1 [53]		0.50 ± 0.2 (ECT, [70])
			1.0 ± 0.2 (AMB, [81])
			0.62 ± 0.2 (AMB, [82])

TABLE 3.1 (*continued*)

Molecule	EA(ECD) (eV)	EA(Lit., absolute) (eV)	EA(Lit., other methods) (eV)
COS	>0.4 ± 0.1 [53]		0.46 ± 0.2 (AMB, [81])
N_2O	0.27 ± 0.2 [54]		0.22 ± 0.1 (Cor, [83]) 0.6 ± 0.2 (ECT, [92])
CH_3NO_2	0.45 ± 0.05 [57]		0.44 ± 0.2 (AMB, [84])
$C_6H_5NO_2$	⩾0.8 ± 0.2 [57] >0.5 ± 0.2 [100]		<1.10 ± 0.04 (TCT, [85]) >0.7 ± 0.2 (TCT, [85]) >0.53 (Theor, [68])
SF_6	⩾0.7 ± 0.2 [56]		0.54 ± 0.2 (AMB, [87]) 0.46 ± 0.2 (AMB, [88]) 0.6 ± 0.1 (TCT, [86]) ⩾0.46 (TCT, [90]) <1.465 (TCT, [90]) 1.05 (ECT, [91]) ⩾1.06 (Cor, [68]) 0.75 ± 0.1 (TCT, [89])
$C_4H_6O_2$	0.63 ± 0.05 [57] 0.6 (single point) [33]		⩾0.41 (Theor, [68])
$C_2H_3O_2$	3.36 ± 0.05 [59]		3.30 (EI, [93]) 3.39 (TCT, [94])
Benzophenone	0.64 ± 0.1 [53]		⩾1.59 (TCT, [95]) ⩽1.89 (TCT, [95]) 0.63 ± 0.2 ($E_{\frac{1}{2}}$, [97])
Naphthoquinone	⩾0.8 ± 0.2 [58]		1.86 ± 0.3 (TCT, [95]) ⩾0.6 (Theor, [68])

by one of the methods discussed earlier and a reference number given in parentheses following the value(s) [69–95]. In Fig. 3.7, the values obtained from the ECD method are plotted versus the literature values. If there is a clearcut "best" literature value, this has been used, otherwise, the average value is used. The line is drawn with a unit slope and a zero intercept and thus has no adjustable parameters.

The general correlation is very good and the results for O_2 and NO are especially important since there are at least two literature values obtained by "absolute" methods and for O_2 there are two independent ECD values. The ECD value for NO_2 agrees well with one of the "absolute" values obtained by photodetachment but does not agree with the others. This could be due to the fact that the "accepted" value could refer to an excited state of the ion and the higher values refer to the ground state. This possibility can not be ruled out since the exact state of the ion is not determined in any of the experimental procedures. Indeed, there are higher estimates for the electron affinities of NO, O_2, and SF_6, as well as NO_2. However, the comparison of the ECD values with the

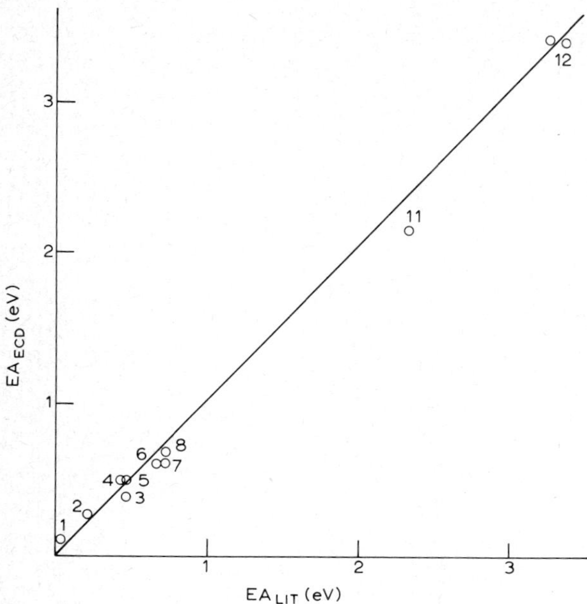

Fig. 3.7. Electron affinity determined by electron capture versus the electron affinity from the literature. (1) (NO, (2) N_2O, (3) COS, (4) O_2, (5) CH_3NO_2, (6) $C_4H_6O_2$, (7) CS_2, (8) SF_6, (11) NO_2, (12) $C_2H_3O_2$.

literature values is still valid as long as the measurements are made on the same state of the negative ion.

The only major discrepancy between the ECD values in Table 3.1 and the literature values is one of the estimates of the electron affinity of benzophenone. From the ECD temperature data obtained from Vessman and Hartvig [96], shown in Fig. 3.6, there is a definite alpha region for benzophenone. This is further substantiated by other ECD temperature data obtained for several substituted benzophenones, which will be discussed in Section 3.5. The electron affinity obtained from these ECD data agrees well with the independent estimate of the electron affinity from ECD data obtained by Gregory and Lovelock [32] and an estimate of the electron affinity obtained from half wave reduction potentials reported by Rashkov et al. [97]. However, based upon thermal charge transfer experiments in an ICR, Rains et al. [95] place the electron affinity of benzophenone greater than the electron affinity of the methoxy radical, 1.59 eV [98]. These experiments also place the electron affinity of maleic anhydride and nitrobenzene above that of the methoxy radical in contradiction to other estimates of EA (nitrobenzene) = 1.1 eV [85,86] and EA (maleic anhydride) = 1.4 ± 0.2 eV [99]. Thus there appears to be a calibration problem in the charge transfer experiments and perhaps the methoxy radical should not be used to set the lower limit of the scale. If nitrobenzene were used as the lower limit then the electron affinity of benzophenone would be less than 1.1 eV in agreement with the ECD and the $E_{\frac{1}{2}}$ estimates.

Also listed in Table 3.1 is the estimate of the electron affinity of biacetyl obtained from a relative capture coefficient determined at a single temperature. In this case there is little doubt that the capture is non-dissociative and this is borne out by the detailed temperature dependence study. However, in general this is not a recommended procedure since there is no confirmation that the capture is non-dissociative and that the data are in the alpha region. If the capture is non-dissociative and the single data point is in the beta region, then the calculated electron affinity will be a lower limit to the electron affinity. If the capture is dissociative the calculated electron affinity will be erroneous. Unfortunately Briegleb [100] has calculated electron affinities for some halogenated compounds using single data points and most certainly these electron affinities are not valid. A high value of the electron affinity can be obtained in the case where the data are for a non-dissociative compound but is referenced to a compound with data in a beta region, as in the case of data for several phthalate compounds presented by Kampars and Neilands [101] and taken from the data of Kuhn et al. [102]. Thus the electron affinities obtained from a single relative capture coefficient reported in the literature must be considered along with these restrictions and should always be verified by the measurement of the temperature dependence.

In order to obtain precise electron attachment rate constants using the ECD, a simultaneous determination of K and $k'_D[\oplus_0]$ must be carried out. This has been done in our laboratory for C_7F_{14}, SF_6, CH_2Cl_2, $CHCl_3$, CCl_4, Cl_2, I_2, and Br_2 by Wentworth et al. [103,104]. Accurate estimates of k_1 can also be obtained from data presented by Van de Wiel and Tommassen [51] and from Freeman [52]. Approximate values of this rate constant can be obtained for nitrobenzene, COS, CS_2, N_2O, NO_2, biacetyl, and naphthoquinone using a nominal value of $k'_D[\oplus_0]$. All of these values are tabulated in Table 3.2

TABLE 3.2

RATE CONSTANTS FOR THERMAL ELECTRON ATTACHMENT

Compound	k_1 (ECD,exact) [51,52,103,104] (cm^3 $molecule^{-1}$ s^{-1})	k_1 (ECD,approximate) [53] (cm^3 $molecule^{-1}$ s^{-1})	k_1 (lit.) [49,105–114]* (cm^3 $molecule^{-1}$ s^{-1})
SF_6	$3.9 \cdot 10^{-7}$		$2.85 \cdot 10^{-7}$
C_7F_{14}	$13.0 \cdot 10^{-8}$		$7.5 \cdot 10^{-8}$
CCl_4	$4.2 \cdot 10^{-7}$		$4.1 \cdot 10^{-7}$
$CHCl_3$	$3.8 \cdot 10^{-9}$		$2.4 \cdot 10^{-9}$
CH_2Cl_2	$2.9 \cdot 10^{-12}$		$4.7 \cdot 10^{-12}$
Cl_2	$2.8 \cdot 10^{-10}$		$3.1 \cdot 10^{-10}$
Br_2	$1.3 \cdot 10^{-10}$		$1 \cdot 10^{-11}$
I_2	$1.4 \cdot 10^{-10}$		$1.8 \cdot 10^{-10}$
O_2	$6.7 \cdot 10^{-12}$		$7 \cdot 10^{-12}$
NO	$2.0 \cdot 10^{-12}$		$2.5 \cdot 10^{-12}$
N_2O		$1 \cdot 10^{-13}$	$1 \cdot 10^{-13}$
COS		$1 \cdot 10^{-11}$	
CS_2		$1 \cdot 10^{-7}$	
NO_2		$1 \cdot 10^{-13}$	$3 \cdot 10^{-13}$
Nitrobenzene		$1 \cdot 10^{-8}$	$1 \cdot 10^{-9}$
Biacetyl		$1 \cdot 10^{-7}$	$4 \cdot 10^{-8}$
1,4-Naphthoquinone		$1 \cdot 10^{-8}$	$7 \cdot 10^{-9}$

*Refs. 68 and 114 offer excellent summaries of rate constants for thermal electron attachment.

along with literature values. The literature values for O_2, NO, and N_2O are three-body rate constants which have been converted to two-body rate constants by considering the 10% methane as the third body with a concentration of $2.4 \cdot 10^{18}$ cm^{-3}. The rate constants are strongly dependent upon the specific third body so that the values for the compounds most closely resembling methane have been chosen for comparison. The values are plotted in Fig. 3.8 as log k_1(ECD) vs. log k_1(LIT). The line has been drawn through zero with a unit slope so that there are no adjustable parameters. Considering the errors in both measurements, the correlation is very good.

Except for NO_2, the activation energy for the rate constants for non-dissociative attachment to these molecules is less than 0.1 eV. Specific values for E_1^* have been determined for SF_6, O_2, and NO and are 0.04 eV, 0.06 eV, and 0.04 eV, respectively.

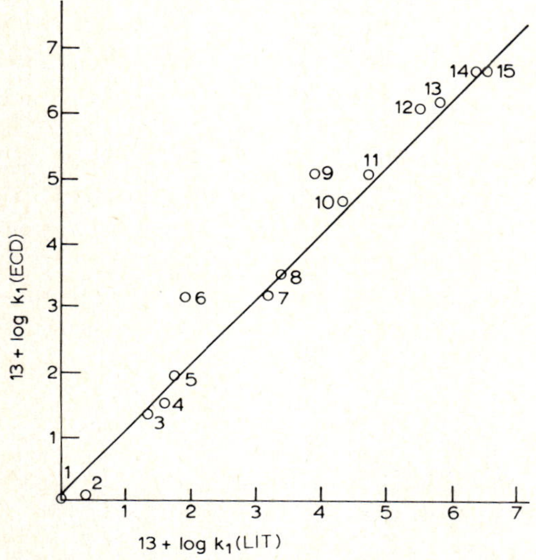

Fig. 3.8. log (10^{13} k_1) determined with the ECD versus log(10^{13} k_1) from the literature. (1) N_2O, (2) NO_2, (3) NO, (4) CH_2Cl_2, (5) O_2, (6) Br_2, (7) I_2, (8) Cl_2, (9) $C_6H_5NO_2$, (10) $CHCl_3$, (11) 1,4-naphthoquinone, (12) $C_4H_6O_2$, (13) C_7F_{14}, (14) SF_6, (15) CCl_4.

3.4.3. Potential energy functions and E^*

There are very few estimates of activation energies for dissociative thermal electron attachment processes reported in the literature. Two examples of comparisons for Mechanism IV are the good agreement between the electron affinities of the acetate radical and NO_2 determined by the ECD and other techniques as shown in Table 3.1 and Fig. 3.7. For Mechanisms II and III, there are little data for comparison so that it is necessary to compare the E^* values obtained with the ECD with E^* values calculated from empirical potential energy curves. These potential energy curves are obtained using some other experimental parameter such as the vertical attachment energy, the adiabatic electron affinity or the cross section for dissociative attachment.

It has been shown [38,39] that two-dimensional potential energy functions can be used to explain the different mechanisms of electron attachment to different molecules.

The Morse function referenced to a zero potential at infinite separation was selected to approximate the potential energy of the neutral molecule. A modification of the Morse potential for the neutral molecule was used to approximate the potential energy function of the negative ion. The modification consisted of (1) the addition of the electron affinity of the negative ion (radical or atom) dissociated from the molecule, and (2) the utilization of a parameter, k_A, to alter the attractive term of the Morse function. For the halogens such functions can be written as:

$$U(X_2) = -2D^0_{X_2} \exp[-\beta(r-r_e)] + D^0_{X_2} \exp[-2\beta(r-r_e)] \tag{3.37}$$

where $U(X_2)$ is the Morse potential energy function.

$$U(X_2^{-*}) = -2k_A D^0_{X_2} \exp[-\beta(r-r_e)] + D^0_{X_2} \exp[-2\beta(r-r_e)] - EA_X \tag{3.38}$$

is the potential energy function of X_2^{-*}. The terms in these functions are defined as

$$D^0_{X_2} = D_{X_2} + \tfrac{1}{2}h\nu_0 \tag{3.39}$$

$$\beta = \left(\frac{2\pi^2\mu}{D^0_{X_2}}\right)^{1/2} \nu_0 \tag{3.40}$$

where $D^0_{X_2}$ is the dissociation energy of X_2, ν_0 is the fundamental vibrational frequency, μ is the reduced mass, r is the internuclear separation, r_e is the equilibrium internuclear distance, and EA_X is the electron affinity of the halogen atom.

In the course of the study of the halogens, it was found that eqn. 3.38 fits the potential energy curve for the excited state X_2^{-*}. However, it does not fit the ground state potential energy function of X_2^-. We have derived a potential energy function that accounts for the ground state of the molecular negative ion. Such a function can be written as:

$$U(X_2^-) = -2D^0_{X_2} \exp[-\beta(r-r_e)] + k_R D^0_{X_2} \exp[-2\beta(r-r_e)] - EA_X \tag{3.41}$$

The only difference between $U(X_2^-)$ and $U(X_2^{-*})$ is the use of k_R, a parameter which modified the repulsive part of the function for the neutral molecule instead of the k_A that modifies the attractive term.

Thus with equations derived from eqns. 3.37, 3.38, and/or 3.41, the quantities k_R [104], and/or k_A [38,39] can be determined from experimental quantities. Then the activation energy, E^*, for thermal electron attachment can be calculated from the value of k_R and/or k_A according to the equations

$$y_1 = \left(\frac{EA_X}{(k_R - 1)D^0_{X_2}}\right)^{1/2} \tag{3.42}$$

$$E^* = -2D^0_{AB} y_1 + k_R D^0_{X_2} y_1^2 - EA_X + D^0_{X_2} - \tfrac{1}{2}h\nu_0 \tag{3.43}$$

$$y_2 = \frac{EA_X}{2(1-k_A)D^0_{X_2}} \tag{3.44}$$

$$E^* = -2k_A D^0_{X_2} y_2 + D^0_{X_2} y_2^2 - EA_X + D^0_{X_2} - \tfrac{1}{2}h\nu_0 \tag{3.45}$$

This has been done for a series of aliphatic and aromatic halogen compounds and the results are shown in Table 3.3 and in Fig. 3.9. The agreement is quite good. The choice of the particular function is difficult and the majority of these data have been calculated from k_A.

TABLE 3.3

ACTIVATION ENERGIES FOR MECHANISMS II AND III [38,40]

Molecule	E^* (calculated) (from potential energy curves) (kcal mole^{-1})	E^* (experimental) (kcal mole^{-1})
CH_3Cl	13.3	12.5 ± 0.4
tert.-C_4H_9Cl	10.3	10.9 ± 0.4
1,2-Dichloroethane	10.0	8.7 ± 0.2
CH_2Cl_2	6.1	7.5 ± 0.4
CH_3Br	5.6	5.7 ± 0.4
$CHCl_3$	5.1	3.4 ± 0.1
CH_2Br_2	0.2	1.2 ± 0.2
CH_3I	1.0	0.6 ± 0.1
$CHBr_3$	0.0	0.2 ± 0.4
CCl_4	0.0	−0.6 ± 0.1
Cl_2	0.8	0.8 ± 0.1
Br_2	0.08	0.2 ± 0.2
I_2	0.8	0.7 ± 0.1
Chlorobenzene	9.76	9.25 ± 0.66
o-Chlorotoluene	11.10	10.26 ± 1.19
o-dichlorobenzene	7.46	6.91 ± 0.29
3′-Chloroacetophenone	5.84	4.60
4′-Chloroacetophenone	5.52	4.80
1-Chloronaphthalene	8.04	9.87 ± 0.29
Bromobenzene	5.53	5.91 ± 0.33
o-Bromotoluene	6.44	6.31 ± 0.34
1-Bromonaphthalene	0.10	1.74 ± 0.43
Iodobenzene	1.20	1.55 ± 0.32

3.5. CORRELATIONS AND PREDICTIONS

3.5.1. Introduction

From the previous section, it is clear that the ECD results obtained from the kinetic model are in basic agreement with data obtained from other techniques for studying the reactions of thermal electrons with molecules. Some fundamental data for a few molecules which have not been studied by the ECD technique are available and can be used for predictive purposes. However, the vast majority of molecules which have been analyzed by the ECD have not been studied with other techniques and the question arises as to whether these results are in agreement with the kinetic model. Also, there are numerous compounds which could be analyzed by the ECD if the response were high

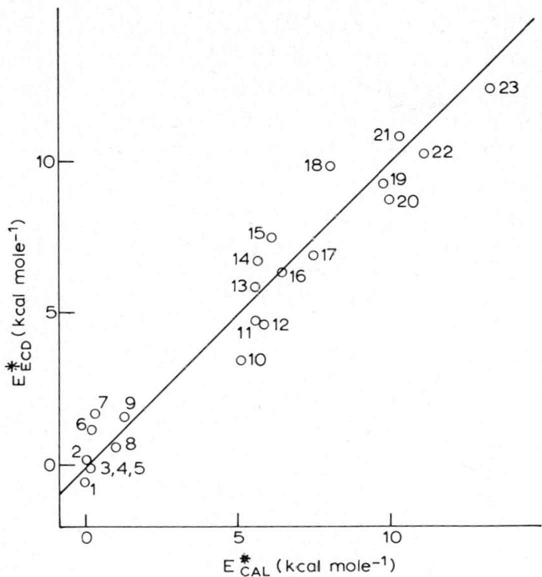

Fig. 3.9. Activation energy, E^*, determined by ECD versus activation energy, E^* calculated with potential energy curves. (1) CCl_4, (2) $CHBr_3$, (3) Cl_2, (4) Br_2, (5) I_2, (6) bromonaphthalene, (7) CH_2Br_2, (8) CH_3I, (9) C_6H_5I, (10) $CHCl_3$, (11) C_6H_5Br, (12) 4′-chloroacetophenone, (13) CH_3Br, (14) 3′-chloroacetophenone, (15) CH_2Cl_2, (16) *o*-bromo-toluene, (17) *o*-dichlorobenzene, (18) 1-chloronaphthalene, (19) C_6H_5Cl, (20) *tert.*-C_4H_9Cl, (21) 1,2-$C_2H_4Cl_2$, (22) *o*-chlorotoluene, (23) CH_3Cl.

enough under optimum conditions. For this reason it would be advantageous if fundamental parameters of the ECD model (electron affinities, activation energies, pre-exponential terms, and rate constants) could be correlated to the structure of the molecule or to some other experimental procedure for which a larger data base exists. Fortunately both of these approaches have been explored in the past and it is a simple matter to test these relationships, as in the previous section, or to examine general trends observed in earlier studies relating structure and response as related to the kinetic model.

The molecular electron affinities can be obtained from correlations with half wave reduction potentials and charge transfer complex absorption spectra [67]. The activation energies for dissociative attachment can be correlated to the exothermicity of the reaction so that if bond dissociation energies and electron affinities of the radicals are available, then E^* can be estimated. In addition, E^* can be estimated from potential energy curves as in the last section. The estimation of the pre-exponential terms and the rate constants for recombination are more difficult but fortunately, the variation is less than plus or minus an order of magnitude so that if the mechanism of the reaction can be determined, then the temperature and concentration dependence can be predicted and an absolute estimate of the response can be made. The correlation of the capture coefficients with the structure of molecules with high response can possibly lead to a rationale for the magnitudes of the attachment rate constants.

3.5.2. Electron affinities and $E_{\frac{1}{2}}$ or E_{CT}

In 1976, Chen and Wentworth examined the correlation between the electron affinities of pi charge transfer complex acceptors determined in the gas phase with $E_{\frac{1}{2}}$ and E_{CT} values measured in solution. At that time, there were fewer than fifteen data points available. At present, there are over fifty compounds with both electron affinities and half wave reduction potential data so that the correlation has been reexamined [115].

The available data are plotted in Fig. 3.10 and are tabulated in Table 3.4 [116–132].

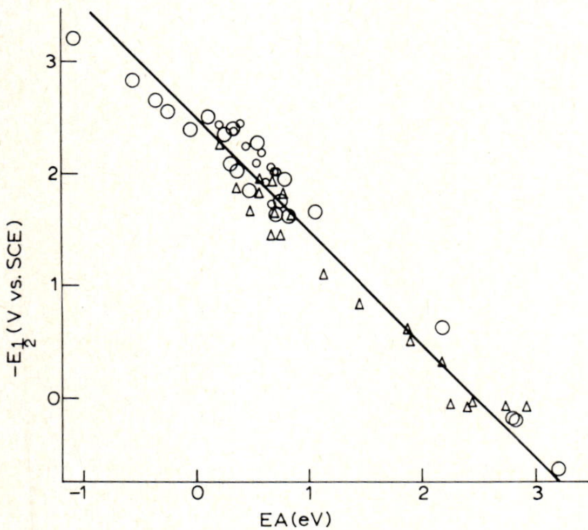

Fig. 3.10. Half wave reduction potential ($-E_{1/2}$) versus gas phase electron affinity. See Table 3.4 for the compounds.

TABLE 3.4

ELECTRON AFFINITIES AND HALF WAVE REDUCTION POTENTIALS

Compound	Electron affinity (eV) [115–127]★	$-(\bar{E}_{\frac{1}{2}})$ [97,128–133] (V vs. SCE)	Deviation (eV) EA − ($E_{\frac{1}{2}}$ + 2.49)
Hexacyanobutadiene	3.24	−0.60	0.15
p-Fluoranil	2.92,2.27	−0.04	0.38,−0.26
Tetracyanoethylene	2.88,2.80	−0.20	0.22,0.11
Tetracyanoquinodimethane	2.83,2.80	−0.16	0.18,0.15
p-Chloranil	2.76,2.40	−0.02	0.25,−0.11
p-Bromanil	2.44	0.00	−0.05
Fluorobenzoquinone	2.16	0.37	0.04
s-Tetracyanobenzene	2.15	0.66	0.32
p-Benzoquinone	1.89	0.53	−0.07
Naphthoquinone	1.86	0.63	0.0
Maleic anhydride	1.40	0.87	−0.22
Nitrobenzene	1.10	1.12	−0.27
o-Dicyanobenzene	1.04	1.68	0.23
Acenaphthylene	0.80	1.65	−0.04

TABLE 3.4 (*continued*)

Compound	Electron affinity (eV) [115–127]*	$-(\bar{E}_{\frac{1}{2}})$ [97,128–133] (V vs. SCE)	Deviation (eV) EA – ($E_{\frac{1}{2}}$ + 2.49)
Cinnamaldehyde	0.80	1.64	–0.02
1,2-Benzpyrene	0.67	2.06	0.25
1-Naphthaldehyde	0.73	1.64	–0.02
1,2-Benzanthracene	0.67	2.06	0.25
Azulene	0.66	1.64	–0.19
Benzophenone	0.64	1.76	–0.09
p-Methylbenzophenone	0.64	1.86	0.11
Dibenz[a,j]anthracene	0.64	2.07	0.22
Dibenz[a,h]anthracene	0.64	2.05	0.23
2-Naphthaldehyde	0.63	1.50	–0.36
Fluoranthene	0.63	1.77	–0.09
Biacetyl	0.63	1.71	–0.15
Cyclooctatetraene	0.60	1.61	–0.28
Pyrene	0.59	2.10	0.20
Anthracene	0.56	1.95	0.02
p-Methoxybenzophenone	0.56	1.88	–0.05
Benzo[*c*]phenanthrene	0.54	2.22	0.27
Diethyl phthalate	0.52	1.87	–0.10
Picene	0.52	2.29	0.22
3,4-Benzpyrene	0.51	2.13	0.15
Benzaldehyde	0.50	1.81	–0.18
s-Tetrachlorobenzene	0.45	1.85	–0.19
Nitromethane	0.44	1.68	–0.37
Chrysene	0.41	2.25	0.19
Acetophenone	0.34	1.91	–0.24
s-Trichlorobenzene	0.34	1.99	–0.16
p-Dichlorobenzene	0.34	2.49	0.34
Phenanthrene	0.31	2.43	0.25
Triphenylene	0.29	2.44	0.24
1-Chloronaphthalene	0.28	2.15	–0.11
Benzonitrile	0.24	2.32	0.07
Methyl benzoate	0.20	2.23	–0.06
Naphthalene	0.15	2.48	0.14
o-Dichlorobenzene	0.09	2.51	0.11
Styrene	–0.25	2.46	–0.25
Biphenyl	–0.37	2.63	–0.23
Pyridine	–0.62	2.71	–0.40
Benzene	–1.15	3.31	–0.33

*Electron affinity values are summarized in refs. 41, 60, 67, 100, 101, 116, and 127.

The line is drawn for the least squares estimate,

$$EA = E_{\frac{1}{2}} + 2.49 \pm 0.20\,(eV) \tag{3.46}$$

The larger symbols are used to represent a data point with only a single $E_{\frac{1}{2}}$ value. The smaller symbols represent the average of multiple $E_{\frac{1}{2}}$ values, but all $E_{\frac{1}{2}}$ values were used to determine the above equation. Also shown in Table 3.4 are the deviations from the correlation line. Except for the last four data points, which represent electron beam experiments, there are no systematic errors. Even including these values, the standard deviation is consistent with the errors in the electron affinity and the $E_{\frac{1}{2}}$ values considering that different solvents and data sources are used. Coincidentally, a value of 2.49 ± 0.26 was obtained for the pi charge transfer complex acceptors alone [67]. The electron affinities from half wave reduction potentials generally agree quite well with values obtained from charge transfer complex absorption spectra data.

Thus the mechanism and the temperature dependence can be established from the electron affinity obtained from the experimentally determined E_{CT} or the $E_{\frac{1}{2}}$ values. If there is a strongly exothermic dissociative process or if the electron affinity is greater than 0.8 eV, the data will be in the beta region and will be temperature independent. The magnitude of the response will be governed by the rate constant for thermal electron attachment which is generally large for most organic compounds. If the electron affinity is between 0.4 eV and 0.8 eV, then the data will exhibit both a beta and an alpha region. At higher temperatures, the response will decrease with increasing temperature and the slope will be governed by the electron affinity. The intercept in the high temperature region at $1/T = 0$ will be 14.7 ± 2.0. If the electron affinity is less than 0.5 eV, then the data will be in the alpha region and will decrease with increasing temperature. The only exceptions to this latter point are molecules with low rate constants for thermal electron attachment, such as O_2, NO, COS, and NO_2, which are in the beta region. In the alpha region, the higher the electron affinity is, the higher the response will be. In the beta region the response depends on the rate constant which is presently unpredictable.

3.5.3. Activation energies and exothermicity

In gas phase ion–molecule reactions, it is generally stated that an exothermic reaction will proceed and an endothermic reaction will not. In the case of thermal electron attachment reactions, the situation is not so clearcut. The activation energy for dissociation is however related to the exothermicity of the reaction. In the case of Mechanism IV, $E^* = \Delta E = D_{AB} - \mathrm{EA_B}$ where ΔE is the exothermicity of the reaction, D_{AB} is the bond dissociation energy, and $\mathrm{EA_B}$ is the electron affinity of the radical, B. For Mechanism II, it has been shown empirically that there is a proportionality between the quantities according to

$$E^* = 15.3 + D_{AB} - \mathrm{EA_B} \tag{3.47}$$

A similar relationship was obtained by Kojima et al. [134]. In the case of Mechanism III, there are not enough data to clearly establish a correlation between E^* and ΔE. Representative data are shown in Fig. 3.11 where it can be clearly seen that the E^* for the compounds reacting according to Mechanism III, the aromatic halides, generally have a smaller activation energy than the corresponding aliphatic compounds. Compare, for example, chlorobenzene with chloromethane.

From the above equation and the mechanisms, it should be clear that the activation energy is not zero for an exothermic reaction and hence the ECD response for a dissociative compound will not be a maximum for an exothermic reaction. Thus in the case of dissociative compounds, the key to the response is the activation energy. If the activation energy is small, i.e. a small bond dissociation energy and/or a large electron affinity of

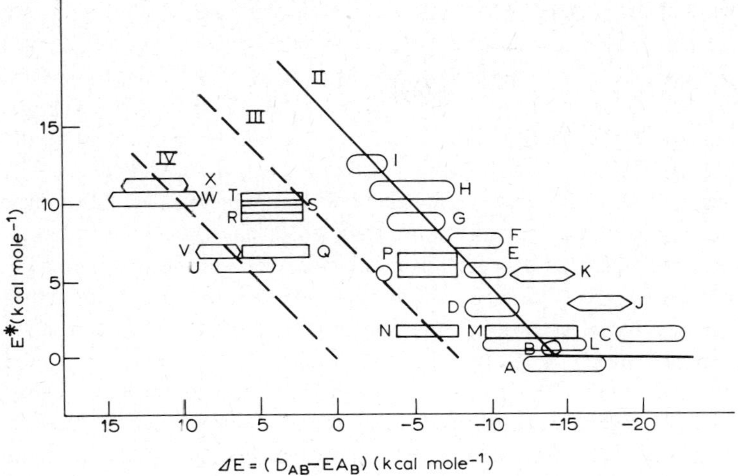

Fig. 3.11. Activation energy, E^*, versus ΔE = (D_{AB} − EA_B). Mechanism II: (A) CCl_4, (B) CH_3I, (C) $CHBr_3$, (D) $CHCl_3$, (E) CH_3Br, (F) CH_2Cl_2, (G) 1,2-$C_2H_4Cl_2$, (H) *tert.*-butyl chloride, (I) CH_3Cl, (J) $C_6H_5CH_2Cl$, (K) allyl chloride, (L) CH_2Br_2. Mechanism III: (M) C_6H_5, (N) 1-bromonaphthalene, (O) C_6H_5Br, (P), *o*-chlorotoluene, (Q) *o*-dichlorobenzene, (R) 1-chloronaphthalene, (S) C_6H_5Cl, (T) *o*-chlorotoluene. Mechanism IV: (U) acetic anhydride, (V) CH_3NO_2, (W) benzyl acetate, (X) ethyl acetate.

the radical, then the compound will have a large ECD response and the response will be relatively temperature independent. If the value of E^* is greater than zero, the response will depend upon the magnitude of E^*. In general, the larger the E^*, the smaller the response at a given temperature. The response will increase with increasing temperature and the slope will be governed by the magnitude of the activation energy.

There are some compounds which exhibit data in the alpha, the beta and the gamma regions. This requires a unique combination of values of E^* and EA. One such compound is chloroacetophenone. At low temperatures the ECD coefficient is temperature independent, at intermediate temperatures it decreases with increasing temperatures, and at high temperatures it increases with increasing temperature.

3.5.4. Relationship of molecular structure to response – Prediction of electron-capture coefficients

The ultimate value of a theory is in prediction. If the pertinent molecular properties are available or can be estimated from correlations, then the response can be predicted as a function of temperature and as a function of concentration. However, in many cases, the only data available for prediction are the structure of the compound and predictions must be made on this basis. In the early studies of electron capture, Maggs et al. [3], Lovelock et al. [135], Zielinski and co-workers [136,137], Clemons and Altshuller [138], and Sullivan [139] investigated this relationship on an empirical basis. These results can be rationalized on the basis of the kinetic model discussed in this section. For example, at a constant temperature the response is in the order I > Br > Cl ≫ F. This is primarily due to the decreasing bond dissociation energy in the series of halogens. The same is true for the sequence tertiary > secondary > primary. Sullivan notes that there is an upper limit to the electron-capture response which would correspond to an electron affinity greater than 0.8 eV or an E^* value of about zero [139]. This is also evidenced by a substantially constant high response for nitrobenzenes due to the fact that the electron affinities are greater than 1.0 eV. Zlatkis and Pettitt [140] examined the relationships between response and structure for a series of derivatizing agents and obtained temperature dependences to establish the capture mechanism. From these data, it is clear that capture to the various derivatives proceeds primarily by Mechanisms I, III, and IV. For example pentafluorobenzamides capture according to Mechanism IV. The addition of multiple electronegative substituents can also be explained on the basis of the kinetic model. Where groups such as acetyl, aldehyde, or nitro are added to a benzene ring, the electron affinity will increase and multiple substitutions will increase the response unless the electron affinity of the monosubstituted species is already above 0.8 eV. In the case of dissociative compounds, substitution of a second electronegative group on the same carbon will decrease the bond dissociation energy and give a higher response, until the activation energy is zero. The addition of a second electronegative group to an adjacent carbon will have less effect on the bond dissociation energy. With an estimate of E^* a predicted capture coefficient can be calculated from equation 3.34 where A_1/k_D can be taken as $5 \cdot 10^{13}$.

For aromatic compounds, both the electron affinity of the aromatic radical and the electron affinity of the molecule will be affected by multiple substitutions. The precise effect of a substituent can be estimated by the use of Hückel calculations and empirical parameters as was done by Wentworth et al. [141]. The effect of multiple substituents is to increase the electron affinity of the molecule and hence to increase the ECD response until saturation is reached. For example, five fluorines on benzene are sufficient to raise the electron affinity above 0.8 eV.

If the estimated $EA < 0.8$, the predicted capture coefficient can be calculated from eqn. 3.32 where the pre-exponential constant can be taken as $2.4 \cdot 10^6$.

Thus in summary, in order to estimate the response from the structure, the mechanism must first be established. This can be done by examining the exothermicity of the possible dissociative reactions and this requires an estimate of the electron affinity of the leaving group and an estimate of the bond dissociation energy. If there are no exothermic pathways and the molecule contains some electronegative functional groups, then capture can possibly occur via Mechanism I and if the electron affinity of the molecule is greater than about 0.8 eV, the response will be high. If a dissociative process can occur, then an estimate of the activation energy must be made based upon the relationship between the structure and the bond dissociation energy and the electron affinity of the leaving group. If the leaving group is a halogen, this is well known. If the leaving group is a carbon-containing radical, some values are available. Bond dissociation energies are known in some cases but frequently they must be estimated using known values of similar type bonds. From these considerations, an estimate of E^* can be obtained and in general if E^* is small the response will be large.

3.5.5. Examples of predictions

In order to illustrate the prediction of EĊD responses, Chen and Wentworth [53] presented a discussion of the response for a number of aromatic hydrocarbons, aldehydes, and ketones studied in our laboratory and several substituted benzophenones studied by Vessman and Hartvig [96]. These data are shown in Figs. 3.12–3.15.

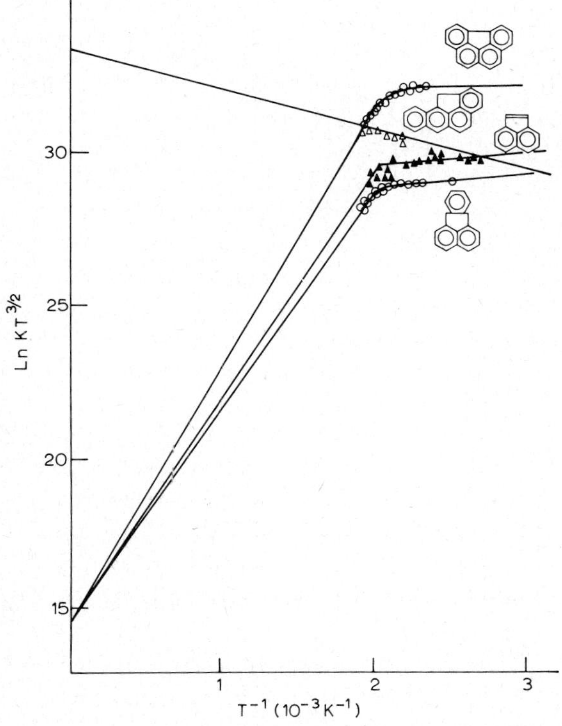

Fig. 3.12. ln $KT^{3/2}$ versus $1/T$ for aromatic hydrocarbons.

The addition of an acetyl group to naphthalene is predicted to increase the electron affinity over that of naphthalene and results in data in both alpha and beta regions. The addition of a second acetyl group to acetophenone and a second aldehyde group to benzaldehyde is expected to result in an electron affinity greater than 1.0 eV. In this case, data are also available for $E_{\frac{1}{2}}$ values which predict electron affinities of 1.08 eV and 1.33 eV, respectively. Thus the response should be in the beta region and should be high and temperature independent. Based on the data for the half wave reduction potentials, the electron affinity of fluoranthene should be 0.72 eV and that for acenaphthylene should be 0.84 eV which predicts that the response should be high and independent of temperature at the low temperatures and, depending upon the specific k_1 values, should enter the alpha regions at high temperatures. By analogy to fluoranthene, benzofluoranthene and cholanthrene should also have a high electron affinity and hence a high ECD response. From Figs. 3.12 and 3.13 it can be seen that the predictions are borne out and the effects

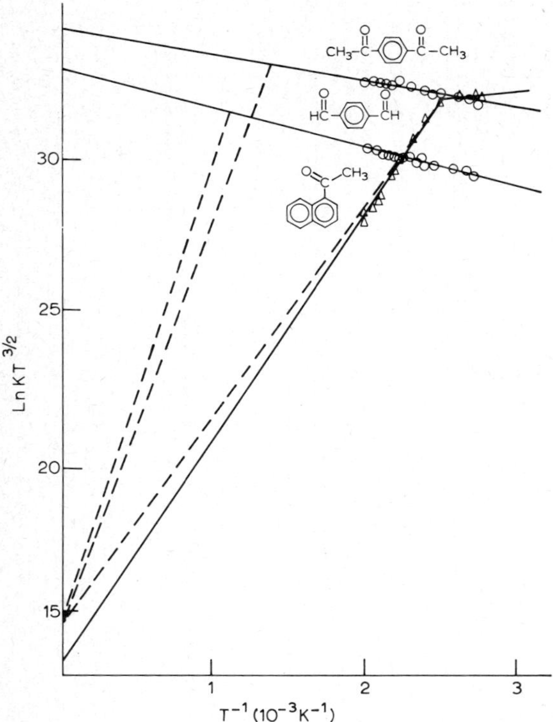

Fig. 3.13. ln $KT^{3/2}$ versus $1/T$ for carbonyl compounds.

of the different k_1 values can be appreciated. The prediction and results are given in Table 3.5.

In the case of the benzophenones, half wave reduction potentials are available for all the substituted compounds except for the nitro, the trifluoromethyl, and the ethyl compounds. The predicted electron affinities vary from 0.61 eV to 1.0 eV and the ECD data agree very well with the predicted responses. At the low temperatures, the response of all the compounds is high and at higher temperatures, the response of the compounds with the lower electron affinities decreases with increasing temperatures. Based upon structural considerations, it is predicted that the electron affinity of the nitro compound will be greater than 1.0 eV, while the electron affinity of the ethyl compound will be about the same as the methyl compound, and the electron affinity of the CF_3 derivative will be higher than the benzophenone value. The results for these three compounds agree with these predictions as can be seen in Table 3.5 and Figs. 3.14 and 3.15.

The temperature dependence of the response for dissociative compounds can be predicted from the exothermicity of the reaction and eqn. 3.46. Kojima et al. [134] have measured the ECD response for several alkylhalides as a function of temperature. The average values obtained by these authors are compared with the calculated values in Table 3.6. They also found that the activation energy decreases for increasing length of the alkyl chain which would be in agreement with a decrease in bond dissociation energy, although the available data are not accurate enough to predict such a trend.

TABLE 3.5

ELECTRON AFFINITIES AND RATE CONSTANTS

Compound	$-E_{\frac{1}{2}}$ [97,132,142] (V vs. SCE)	EA(calc.)* (eV)	EA(ECD) [115] (eV)	k_1 (ECD) [115] (approximate) (cm^3 $molecule^{-1}$ s^{-1})
Acetonaphthone-1			0.61 ± 0.1	$1\cdot10^{-7}$
p-Diacetylbenzene	1.41	1.08	0.8 ± 0.2	$1\cdot10^{-7}$
Terephthalaldehyde	1.16	1.33	⩾0.8 ± 0.2	$3\cdot10^{-8}$
Fluoranthene	1.77	0.72	⩾0.63 ± 0.2	$3\cdot10^{-8}$
Benzo[*ghi*]fluoranthene			0.68 ± 0.1	$1\cdot10^{-7}$
Cholanthrene			0.8 ± 0.2	$5\cdot10^{-8}$
Acenaphthylene	1.65	0.84	⩾0.7 ± 0.2	$3\cdot10^{-8}$
Benzophenone	1.75	0.73	0.64 ± 0.1	
4-CH_3	1.79	0.64	0.64 ± 0.1	
4-C_2H_5			0.64 ± 0.1	
4-CH_3O	1.86	0.61	0.56 ± 0.1	
4-CF_3			0.8 ± 0.2	$7\cdot10^{-8}$
4-NO_2			>1 ± 0.2	$2\cdot10^{-7}$
4-F	1.74	0.75	0.66 ± 0.1	$5\cdot10^{-8}$
4-Cl	1.62	0.78	0.8 ± 0.2	$2\cdot10^{-8}$
4-Br	1.60	0.89	0.9 ± 0.2	$7\cdot10^{-8}$
4-I	1.49	1.00	1 ± 0.2	$3\cdot10^{-7}$
3-F	1.61	0.88		
3-Cl	1.59	0.9		
3-Br	1.59	0.93		

*The values of EA (calc.) are obtained from EA = $E_{\frac{1}{2}}$ + 2.49 ± 0.2 eV.

TABLE 3.6

ACTIVATION ENERGIES FOR ALKYL HALIDES

Compound	E^* (predicted) ($kcal\ mole^{-1}$)	E^* (experimental) ($kcal\ mole^{-1}$) [134]
Primary RCl	12.5	11.3
Secondary RCl	11.1	10.6
Tertiary RCl	10.9	9.7
Primary RBr	5.7	6.0
Secondary RBr	4.0	4.6
Tertiary RBr	2.0	2.0
Primary RI	0.6	2.1

From these examples, it is clear that in order to obtain a meaningful correlation of response and chemical structure, the temperature dependence of the response must be determined. The significance of the temperature data was not recognized by the early workers and in fact the temperature was not often reported [32–34, 135–138]. More recently, workers such as Zlatkis and Pettitt [140], Poole et al. [143], Poole [144],

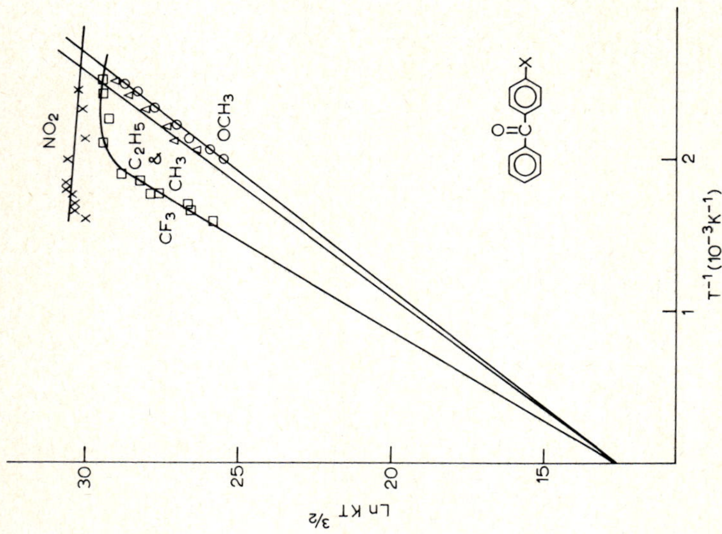

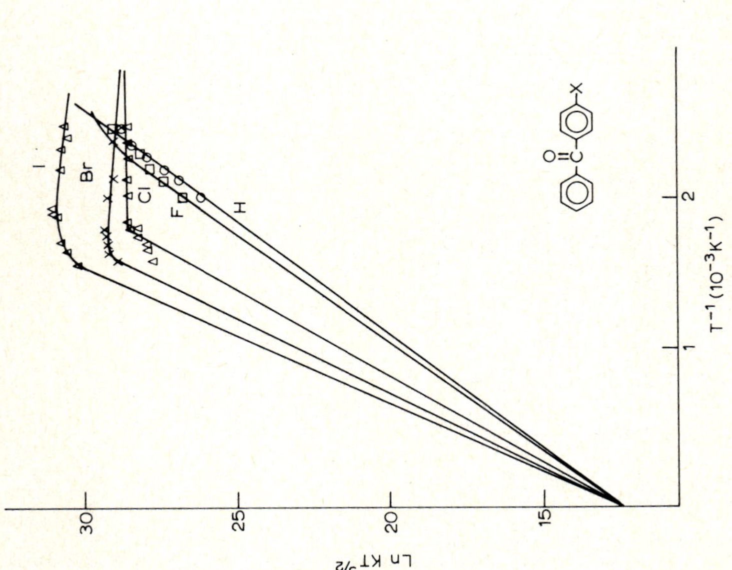

Fig. 3.14. ln $KT^{3/2}$ versus $1/T$ for halo-substituted benzophenones.

Fig. 3.15. ln $KT^{3/2}$ versus $1/T$ for substituted benzophenones.

Simmonds [30], Hattori et al. [126], Kojima et al. [134,145], and Vessman and Hartvig [96] have obtained the necessary data as a function of temperature. However, the exact ECD system also must be considered. For example Hattori et al. [146] used a d.c. system to study the temperature dependence of several substituted benzenes and found results different from that obtained with a pulsed system. The theory discussed in this chapter would not necessarily apply to such a system, but there may be some similarities.

REFERENCES

1 W.E. Wentworth, E.C.M. Chen and J.E. Lovelock, *J. Phys. Chem.*, 70 (1966) 445.
2 J.E. Lovelock and S.R. Lipsky, *J. Amer. Chem. Soc.*, 82 (1960) 431.
3 R.J. Maggs, P.L. Joynes, A.J. Davies and J.E. Lovelock, *Anal. Chem.*, 43 (1971) 1966.
4 D.C. Fenimore, P.R. Loy and A. Zlatkis, *Anal. Chem.*, 43 (1971) 1972.
5 G.R. Shoemake, D.C. Fenimore and A. Zlatkis, *J. Gas Chromatogr.*, 3 (1965) 287.
6 J.H. Beckinski and J.C. Sternberg, *Abstracts 148th National Meeting ACS, Chicago, IL, Sept. 1964, No. 52, p. 22A.*
7 J.J. Sullivan, *Hewlett Packard Technical Paper (30th Pittsburgh Conference, Cleveland, OH, March 1979, Paper 256).*
8 W.E. Wentworth, A. Tishbee, C.F. Batten and A. Zlatkis, *J. Chromatogr.*, 112 (1975) 229.
9 J.E. Lovelock, in C.L.A. Harbourn (Editor), *Gas Chromatography 1968*, Elsevier, Amsterdam, 1969.
10 J.E. Lovelock, D.C. Fenimore and A. Zlatkis, *J. Gas Chromatogr.*, 5 (1967) 392.
11 M.J. Cohen and F.W. Karasek, *J. Chromatogr. Sci.*, 8 (1970) 330.
12 E.C. Horning, D.I. Carroll, I. Dzidic, S.-N. Lin, R.N. Stillwell and J.-P. Thenot, *J. Chromatogr.*, 142 (1977) 481.
13 E.P. Grimsrud and S.H. Kim, *Anal. Chem.*, 51 (1979) 537.
14 S. Kapila and W.A. Aue, *J. Chromatogr.*, 148 (1978) 343.
15 M.P. Phillips, P.D. Goldan, W.C. Kuster, R.E. Sievers and F.C. Fehsenfeld, *Anal. Chem.*, 52 (1980) 45.
16 E.P. Grimsrud and R.G. Stebbins, *J. Chromatogr.*, 155 (1978) 19.
17 W.A. Aue and S. Kapila, *J. Chromatogr.*, 188 (1980) 1.
18 W.A. Aue and K.W.M. Siu, *J. Chromatogr.*, 203 (1981) 237.
19 J. Rosiek, I. Śliwka and J. Lasa, *J. Chromatogr.*, 137 (1977) 245.
20 I. Śliwka, J. Lasa, J. Rosiek and A. Korus, *J. Chromatogr.*, 150 (1978) 1.
21 M.W. Siegel and M.C. McKeown, *J. Chromatogr.*, 122 (1976) 397.
22 I. Śliwka, J. Lasa and J. Rosiek, *J. Chromatogr.*, 172 (1979) 1.
23 E. Broś and F.M. Page, *J. Chromatogr.*, 126 (1976) 271.
24 P.L. Patterson, *J. Chromatogr.*, 134 (1977) 25.
25 J.A. Ayala, W.E. Wentworth and E.C.M. Chen, *J. Chromatogr.*, 195 (1980) 1.
26 E.P. Grimsrud, S.H. Kim and P.L. Gobby, *Anal. Chem.*, 51 (1979) 223.
27 P.L. Gobby, E.P. Grimsrud and S.W. Warden, *Anal. Chem.*, 52 (1980) 473.
28 D.A. Miller and E.P. Grimsrud, *Anal. Chem.*, 51 (1979) 851.
29 E.C.M. Chen, W.E. Wentworth and J.A. Ayala, *J. Chromatogr.*, 188 (1980) 89.
30 P.G. Simmonds, *J. Chromatogr.*, 166 (1978) 593.
31 J.E. Lovelock, *Nature (London)*, 189 (1961) 729.
32 N.L. Gregory and J.E. Lovelock, *Proceedings International Gas Chromatography Symposium, Lansing, MI, June 1961*, Academic Press, New York, p. 151.
33 J.E. Lovelock, *Anal. Chem.*, 33 (1961) 162.
34 W.E. Wentworth and R.S. Becker, *J. Amer. Chem. Soc.*, 84 (1962) 4263.
35 R.S. Becker and W.E. Wentworth, *J. Amer. Chem. Soc.*, 85 (1963) 2210.
36 R.S. Becker and E.C.M. Chen, *J. Chem. Phys.*, 45 (1966) 2403.

37 W.E. Wentworth, R.S. Becker and R. Tung, *J. Phys. Chem.*, 71 (1967) 1652.
38 J.C. Steelhammer and W.E. Wentworth, *J. Chem. Phys.*, 51 (1969) 1802.
39 W.E. Wentworth and J.C. Steelhammer, *Advan. Chem. Ser. Radiation Chemistry*, ACS Publications, Washington, DC, 1968, Ch. 4.
40 W.E. Wentworth, R. George and H. Keith, *J. Phys. Chem.*, 51 (1969) 1791.
41 W.E. Wentworth and E.C.M. Chen, *J. Gas Chromatogr.*, 5 (1967) 170.
42 W.E. Wentworth and E.C.M. Chen, *J. Chromatogr.*, 186 (1979) 99.
43 J. Connor, *J. Chromatogr.*, 200 (1980) 15.
44 J.E. Lovelock, unpublished results.
45 J.E. Lovelock and A.J. Watson, *J. Chromatogr.*, 158 (1978) 123.
46 P. Devaux and G. Guiochon, *J. Gas Chromatogr.*, 5 (1967) 341.
47 C.C. Han, *Master's Thesis*, University of Houston, Houston, TX, 1970.
48 M. Scolnick, *J. Chromatogr. Sci.*, 7 (1969) 300.
49 B.H. Mahan and C.E. Young, *J. Chem. Phys.*, 44 (1966) 2182.
50 W.E. Wentworth, in I.I. Damsky and J.A. Perry (Editors), *Recent Advances in Gas Chromatography*, Marcel Dekker, New York, 1971, Ch. 5.
51 H.J. van de Wiel and H. Tommassen, *J. Chromatogr.*, 71 (1972) 1.
52 R.R. Freeman, *Doctoral Dissertation*, University of Houston, Houston, TX, 1971.
53 E.C.M. Chen and W.E. Wentworth, in preparation.
54 W.E. Wentworth, R.R. Freeman and E.C.M. Chen, *J. Chem. Phys.*, 55 (1971) 2075.
55 M.E. Morrison and W.H. Corcoran, *Anal. Chem.*, 39 (1967) 255.
56 E.C.M. Chen, W.E. Wentworth and R.D. George, *J. Chem. Phys.*, 49 (1967) 1973.
57 J.C. Steelhammer, *Master's Thesis*, University of Houston, Houston, TX, 1969.
58 E.C.M. Chen, *Doctoral Dissertation*, University of Houston, Houston, TX, 1966.
59 W.E. Wentworth, J. Steelhammer and E.C.M. Chen, *J. Phys. Chem.*, 72 (1968) 2671.
60 F.M. Page and G.C. Goode, *Negative Ions and the Magnetron*, Wiley-Interscience, New York, 1969.
61 J.L. Packs and A.V. Phelps, *J. Chem. Phys.*, 45 (1966) 4316.
62 M. McFarland, D.B. Dunkin, F.C. Fehsenfeld, A.L. Schmeltekopf and E.E. Ferguson, *J. Chem. Phys.*, 56 (1972) 2358.
63 R.J. Celotta, R.A. Bennett, J.L. Hall, M.W. Siegel and J. Levine, *Phys. Rev.*, A6 (1972) 631.
64 M.W. Siegel, R.J. Celotta, R.J. Hall and R.A. Bennett, *Phys. Rev.*, A6 (1972) 607.
65 E. Herbst, J.A. Patterson and W.C. Lineberger, *J. Chem. Phys.*, 61 (1974) 1301.
66 P. Warnek, *Chem. Phys. Lett.*, 3 (1969) 532.
67 E.C.M. Chen and W.E. Wentworth, *J. Chem. Phys.*, 63 (1975) 3183.
68 L.G. Christophorou, *Advan. Electron. Electron Phys.*, 46 (1978) 55.
69 J. Berkowitz, W.A. Chupka and D. Gutman, *J. Chem. Phys.*, 55 (1971) 2733.
70 B.M. Hughes, C. Lifshitz and T.O. Tiernan, *J. Chem. Phys.*, 59 (1973) 3162.
71 S.J. Nalley, R.N. Compton, H.C. Schweinler and V.E. Anderson, *J. Chem. Phys.*, 59 (1973) 4125.
72 F.C. Fehsenfeld, E.E. Ferguson and A.L. Schmeltekopf, *J. Chem. Phys.*, 45 (1966) 1844.
73 T.O. Tiernan, B.M. Hughes and C. Lifshitz, *J. Chem. Phys.*, 55 (1971) 5692.
74 A.P.M. Baede, *Physica (Utrecht)*, 59 (1972) 541.
75 S.J. Nalley and R.N. Compton, *Chem. Phys. Lett.*, 9 (1971) 529.
76 D.B. Dunkin, F.C. Fehsenfeld and E.E. Ferguson, *Chem. Phys. Lett.*, 15 (1972) 257.
77 R.K. Curran, *Phys. Rev.*, 125 (1962) 910.
78 E.E. Ferguson, F.C. Schmeltekopf and F.C. Fehsenfeld, *Advan. Chem. Ser.*, 80 (1969) 83.
79 K.M.A. Refaey, *Int. J. Mass Spectrom. Ion Phys.*, 21 (1976) 21.
80 C.B. Leffert, W.M. Jackson and E.W. Roth, *J. Chem. Phys.*, 58 (1973) 5801.
81 R.N. Compton, P.W. Reinhardt and C.D. Cooper, *J. Chem. Phys.*, 63 (1975) 3821.
82 H. Dispert and K. Lachmann, Hagn-Meitner Institut für Kernforschung, Berlin, HMI-B198 (1975) 34.
83 D.G. Hooper, A.C. Wahl, R.L.C. Wu and T.O. Tiernan, *J. Chem. Phys.*, 65 (1976) 5484.
84 R.N. Compton, P.W. Reinhardt and C.D. Cooper, *J. Chem. Phys.*, 68 (1978) 4360.
85 A. Henglein and G.A. Muccini, *J. Chem. Phys.*, 39 (1959) 1426.

86 C. Lifshitz, T.O. Tiernan and B.M. Hughes, *J. Chem. Phys.*, 59 (1973) 3182.
87 R.N. Compton and C.D. Cooper, *J. Chem. Phys.*, 59 (1973) 4140.
88 R.N. Compton, P.W. Reinhardt and C.D. Cooper, *J. Chem. Phys.*, 68 (1978) 2023.
89 C.B. Leffert, S.Y. Tang, E.W. Rothe and T.C. Cheng, *J. Chem. Phys.*, 61 (1974) 4929.
90 F.C. Fehsenfeld, *J. Chem. Phys.*, 54 (1971) 438.
91 K.M.A. Refaey and J.L. Franklin, *Int. J. Mass Spectrom. Ion Phys.*, 26 (1978) 125.
92 T.O. Tiernan and R.P. Clow, *Advan. Mass Spectrom.*, 6 (1974) 295.
93 S. Tsuda and W.H. Hamill, *Advan. Mass Spectrom.*, 2 (1966) 249.
94 R. Yamdgani and P. Kebarle, *J. Amer. Chem. Soc.*, 96 (1973) 5940.
95 L.J. Rains, H.W. Moore and R.T. McIver, *J. Chem. Phys.*, 60 (1978) 3709.
96 J. Vessman and P. Hartvig, *Acta Pharm. Suecica*, 9 (1972) 463.
97 I.B. Rashkov, Tz.G. Popov, Gr.S. Michailov, I.M. Panayotov and A.Z. Trifonov, *Monatsh. Chem.*, 101 (1970) 1797.
98 K.J. Reed and J.I. Braumann, *J. Amer. Chem. Soc.*, 97 (1975) 1625.
99 R.N. Compton, P.W. Reinhardt and C.D. Cooper, *J. Chem. Phys.*, 60 (1974) 2953.
100 G. Briegleb, *Angew. Chem. Int. Ed.*, 3 (1964) 617.
101 V. Kampars and O. Neilands, *Russ. Chem. Rev.*, 46 (1977) 503.
102 W.F. Kuhn, R.J. Levins and A.C. Lilly, *J. Phys. Chem.*, 49 (1968) 5550.
103 W.E. Wentworth, E.C.M. Chen and J.A. Ayala, *J. Phys. Chem.*, in press.
104 W.E. Wentworth, E.C.M. Chen and J.A. Ayala, *J. Phys. Chem.*, submitted for publication.
105 J.M. Warman and M.C. Sauer, *Int. J. Radiat. Phys. Chem.*, 3 (1971) 273.
106 E. Schultes, A.A. Christodoulides and R.N. Schindler, *Chem. Phys.*, 8 (1975) 354.
107 R.N. Fessenden and K.M. Bansal, *J. Chem. Phys.*, 53 (1970) 3468.
108 A.A. Christodoulides, R. Schumacher and R.N. Schindler, *J. Phys. Chem.*, 79 (1975) 1904.
109 G.D. Sides, T.O. Tiernan and R.J. Hanrahan, *J. Chem. Phys.*, 65 (1976) 1966.
110 F.K. Truby, *Phys. Rev.*, A4 (1971) 613.
111 L. Bouby, F. Fiquet-Fayard and Y. LeCoat, *Int. J. Mass Spectrom. Ion Phys.*, 3 (1970) 439.
112 D.A. Parkes, *J. Chem. Soc. Faraday Trans.*, 11 (1972) 2121.
113 A.V. Phelps and R.E. Vorshall, *J. Chem. Phys.*, 49 (1968) 3246.
114 G.E. Caledonia, *Chem. Rev.*, 75 (1975) 333.
115 E.C.M. Chen and W.E. Wentworth, *J. Chromatogr.*, submitted for publication.
116 B.K. Janousek and J.I. Brauman, in M.T. Bowers (Editor), *Gas Phase Ion Chemistry*, Vol. 2, Academic Press, New York, 1979, p. 53.
117 C.D. Cooper, W.F. Frey and R.N. Compton, *J. Chem. Phys.*, 69 (1978) 2367.
118 V.A. Nazarenko and V.D. Pokhodenko, *Int. J. Mass Spectrom. Ion Phys.*, 31 (1979) 381.
119 C.E. Klots, R.N. Compton and V.F. Raaen, *J. Chem. Phys.*, 60 (1974) 1177.
120 R.N. Compton and C.D. Cooper, *J. Chem. Phys.*, 66 (1977) 4325.
121 I. Nenner and G.J. Schultz, *J. Chem. Phys.*, 62 (1975) 1747.
122 W.E. Wentworth and E.C.M. Chen, *J. Phys. Chem.*, 71 (1967) 1929.
123 W.E. Wentworth and W.T. Ristau, *J. Phys. Chem.*, 73 (1969) 2126.
124 J. Michl, *J. Mol. Spectrosc.*, 30 (1969) 30.
125 P.D. Burrow, J.A. Michejda and K.D. Jordan, *J. Amer. Chem. Soc.*, 98 (1976) 6392.
126 Y. Hattori, Y. Kuge and S. Asada, *Nippon Kagaku Kaishi*, 12 (1978) 1102.
127 H.M. Rosenstock, K. Draxl, B.W. Steiner and J.T. Herron, *J. Phys. Chem. Ref. Data*, 6 (1977) Suppl. No. 1.
128 T. Shimokawa and T. Sawai, *Bull. Chem. Soc. Jap.*, 50 (1977) 366.
129 F. Gerson, H. Ohya-Nishiguchi and C. Wydler, *Angew. Chem. Int. Ed. Engl.*, 15 (1976) 552.
130 G.W.C. Milner, *The Principles and Applications of Polarography*, Longmans, Green and Co., New York, 1954.
131 L. Meites, *Polarographic Techniques*, Wiley Intersciende, New York, 1965.
132 C.K. Mann and K.K. Barnes, *Electrochemical Reactions in Non-Aqueous Systems*, Marcel Dekker, New York, 1970.
133 L. Meites, P. Zuman, W.J. Scott, B.H. Campbell, A. Dubinsky, L. Lampugnani and R. Zuman, *Electrochemical Data*, Wiley-Interscience, New York, 1974.

134 T. Kojima, Y. Tanaka and M. Satouchi, *Anal. Chem.*, 48 (1976) 1760.
135 J.E. Lovelock, A. Zlatkis and R.S. Becker, *Nature (London)*, 193 (1962) 540.
136 W.L. Zielinski, Jr., L. Fishbein and R.O. Thomas, *J. Chromatogr.*, 30 (1967) 177.
137 W.L. Zielinski, Jr. and L. Fishbein, *J. Chromatogr.*, 28 (1967) 293.
138 C.A. Clemons and A.P. Altshuller, *Anal. Chem.*, 38 (1966) 133.
139 J.J. Sullivan, *J. Chromatogr.*, 87 (1973) 9.
140 A. Zlatkis and B.C. Pettitt, *Chromatographia*, 2 (1969) 484.
141 W.E. Wentworth, R.S. Becker and L.W. Kao, *J. Phys. Chem.*, 79 (1975) 1161.
142 A. Das Gupta, S. Chatterjee and N.K. Das Gupta, *Bull. Chem. Soc. Jap.*, 52 (1979) 3070.
143 C.F. Poole, S. Singhawangcha and A. Zlatkis, *J. Chromatogr.*, 158 (1978) 33.
144 C.F. Poole, *J. Chromatogr.*, 118 (1976) 280.
145 T. Kojima, M. Satouchi and Y. Tanaka, *Bull. Chem. Soc. Jap.*, 52 (1979) 277.
146 Y. Hattori, Y. Kuge and S. Nakagawa, *Bull. Chem. Soc. Jap.*, 51 (1978) 2249.

Chapter 4

Selective electron-capture sensitization

FRED C. FEHSENFELD, PAUL D. GOLDAN, MICHAEL P. PHILLIPS and ROBERT E. SIEVERS

CONTENTS

4.1. INTRODUCTION

The widespread use of the electron-capture detector (ECD) is due to its exceptional sensitivity in detecting compounds that rapidly attach electrons such as halogen-containing pesticides, polychlorinated biphenyls, and refrigerants. On the other hand, the ECD is relatively insensitive to compounds such as the alkanes that do not rapidly attach electrons. The sensitivity of the ECD and the resulting detection limit for a given compound are strongly influenced by many factors: the purity of the carrier gas, the detector temperature, column bleed, septum bleed, and leaking column fittings. Usually, the impurities present in the carrier gas act adversely to reduce the sensitivity of detection of compounds by the ECD. In some cases, however, the presence of these impurities can enhance the sensitivity of the ECD to specific compounds. For example, Van de Wiel and Tommassen [1] discovered an increased sensitivity for butyl bromide when oxygen was added to the nitrogen carrier gas in their fixed-frequency, pulsed ECD. Grimsrud and co-workers [2–6] demonstrated that oxygen purposely added to the nitrogen carrier gas can be used to increase the sensitivity of the ECD to certain organic compounds. Simmonds [7] has observed that oxygen added to the carrier gas enhances the sensitivity of the ECD to carbon dioxide. In each of these studies it was proposed that at elevated temperatures the collisional detachment of electrons from O_2^- competes with the attachment process so that, in the presence of O_2, O_2^- and electrons are in thermodynamic equilibrium. As a consequence, any reaction which converts O_2^- to a stable negative ion results in electron removal from the ECD plasma and generates a detector response even in the absence of direct electron attachment by the analyte. This is discussed in detail in Chapter 5.

Obviously, other schemes involving the negative ion chemistry of the ECD can be devised to increase the response of the detector to compounds that do not rapidly attach electrons. Among these is a mechanism based on the following steady-state negative ion chemistry of N_2O in a N_2 carrier gas [8].

$$e + N_2O \xrightarrow{k_1} O^- + N_2 \qquad (4.1)$$

$$O^- + N_2O \xrightarrow{k_2} NO^- + NO \qquad (4.2)$$

$$NO^- + N_2 \xrightarrow{k_3} NO + N_2 + e \qquad (4.3)$$

in which electrons are attached (eqn. 4.1) and released (eqn. 4.3). The concentrations of electrons, O^-, and NO^- in the ECD depend on the rate constants for eqns. 4.1–4.3, the concentration of N_2O in the N_2 carrier gas, and the operating temperature of the detector. Thus, any compound (analyte, A) that can react with O^- or NO^- to form a stable negative ion

$$O^- + A \xrightarrow{k_4} \text{stable negative ion} \qquad (4.4)$$

will interrupt the reaction cycle, causing a reduction in the electron density and a detector response. This is particularly significant because O^- reacts with many compounds that do not directly capture electrons and are not normally sensitively detected by the ECD This technique has been termed "selective electron-capture sensitization" (SECS) because of the specific chemical nature of the enhancement process.

4.2. THEORY

The steady-state reaction sequence, represented by eqns. 4.1–4.4, is illustrated in Fig. 4.1. The reactions that comprise this chemistry are relatively well understood [9–15]. Three-body reactions that serve to convert electrons to O^- [13,14] and O^- to NO^- [13] can be neglected compared to reactions 4.1 and 4.2 in the detector temperature range normally used (i.e., 100–400°C). Rate constants for eqns. 4.1–4.3 are

$k_1 = 7.3 \cdot 10^{-8} \exp(-4800/T)\ \text{cm}^3\text{s}^{-1}$ [10,11]

$k_2 = 2.2 \cdot 10^{-10}\ \text{cm}^3\text{s}^{-1}$ [9,13]

$k_3 = 10^{-11} \exp(-700/T)\ \text{cm}^3\text{s}^{-1}$ (estimated from ref. 12)

Because NO^- is destroyed by reaction with the N_2 carrier gas at a much faster rate than it is produced by the reaction of O^- with N_2O, the expected concentration of NO^- is

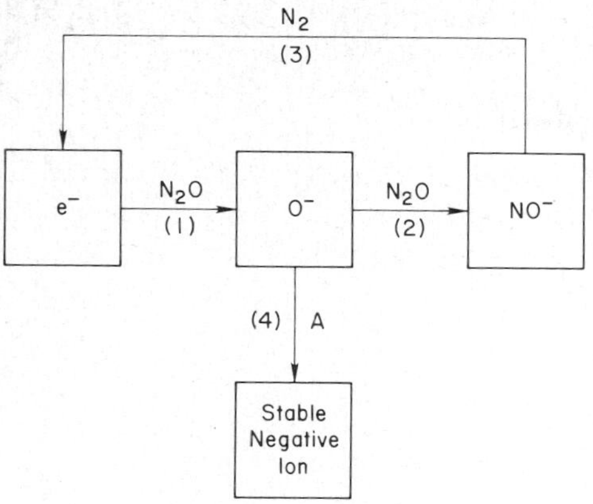

Fig. 4.1. Illustration of the steady-state negative ion chemistry used to explain the N_2O-induced selective electron capture sensitization. (From ref. 8.)

much smaller than that of electrons or O^- and, therefore, reactions of NO^- are not significant and will be omitted from further discussion.

The time required for establishing a reactive steady state depends on the time constant for the slowest reaction. For any reasonable combinations of N_2O concentrations in the carrier gas and detection temperatures below 400°C, the constraint on the approach to steady state will be imposed by eqn. 4.1. For a detector cell pressure of 760 Torr, the lifetime associated with eqn. 4.1 will be $\tau_1 = 2.2 \cdot 10^{-8}/R_{N_2O}$ (s) at 350°C, where R_{N_2O} is the N_2O mole fraction in the carrier gas. For N_2O mole fractions, $R_{N_2O} \geqslant 10^{-7}$, τ_1 will be short compared to the passage time of the carrier gas mixture through the detector (ca. 1 s).

As the concentration of N_2O in the carrier gas increases, the rate of formation of O^- through eqn. 4.1 increases. The O^- production through eqn. 4.1 is balanced by the destruction through both eqn. 4.2 and conversion to stable ions through eqn. 4.4. These two processes produce a steady-state concentration of O^- relative to electrons which, for $k_2[N_2O] \gg k_4[A]$, is given by

$$\frac{[O^-]}{[e]} \approx \frac{k_1}{k_2} = 331 \exp(-4800/T) \tag{4.5}$$

At this point the enhancement of the detector response associated with O^- secondary reactions becomes saturated. The approach to this saturation for H_2, CH_4, and CO_2 is shown in Fig. 4.2 by the flattening of the detector response curves at N_2O mole fractions greater than 10 ppm (v/v) (throughout this chapter mixing ratios or concentrations are expressed as volume/volume units, except where indicated otherwise). Even though H_2, CH_4, and CO_2 do not directly attach electrons, all three compounds are observed to induce a small response in the ECD even when using a N_2 carrier gas with no intentionally added N_2O. If the measured response to a fixed sample size of these analytes with no N_2O added to the carrier is designated as F_0, the increase in the response attributable to the presence of N_2O is given by $F_{N_2O} = F - F_0$. The response increase, F_{N_2O}, is plotted as a function of the N_2O mole fraction, R_{N_2O}, in Fig. 4.2. The N_2O mole frac-

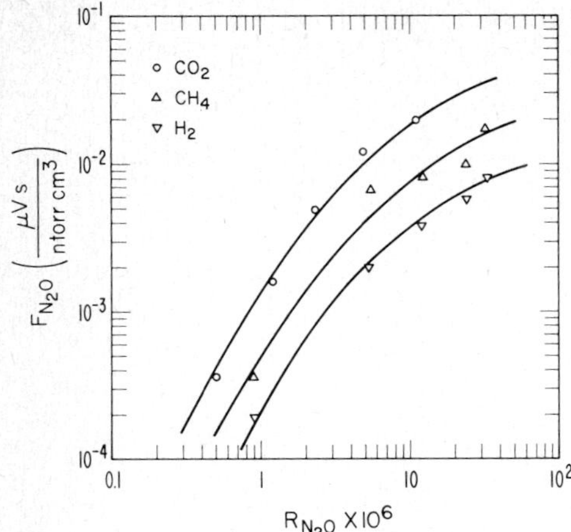

Fig. 4.2. Plot of F_{N_2O} for CO_2, CH_4, and H_2 as a function of the N_2O mole fraction, R_{N_2O}. The experimental data for CO_2 are plotted as open circles (○), for CH_4 as open triangles (△), and for H_2 as inverted triangles (▽). (From ref. 8.)

tion at which saturation occurs depends on the details of the high-pressure secondary reactions of O^- with N_2O, and the amount of impurities that react with O^- in the ECD. The response, F_{N_2O}, at saturation depends on the rate at which O^- is destroyed by the test compounds, and the ratio of O^- to electrons as deduced from eqn. 4.5.

Although the concentration of O^- relative to free electrons in eqn. 4.5 does not depend on R_{N_2O}, it does depend strongly on temperature, as indicated by the plot of F_{N_2O} for CO_2, CH_4, and H_2 as a function of temperature shown in Fig. 4.3. This trend of increasing F_{N_2O} with increasing temperature is expected to continue until $[O^-]/[e]$ is greater than unity. According to eqn. 4.5, at a detector temperature of 350°C, $[O^-]/[e] = 0.15$. Experimental results, obtained using a detector operating at 350°C, show that the N_2O mole fraction in the N_2 carrier gas stream may be raised from $\ll 1$ ppb* to 20 ppm (i.e., the saturation point) with a change in the detector standing current of approximately 20%, in good agreement with the value predicted by eqn. 4.5 [16].

Details of the experimental curves shown in Fig. 4.3 can be explained by considering the known ion chemistry of CH_4, CO_2, and H_2 with O^-. For example, the CH_4 reaction is

$$O^- + CH_4 \xrightarrow{k_6} OH^- + CH_3 \qquad (4.6)$$

with $k_6 = 8 \cdot 10^{-11}$ $cm^3 s^{-1}$ [17]. This reaction exhibits little energy (temperature) dependence. As a result, the temperature dependence of F_{N_2O} for CH_4 almost entirely reflects the temperature dependence of eqn. 4.1 as indicated in eqn. 4.5. The slope of F_{N_2O} for CH_4 in Fig. 4.3 implies an activation energy for k_1 of 0.44 eV, in excellent agreement with the value of 0.46 eV measured by Warman et al. [10], and the value of 0.45 eV measured by Wentworth et al. [11].

*Throughout this chapter the American billion (10^9) and trillion (10^{12}) are meant.

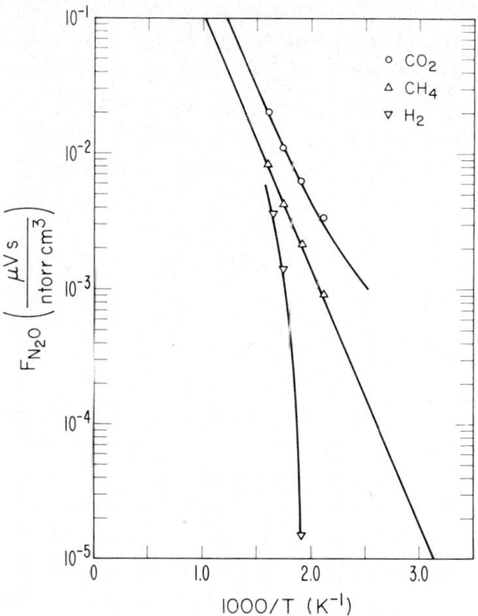

Fig. 4.3. Plot of F_{N_2O} for CO_2, CH_4, and H_2 as a function of temperature. The experimental data for CO_2 are plotted as open circles (○), for CH_4 as open triangles (△), and for H_2 as inverted triangles (▽). The N_2O mole fraction in the carrier gas was 12 ppm for these measurements. (From ref. 8.)

The reaction of CO_2 with O^-, on the other hand, is a three-body process,

$$O^- + CO_2 + M \xrightarrow{k_7} CO_3^- + M \tag{4.7}$$

The rate constant for this reaction is $k_7 = 3.1 \cdot 10^{-28}$ $cm^3 s^{-1}$ [18] at room temperature with O_2 acting as the third body. If it is assumed that this reaction rate is approximately the same with N_2 as the third body, M, and with $[N_2] = 1.2 \cdot 10^{19}$ molecules cm^{-3} in the ECD at atmospheric pressure and 350°C, reaction 4.7 should proceed more rapidly than reaction 4.6; consequently, F_{N_2O} $(CO_2) > F_{N_2O}$ (CH_4). The temperature dependence of F_{N_2O} (CO_2) does not yield a straight line in the Arrhenius plot of Fig. 4.3 as in the case of CH_4. The slight upward curvature observed for F_{N_2O} (CO_2) is explained by the negative temperature dependence of eqn. 4.7.

Finally, the reaction of H_2 with O^- has two channels at low energy

$$O^- + H_2 \rightarrow OH^- + H \tag{4.8a}$$

$$O^- + H_2 \rightarrow H_2O + e \tag{4.8b}$$

The rate constant for this reaction is $k_8 = 6.4 \cdot 10^{-10}$ $cm^3 s^{-1}$ [19]. At low temperature, however, the associative detachment channel is dominant, while the production of OH^- increases rapidly with increasing temperature. The competition of these two processes explains the sharp downward curvature of F_{N_2O} (H_2) in Fig. 4.3 at lower temperatures. In fact, the response of the ECD to H_2 becomes negative at detector temperatures less than 200°C. When the associative detachment process, reaction 4.8b, dominates, electrons are produced from the reaction of O^- with H_2, accounting for the negative response.

The presence of impurities in the ECD has a significant effect on the degree of chemical sensitization that may be achieved using N_2O doping. The source of these impurities may be the carrier gases used, transfer lines, regulators, valves, or other equipment used in conjunction with the gas chromatograph. Column bleed or plumbing and valving leaks can be very significant sources of contamination in the ECD.

In order to understand the role of impurities in the chemical sensitization process, one must consider the rate equations for [e], $[O^-]$, and $[NO^-]$. These rate equations are obtained from eqns. 4.1–4.4 in Section 4.1 along with eqn. 4.9 which includes reactions of O^- with carrier gas impurities (Imp.) not intentionally added:

$$O^- + \text{Imp.} \xrightarrow{k_9} \text{stable negative ion} \tag{4.9}$$

The rate equations for [e], $[O^-]$, and $[NO^-]$ are given as eqns. 4.10–4.12:

$$\frac{\partial[\text{e}]}{\partial t} = S + k_3[NO^-][N_2] - k_1[\text{e}][N_2O] - L[\text{e}] \tag{4.10}$$

$$\frac{\partial[O^-]}{\partial t} = k_1[\text{e}][N_2O] - k_2[O^-][N_2O] - k_4[O^-][\text{A}] - k_9[O^-][\text{Imp.}] \tag{4.11}$$

$$\frac{\partial[NO^-]}{\partial t} = k_2[O^-][N_2O] - k_3[NO^-][N_2] \tag{4.12}$$

In these equations, t is time, S is the electron ionization rate per unit volume and L represents, with the exception of reactions 4.1 and 4.4, all electron loss mechanisms which are linearly dependent on the electron density, [e], such as diffusion, mass flow out of the detector, and reaction with other trace species in the carrier gas.

In steady state, [e] is given by

$$[\text{e}] = [\text{e}]_0 \left\{L + \frac{k_1[N_2O]k_9[\text{Imp.}]}{k_2[N_2O] + k_9[\text{Imp.}]}\right\}\left\{L + \frac{k_1[N_2O](k_4[\text{A}] + k_9[\text{Imp.}])}{k_2[N_2O] + k_4[\text{A}] + k_9[\text{Imp.}]}\right\}^{-1} \tag{4.13}$$

where $[\text{e}]_0 = [\text{e}]$ when $[\text{A}] = 0$. For $k_2[N_2O] \gg k_4[\text{A}]$ and $k_2[N_2O] \gg k_9[\text{Imp.}]$, eqn. 4.13 reduces to

$$[\text{e}] \approx [\text{e}]_0 \left\{\frac{Lk_2 + k_1k_4[\text{A}] + k_1k_9[\text{Imp.}]}{Lk_2 + k_1k_9[\text{Imp.}]}\right\}^{-1} \tag{4.14}$$

Eqn. 4.14 predicts that the detection limit for A using the N_2O-sensitized ECD increases with increasing L or [Imp.], i.e.,

$$\left.\frac{\partial[\text{e}]}{\partial[\text{A}]}\right|_{\lim[\text{A}]\to 0} = -[\text{e}]_0 \left\{\frac{k_1k_4}{Lk_2 + k_1k_9[\text{Imp.}]}\right\} \tag{4.15}$$

This means that the sensitivity of the chemically enhanced ECD decreases as the electron loss mechanisms mentioned above and/or reactions of O^- with impurities in the carrier gas increase.

The effect of impurities on the response of a N_2O-doped ECD has been investigated using a controlled environment gas chromatograph in which the entire system, plumbing, gas sampling valve, column, and detector, were housed in a sealed box which was flushed by nitrogen (the column effluent) [16,20]. This precaution was undertaken to avoid contamination due to atmospheric O_2 and CO_2 diffusing into the system through small leaks in valves and fittings. When the sealed case of the gas chromatograph was opened, or flushed with air instead of N_2, allowing O_2 and CO_2 to be entrained into the carrier gas stream through the inevitable small leaks in the system valves and plumbing, the minimum detectable level for most compounds increased by a factor of two. This is due to the reaction, within the detector, of electrons with inadvertently present O_2, to form O_2^-, an electron loss mechanism that can be represented formally by an increase in L, and of O^- with O_2 to form O_3^- or O^- with CO_2 to form CO_3^- (i.e., [Imp.] has increased).

The use of a controlled environment gas chromatograph is by no means essential for the successful application of the N_2O-doping technique. For comparison, typical detection limits obtained using standard, commercially available gas chromatographs operated in the N_2O-sensitized mode are about two times higher than those obtained using the controlled environment instrument. It is essential, however, that impurities entering the carrier gas be minimized in order to take maximum advantage of this chemical sensitization technique.

Other factors, in addition to carrier gas impurities, can also have an effect on the response of the N_2O-doped ECD. The detector response to carbon monoxide is an interesting example. When carbon monoxide is present in the detector, the following reactions may be expected to occur:

$$CO + O^- \xrightarrow{k_{16}} CO_2 + e \tag{4.16}$$

$$CO + \text{wall} \xrightarrow{\tau_W} CO_2 + \text{wall} \tag{4.17}$$

$$CO_2 + O^- + M \xrightarrow{k_7} CO_3^- + M \tag{4.7}$$

Considering only the associative detachment reaction of O^- with CO (eqn. 4.16, k_{16} = $4.5 \cdot 10^{-10}$ $cm^3\,s^{-1}$ at 350°C [19]), one would expect no response to CO because for each electron lost by the interruption of the cycle represented by eqns. 4.1–4.3, an electron is regenerated in reaction 4.16. The observed positive response to CO in a N_2O-doped ECD may be explained by the further reaction of the CO_2 generated in reaction 4.16 with another O^- ion as shown earlier in reaction 4.7. Alternatively, carbon monoxide in the sample may be oxidized to CO_2 by the hot metal surfaces inside the ECD at 350°C. Eqn. 4.17 represents this oxidation of CO on the hot walls of the detector with τ_W being the associated time constant.

Calculations and experimental results have shown that reaction 4.16 may be neglected with respect to reaction 4.17 at 350°C [21]. Thus, the N_2O-doped ECD appears to act as a catalytic converter of CO to CO_2 with a response that is dependent on the reaction of O^- with CO_2 in reaction 4.7. The oxidation of CO on the detector walls proceeds rapidly enough with respect to the residence time of the CO in the detector so that the CO elution rate from the column is the sensitivity limiting factor, and no significant improvement in sensitivity would be expected from intentionally oxidizing the CO prior to entering the detector.

Finally, it should be noted that the result of summing the steady-state ion chemistry outlined in eqns. 4.1–4.3 represents the electron catalyzed conversion of N_2O to N_2 and NO,

$$2N_2O \xrightarrow{e} N_2 + 2NO \qquad (4.18)$$

This cycle suggests that NO should be formed in the ECD when N_2O is present. The rate of production of NO in the carrier gas expected from eqn. 4.18 may be calculated from

$$\frac{\partial [NO]}{\partial t} = 2k_1 [e] [N_2O] \qquad (4.19)$$

This equation is derived from the rate equation for the production of NO, eqn. 4.11 ([A] = 0, [Imp.] = 0) and eqn. 4.12. At 350°C, $k_1 = 3.3 \cdot 10^{-11}$ cm^3 s^{-1}. The average concentration of electrons in the ECD cell may be calculated using $[e] = I/ef_0 V$, where I = standing current, e = electron charge, V = detector cell volume, and f_0 = pulse frequency. The assumption is made that all of the electrons in the cell are collected with each pulse. For the ECD used in these studies $I \approx 0.8$ nA, $e = 1.602 \cdot 10^{-19}$ C, $V \approx 1$ cm^3, and $f_0 = 10^3$ Hz yielding an electron concentration of approximately $5 \cdot 10^6$ electrons cm^{-3}. A nitrous oxide concentration of 20 ppm at 760 Torr and 350°C implies that $[N_2O] = 2.4 \cdot 10^{14}$ molecules cm^{-3}. Using these values in eqn. 4.19, the rate of production of NO is ca. $8 \cdot 10^{10}$ molecules cm^{-3} s^{-1}. The concentration of N_2 is $1.2 \cdot 10^{19}$ molecules cm^{-3} in the ECD at atmospheric pressure and 350°C. For a carrier gas flow-rate of 0.5 std. cm^3 s^{-1} through the 1-cm^3 cell (i.e., residence time ≈ 1 s), the calculated rate of production of NO implies that the expected mixing ratio of NO in the ECD effluent carrier gas should be ≈ 6 ppb.

In order to determine if NO is actually produced in the ECD in practice, the effluent of the gas chromatograph was connected to the inlet of a chemiluminescence detector that has been described in the literature [22–24]. In this way, the NO level was measured in the carrier gas containing 20 ppm N_2O both at a point prior to the ECD and immediately after the carrier gas exited the ECD [20]. The level of NO found before the ECD was $\leqslant 0.5$ ppb. The level of NO found in the carrier gas after it exited the ECD was 5.5 ppb, indicating that about $6 \cdot 10^{10}$ NO molecules cm^{-3} s^{-1} were being generated inside the ECD at 350°C, in good agreement with the predicted $8 \cdot 10^{10}$ molecules cm^{-3} s^{-1}.

As an additional check, the ECD was cooled and the NO level measured. At a detector temperature of 55°C, the level of NO in the ECD effluent was found to be below the detection limit, $\leqslant 0.5$ ppb, as it was in the carrier before passing into the ECD. Raising the temperature of the detector back up to 350°C restored the 5 ppb difference in the NO level between the effluent and inlet of the ECD.

4.3. EQUIPMENT AND EXPERIMENTAL PRECAUTIONS

Selective electron-capture sensitization can be achieved with very minor modifications of commercially available electron-capture gas chromatographs. Selection of instrumentation accessories and operating conditions must be guided with careful attention to the following requirements arising from the theory of sensitization:

(a) Care must be taken to minimize system leaks and contaminants in the carrier gas and throughout the flow system. This includes the elimination of elastomeric materials in the flow system to the extent possible. We have also found it necessary to rinse all metal tubing that comes in contact with the carrier gas with reagent grade hexane, then acetone, and finally ethanol, followed by drying at elevated temperature.

(b) The detector must be operated at temperatures of approximately 300–350°C for maximum sensitivity.

While in some of our studies a hermetically sealed chromatograph, in which the flow system is bathed in nitrogen to reduce air leaks into the flow system, was used, ordinary commercially available instrumentation has been employed for most experiments. The carrier gas must be purified by passing it through a Molecular Sieve 13X cartridge before entering the chromatograph. Even "ultrahigh purity" grade nitrogen contains impurities that must be removed. In some instances the highest grade nitrogen cylinders contained impurities more objectionable than lower grade cylinders [16].

The nitrous oxide can be introduced into the carrier gas stream between the analytical column outlet and the ECD by means of a permeation device [16,20,21]. The optimum concentration range is from 15 to 70 ppm N_2O in N_2. The device consists of a PTFE permeation membrane and a compression fitting "cross". The use of a compression fitting "cross" allows continuous flushing of the permeation tube with N_2O and also makes rapid changing of dopant gases convenient. In the earliest experiments nitrous oxide was mixed with nitrogen in carrier gas reservoirs, but this practice has the disadvantage that it is difficult to purify the carrier gas without also removing amounts of nitrous oxide which change with time.

For most purposes commercially available instruments with constant-current electron-capture detectors containing ^{63}Ni high-temperature foils are suitable. Care should be taken to minimize impurities, column bleed, etc., because the net effect of sensitization is to convert a relatively specific detector sensitive to a limited number of compounds into a detection system responding sensitively to a much wider range of compounds, including many kinds of impurities, contaminants and artifacts. Occasionally "negative" peaks or W-shaped peaks are observed when nitrous oxide sensitization is attempted. Usually these undesirable peak shapes can be avoided by eliminating impurities in the system.

4.4. APPLICATIONS

4.4.1. Vinyl chloride

4.4.1.1. Introduction

As a result of the recognition that occupational exposure to vinyl chloride (C_2H_3Cl)

poses significant health hazards, federal regulations governing the use of this material have been established. Standards promulgated in 1974 [25] call for an action level of 0.5 ppm time weighted average exposure. If this level of vinyl chloride is exceeded, an extensive personal monitoring system must be implemented. This directive permits a maximum 8-h time weighted average exposure of 1 ppm and a maximum 15 min exposure of 5 ppm.

The continuing reduction in the permissible levels of human exposure to C_2H_3Cl as the nature of the hazard becomes more fully understood, has led to a search for improved methods for the detection and monitoring of this common industrial chemical. The method currently recommended by the National Institute of Occupational Safety and Health [26] is the adsorption of substrates from whole air samples in activated charcoal "traps" followed by desorption using CS_2 and gas chromatographic analysis using a flame ionization detector (FID). This technique is very readily adaptable to the time weighted average requirement simply by drawing ambient air over a suitable trap at low flow-rates for the requisite 8 h. The technique does, however, have a number of serious drawbacks. Backup traps are necessary for each sample to ensure that vinyl chloride breakthrough has not occurred, making the analysis more time consuming. Trap sealing, transport, and storage procedures may adversely affect the reproducibility and efficiency of the analysis. The use of CS_2 for desorption of the traps is not only cumbersome and time consuming, but also represents safety hazards during analysis (flammability and toxicity) [27]. This last difficulty may be circumvented by the use of thermal desorption techniques which allow some simplification of operation while requiring programming of the gas chromatographic column temperature. The ultimate detection limit and reproducibility of this approach appear to be approximately 10^{12} molecules (ca. 100 pg) of vinyl chloride and $\pm 10\%$, respectively [28].

A moderately priced detection system for vinyl chloride that is capable of either direct, in situ air sampling or grab sampling for laboratory analysis at low level (ppb) would avoid most of the above difficulties. One of the most sensitive methods for detecting vinyl chloride is mass spectrometry in the specific ion monitoring mode (SIM-MS). This technique offers excellent selectivity, a significant advantage over the nonspecific FID, and a detection limit of $1-2 \cdot 10^{11}$ molecules (10–20 pg), corresponding to a mixing ratio of 4–8 ppb for a sample size of 1 std. cm^3 of air [29,30]. Although this detection limit is adequate, the expense of SIM-MS places it out of reach as a general monitoring tool in many laboratories. Infrared absorption systems with folded optical paths yield detection limits of about 1 ppm operating with the 941 cm^{-1} or 917 cm^{-1} absorption bands [31,32]; however, many substances can interfere with this nonspecific analysis. Work with a photoionization detector has yielded a detection limit for C_2H_3Cl of $5-10 \cdot 10^{13}$ molecules per sample using a porous polymer packed column [33]. A chemiluminescence detector based upon the chemiluminescence of the excited reaction products of olefins and ozone has also been developed [34–36]. For a 25-ml sample size the noise equivalent concentration based on peak height is 50 ppb [35]. This corresponds to $3.4 \cdot 10^{13}$ molecules of C_2H_3Cl (3.5 ng of C_2H_3Cl) at standard temperature and pressure. The sensitivity of each of the latter three methods is less than that of the flame ionization detector approach.

The response of a conventionally operated ECD to vinyl chloride is also too low to be of practical use. Lande [28] has quoted a detection limit of approximately $2 \cdot 10^{13}$ molecules per sample (2 ng) corresponding to a mixing ratio of 0.7 ppm for a 1 std. cm^3 air sample. However, Miller and Grimsrud [4], using O_2 as a dopant in their carrier gas at the 0.2% level, found a sensitivity enhancement for C_2H_3Cl of approximately a factor of 110 at a detector temperature of 250°C. The increase in signal-to-noise ratio, and thus the lowering of the detection limit, for C_2H_3Cl is about a factor of 7 [16].

By contrast, the detection limit for C_2H_3Cl, using an otherwise standard ECD, can be lowered about three orders of magnitude to $1.4 \cdot 10^{10}$ molecules (ca. 1.4 pg) by adding N_2O to the carrier gas stream [16]. This detection limit corresponds to a mixing ratio of about 0.5 ppb for a 1 std. cm^3 air sample. The application of this technique to the monitoring of C_2H_3Cl obviates all of the complex handling, storage, and desorption procedures associated with the carbon-trap FID analysis. The achievable sensitivity would allow the continuous in situ monitoring for C_2H_3Cl at levels down to 1 ppb. An air sample slowly collected through a critical orifice into an evacuated container would suffice for time weighted average measurements. Only about 2 std. cm^3 of air are required for each vinyl chloride measurement; therefore, a cigarette pack size evacuated sampler that can be easily worn by personnel in critical areas would provide for several replicate samples. The removal of samples from an ambient or a subambient pressure container presents no serious technical problems and can be routinely accomplished using an evacuated sample inlet system on a standard commercially available gas chromatograph [37].

4.4.1.2. Analysis of vinyl chloride in air

A sample chromatogram taken with a nitrous oxide mixing ratio of 16 ppm in nitrogen is shown in Fig. 4.4. The chromatogram shows the detector response to a 1.50 std. cm^3 sample of N_2 containing C_2H_3Cl at a mole fraction of 4.18 ppb (i.e., $1.68 \cdot 10^{11}$ molecules or 17.8 pg). The sample chromatogram was obtained using a 3.4 m $\times$ 1/8 in. O.D. stainless-steel column packed with 80–100 mesh Porasil C and operated at 100°C with a flowrate of 25 std. cm^{-3} min^{-1}. The ECD temperature was 350°C. Under optimum operating conditions, when the N_2 carrier gas contains 20 ppm N_2O and the detector temperature is 350°C, the mole fraction of C_2H_3Cl resulting in a signal-to-noise ratio of 2 for a 1.5 std. cm^3 air sample is 350 ppt, corresponding to $1.4 \cdot 10^{10}$ molecules (1.5 pg). This detection limit is a factor of 760 lower than the detection limit observed for the same instrument with no N_2O ($<$1 ppb) present in the carrier gas. A plot of the peak height response of the N_2O-doped ECD to C_2H_3Cl as a function of the concentration of C_2H_3Cl injected is shown in Fig. 4.5, indicating that the detector operated in this mode gives a linear response over two orders of magnitude (1.5–150 pg). Quantitation over at least three orders of magnitude is easily achievable and the linear dynamic range can be extended by operating the detector in a fixed-current variable pulse frequency mode rather than the fixed-frequency variable current mode used for collecting the data in Fig. 4.5.

The response of the N_2O-doped ECD to C_2H_3Cl as a function of detector temperature is shown in Fig. 4.6. The response is greatest at the maximum operating temperature of the ECD, 350°C, and decreases with decreasing temperature so that the minimum detectable level of C_2H_3Cl increases by a factor of ten at a detector temperature of 190°C. The ECD response to C_2H_3Cl as a function of N_2O concentration in the carrier gas is shown in Fig. 4.7. The lower curve is the detector response showing the increase in sensitivity from the zero N_2O level, and the upper curve shows the signal-to-noise ratio achieved on the Porasil C column mentioned previously for injections of 1.5 std. cm^3 gas

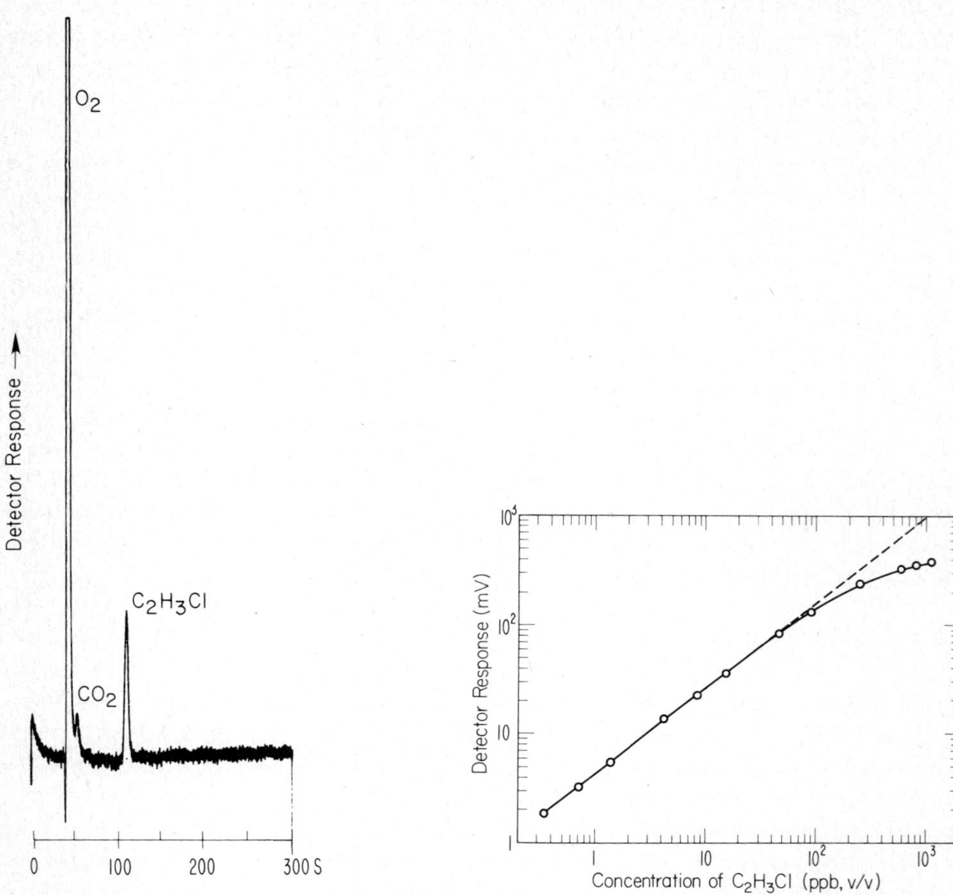

Fig. 4.4. Sample chromatogram obtained with a 3.4 m Porasil C column at 100°C showing the sensitized ECD response to a 1.50 std. cm^3 sample containing C_2H_3Cl at a mole fraction of 4.18 ppb (i.e., $1.68 \cdot 10^{11}$ molecules or 17.8 pg). ECD temperature, 350°C. (From ref. 16.)

Fig. 4.5. Peak height response of the N_2O-doped ECD to C_2H_3Cl as a function of the C_2H_3Cl concentration in a 1.50 std. cm^3 sample. One ppb corresponds to a sample containing $4.03 \cdot 10^{10}$ molecules (4.18 pg). (From ref. 16.)

samples containing $4.0 \cdot 10^{10}$ molecules (4.2 pg) of C_2H_3Cl. The observed increase in sensitivity results in an increased signal-to-noise ratio of a factor of 760 as the level of N_2O is increased from $\ll 1$ ppb to approximately 20 ppm. Since the signal-to-noise ratio is essentially independent of the N_2O mole fraction between the levels of 16 and 70 ppm, this is not a critical system parameter; the N_2O concentration need only be kept constant at some convenient mixing ratio in the 20- to 50-ppm region.

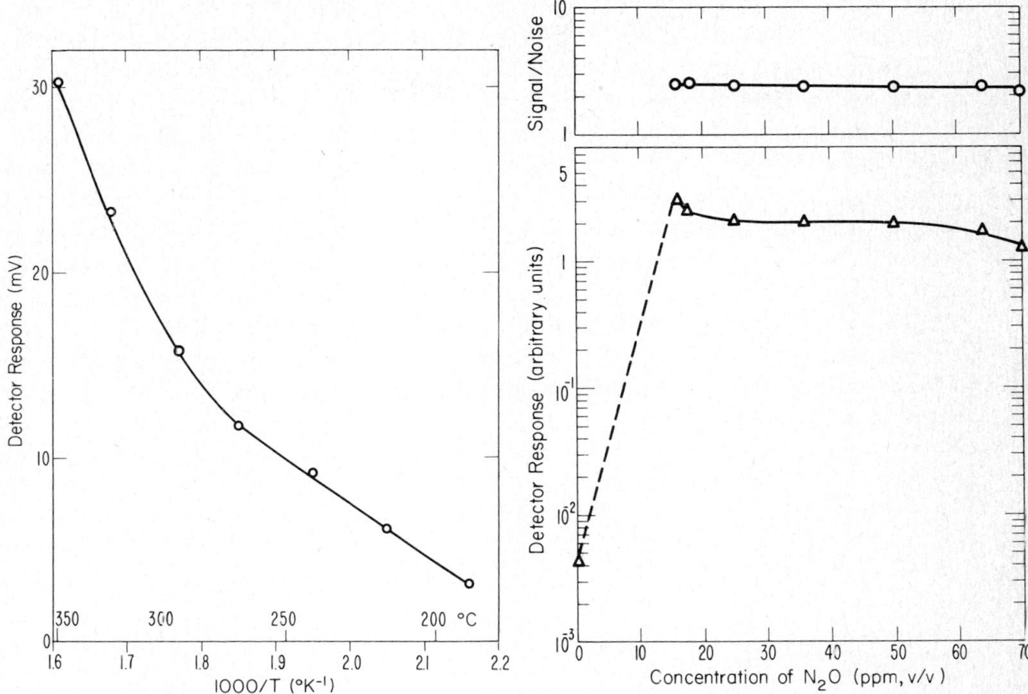

Fig. 4.6. Peak height response to C_2H_3Cl as a function of the detector temperature with a N_2O mole fraction of 17 ppm in the N_2 carrier gas stream. (From ref. 20.)

Fig. 4.7. Peak height response of the sensitized ECD to C_2H_3Cl and the signal-to-noise ratio as a function of N_2O dopant mixing ratio obtained with a 3.4 m Porasil C column at 100°C. (From ref. 16.)

Both a 3.4 m long Porasil C column and a 2.4 m long Porapak Q column have been evaluated for use in the analysis of vinyl chloride in air [16]. Both of these columns, when operated isothermally at 100°C with a carrier flow-rate of 25 std. cm^3 min^{-1}, give entirely adequate separation of vinyl chloride from any other species, including O_2, present in the artificial C_2H_3Cl and N_2 mixture shown in Fig. 4.4. The retention times for C_2H_3Cl on both of these columns were in good agreement with available literature values [38,39] and both columns yielded reasonable efficiencies for vinyl chloride. Neither of these columns is, however, recommended for actual atmospheric sampling since the Porasil C material suffers from severe overloading by CO_2 and poor separation of C_2H_3Cl from $CFCl_3$ and the Porapak Q from potential interference with the C_2H_3Cl analysis by acetaldehyde and ethylene oxide [39]. A much more suitable column for air analysis, consisting of an 80–100 mesh mixture of Porapak S–Porapak T (80:20, w/w) as suggested by Krockenberger et al. [40], was evaluated [16]. The column was 1.5 m long and was operated isothermally at 80°C. Although this column had a lower efficiency than either the Porasil C or the Porapak Q columns, it gave entirely adequate resolution of C_2H_3Cl from all components found in the dry whole air samples we investigated, and it is reportedly free from interference by acetaldehyde and ethylene oxide [40]. Fig. 4.8 shows a sample chromatogram taken using this latter column. The sample, 1.5 std. cm^3 of air "spiked" with 22 ppb of vinyl chloride, was separated using a N_2 carrier doped with approximately 20 ppm of N_2O and a detector temperature of 350°C. The vinyl chloride is clearly separated from the O_2, CO_2, and other components such as CF_2Cl_2 and $CFCl_3$ present in the whole air sample. The lower column efficiency raises the minimum detectable level of vinyl chloride to approximately $4.4 \cdot 10^{10}$ molecules per sample (ca. 4.6 pg) corresponding to a minimum detectable mixing ratio of 1.1 ppb for a 1.5 std. cm^3 air sample. Since the ECD is sensitive to water that elutes just before the vinyl chloride on this column, whole air samples may need to be dried if taken in a humid environment [41–43].

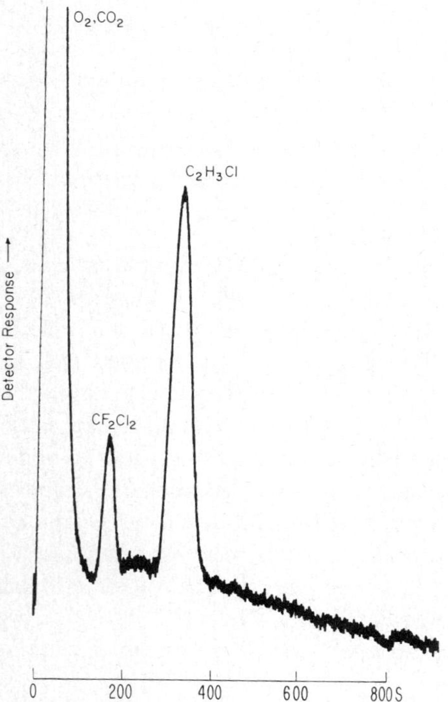

Fig. 4.8. Chromatogram of a 1.5 std. cm^3 whole air sample spiked with 22 ppb C_2H_3Cl obtained with a 1.5 m Porapak S and T mixed sorbent column. (From ref. 16.)

4.4.2. Carbon monoxide, hydrogen, and methane determinations

4.4.2.1. Comparison of techniques used for the analysis of carbon monoxide in air

Many methods have been developed for the analysis of carbon monoxide in air because of its importance in the reactions producing photochemical smog. However, only some of these methods are useable for measurements of CO at trace levels, and each requires a dedicated or specialized instrument not often available in most laboratories. Non-dispersive infrared instruments are widely used for measurements of CO in polluted cities, but suffer from poor sensitivity and susceptibility to interferences. One technique takes advantage of the chemical reaction between CO and HgO at 210°C. The resulting mercury vapor is measured by its absorption of resonance radiation at 2537 Å provided by a mercury lamp [44]. Although allowing very low detection limits for CO of 2–3 ppb [45,46], this technique is also sensitive to other reducing gases such as H_2, SO_2, olefins, and aldehydes that also undergo chemical reactions with hot HgO and can cause erroneously high CO measurements if not removed prior to analysis. Another method for the detection of trace quantities of CO is the helium ionization detector. Bruner et al. [47] report a detection limit of 70 ppb CO in air, but the CO peak appears as a small shoulder on a large, tailing N_2 peak. In addition, the mechanism by which the detector operates is not fully understood and no satisfactory explanation has been given for the negative and bipolar detector responses for some gases or the changes in polarity of the detector signal with changing purity of the helium carrier gas [48,49]. Although a detection limit of 20 ppb CO in a 100-μl He sample ($5.4 \cdot 10^{10}$ CO molecules) has been achieved [49], the difficulties mentioned above prevent the widespread use of this detector for routine CO analysis in whole air samples. Another method used for local measurements of small quantities of CO is the catalytic reduction of CO to CH_4 and detection with a FID [50–55]. The amount of CO producing a signal-to-noise ratio of 2 using a molecular sieve column under ideal laboratory conditions is $4 \cdot 10^{11}$ molecules (18 pg) per sample, corresponding to a mole fraction of 15 ppb for a 1 std. cm^3 sample.

When a N_2O-doped ECD is used [20,21] a signal-to-noise ratio of 2 can be achieved with a whole air sample containing $3.4 \cdot 10^{11}$ CO molecules (16 pg). This detection limit corresponds to a mole fraction of 13 ppb for a 1 std. cm^3 air sample, which is about the same sensitivity as that achieved under optimal conditions using the catalytic conversion/FID method mentioned previously. The elimination of the flammable compressed H_2 required for the catalytic CO converter and the open flame, which is sometimes inadvertently extinguished, in the FID, together with the convenience of needing only one compressed carrier gas cylinder, make this use of the ECD extremely attractive for portable gas chromatographs intended for monitoring CO at background levels. Using a molecular sieve analytical column, one also has the added bonus of simultaneous measurement of H_2 along with the usual CH_4 determination achieved with an FID.

4.4.2.2. Simultaneous measurements of carbon monoxide, hydrogen, and methane in air

The response of the N_2O-doped ECD to CO as a function of the N_2O dopant concentration is similar to that mentioned previously for vinyl chloride in Section 4.4.1.2. The signal-to-noise ratio is essentially independent of the N_2O mole fraction between the levels of about 16 and 70 ppm so that the N_2O concentration need only be set at some convenient level in this range. The detector response to CO as a function of detector temperature is also similar to that reported for several gases [8,16,20,21,56]; that is, the response is greatest at the maximum operating temperature of the detector, 350°C, and decreases by approximately a factor of 10 as the detector temperature is reduced to 200°C.

The response of the N_2O-doped ECD to CO has been found to be linear within the 1% precision of the measurement up to $5 \cdot 10^{13}$ CO molecules (2.3 ng) per sample, corresponding to a mole fraction of 1.9 ppm for a 1 std. cm^3 sample. Departures from linearity do not exceed 10% up to $10 \cdot 10^{13}$ CO molecules per sample, corresponding to a CO mole fraction of 4 ppm for a 1 std. cm^3 sample. Quantitation outside of this range can be accomplished by the use of appropriate calibration data.

A chromatogram of 2.2 std. cm^3 whole air sample is shown in Fig. 4.9. Because of the sample pre-cut configuration used [20,21], the chromatogram consists of only four peaks, corresponding successively to H_2, O_2, CH_4, and CO. The mole fractions of H_2, CH_4, and CO are 0.60, 1.67, and 0.39 ppm respectively. As may be seen, all four compounds are readily quantitated even at these low levels. It should be emphasized that the measurement of atmospheric concentrations of H_2 along with the CO and CH_4 cannot be achieved with the catalytic methanization/FID technique mentioned in Section 4.4.2.1, while it is relatively easily accomplished with SECS.

Atmospheric mixing ratios of CO, CH_4, and H_2 are currently being monitored at a mountain research station west of Boulder, CO, U.S.A., jointly maintained by the National Oceanic and Atmospheric Administration and the University of Colorado. Located on Niwot Ridge at an altitude of 3050 m (longitude 105°32′, latitude 40°30′), the site is usually swept by moderate to strong winds from the west. As no large anthropogenic sources are located west of the site for some 500 miles, Salt Lake City being the nearest, it is thought that concentrations of minor atmospheric constituents measured under conditions of westerly flow should be representative of background levels over the North American continent. Occasionally, "upslope" conditions exist during which air from the heavily populated Denver-Boulder metropolitan area is swept upward from the east into the Rocky Mountains to altitudes exceeding 3000 m. Under these conditions, the Niwot Ridge site affords an opportunity to sample "aged" polluted air under conditions of high UV solar radiation.

Carbon monoxide mixing ratios found during the onset of just such an episode on September 22, 1980 are shown in Fig. 4.10. The morning was overcast, foggy, and calm. At about 1:00 p.m. the cloud cover broke up and a slight breeze from the southeast developed. By 2:00 p.m. the wind was from the east at 5 to 10 miles/h and a strong "upslope" easterly wind condition existed. Fig. 4.10 clearly shows the increase in the CO mole fraction from about 250 ppb to more than 400 ppb during this period. Concurrent measurements showed that the ozone concentrations doubled, increasing from 31 to 59 ppb, during the same period.

Carbon monoxide mixing ratios have been monitored intermittently at this site with the sensitized ECD technique since August 2, 1980. From August 2 to September 22 the

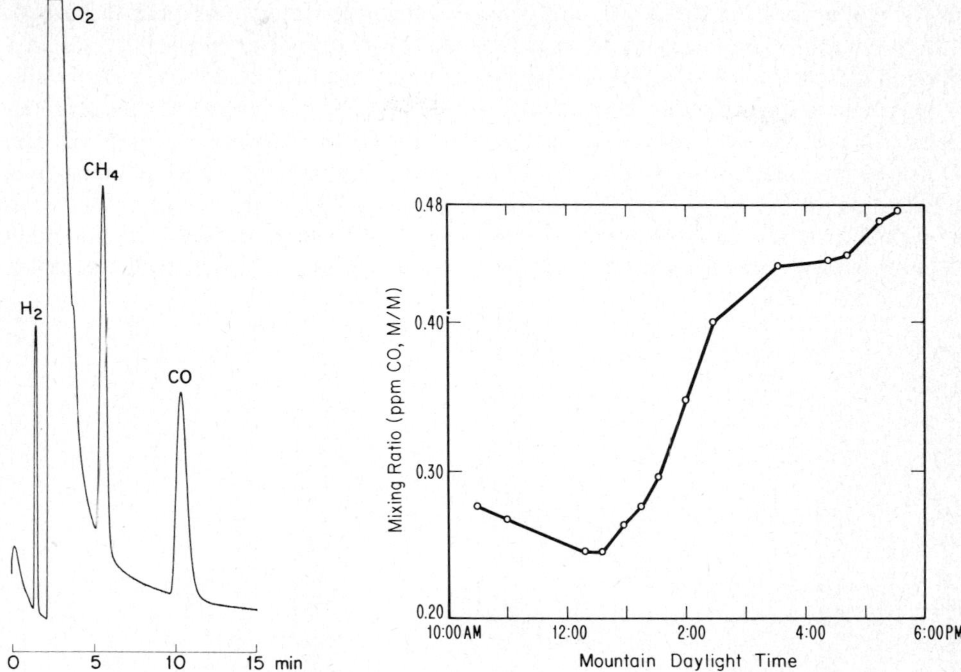

Fig. 4.9. Chromatogram of a 2.2 std. cm^3 mountain air sample obtained using a 2.4 m stainless steel column, packed with 100–120 mesh molecular sieve 5A and operated at 80°C. A 2.4 m long stainless-steel column, packed with 100–120 mesh Porapak Q and operated at 50°C, was used as a sample pre-cut column to separate H_2, O_2, CH_4, and CO from the other minor components in the air sample. The N_2O mole fraction in the nitrogen carrier gas (added post-column with a permeation device) was approximately 30 ppm. Carrier flow-rate was 42 std. cm^3 min^{-1}. (From refs. 20 and 21.)

Fig. 4.10. Carbon monoxide mixing ratios over a 7-h period, at Niwot Ridge, Colorado. M/M refers to CO mole fraction. (From refs. 20 and 21.)

concentrations ranged from 160 to 473 ppb, with mixing ratios in excess of 300 ppb being observed only during the strong "upslope" on September 22. Analysis of these data showed a CH_4 mole fraction of about 1.7 ppm and an average H_2 mole fraction of 0.6 ppm with diurnal variations of ±5% and ±15%, respectively.

4.4.2.3. Human breath analysis

A N_2O-doped ECD has been used to measure hydrogen and methane in samples of human breath. Data from these experiments indicate that the concentration of hydrogen may be up to 20 times higher in exhaled air than in ambient air. The change in methane concentration in a small number of healthy subjects sampled was negligible. A chromatogram obtained from the air exhaled by an adult male is presented in Fig. 4.11. The peak at 0.3 min arises from about 6 pg of hydrogen. This chromatogram was obtained using a 1.8 m × 1/8 in. O.D. stainlèss-steel column packed with 100–120 mesh molecular sieve 5A and operated at 50°C with a nitrogen carrier flow-rate of 40 ml min^{-1}. The sample loop volume was 90 µl. This method may prove valuable for determining hydrogen levels in biological samples [57,58]. Because hydrogen diffuses very rapidly, it is expected that hydrogen generated by bacteria or by processes in the body should be rapidly transported by the blood to the lungs where it can be readily partitioned into the air and exhaled. It is expected that measurements such as these coupled with determinations of many other compounds will provide useful diagnostic information for the detection of disease states.

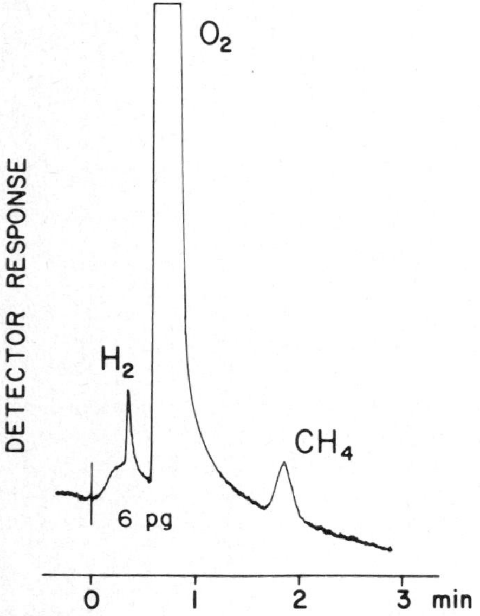

Fig. 4.11. Measurement of H_2 and CH_4 in expired human breath. Data were obtained using a 1.8 m molecular sieve 5A column at 50°C. Sample loop volume was 90 µl. Peak at 0.3 min arises from about 6 pg of H_2. (From ref. 56.)

4.4.3. Sensitization of detection of other compounds

Determinations have been made of the relative ECD responses to a variety of compounds [8,16,20,21,56] that are listed in Table 4.1. Enhancements in the ECD response to ethane, *n*-propane, *n*-butane, *n*-pentane, and *n*-hexane are included in the table. These hydrocarbon measurements were made using a 1.5 m × 1/8 in. O.D. stainless-steel column packed with 100–150 mesh Porasil C and operated isothermally at 100°C with a carrier flow-rate of 40 std. cm^3 min^{-1}. The ECD temperature was 350°C and the carrier gas contained 20 ppm N_2O. Fig. 4.12 shows chromatographic traces for a mixture of alkanes analyzed with and without N_2 in the carrier gas. Detection of picogram levels of these hydrocarbons has been achieved when N_2O is present in the carrier gas [56].

TABLE 4.1

ELECTRON-CAPTURE DETECTOR RESPONSE ENHANCEMENT FACTORS WHEN NITROUS OXIDE IS ADDED TO THE CARRIER GAS

Compound	Formula	Enhancement factor*
Hydrogen	H_2	40
Carbon dioxide	CO_2	10,000
Methane	CH_4	13
Ethane	C_2H_6	27
n-Propane	C_3H_8	31
n-Butane	C_4H_{10}	34
n-Pentane	C_5H_{12}	29
n-Hexane	C_6H_{14}	11
Dichlorodifluoromethane	CCl_2F_2	0.23
Dichlorotetrafluoroethane	$CClF_2CClF_2$	0.29
Trichlorofluoromethane	CCl_3F	0.40
1,1,1-Trichloroethane	CH_3CCl_3	0.51
Trichlorotrifluoroethane	CCl_2FCClF_2	0.33
Carbon tetrachloride	CCl_4	0.43
Dichloromethane	CH_2Cl_2	0.32
Chloroform	$CHCl_3$	0.40
Tetrachlorothene	C_2Cl_4	0.45
Carbon sulfide	COS	1
Methyl chloride	CH_3Cl	1
Chlorodifluoromethane	$CHClF_2$	63
Methyl fluoride	CH_3F	782
Acetaldehyde	C_2H_4O	11
Vinyl chloride	C_2H_3Cl	760
Carbon monoxide	CO	100

$$^{\star}\text{Enhancement factor} = \frac{\text{detection limit with 20 ppm } N_2O \text{ in } N_2 \text{ carrier}}{\text{detection limit without } N_2O \text{ in } N_2 \text{ carrier}}$$

Some representative halocarbons that are normally sensitive to electron-capture detection are also listed in Table 4.1. The detection limits for these compounds are raised by a factor of 2 to 4 when N_2O is present in the carrier gas simply as a result of the increased baseline noise. A 4 m $\times$ 1/8 in. O.D. stainless-steel column, packed with 100–150 mesh Porasil C and operated isothermally at 100°C with a carrier flow-rate of 60 std. cm^3 min^{-1} was used for the halocarbon measurements.

Both carbonyl sulfide (COS) and methyl chloride (CH_3Cl) are important atmospheric constituents. The ECD is reasonably sensitive to both of these compounds. As can be seen in Table 4.1, no enhancement was observed for either of these compounds at a detector temperature of 350°C and 20 ppm N_2O in the carrier gas. The CH_3Cl and COS measurements were made on the 4 m Porasil C column used for the halocarbon measurements. Column temperatures used were 100°C for CH_3Cl and 50°C for COS.

The ECD shows an enhanced response for chlorodifluoromethane ($CHClF_2$), methyl fluoride (CH_3F), acetaldehyde (C_2H_4O), and carbon dioxide when 20 ppm of N_2O are added to the carrier gas. The response enhancement factors for these compounds range

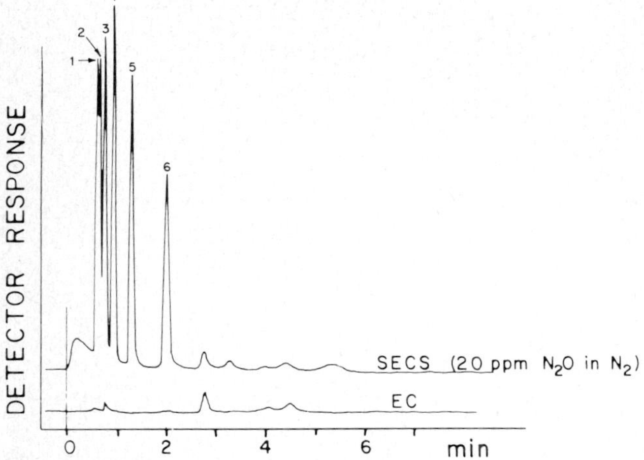

Fig. 4.12. SECS detection of a mixture of *n*-alkanes. Detector temperature was 350°C with a carrier flow-rate of 40 std. cm^3 min^{-1}. The analysis was done using a 1.5 m long stainless-steel column packed with Porasil C and operated at 100°C. Sample size was 0.59 ml and the concentration of each *n*-alkane in the N_2 gas standard was 15 ppm. Peaks: 1 = methane; 2 = ethane; 3 = propane; 4 = butane; 5 = pentane; 6 = hexane. (From ref. 56.)

from 11 for acetaldehyde up to 10,000 for carbon dioxide and are shown in Table 4.1. A 3.4 m X 1/8 in. O.D. stainless-steel column, packed with 80–100 mesh Porasil C and operated at 50°C with a carrier flow-rate of 30 std. cm^3 min^{-1}, was used for the $CHClF_2$ and CH_3F measurements. A 2.4 m X 1/8 in. O.D. stainless-steel column, packed with 100–120 mesh Porapak Q and operated at 100°C and a carrier flow-rate of 60 std. cm^3 min^{-1}, was used for the acetaldehyde measurements. Finally, a 3.7 m X 1/8 in. O.D. stainless-steel column, packed with 80–100 mesh Porapak Q and operated at 50°C with a flow-rate of 60 std. cm^3 min^{-1}, was used for the CO_2 determinations.

It should be emphasized that the response enhancements given in Table 4.1 represent increases or decreases in signal-to-noise ratios when N_2O is added to the carrier gas. That is, the enhancement factor is the ratio of the detection limit (signal-to-noise ratio of 2) with N_2O added to the detection limit without N_2O. Therefore, increases or decreases in noise are taken into account and the enhancement factors are not simply changes in sensitivity.

4.5. CONCLUSIONS

These results indicate that the presence of 1 to 70 ppm of N_2O in the ECD cell acts to reduce the detection limit of the ECD to compounds which do not rapidly attach electrons but which do react with O^- to form stable negative ions. A steady-state negative ion chemistry involving N_2O has been proposed to account for this effect. This scheme quantitatively accounts for a wide range of observations, checks and tests, which are outlined in the preceding sections. At present the N_2O-enhanced gas chromatography–ECD is being used in the field for routine trace gas analysis.

Since the results of the previous applications section indicate that lower detection limits are observed only for analytes that react more rapidly with O^- than directly with electrons, and the detection limits for compounds such as $CFCl_3$ or CCl_4 that rapidly convert electrons to stable negative ions are raised slightly upon addition of N_2O to the carrier because of increased baseline noise, advantage may be taken of this equalization phenomenon in the frequently encountered instances in which two or more analytes of interest have markedly different response factors, necessitating repeated independent analyses. In addition, the relative response factors observed for the various classes of compounds when N_2O is added to the carrier gas can be coupled with retention indices to aid in the confirmation of the identities of unknown peaks in a chromatogram. For example, an atmospheric sample chromatographed in the absence of N_2O may give a complex chromatogram consisting of peaks due to chlorofluorocarbons, other halogenated hydrocarbons, and numerous aliphatic and aromatic hydrocarbons. When the same sample is chromatographed in the presence of N_2O, one would expect the aliphatic and aromatic hydrocarbon peaks to be much larger, while the peaks due to halogenated compounds will not be enhanced. From a comparison of these two chromatograms, specifically the enhancement data coupled with retention data, one can make rapid tentative assignments of compound classes, and perhaps identify the specific compounds. Consequently, the practical effect of adding N_2O or other dopant gases to the carrier gas is to convert the relatively selective ECD into a much less specific, general-purpose detector, with no change in instrumentation, simply by switching from one carrier gas mixture to another.

REFERENCES

1 H.J. van de Wiel and P. Tommassen, *J. Chromatogr.*, 71 (1972) 1–7.
2 E.P. Grimsrud and D.A. Miller, *Anal. Chem.*, 50 (1978) 1141–1145.
3 E.P. Grimsrud and R.G. Stebbins, *J. Chromatogr.*, 155 (1978) 19–34.
4 D.A. Miller and E.P. Grimsrud, *Anal. Chem.*, 51 (1979) 851–859.
5 E.P. Grimsrud and D.A. Miller, *J. Chromatogr.*, 192 (1980) 117–125.
6 E.P. Grimsrud, D.A. Miller, R.G. Stebbins and S.H. Kim, *J. Chromatogr.*, 197 (1980) 51–58.
7 P.G. Simmonds, *J. Chromatogr.*, 166 (1978) 593–598.
8 M.P. Phillips, R.E. Sievers, P.D. Goldan, W.C. Kuster and F.C. Fehsenfeld, *Anal. Chem.*, 51 (1979) 1819–1825.
9 R. Marx, G. Mauclaire, F.C. Fehsenfeld, D.B. Dunkin and E.E. Ferguson, *J. Chem. Phys.*, 58 (1973) 3267–3273.
10 J.M. Warman, R.W. Fessenden and G. Bakale, *J. Chem. Phys.*, 57 (1972) 2702.
11 W.E. Wentworth, E. Chen and R. Freeman, *J. Chem. Phys.*, 55 (1971) 2075.
12 M. McFarland, D.B. Dunkin, F.C. Fehsenfeld, A.L. Schmeltekopf and E.E. Ferguson, *J. Chem. Phys.*, 56 (1972) 2358–2364.
13 D.A. Parkes, *J. Chem. Soc., Faraday Trans. 1*, (1972) 2103.
14 H. Shimamori and R.W. Fessenden, *J. Chem. Phys.*, 68 (1978) 2757–2766.
15 H. Shimamori and R.W. Fessenden, *J. Chem. Phys.*, 69 (1978) 4732–4742.
16 P.D. Goldan, F.C. Fehsenfeld, W.C. Kuster, M.P. Phillips and R.E. Sievers, *Anal. Chem.*, 52 (1980) 1751–1754.
17 W. Lindinger, D.L. Albritton, F.C. Fehsenfeld and E.E. Ferguson, *J. Chem. Phys.*, 63 (1975) 3238.
18 F.C. Fehsenfeld and E.E. Ferguson, *J. Chem. Phys.*, 61 (1974) 3181–3193.
19 M. McFarland, D.L. Albritton, F.C. Fehsenfeld, E.E. Ferguson and A.L. Schmeltekopf, *J. Chem. Phys.*, 59 (1973) 6629–6635.

20 M.P. Phillips, Ph.D. Dissertation, University of Colorado, Boulder, CO, 1980.
21 F.C. Fehsenfeld, P.D. Goldan and M.P. Phillips, in preparation.
22 B.A. Ridley and L.C. Howlett, *Rev. Sci. Instrum.*, 45 (1974) 742–746.
23 O.C. Zafiriou and M. McFarland, *Anal. Chem.*, 52 (1980) 1662–1667.
24 M. McFarland, D. Kley, J.W. Drummond and D.L. Albritton, *Anal. Chem.*, submitted for publication.
25 Fed. Reg., 39: 194 (1974) 35890.
26 J.V. Crable and D.G. Taylor, *NIOSH Manual of Analytical Methods*, U.S. Government Printing Office, Washington, DC, 1974.
27 N.I. Sax, *Dangerous Properties of Industrial Materials*, Van Nostrand Reinhold, New York, 5th ed., 1979.
28 S.S. Lande, *Amer. Ind. Hyg. Assoc. J.*, 40 (1979) 96–107.
29 R.C. Lao, R.S. Thomas and J.L. Monkman, *Amer. Ind. Hyg. Assoc. J.*, 37 (1976) 1.
30 J.D. Rosen, J.R. Morano, S.R. Pareles, J.R. Giacin and S.G. Gilbert, *J. Ass. Offic. Anal. Chem.*, 58 (1975) 700.
31 M.V. Zeller, *Infrared Determination of Detection Limits, Part II, Vinyl Chloride; Perkin-Elmer Infrared Bulletin 46*, Perkin-Elmer Corp., Norwalk, CT, May 1974.
32 D.S. Lavery and P.A. Wilks, *Amer. Lab.*, 6, No. 10 (1974) 53.
33 P.D. Goldan and F.C. Fehsenfeld, Aeronomy Laboratory, Environmental Research Laboratories, National Oceanic and Atmospheric Administration, unpublished results.
34 S.A. Myers, H.J. Quinn and W.C. Zook, *Amer. Ind. Hyg. Assoc. J.*, 36 (1975) 332.
35 W.A. McClenny, B.E. Martin, R.E. Baumgardner, Jr., R.K. Stevens and A.E. O'Keefe, *Environ. Sci. Technol.*, 10 (1976) 810–813.
36 S.J. Arnold, G.H. Kimbell and D.R. Snelling, *Can. J. Chem.*, 52 (1974) 2608.
37 P.D. Goldan, W.C. Kuster, D.L. Albritton and A.L. Schmeltekopf, *J. Geophys. Res.*, *C1*, 85 (1980) 413–423.
38 D.T. Williams and W.R. Miles, *J. Ass. Offic. Anal. Chem.*, 58 (1975) 272–275.
39 A. Krishen and R.G. Tucker, *Anal. Chem.*, 48 (1976) 455–456.
40 D. Krockenberger, H. Lorkowski and L. Rohrschneider, *Chromatographia*, 12 (1979) 787–789.
41 E. Burghardt, R. Jeltes, H.J. van de Wiel and E.J. Oranje, *Atmos. Environ.*, 13 (1979) 1057–1060.
42 J. Krasnec, *Atmos. Environ.*, 14 (1980) 391.
43 E. Burghardt, H.J. van de Wiel, E.J. Oranje and R. Jeltes, *Atmos. Environ.*, 14 (1980) 391–392.
44 R.C. Robbins, K.M. Borg and E. Robinson, *J. Air Pollution Control Ass.*, 18 (1968) 106.
45 L.A. Cavanagh, C.F. Schadt and E. Robinson, *Environ. Sci. Technol.*, 3 (1969) 251–257.
46 W. Seiler and C. Junge, *J. Geophys. Res.*, 75 (1970) 2217–2226.
47 F. Bruner, P. Ciccioli and R. Rastelli, *J. Chromatogr.*, 77 (1973) 125–129.
48 F.F. Andrawes and E.K. Gibson, *Anal. Chem.*, 52 (1980) 846–851.
49 F.F. Andrawes, R.S. Brazell and E.K. Gibson, *Anal. Chem.*, 52 (1980) 891–896.
50 F.W. Hightower and A.H. White, *Ind. Eng. Chem.*, 20 (1928) 10.
51 K. Porter and D.H. Volman, *Anal. Chem.*, 34 (1962) 748–749.
52 F.W. Williams, F.J. Woods and M.E. Umstead, *J. Chromatogr. Sci.*, 10 (1972) 570–572.
53 L. Dubois and J.L. Monkman, *Anal. Chem.*, 44 (1972) 74–76.
54 A. Volz, D.H. Ehhalt, L.E. Heidt and W. Pollock, in K.H. Grasnick (Editor), *Proc. Joint Symp. on Atmospheric Ozone, Dresden, 1976*, published under the auspices of the Int. Commission on Atmospheric Ozone and on Atmospheric Chemistry and Global Pollution of the Int. Assoc. of Meteorology and Atmospheric Physics in collaboration with the World Meteorology Organization and the Int. Assoc. of Geomagnetism and Aeronomy, 1977, pp. 219–226.
55 K. Suzuki, K. Horiuchi, Y. Kasahara, T. Shirai and S. Yanagisawa, *Bunseki Kagaku*, 29 (1980) 152–155; *C.A.*, 92 (1980) 208562s.
56 R.E. Sievers, M.P. Phillips, R.M. Barkley, M.A. Wizner, M.J. Bollinger, R.S. Hutte and F.C. Fehsenfeld, *J. Chromatogr.*, 186 (1979) 3–14.
57 M.D. Levitt, *N. Engl. J. Med.*, 281 (1969) 122.
58 H.L. Gearhart, D.P. Boxe, C.A. Smith, R.D. Morrison, J.D. Welsh and T.K. Smalley, *Anal. Chem.*, 48 (1976) 393.

Chapter 5

Oxygen-doping of the carrier gas in electron-capture detection

ERIC P. GRIMSRUD

CONTENTS

5.1. INTRODUCTION

During the past twenty-five year period in which the development of various forms of the electron-capture detector (ECD) for gas chromatography (GC) has occurred, several different gases or gas mixtures have been investigated for their use as the bulk carrier gas in which gaseous electron-capture processes can be measured. It has been noted that by changing the carrier gas, the ECD can be made to behave in a variety of ways. These differences have generally been caused by affecting various properties of the electrons, such as their thermalization rate or drift velocity [1]. Also, the choice of carrier gas can affect the concentration of electronically excited metastables, which can cause ionization of molecules thereby complicating the ECD signal [2]. Nitrogen and argon–methane mixtures have become favored carrier gases because with these, conditions within the detector facilitate the observation of electron-capture processes alone. In this search for appropriate carrier gases, gases which have electron affinities sufficient to cause the capture of electrons themselves have been avoided. This is entirely understandable since such carrier gases will reduce the population of the essential reagent of the ECD, the electron. Therefore, until recently any consideration given to electronegative components of the carrier gas has been primarily directed toward the means by which they might be eliminated. During the last few years, however, it has become apparent that when certain electron-attaching molecules, such as oxygen or nitrous oxide, are made a component part of the carrier, very interesting and analytically useful effects are observed. With these carrier gases an increase in the ECD's response to numerous compounds can occur by reactions of the negative ions which are then present. The ion–molecule reaction between a reagent negative ion, N^-, and the sample component, A, which provides a portion of this response is shown generally as eqn. 5.1.

$$N^- + A \rightarrow \text{products} \tag{5.1}$$

A special requirement of the negative ions used in this way for electron-capture detection is that their instantaneous concentration within the cell must be coupled to the electron density, since the above reaction alone cannot produce an ECD response. The coupling of the electron and reagent negative ion concentration is represented generally as eqn. 5.2.

$$e^- + N \rightleftharpoons N^- \tag{5.2}$$

The chemical sum of eqns. 5.1 and 5.2 is equivalent to direct electron capture. However, an ion–molecule reation, rather than direct electron capture, determines the magnitude of this response. In this chapter, the use of oxygen in the carrier gas to produce O_2^- and an altered ECD response by this general mechanism will be described.

Because oxygen has long been recognized as a common carrier gas impurity, several investigations have attempted to characterize its detrimental effects upon analyses by the ECD [2–5]. Of particular interest to the theme of this chapter is a report in 1972 by Van de Wiel and Tommassen [6] in which the primary observable effect of oxygen, the lowering of the standing current and attendant loss of electrons, is described. One can be certain that the response of a fixed frequency, pulsed ECD (used in their study) to strongly electron-attaching compounds will be diminished if oxygen contamination of the carrier gas has significantly reduced the electron population within the cell. It was, therefore, interesting that Van de Wiel and Tommassen noted that with severe oxygen contamination, the response of their instrument to a weakly responding compound, *n*-butyl bromide, actually increased somewhat. They suggested that this response increase was due to assistance provided by O_2^- in the overall negative charge transfer to this sample molecule. This chapter describes the extension and development of this fascinating observation made by Van de Wiel and Tommassen. The instrument used by these researchers had a pulsed, fixed-frequency ECD. As will be shown later, if they had had available a constant-current ECD [7] (which was first described in the literature only the previous year), they would undoubtedly have observed a much larger oxygen-induced response enhancement to *n*-butyl bromide. Following the commercial availability of the constant-current ECD, Grimsrud and Stebbins [8] reported in 1978 the effects of oxygen on the response of numerous EC-active compounds using this detector. The absolute response of this ECD to the two most weakly responding compounds in that study, *n*-butyl chloride and anthracene, was increased with oxygen by almost two orders of magnitude. This measurement and similar observations by Simmonds and also Rasmussen, made independently at about the same time, clearly indicated that the effects of oxygen on an ECD would be analytically useful. In the few years since 1978, we have investigated the general applicability and the fundamentals of the oxygen-doped ECD technique [9–13] and Simmonds [14] and also Rasmussen and co-workers [15, 16] have reported specific applications of the O_2-doped ECD to atmospheric analyses. Kallos [17] has developed an automated GC–ECD system with O_2-doped carrier for monitoring bis-chloromethyl ether in an industrial atmosphere. Clearly, the number of research groups who have reported their experiences with the oxygen-doped ECD is, so far, quite small, and the ultimate importance of the method to the broad scope of chemical analysis is yet to be determined. In this chapter a summary of our present knowledge of this promising new twist on electron-capture detection will be given in the hope that it will stimulate its further exploration. For more detail than is possible to relate here, the interested reader is encouraged to consult the references cited.

5.2. INSTRUMENTATION FOR THE OXYGEN-DOPED ELECTRON-CAPTURE DETECTOR

Most of the data presented for discussion in this chapter have been obtained in our lab using a Varian 3700 gas chromatograph equipped with a constant-current, pulsed ECD. This particular instrument has been described in detail by Patterson [18]. Improved ECD analyses by O_2-doping have also been reported by Rasmussen and co-workers using Hewlett-Packard 5710 [16] and Perkin Elmer 3920-B [15] gas chromatographs and by Simmonds using a Pye 104 [14]. In our lab we have observed oxygen-caused responses with a variety of homemade ECDs [10, 13]. Clearly, the successful use of O_2-doping is not limited to any one detector design. It will be shown later in this chapter, however, that for O_2-doping the constant-current (frequency-modulated) mode of signal processing appears to offer significant advantages over the fixed-frequency mode. While the direct-current mode of signal processing has not yet been carefully investigated in conjunction with O_2-doping, there is no apparent reason to expect from it any advantages over the pulsed modes. Of the ECD ionizers presently available, ^{63}Ni is the only one with which O_2-doping experiments have been reported. These ionizers can be operated at high temperatures (up to 400°C) and appear to suffer no detrimental effects from exposure to oxygen.

The only alteration of normal GC–ECD procedures required for O_2-doping is the addition of oxygen to the carrier gas. For routine analytical work this is probably done most conveniently by the addition of an appropriate makeup gas containing oxygen to the carrier flow system immediately ahead of the detector and after the column. If oxygen will not damage the chromatographic column in use, a single oxygen-containing gas can be used as the GC carrier gas. Both nitrogen and argon–methane have been used successfully as the major gas into which the oxygen or air is added.

5.3. SOME TYPICAL CHROMATOGRAMS

Perhaps a good way to start this account of O_2-doping is by showing some raw chromatograms of test substances and real samples run with and without oxygen added to the carrier gas. In Fig. 5.1 is shown the repeated analysis of a gaseous sample containing various amounts of the chlorinated methanes. The compound in least abundance, CCl_4, responds most strongly under the normal O_2-free condition shown by chromatogram a. In chromatogram b with 2.0‰ oxygen added, the response to CCl_4 is almost unaffected (is increased slightly). In considering the other three compounds, it is seen that as the chlorine number per molecule is decreased, the normal response decreases sharply. Meanwhile, in chromatogram b the O_2-doped responses to the normally poorly responding compounds is seen to be impressively enhanced relative to those of chromatogram a. The response to CH_3Cl, which is the most spectacular case, has increased about 100 times for these conditions.

The GC–ECD response of a 1,2-benzanthracene sample is shown in Fig. 5.2 for various amounts of O_2 present in the carrier gas. It is seen that the response to this compound increases continuously with added oxygen as progressively larger attenuations are required to record this peak. With no O_2 added, the two impurities (probably oxygenated polycyclic aromatic hydrocarbons which typically have very high EC sensitivity) produce peaks as large as the major component of this sample. The response to these impurities is

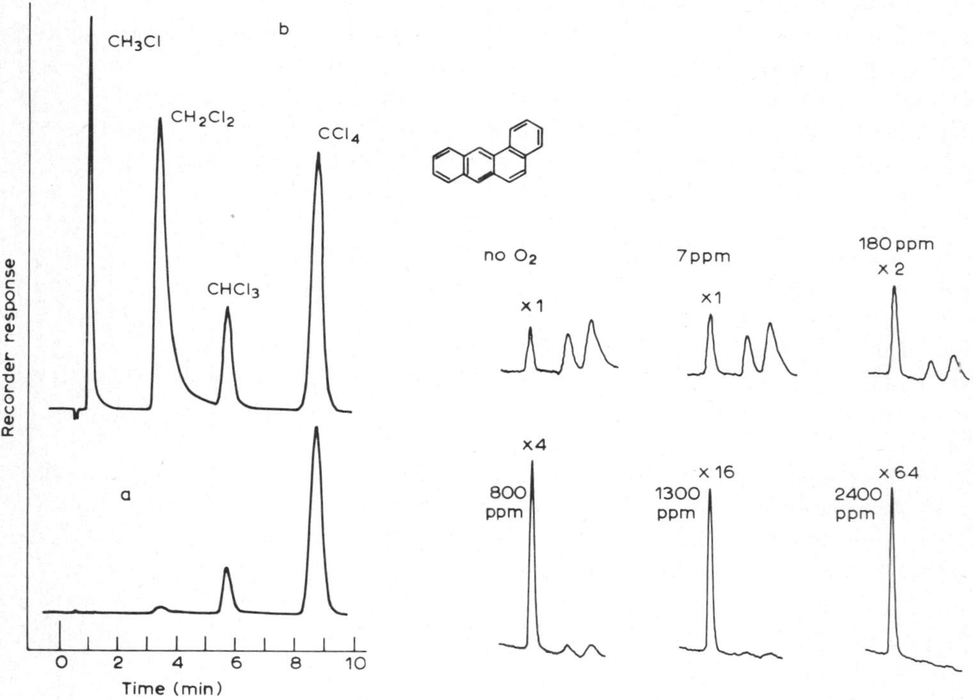

Fig. 5.1. Typical chromatograms in studies of the effects of oxygen on ECD responses to the chloromethanes. (a) Normal ECD conditions of pure nitrogen carrier gas. (b) Same standard run with 2.0‰ oxygen in the carrier gas. Detector temperature is 300°C. Sample loop 2 ml. Column: 10 ft. × 1/8 in. stainless steel packed with 10% SF-96 on Chromosorb W, 26°C. Concentrations are 620 ppb CH_3Cl, 960 ppb CH_2Cl_2, 68 ppb $CHCl_3$, and 17 ppb CCl_4. (Reproduced from ref. 9.)

Fig. 5.2. ECD response to 20 ng 1,2-benzanthracene sample with various amounts of oxygen added to the carrier gas. The second and third peaks observed in each analysia are due to impurities. Detector 250°C. Column: 1.5 ft. × 1/8 in. stainless steel packed with 3% OV-101 on Chromosorb W, 150°C. (Reproduced from ref. 11.)

not strongly affected by oxygen, however, so that with 2.4‰ oxygen added to the carrier gas, the chromatogram is unfavorably biased against the impurities relative to 1,2-benzanthracene. (The increasingly sloped baseline accompanying oxygen addition seen in this example is not inherent in the method, but is due to the device, an exponential dilutor, used in this experiment for varying the oxygen concentration in the carrier gas.)

Finally, in Fig. 5.3 is shown an application of O_2-doping to the analysis of CH_3Cl in a real atmospheric sample. A 20-ml whole air sample is introduced to the gas chromatograph and the components are frozen onto the head of the column using an initial sub-ambient temperature as described in the figure caption. Chromatograms a, b and c were obtained by the normal ECD method, without oxygen added to the carrier gas. A sharp negative spike and the first time marker indicate the point at which the 20-ml sample loop is opened. About 45 s is allowed for the sample oxygen to pass through the column during which the disturbance it causes is clearly seen in chromatogram a of a real air sample. At the point of the second time marker, the column temperature is rapidly raised to 10°C. Several peaks then appear, the two largest of which in chromatogram a are due to CF_2Cl_2 and $CFCl_3$. Chromatogram b shows the methyl chloride peak from a standard of concentration 7.7 ppb (parts per billion, volume ratio). The third time marker at about 1.5 min denotes the expected retention time of CH_3Cl. The air sample (chromatogram a) has a peak very near the same retention time. If this peak were assumed to be due to

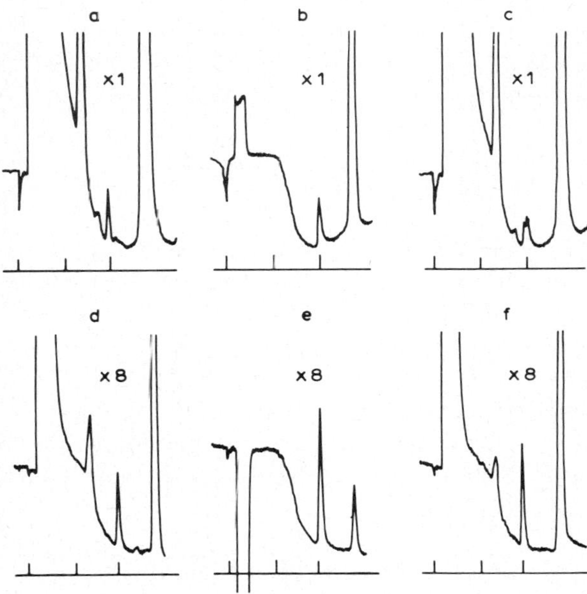

Fig. 5.3. Chromatograms of real air samples and CH_3Cl standards. a–c are under normal ECD conditions of pure nitrogen carrier gas. d–f are with 2.0‰ oxygen added to the carrier gas. The detector temperature is 300°C. Same column as in Fig. 5.1. The time markers indicate the following events: first, the sample valve is opened; second, 45 s later the rapid temperature program of the GC oven from −50°C to 10°C is commenced; and third, methyl chloride elutes at about 1.5 min. Samples: (a) ambient air, (b) standard with 7.7 ppb CH_3Cl, (c) equivolume mixture of a and b, (d) ambient air, (e) standard with 1.5 ppb CH_3Cl, (f) equivolume mixture of d and e. (Reproduced from ref. 9.)

CH_3Cl, however, approximately 8 ppb CH_3Cl in the air sample would be indicated, about ten times higher than the expected value of about 0.70 ppb [15, 19, 20]. Based on the sensitivity demonstrated by chromatogram b, only a barely perceivable peak for CH_3Cl might have been expected from the air sample. Chromatogram c shows that the peak under consideration in the air sample is not, in fact, CH_3Cl since a doublet is observed upon mixing the air sample with an equal volume of the CH_3Cl standard.

Chromatograms d, e and f were obtained with 2.0‰ oxygen in the carrier gas (temperature = 300°C). A recorder attenuation of eight times greater is now necessary to keep a higher baseline and peak responses on scale. Under this condition a peak of high signal-to-noise quality is again observed in the air analysis, chromatogram d, at the approximate retention time of CH_3Cl. In this case, however, chromatogram f of the mixture of standard and sample shows only one peak, suggesting that in this instance the peak in the air sample is due to CH_3Cl. The peak observed previously in chromatogram a near the retention time of CH_3Cl has been essentially lost in the attenuation increase because the response to that unknown substance is not amplified by O_2-doping. The concentration of CH_3Cl in the atmospheric sample deduced by comparison with chromatogram e (which is now of a 1.5 ppb CH_3Cl standard) is 0.80 ppb. This value is in very reasonable agreement with the generally accepted value for background air.

It is evident just from the chromatograms shown above that this simple procedure for altering ECD response will increase the power of the ECD for chemical analysis. Prior to discussing other applications of this technique, however, this account of O_2-doping will first turn to its more fundamental aspects which, hopefully, will provide some understanding of why and where the oxygen-doped ECD can be gainfully used.

5.4. EXPERIMENTAL CHARACTERIZATION AND THEORY

In order to determine optimum experimental conditions for taking advantage of the enhancements of response such as those illustrated in Figs. 5.1–5.3, some effort has been directed toward characterizing the effects of the variables at hand. We have also developed a theoretical model for O_2-doping which appears to agree quite well with experimental tests of the method. This theory (in somewhat abbreviated form relative to that given in ref. 10) and a characterization of the experimental parameters will, therefore, be described together. The chlorinated methanes will serve as examples in this section to illustrate typical experimental responses. Similar experiments applied to other molecules can be found in refs. 10 and 11.

5.4.1. Oxygen-doping with the constant-current electron-capture detector

The following events are expected to influence the electron concentration in an oxygen-doped ECD:

$$\text{radiation} \xrightarrow{S} P^+ + e \tag{5.3a}$$

$$e + A \xrightarrow{k_{eA}} B^- + \text{neutrals} \tag{5.3b}$$

$$e + O_2 \underset{k_{-0}}{\overset{k_0}{\rightleftharpoons}} O_2^- \qquad K_0 = \frac{k_0}{k_{-0}} \tag{5.3c}$$

$$O_2^- + A \xrightarrow{k_{0A}} C^- + \text{neutrals} \tag{5.3d}$$

$$P^+ + (e, B^- \text{ or } C^-) \xrightarrow{R} \text{neutrals} \tag{5.3e}$$

Eqn. 5.3a represents the ^{63}Ni ionization of the carrier gas to positive ions and electrons. Eqn. 5.3b is the direct electron capture by the sample molecule, A. Eqn. 5.3c is electron attachment and detachment with oxygen. If a sufficient quantity of oxygen is present, the forward and reverse rates of eqn. 5.3c will often be faster than the other chemical reactions, allowing the concentrations of the species coupled by eqn. 5.3c to be related by an equilibrium constant, K_0. Eqn. 5.3d is an ion–molecule reaction between O_2^- and A. Eqn. 5.3e is positive ion recombination with each of the negative species present. In view of recent articles dealing with the fundamentals of the ECD [21–23], it appears reasonable to assume that one recombination rate, R, applies reasonably well to each negative species listed in eqn. 5.3e. Furthermore, the positive ion concentration in the ECD will remain relatively constant at a level approached by the sum of the negative charges, $\eta_e + \eta_{C_2^-} + \eta_{B^-} + \eta_{C^-}$ (where n_i refers to the number density of each species i) the cell approaches equilibrium after each pulse [10]. In addition to essential eqns. 5.3a–e, others have been considered but are not included here for the sake of simplicity. Further clustering of O_2^- to form O_4^- or $O_2(H_2O)^-$ have been shown to be unimportant for the usual conditions of the detector [8]. The reactions of carrier gas impurities such as column bleed will often be significant under analysis conditions and probably constitute the major cause of departures to be expected from the model comprised of eqns. 5.3a–e.

The following expressions describe the changes in η_e and $\eta_{O_2^-}$ expected from eqns. 5.3a–e during the periods between pulses.

$$\frac{d\eta_e}{dt} = S - k_{eA}\eta_e\eta_A - k_0\eta_e\eta_{O_2} + k_{-0}\eta_{O_2^-} - R\eta_+\eta_e \tag{5.4a}$$

$$\frac{d\eta_{O_2^-}}{dt} = k_0\eta_e\eta_{O_2} - k_{-0}\eta_{O_2^-} - k_{0A}\eta_{O_2^-}\eta_A - R\eta_+\eta_{O_2^-} \tag{5.4b}$$

If sufficient O_2 is present in the carrier gas it is reasonable to assume a steady-state condition is approached at the end of each field-free period between pulses where each of the above derivatives approaches zero. Then the two equations can be combined and an expression is obtained for η_e with or without sample present. For the frequency-modulated or constant-current ECD, the pulse frequency, f, is related inversely to the population of electrons existing at the instant of pulsing [7, 23] so that

$$f = \frac{K}{\eta_e} \tag{5.5}$$

where K is a proportionality constant depending on various preselected experimental factors (such as the level of current demanded by the constant-current ECD). Considering the no-sample condition first, the expression for the baseline frequency of the O_2-doped ECD becomes

$$f_0 = \frac{K}{S}(R\eta_+) + \frac{K}{S}\frac{R\eta_+ k_0}{k_{-0} + R\eta_+}\eta_{O_2} \tag{5.6}$$

It has been shown [10] that $k_{-0} \gg R\eta_+$ at temperatures of 250°C or above. This allows the baseline frequency to be expressed simply as:

$$f_0 = \frac{K}{S}(R\eta_+) + \frac{K}{S}(R\eta_+ K_0\, \eta_{O_2}) \tag{5.7}$$

where $K_0 = k_0/k_{-0}$ is the equilibrium constant for eqn. 5.3c. In Fig. 5.4 the baseline frequency of a constant-current ECD measured as a function of O_2 concentration is shown for temperatures from 200°C to 400°C. As predicted by eqn. 5.7, f_0 increases linearly with η_{O_2} in each case. This instrument has a linear response range up to 100 kHz. It is seen that only for the 200°C case in Fig. 5.4 has the baseline frequency been raised by the oxygen levels used above this maximum acceptable value. From eqn. 5.7 we might expect that the slopes for each temperature in Fig. 5.4 divided by the intercepts would yield a value equal to K_0 for each temperature. The slope/intercept ratio of 15 different plots similar to those of Fig. 5.4 are 3.76 (±0.3) · 10^{-16} ml, 1.5 (±0.1) · 10^{-16} ml and 8.8 (±0.4) · 10^{-17} ml for 250°C, 300°C and 350°C, respectively (1.0‰ oxygen = 2.7 · 10^{16} · 273/T molecules per ml). The rate constants, k_0 and k_{-0}, and their temperature depen-

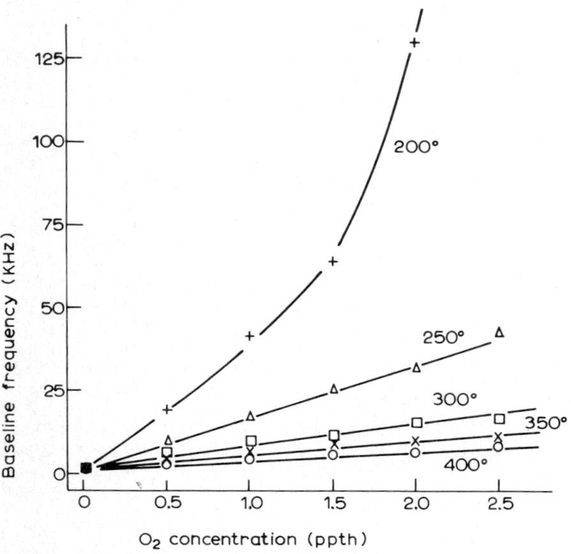

Fig. 5.4. The effect of oxygen doping on the ECD chromatogram baseline at several detector temperatures. (Reproduced from ref. 9.)

dences have been determined by others [24] and from these, an expression for K_0 has been deduced [10] shown as eqn. 5.8, where T is in degrees Kelvin.

$$K_0 = 2.6 \cdot 10^{-16} \cdot T^{-3/2} \exp \frac{4990}{T} \text{ ml} \tag{5.8}$$

Values for K_0 of $3.1 \cdot 10^{-16}$ ml, $1.2 \cdot 10^{-16}$ ml and $5.2 \cdot 10^{-17}$ ml at 250°C, 300°C and 350°C, respectively, are predicted by this equation. These expected values compare reasonably well with those obtained from Fig. 5.4.

To obtain an expression for the response to sample with oxygen-doping, it is necessary to again solve the rate equations 5.4a and 5.4b, this time with sample, A, present. An expression for f (frequency with sample) is obtained which can again be simplified by the reasonable assumption that $k_{-0} \gg R\eta_+$ and also, for this case, $k_{-0} \gg k_0\eta_A$ [10]. The net response to sample, $f - f_0$, is then given by eqn. 5.9.

$$f - f_0 = \frac{K}{S} (k_{eA} + K_0 \eta_{O_2} k_{0A}) \eta_A \tag{5.9}$$

That the oxygen-enhanced response measured with a constant-current ECD is linearly related to the sample concentration as predicted by eqn. 5.9 is shown in Fig. 5.5. The response is linear for the nearly three orders of magnitude change in sample concentration from the detection limit to the 100 kHz limit imposed by the instrument itself. It might be mentioned at this point that using oxygen-free nitrogen carrier gas and the same instrument, we observe a very non-linear response to CH_3Cl [25]. (The reason for this is not

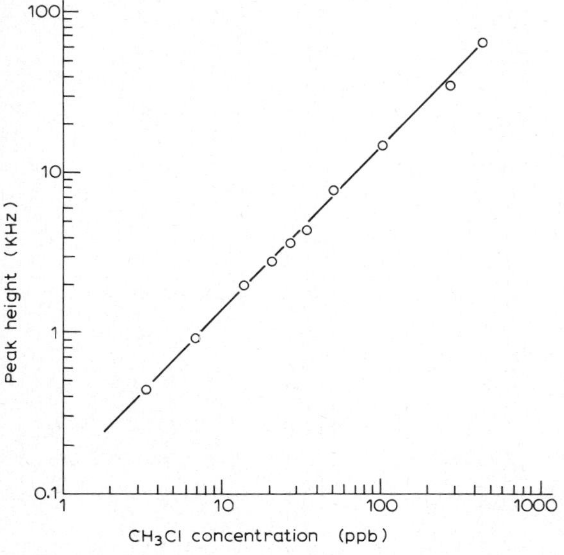

Fig. 5.5. ECD calibration curve for methyl chloride using oxygen-doped carrier gas. Detector temperature 300°C. Oxygen concentration in nitrogen carrier gas is 2.0‰.

well understood.) Therefore, for the case of CH_3Cl and other slightly chlorinated hydrocarbons, both sensitivity and linearity of response are improved by O_2-doping.

Since the normal response of this detector to a compound is given by $f - f_0 = K/S$ $(k_{eA})\ \eta_A$, the expected response enhancement, RE, caused by O_2 to substance A is given by

$$RE = 1 + K_0 \eta_{O_2} \frac{k_{oA}}{k_{eA}} \tag{5.10}$$

For all compounds studied in our laboratory, the dependence of the relative response on oxygen concentrations has been near-linear as predicted by eqn. 5.10 [9–11]. This is demonstrated for the chlorinated methanes in Fig. 5.6.

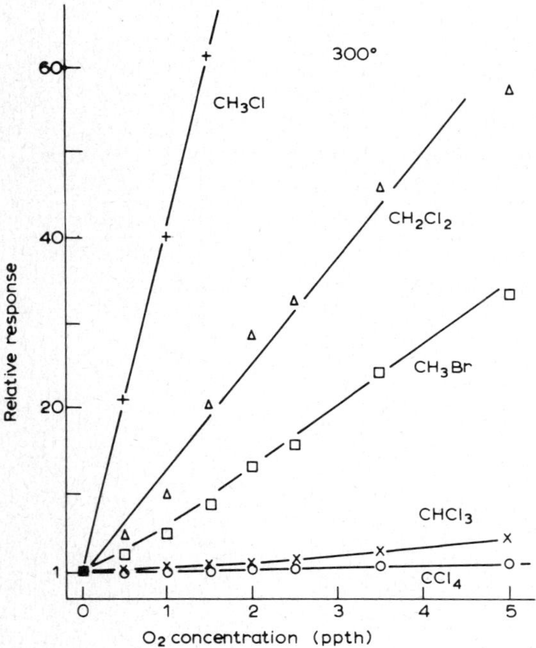

Fig. 5.6. The effect of oxygen-doping on the ECD response to several halogenated methanes at a detector temperature of 300°C.

Since K_0 can be evaluated by eqn. 5.6, a reasonable prediction of the product, $K_0\eta_{O_2}$, at any temperature and oxygen concentration can be obtained. For example, for η_{O_2} = 2.0‰ ($\eta_{O_2} = 5.4 \cdot 10^{16} \cdot 273/T\ \text{ml}^{-1}$),

$$RE = 1 + 8.5 \frac{k_{oA}}{k_{eA}} \text{ at } 250^\circ\text{C} \tag{5.11a}$$

$$RE = 1 + 3.0\,\frac{k_{oA}}{k_{eA}} \text{ at } 300^{\circ}\text{C} \tag{5.11b}$$

$$RE = 1 + 1.2\,\frac{k_{oA}}{k_{eA}} \text{ at } 350^{\circ}\text{C} \tag{5.11c}$$

$$RE = 1 + 0.54\,\frac{k_{oA}}{k_{eA}} \text{ at } 400^{\circ}\text{C} \tag{5.11d}$$

If it is assumed for the moment that the ratio k_{oA}/k_{eA} is relatively invariant with temperature [10], the temperature dependence of the oxygen-enhanced responses to methyl chloride shown in Fig. 5.7 can be largely understood in terms of eqns. 5.11a–d. Thus, the larger enhancements observed at the lower temperatures are due primarily in this case to the larger oxygen–electron equilibrium constant. It should be noted that eqns. 5.11a–d predict that the magnitude of oxygen-response enhancements measured with the constant-current ECD are not limited to the value of k_{oA}/k_{eA}, as one might naively predict, but increase continuously beyond that value with added O_2. Thus, for the case of methyl chloride at 250°C, an *RE* value of about 200 with 2.0‰ is achieved which, according to eqns. 5.11a, results from a ratio for k_{oA}/k_{eA} of approximately 23.

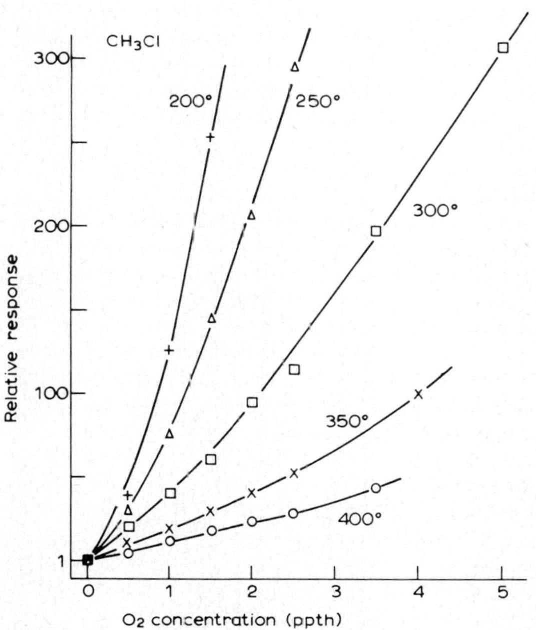

Fig. 5.7. The effect of detector temperature on the oxygen-induced response enhancements of methyl chloride. (Reproduced from ref. 9.)

5.4.2. Oxygen-doping with the fixed-frequency electron-capture detector

While the majority of ECDs being purchased today are of the constant-current type, many are still in use which have their frequency of pulsing fixed and the signal is derived from a simple decrease in sample current. Because of the much smaller linear response range of these instruments as compared with the constant-current ECDs, we have not expected the fixed-frequency ECDs to be as readily adaptable to oxygen-doping. Furthermore, in a comparative study using methyl chloride as the sample [13], we have found another more encompassing reason why the oxygen-doped, fixed-frequency ECD may not work as well, even when the sample sizes are kept small and within the linear range of either instrument. In Fig. 5.8 are shown the *RE* values measured for methyl chloride as a function of oxygen concentration using a fixed-frequency ECD. It is seen that the measured response increases sharply with the initial addition of oxygen, but after about 0.3‰ O_2 the response is weakened by additional oxygen. Furthermore, the magnitude of the *RE* value is much smaller, only 1.7 with 2.0‰ O_2 and reaches only 4, the maximum value, with 0.30‰ O_2. At this detector temperature it will be recalled (Fig. 5.7) that an *RE* value of about 200 is obtained with 2.0‰ O_2 using the constant-current ECD.

An understanding of why this detector provides response enhancements very different from those obtained with the constant-current ECD can be obtained by again considering eqns. 5.3a–e. For the fixed-frequency mode, the response is taken as the decrease in the measured current, and for the small-sample and oxygen-free condition the response can

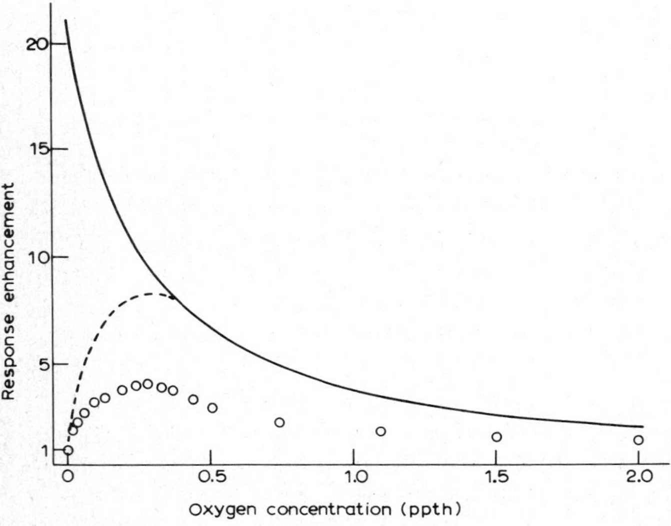

Fig. 5.8. Response of fixed-frequency ECD to a 12 ppb methyl chloride sample as a function of oxygen concentration in the carrier gas. Measured relative responses (points) and predicted responses (continuous curves) are shown.

be expressed by eqn. 5.12:

$$R_e = Q\,k_{eA}\eta_e^0\eta_A \tag{5.12}$$

where Q is a proportionality constant and η_e^0 denotes the electron density in an oxygen-free ECD. With oxygen present in the detector, the response will be

$$R_{O_2} = Q\,(k_{eA}\eta_e\eta_A + \phi\,k_{oA}\eta_{O_2^-}\eta_A) \tag{5.13}$$

where $\phi = d\eta_e/d\eta_{O_2^-}$ is a constant which relates the loss of O_2^- via eqn. 5.3d to the attendant loss of electrons by eqn. 5.3c. Assuming eqn. 5.3c to be in an equilibrated state (the assumption is reasonable when relatively large amounts of O_2 are present), the concentrations of the species in this reaction are related by $K_0 = \eta_{O_2^-}/\eta_e\eta_{O_2}$. Therefore, $\phi = 1/K_0\eta_{O_2}$ and, also, $\phi = \eta_e/\eta_{O_2^-}$. Eqn. 5.13 then becomes

$$R_{O_2} = Q\,(k_{eA} + k_{oA})\,\eta_e\eta_A \tag{5.14}$$

Dividing this by the normal response given in eqn. 5.12, the response enhancement expected is

$$RE = \frac{\eta_e}{\eta_e^0}\left(1 + \frac{k_{oA}}{k_{eA}}\right) \tag{5.15}$$

Since the sum of negative charges in the O_2-doped ECD will tend to remain constant [10], $\eta_e^0 \approx \eta_e + \eta_{O_2^-}$ for the small-sample condition (where η_{B^-} and η_{C^-} are negligible). With this equality and the expression for K_0 given above, eqn. 5.16 is obtained.

$$RE = \frac{1 + \dfrac{k_{oA}}{k_{eA}}}{1 + K_0\eta_{O_2}} \tag{5.16}$$

Eqn. 5.16 is in a form which can be compared with eqn. 5.10 obtained previously for the constant-current ECD. Eqn. 5.16 indicates that the *RE* value measured by the fixed-frequency mode of signal processing can never exceed the value of $k_{oA}/k_{eA} + 1$. Furthermore, as more oxygen is added the *RE* value should decrease, not increase as with the constant-current ECD. Eqn. 5.16 has also been plotted (solid curve) in Fig. 5.8 where values of $k_{oA}/k_{eA} = 20$ and $K_0 = 4.25‰^{-1}$ (determined by fitting the constant-current ECD data for methyl chloride shown in Fig. 5.7 to eqn. 5.10) were chosen. That this curve approaches a maximum value equal to $1 + k_{oA}/k_{eA}$ as O_2 approaches zero is an artifact of our treatment which assumed enough O_2 is always present to maintain eqn. 5.1d at equilibrium. As η_{O_2} approaches zero, eqn. 5.16 fails because eqn. 5.3c is no longer fast in both directions and the concentrations of the species involved are no longer held in a ratio fixed by K_0 as sample enters the detector. The relative response at $\eta_{O_2} = 0$ must be 1.0 by definition. Therefore, with the dashed line in Fig. 5.8 the expected intercept of unity is joined to the solid curve at the point corresponding to about 0.3‰ O_2. This somewhat arbitrary choice is made because the maximum in the experimental data occurred near this point. Other similar curves could be drawn to approximate the expected behavior of this detector in the small oxygen concentration range, but in any case it is clear that the model provides an adequate account of the form of O_2 dependence observed experimentally. The maximum *RE* observed in Fig. 5.8 was only about 4 while a value of approximately 8 might have been predicted from the dashed curve. This difference probably reflects a variation in the oxygen-free response measured with the completely different GC–ECD systems used for this comparative study. (The normal responses will be somewhat dependent on details of each instrument, such as detector design and carrier gas purity.)

In summary, it presently appears that analytical advantages offered by oxygen-doping of the carrier gas may be realized more completely if a constant-current rather than a fixed-frequency ECD is used. For certain instances, if the ratio k_{0A}/k_{eA} for a given substance is very large, the fixed-frequency ECD with a lightly O_2-doped carrier gas will provide significant enhancements. Also, in terms of signal-to-noise responses or ultimate detection limits with O_2-doping, the two instruments may provide similar responses. Nevertheless, the greater linear dynamic range of the constant-current ECD, the relative simplicity of the dependence of the response enhancement on oxygen concentration, and the ability to cause very large enhancements by using large amounts of oxygen all favor the constant-current ECD for O_2-doping.

5.4.3. Signal-to-noise and detection limits

Returning to the constant-current ECD, it is important to recognize that the signal-to-noise response to a given compound will not be increased by O_2-doping as much as is the absolute response $(f - f_0)$. Unfortunately, with the addition of O_2 to the detector an increased baseline noise must be accepted along with the improved response. An increased signal-to-noise response will occur only if the response enhancement exceeds this attendant increase in baseline noise. In Fig. 5.9 is shown the measured peak-to-peak noise of the measured baseline frequency as a function of oxygen concentration and detector temperature. It is seen that the dependence of noise on oxygen concentration very closely resembles the accompanying increase in baseline frequency shown previously in Fig. 5.4. The effect of oxygen on the relative signal-to-noise response to a given compound will be equal to the ratio of its *RE* value and relative noise at each oxygen level. Again, using the chlorinated methanes as an example, the oxygen dependence of the signal-to-noise response to these are shown in Fig. 5.10 at 250°C. For the analysis of CH_3Cl, oxygen doping might be expected to increase the signal-to-noise response and improve the detection limit by about 16 times. (Recall that at 250°C the relative absolute response shown in Fig. 5.7 increases continuously with O_2 well past *RE* = 200.) For compounds which do not enhance greatly, little or no signal-to-noise improvement is to be expected. In fact, if the *RE* value of a compound is very low, such as that of CCl_4, a deterioration of the signal-to-noise response will occur. This effect can be used to advantage to reduce or eliminate uninteresting peaks in a chromatogram as was seen in the air analysis shown in Fig. 5.3.

It has previously been shown by eqns. 5.11a–d and in Fig. 5.7 that greater absolute response enhancements will be observed using lower detector temperatures. In terms of signal-to-noise responses, however, this advantage of lower temperatures may be negated

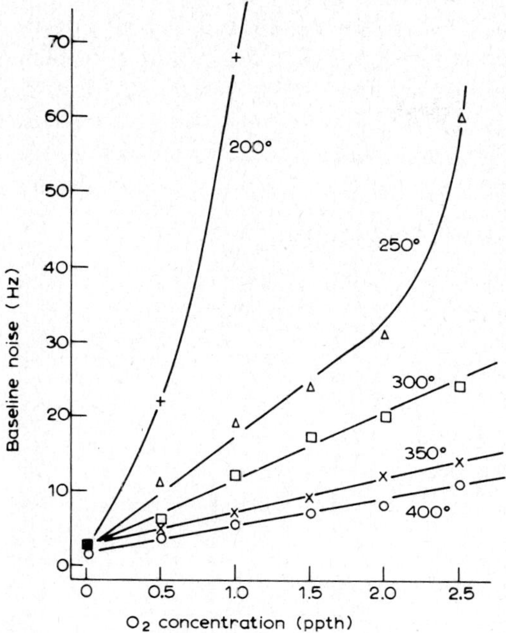

Fig. 5.9. The effect of oxygen-doping on baseline noise (peak-to-peak) at several detector temperatures. (Reproduced from ref. 9.)

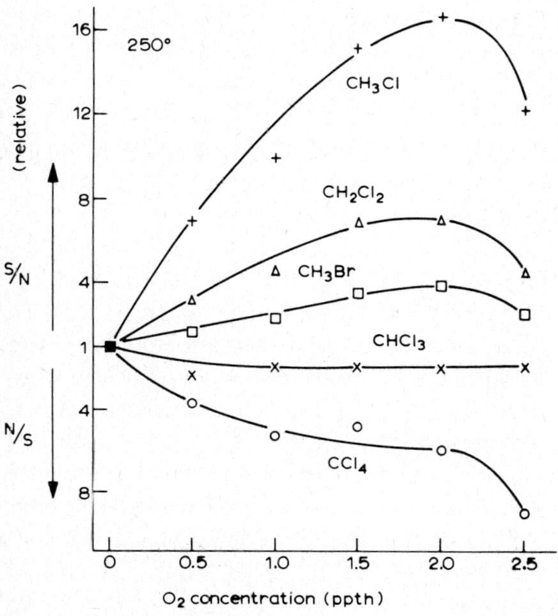

Fig. 5.10. Relative signal-to-noise (S/N) or noise-to-signal (N/S) ratio (whichever is >1) for halogenated methanes as a function of oxygen-doping at 250°C. (Reproduced from ref. 9.)

by the greater noise also expected. In Fig. 5.11 is shown the relative signal-to-noise response to CH_3Cl as a function of temperature and oxygen concentration. It is seen here that the signal-to-noise response to CH_3Cl will be increased approximately one order of magnitude for each of the intermediate temperatures used. At 200°C the signal-to-noise response is lower due to a greater oxygen-caused noise level. At 400°C the signal-to-noise increase is not as large because the absolute response enhancements are considerably smaller.

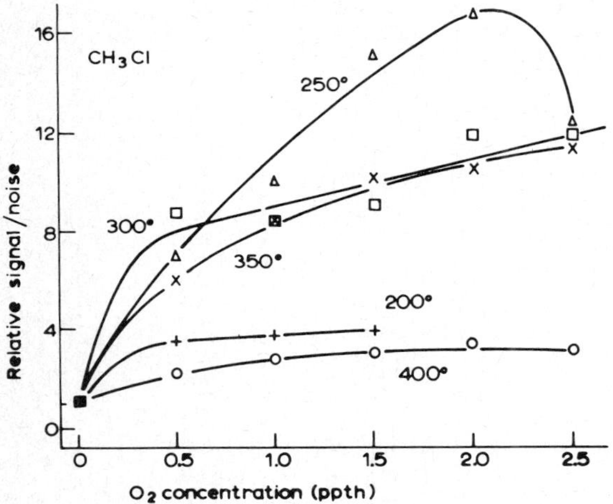

Fig. 5.11. Relative signal-to-noise ratio expected for the ECD analysis of methyl chloride as a function of oxygen-doping and detector temperature. (Reproduced from ref. 9.)

5.5. SURVEY OF ANALYTE RESPONSES WITH THE OXYGEN-DOPED ELECTRON-CAPTURE DETECTOR

5.5.1. Halogenated hydrocarbons

The behavior of a range of ECD-active chemicals in the oxygen-doped ECD have been investigated in our laboratory. The results for a group of halogenated compounds are reported in Table 5.1. For all of them, the *RE* value increases continuously in a near-linear fashion with added oxygen as was shown in Fig. 5.6 for the chlorinated methanes. To describe the behavior of each compound, therefore, it is sufficient to report data for a standard condition of oxygen concentration. Table 5.1 indicates the normal responses and the oxygen-induced response enhancements of 33 simple halogenated hydrocarbons at three detector temperatures with 2.0‰ oxygen in the detector. Under the last column in Table 5.1, entitled R_{O_2}, are listed the relative contributions to the ECD response of the O_2^- reaction, alone. These values provide a clearer measure of the relative rates of eqn. 5.3d for each molecule than do the *RE* values because the latter are a ratio

TABLE 5.1

ELECTRON-CAPTURE DETECTOR RESPONSES AND RESPONSE ENHANCEMENTS CAUSED BY 2.0‰ OXYGEN IN CARRIER GAS

ECD responses under the normal condition of clean carrier gas are the first values listed under each detector temperature. These are relative molar responses normalized with respect to the case of CH_3Cl at 300°C. Oxygen-induced response enhancements are listed in parentheses under each detector temperature. The Relative oxygen-caused response, R_{O_2}, is the contribution to an overall response provided by 2.0‰ oxygen. These values have been calculated from eqn. 5.17 using the ECD data at 300°C, and have been normalized with respect to the case of CH_3Cl. (Reproduced from ref. 10.)

No.	Compound	$R_e(RE)$			R_{O_2}
		250°C	300°C	350°C	
1	CH_3Cl	1.6 (189)	1.0 (113)	1.0 (56)	1.0
2	CH_2Cl_2	8.3 (108)	10.0 (32)	18 (10.9)	2.8
3	$CHCl_3$	459 (4.8)	662 (1.71)	815 (1.42)	4.1
4	CCl_4	9,100 (1.90)	10,500 (1.20)	11,400 (1.15)	18
5	CH_3CH_2Cl	2.1 (228)	1.5 (135)	1.4 (57)	1.8
6	$ClCH_2CH_2Cl$	4.6 (161)	5.1 (69)	5.2 (22)	3.1
7	$CH_3CH_2CH_2Cl$	2.0 (201)	1.3 (127)	1.2 (57)	1.4
8	$CH_3CHClCH_3$	2.0 (195)	1.4 (113)	1.2 (50)	1.4
9	$ClCH_2CH_2CH_2Cl$	2.2 (180)	3.6 (81)	5.1 (34)	2.6
10	$CH_3CH_2CH_2CH_2Cl$	2.3 (145)	1.6 (84)	1.4 (34)	1.2
11	$CH_3CHClCH_2CH_3$	3.5 (132)	2.2 (65)	2.3 (25)	1.2
12	$(CH_3)_3CCl$	1.7 (95)	1.3 (47)	1.8 (13)	0.53
13	$ClCH_2CH_2CH_2CH_2Cl$	3.4 (197)	3.2 (94)	2.9 (53)	2.6
14	$CH_2{=}CHCl$	0.0068 (107)	0.0068 (69)	0.013 (29)	0.0041
15	$CH_2{=}CCl_2$	19 (1.78)	37 (1.13)	67 (1.02)	0.033
16	*trans*-$ClCH{=}CHCl$	1.7 (17)	3.7 (3.9)	8.3 (1.59)	0.095
17	*cis*-$ClCH{=}CHCl$	1.2 (20)	2.3 (5.5)	4.9 (2.0)	0.092
18	$ClCH{=}CCl_2$	505 (2.9)	732 (1.42)	1070 (1.11)	2.6
19	$Cl_2C{=}CCl_2$	4000 (1.79)	4880 (1.16)	6160 (1.02)	7.0
20	$ClCH{=}CHCH_3$	0.0036 (161)	0.0031 (91)	0.0056 (33)	0.0024
21	$CH_2{=}CClCH_3$	0.180 (3.0)	0.190 (1.88)	0.21 (1.11)	0.0015
22	$CH_2{=}CHCH_2Cl$	5.7 (153)	4.9 (61)	6.3 (20)	2.6
23	*trans*-$ClCH_2CH{=}CHCH_2Cl$	380 (8.5)	500 (2.9)	580 (1.69)	8.4
24	C_6H_5-CH_2Cl	42 (44)	67 (10.3)	97 (4.5)	5.5
25	C_6H_5-Cl	0.029 (14.7)	0.068 (3.7)	0.16 (1.84)	0.0017
26	*ortho* Cl-C_6H_4-Cl	20 (3.9)	43 (1.52)	77 (1.26)	0.18
27	*meta* Cl-C_6H_4-Cl	29 (10.2)	60 (2.4)	103 (1.39)	0.75
28	*para* Cl-C_6H_4-Cl	10.6 (8.8)	26 (2.0)	47 (1.25)	0.23
29	CH_3Br	18 (55)	24 (13.3)	39 (5.0)	2.6
30	CF_3Cl	6.9 (2.7)	9.3 (1.11)	12.8 (1.00)	0.0083
31	CHF_2Cl	2.0 (190)	1.35 (90)	0.67 (62)	1.1
32	CF_2Cl_2	174 (3.4)	253 (1.52)	361 (1.10)	1.1
33	$CFCl_3$	4450 (2.2)	5210 (1.31)	5850 (1.03)	14

of two responses and are dependent on the normal ECD response, also. The values of R_{O_2} are determined from the relationship

$$R_{O_2} \propto R_e (RE - 1) \tag{5.17}$$

where R_e and RE values measured at 300°C have been used for Table 5.1, and all R_{O_2} values thus determined have been normalized with respect to the case of CH_3Cl. In considering the data of Table 5.1, many comparisons for illuminating structural effects can be made. The temperature dependence of the measured RE is always negative as was predicted by eqns. 5.11a–d, and the structural effects are the same within each set of measurements at a given temperature. Examples for discussion here will generally be taken from the experiments performed at 300°C.

The number of chlorine atoms per molecule has a large effect on the values of R_e and RE. Using the chlorinated methanes as an example, the ECD responses increase by a factor of about 10^4 in going from CH_3Cl to CCl_4, while the RE values decrease from 113 to 1.20. The relative oxygen-caused contribution to the response, R_{O_2}, increases with chlorine number from 1.0 ± 0.3 for CH_3Cl to 18 ± 9 for CCl_4. The small RE value for CCl_4 is not, therefore, due to a slow reaction with O_2^- but is due simply to its very large normal response.

All of the monochlorinated alkanes in Table 5.1 show relatively large enhancements. For these, the effects of molecular size or branching appear to be relatively unimportant. RE values are generally greater than 100 at 300°C and R_{O_2} values generally exceed 1.0. An exception is noted for *tert.*-butyl chloride where a lower RE of 47 and a lower R_{O_2} of 0.53 are observed. The case of *tert.*-butyl chloride is one of several instances where the absence of a hydrogen atom on the carbon to which the chlorine atom is attached is accompanied by a comparatively low RE value for the monochlorohydrocarbons. Other examples of the apparent importance of an α-hydrogen atom can be seen by comparing $CH_2{=}CClCH_3$ (RE = 1.9) to $CH_2{=}CHCl$ (RE = 69) or CF_3Cl (RE = 1.1) to CHF_2Cl (RE = 90). One might speculate that for these normally weakly responding chlorocarbons, reaction with O_2^- will be faster if products such as the following are possible:

$$O_2^- + RHCl \rightarrow RO\cdot + \cdot OH + Cl^- \tag{5.18}$$

The dichloroethylenes provide a test of the dependence of RE on isomeric variations about a double bond. It is seen that the *trans*-isomer is enhanced by 3.9, the *cis*--isomer by 5.5 and the 1,1-dichloro isomer by only 1.13 at 300°C. The dichlorobenzenes also show small but measurable differences in their RE values. The *ortho, meta* and *para* values of RE are 1.5, 2.4 and 2.0, respectively. It will be seen in the ensuing discussion of the isomers of chloroanthracene that as the size of the aromatic hydrocarbon skeleton is increased, the sensitivity of the measured RE to differences within a set of substitution isomers is increased.

The presence of a double bond, one carbon removed from a carbon–chlorine bond, causes both an increased R_e and R_{O_2} value. This is seen by considering compounds 22 and 23 in Table 5.1. Also, benzyl chloride (compound 24) has a relativey large R_e value and an R_{O_2} value which is among the largest measured here. As in the solution phase these compounds appear to be readily attacked by gas-phase nucleophiles.

5.5.2. Polycyclic aromatic hydrocarbons

In Table 5.2 are shown the normal responses, the response enhancements and the relative O_2^- contribution to the response of 30 polycyclic aromatic and related hydrocarbons. Data for these are reported only for 250°C because at higher temperatures the responses, both with and without oxygen, are usually much weaker and are not useful for analysis. In Table 5.2 the R_e and R_{O_2} values again are normalized, in this case relative to the case of phenanthrene. Inspection of all of the molecules studied indicates that 13 compounds have *RE* values greater than 100 and only 11 have *RE* values less than 50. Therefore, this general class of molecules is quite responsive to O_2-doping and improved signals to many of these can be expected. The other aspect of O_2-doping which seems particularly interesting for this class of molecules is its ability to provide qualitative information concerning each polynuclear aromatic hydrocarbon (PAH) isomer. The combined use of the flame ionization detector and a normal ECD has already been shown to assist in the identification of components in a PAH mixture [26] and the addition to this detection scheme of a configuration capable of also measuring oxygen-induced *RE* values is also worthy of consideration. As can be seen in Table 5.2 the measured *RE* values are influenced measurably by isomeric variations of various types. For example, the five-ringed PAHs have *RE* values which spread continuously from 15 for perylene to 400 for benzo[*e*]pyrene. (Note that benzo[*a*]pyrene has an *RE* of only 29.) A GC analysis scheme which measures responses with and without oxygen would provide information in this case which (along with the GC retention time) would greatly assist in the identification of these isomers in a complex mixture. To appreciate the difficulty normally confronted in this type of determination, it might be recalled that the conventional forms of mass spectrometry would be unable to provide spectra unique to each isomer in many of these cases. Even the electron impact or methane chemical ionization spectra of phenanthrene and anthracene, an example pair which might be considered relatively easy to delineate, are nearly identical [27].

Seven compounds in Table 5.2 contain heteroatoms in the ring structure. It is seen that these are all EC-active and have relatively large and characteristic *RE* values. The two isomers of benzoquinoline produce similar, but easily distinguished, *RE* values of 50 and 70.

The last 11 compounds of Table 5.1 were studied to determine the sensitivity of *RE* values to the more subtle structural variations accompanying the substitution of chloro or methyl groups on the fused rings. It is seen that the chloroanthracenes are, indeed, very sensitive to the position of chlorine atom placement. The 1-, 2- and 9-chloroanthracenes have *RE* values of 50, 8.9 and 4.7, respectively. Since the R_e values of these are nearly the same, it is seen that the differences result mainly from a variation in the reaction rate of each molecule with O_2^-. While 9-chlorophenanthrene has a larger normal response than the chloroanthracenes, its *RE* is unusually small, 1.7.

The effect of structural variations for the methyl-substituted anthracenes is, unfortunately, less pronounced than those of the chloroanthracenes. The *RE* values of 2- and 7-methylanthracene, 84 and 130, are nevertheless readily distinguished. Two of the three monomethylbenz[*a*]anthracenes could be distinguished by their *RE* values. The two dimethylbenz[*a*]anthracenes have extremely different *RE* values. The uniquely low *RE*

TABLE 5.2

ELECTRON-CAPTURE DETECTOR RESPONSES AND RESPONSE ENHANCEMENTS CAUSED BY 2.0‰ OXYGEN IN THE CARRIER GAS AT 250°C

(Reproduced from ref. 12.)

Compound	R_e*	RE**	R_{O_2}***
Phenanthrene	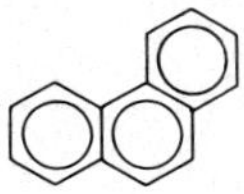 1.0	4.6	1.0
Anthracene	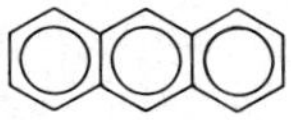 40	62	680
Triphenylene	 1.5	3.0	0.8
Chrysene	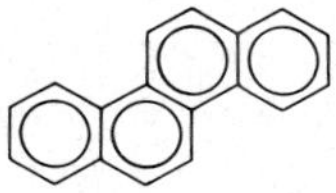 13	140	500
Pyrene	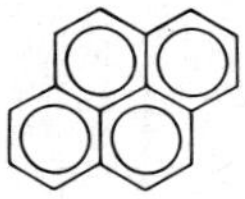 20	185	1000
Benz[*a*]anthracene	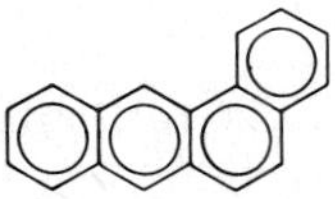 130	165	5900
Tetracene	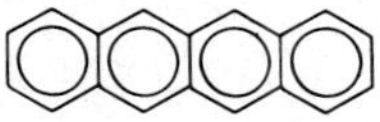 4000	7.4	7100
Benzo[*e*]pyrene	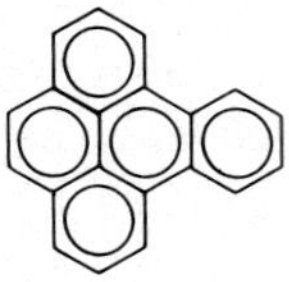 150	400	16,000
Benzo[*a*]pyrene	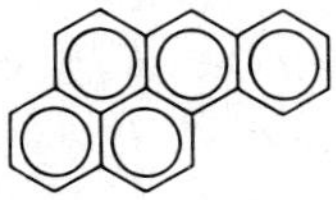 2300	29	18,000
Perylene	 1000	15	3900

TABLE 5.2 (*continued*)

Compound	R_e*	*RE***	R_{O_2}***
Dibenz[*a,c*]anthracene	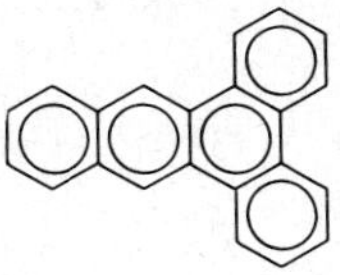 500	190	26,000
Dibenz[*a,h*]anthracene	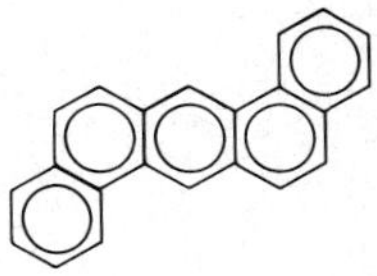 640	67	12,000
Carbazole	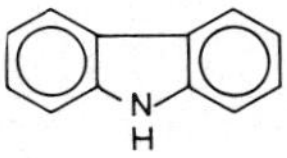 110	200	6100
Dibenzofuran	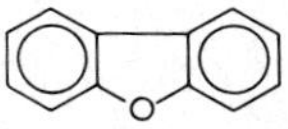 1.6	175	77
Dibenzothiophene	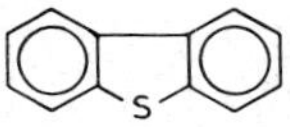 0.9	15	3.5
Acridine	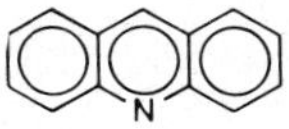 1100	33	9800
Xanthene	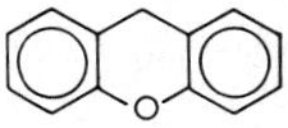 14	130	500
7,8-Benzoquinoline	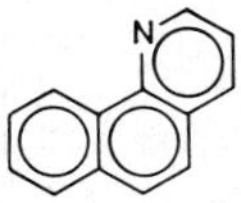 12	50	160
5,6-Benzoquinoline	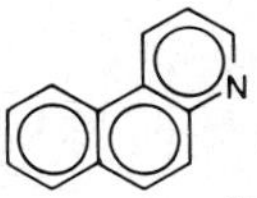 10	70	190
1-Chloroanthrane	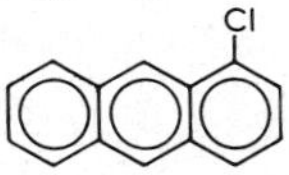 2900	4.7	3000
2-Chloroanthracene	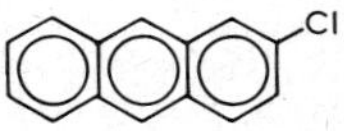 3500	8.9	7700
9-Chloroanthracene	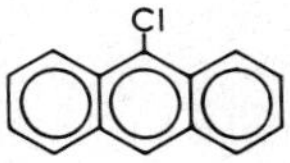 2600	50	34,000

(*Continued on p. 112*)

TABLE 5.2 (*continued*)

Compound		R_e*	RE**	R_{O_2}***
9-Chlorophenanthrene	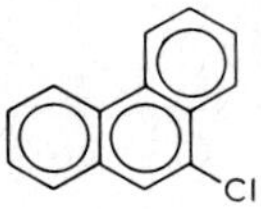	4300	1.7	850
2-Methylanthracene	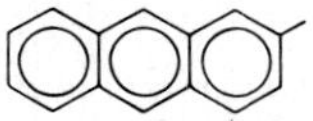	80	85	1900
9-Methylanthracene	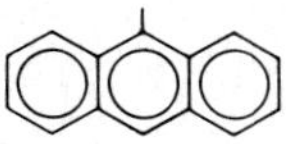	45	130	1600
2-Methylbenz[*a*]anthracene	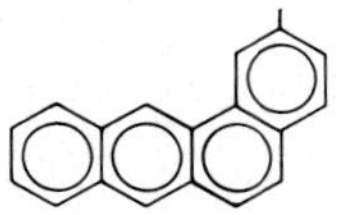	70	140	6700
7-Methylbenz[*a*]anthracene	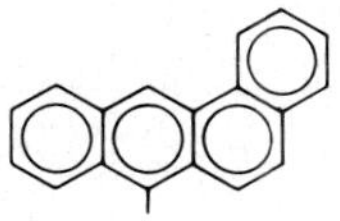	500	100	14,000
12-Methylbenz[*a*]anthracene	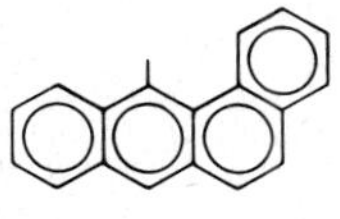	190	150	8100
6,8-Dimethylbenz[*a*]anthracene	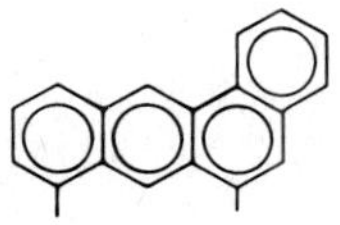	230	140	8700
7,12-Dimethylbenz[*a*]anthracene	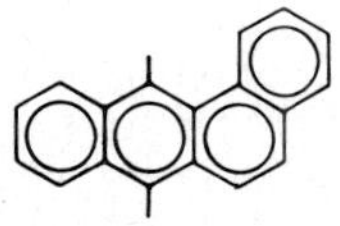	8500	8.9	19,000

*Normal ECD molar responses determined from peak areas all normalized with respect to the case of phenanthrene.
**Response enhancement induced by 2.0‰ oxygen, determined by ratio of peak heights obtained with and without oxygen in the carrier gas.
***Relative rate of reaction with O_2^- normalized to the case of phenanthrene, calculated from the relationship $R_{O_2} \propto R_e(RE - 1)$.

value of 7,12-dimethylbenz[*a*]anthracene of 8.9 appears to be due to an unusually large normal response, about 35 times greater than 6,8-dimethylbenz[*a*]anthracene. The potential of an oxygen-doped detection scheme to sense even these subtle isomeric differences in an analysis by gas chromatography is clearly indicated by these data.

5.5.3. Relative reaction rates with O_2^-

For all of the compounds listed in Tables 5.1 and 5.2, relative rates of reaction with O_2^- have been calculated from the R_e and RE values and have been listed under R_{O_2}. The theoretical development of the oxygen-enhanced response leading to eqn. 5.10 supports our present belief that the R_{O_2} values listed can be taken to reflect the relative values of k_{0A} in eqn. 5.3d for each molecule. If accurate, these measurements and the simple method used to obtain them should provide the field of ion–molecule reaction chemistry with a means of expanding its knowledge of O_2^- reactions. For comparison with our values, we have searched the literature for rate constants of the reactions of O_2^- with halocarbons and found only the measurements of Fehsenfeld et al. [28] for CF_2Cl_2 and $CFCl_3$ under simulated atmospheric conditions. (Apparently, the instrumental difficulties associated with passing oxygen into a mass spectrometer ion source have discouraged the study of O_2^- ion–molecule reactions.) These researchers have informed us, however, that they have also measured the rates of CH_3Cl, $CHCl_3$ and CCl_4 [29]. Their rate constants normalized with respect to CH_3Cl are in the following order: CF_2Cl_2 (0.84), CH_3Cl (1.00), $CHCl_3$ (2.6), $CFCl_3$ (3.0) and CCl_4 (5.6). Their measurements, performed at 27°C and at pressures of about 1 Torr, have an estimated uncertainty of about ±30%. The data in Table 5.1 indicate the following relative rates at 300°C and 1 atm pressure: CH_3Cl (1.0 ± 0.3), CF_2Cl_2 (1.1 ± 0.5), $CHCl_3$ (4.1 ± 1.6), $CFCl_3$ (14 ± 7) and CCl_4 (18 ± 9). While acknowledging the large difference in temperature between the two sets of data, the degree of agreement is considered encouraging to the existing interpretation of oxygen-doping. That the relative rates of O_2^- with CH_3Cl and CF_2Cl_2 are found to be nearly equal by the ECD method is particularly supportive considering the very different ECD sensitivities and R_E values these two molecules have. It presently seems reasonable to consider the R_{O_2} values listed in Tables 5.1 and 5.2 to constitute a set of at least semi-quantitative indicators of reactivity of each molecule with O_2^-.

5.6. APPLICATIONS OF THE OXYGEN-DOPED ELECTRON-CAPTURE DETECTOR

Relatively little time has passed since it was first realized that oxygen can alter the response characteristics of an ECD and, therefore, few laboratories have had the opportunity to apply this method to their problems. The few applications of the O_2-doped ECD which have been reported to date have been concerned with the atmospheric analysis of small molecules where oxygen served to increase the ECD detection limit.

Simmonds [14] reported the use of O_2-doping for improving the analysis of CO_2 in atmospheric samples. In Fig. 5.12 is shown the very impressive response increase to CO_2 observed with 100 ppm oxygen added to nitrogen carrier gas. This application is a particularly interesting one in that the sample compound is quite different from those previously discussed here, and also because the detector used is of the fixed-frequency type. In a previous discussion in this chapter it was pointed out that lower response enhancements are to be expected with this detector as compared to those obtainable with a constant-current ECD. In view of the large enhancement to CO_2 clearly demonstrated in Fig. 5.12, it would appear that for CO_2 the ratio k_{0A}/k_{eA}, which is the upper limit for O_2-induced

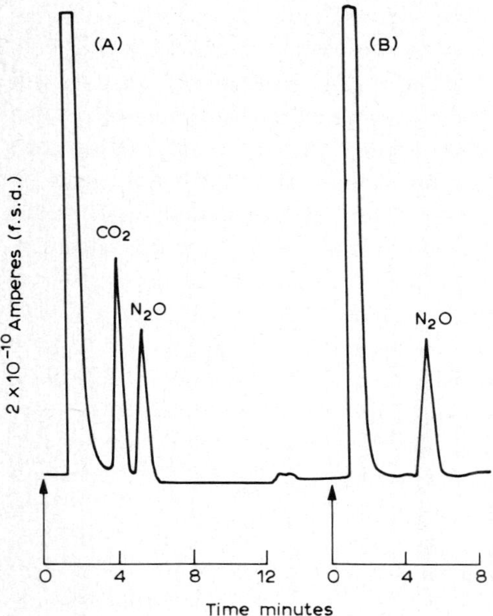

Fig. 5.12. Response of a fixed-frequency ECD to a 3-ml air sample. (A) Nitrogen carrier gas doped with 100 ppm of oxygen; (B) nitrogen carrier gas. Detector temperature 270°C. Column: 5 ft. × 1/4 in. O.D. aluminium tubing packed with 80–100 mesh Porapak QS (Waters Assoc., Milford, MA, U.S.A.). (Reproduced from ref. 14.)

response enhancement using a fixed-frequency ECD (eqn. 5.16), is quite large. Simmonds found that with this instrument a relatively small amount of oxygen (0.10‰) produced the most favorable signal to CO_2. This observation parallels that made here for CH_3Cl using the fixed-frequency ECD (shown in Fig. 5.8).

Rasmussen and co-workers [15] have made extensive use of the O_2-doped ECD. In a report of the global distribution of methyl chloride in the atmosphere, an O_2-doped ECD enabled them to measure CH_3Cl in whole air samples of 5-ml volume directly, without preconcentration. The chosen conditions for operating their constant-current ECD (275°C and 2.0‰ O_2) would be considered optimum for CH_3Cl using the completely different constant-current ECD in our laboratory. Rasmussen and Khalil [16] also report the atmospheric content of $CHClF_2$ (a compound also shown in Table 5.1 to enhance strongly with oxygen). Using a freeze-out loop for preconcentration, 100-ml whole air samples of background air were found to contain 0.045 ppb $CHClF_2$. Rasmussen [30] indicates that in each of the above cases, oxygen-doping increases the signal-to-noise sensitivity by about 10 times. Finally, Rasmussen indicates that the ECD response to CS_2 (which he has found in air over Mt. St. Helens volcanic ash) is similarly enhanced.

Another application of the O_2-doped ECD, just recently reported by Kallos at Dow Chemical Company (Midland, MI, U.S.A.) [17], is for the continuous analysis of bis-chloromethyl ether in an industrial atmosphere. With this procedure, whole air samples will be automatically taken and analyzed every five minutes, 24 hours a day. An oxygen-containing make-up gas provides a response sufficient to detect 1 ppb of this compound in air. This simple and automated system will replace an alternative method which involved less frequent samples taken by the preconcentration of several liters of air on solid adsorbents, analysis by GC–mass spectrometry and a technician dedicated to the task. The oxygen-doped ECD system was operated unattended with exceptional stability and reproducibility for several months. It appeared that a desirable side-effect of O_2-doping in this case was the maintenance by O_2 of a clean detector throughout the test period.

5.7. FUTURE USES AND IMPROVEMENTS

It has been seen in this chapter that O_2-doping can be used to increase ECD response to many compounds. Compounds which enhance are generally those which already have an ECD response (although often a weak one). No enhancement is generally observed for compounds which have no normal ECD response (such as aliphatic hydrocarbons). Thus, O_2^- is a reagent which retains much of the selectivity of the ECD and does not generally produce signals where absolutely none existed previous to doping. The method provides a means of enhancing certain signals that may already be faintly evident in preference to the signals of the strongly responding compounds which normally dominate the chromatogram. For most of the favorable cases examined so far, detection limits are improved by about an order of magnitude. To date, the only classes of compounds which have been studied in detail with the O_2-doped ECD are the halogenated and highly aromatic hydrocarbons. Perhaps more spectacular sensitivity increases will be found for other classes not yet studied. The exceptionally large increase in signal-to-noise response to CO_2 noted by Simmonds [14] indicates that such instances do exist.

From previous considerations of the behavior of a wide range of chemicals in the O_2-doped ECD, it appears that a GC analysis scheme which measures *RE* values of each eluent may be useful for providing qualitative information concerning the sample. The data for the PAHs discussed here suggest that such a detector would be capable of sensing isomeric variations which are often indistinguishable by mass spectrometry. This ability of the ECD to provide subtle structural information by measuring the relative reaction rates of the electron and O_2^- clearly warrants further study. The detector configurations which might best accomplish this are yet to be determined. In addition to simply obtaining separate, paired analyses with and without oxygen or by using two ECDs and an effluent splitter, several other more elegant possibilities come to mind. Tandem ECDs with oxygen added to the carrier stream after the first might work well. Alternatively, oxygen-containing and oxygen-free makeup gas might be added to a single detector in alternate pulses of a few seconds duration. A simultaneous flame or photo-ionization response would also be desirable.

In our studies of O_2-doping, especially when high column temperatures have been required for the separation of large molecules, it has become clear to us that any problems due to the ECD's unwanted response to column bleed materials may be magnified by the presence of oxygen. Apparently, oxygen and O_2^- enhance the response of the detector to some of these just as they do to many sample components of interest. This effect is shown in Fig. 5.13 where an unconditioned column packed with OV-17 has been used with an unconditioned injection septum to illustrate this point. For comparison, a cleaner column packed with SF-96 is also shown. With both columns at 40°C and no oxygen in the carrier gas, the difference in baseline frequency is easily seen but is not alarmingly different. As oxygen is added to the carrier gas, however, the difference in baseline quality of the two columns increases very substantially and with 2.0‰ oxygen, the baseline with the OV-17 is almost intolerable. In this experiment oxygen has been added as a makeup gas after the column so that the effects observed are clearly due to detector chemistry and are not due to any change of the amount of column bleed. With higher column temperatures the problem worsens for the OV-17 column as the baseline frequency then exceeds the 100-kHz limit for linear response of our instrument. Without oxygen added, this column is quite usable at all column temperatures. Although the column and system used to illustrate this point were intentionally chosen to represent the worst case, problems of this type are noticeable even with the better columns we have used at high oven temperatures. Therefore, in order to maintain the ECD baseline frequency as low as possible (determined ideally by the electron–O_2^- equilibrium), it will be desirable to determine the stationary phases and column technologies which produce the least amounts of O_2^- reactive impurities. Capillary columns, which at the time of this writing have not been used in conjunction with the O_2-doped ECD, are known to provide less column bleed than packed columns and are anticipated to work well.

Finally, another dimension of the general technique described here which seems worthy of future attention is the search for new negative ion reagents, other than O_2^-, which are coupled to the cell's electron density and will, therefore, also provide a measurable response upon reaction with sample molecules. One such example, O^-

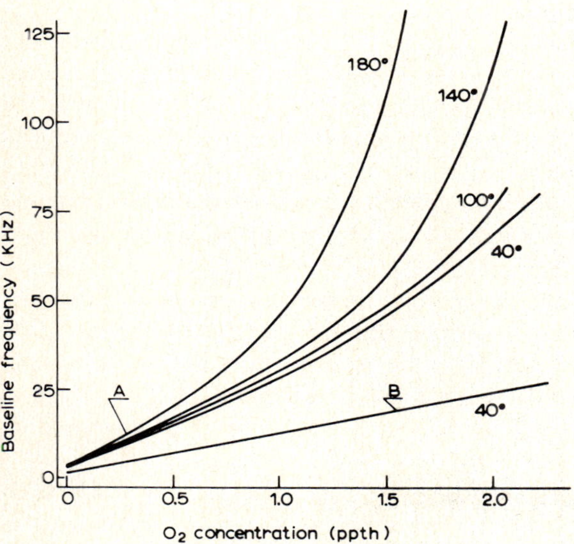

Fig. 5.13. ECD baseline frequency as a function of oxygen concentration in carrier gas using two different chromatographic columns and various oven temperatures. Column A is unconditioned and is packed with 3% OV-17. Column B is well-conditioned and is packed with 10% SF-96. Detector temperature is 275°C.

generated by N_2O-doping [31, 32] has already been discovered and is the subject of Chapter 4. As might be expected of O^- as compared to O_2^- or the electron, a whole new and different set of response characteristics are caused by it. Possibly other electron-active carrier gas reagents capable of causing yet additional sets of response characteristics will be found.

REFERENCES

1 W.E. Wentworth, E. Chen and J.E. Lovelock, *J. Phys. Chem.*, 70 (1966) 445.
2 J.E. Lovelock, *Anal. Chem.*, 35 (1963) 474.
3 J.E. Lovelock, *Anal. Chem.*, 35 (1963) 475.
4 G.G. Guilbault and C. Herrin, *Anal. Chim. Acta*, 36 (1966) 255.
5 F.W. Karasek and D.M. Kane, *Anal. Chem.*, 45 (1973) 576.
6 H.J. van de Wiel and P. Tommassen, *J. Chromatogr.*, 71 (1972) 1.
7 R.J. Maggs, P.L. Joynes, A.J. Davies and J.E. Lovelock, *Anal. Chem.*, 43 (1971) 1966.
8 E.P. Grimsrud and R.G. Stebbins, *J. Chromatogr.*, 155 (1978) 19.
9 E.P. Grimsrud and D.A. Miller, *Anal. Chem.*, 50 (1978) 1141.
10 D.A. Miller and E.P. Grimsrud, *Anal. Chem.*, 51 (1979) 851.
11 E.P. Grimsrud, D.A. Miller, R.G. Stebbins and S.H. Kim, *J. Chromatogr.*, 197 (1980) 51.
12 D.A. Miller, K. Skogerboe and E.P. Grimsrud, *Anal. Chem.*, submitted for publication.
13 E.P. Grimsrud, S.W. Warden and R.G. Stebbins, *Anal. Chem.*, submitted for publication.
14 P.G. Simmonds, *J. Chromatogr.*, 166 (1978) 593.
15 R.A. Rasmussen, L.E. Rasmussen, M.A.K. Khalil and R.W. Dalluge, *J. Geophys. Res.*, submitted for publication.
16 S.A. Penkett, N.J. Prosser, R.A. Rasmussen and M.A.K. Khalil, *Nature (London)*, 286 (1980) 793.
17 G.J. Kallos, *Anal. Chem.*, submitted for publication.
18 P.L. Patterson, *J. Chromatogr.*, 134 (1977) 25.
19 E.P. Grimsrud and R.A. Rasmussen, *Atmos. Environ.*, 9 (1975) 1014.
20 D.R. Cronn and D.E. Harsch, *Anal. Lett.*, 9 (1976) 1015.
21 M.W. Siegel and M.C. McKeown, *J. Chromatogr.*, 122 (1976) 397.
22 E.P. Grimsrud, S.H. Kim and P.L. Gobby, *Anal. Chem.*, 51 (1979) 223.
23 P.L. Gobby, E.P. Grimsrud and S.W. Warden, *Anal. Chem.*, 52 (1980) 473.
24 M.A. Biondi, M.H. Bortner and T. Bauer (Editors), *Defense Nuclear Agency Reaction Rate Handbook*, 4th rev., 1975, Ch. 12.
25 E.P. Grimsrud and D.A. Miller, *J. Chromatogr.*, 192 (1980) 117.
26 A. Bjorseth and G. Eklund, *J. High Resolut. Chromatogr. Chromatogr. Commun.*, 2 (1979) 22.
27 M.L. Lee, D.L. Vassilaros, W.S. Pipkin and W.L. Sorenson, *Proceedings of the 9th Materials Research Symposium, Gaitherburg, MD, April 10–13, 1978*, National Bureau of Standards Special Publication 519, p. 731.
28 F.C. Fehsenfeld, P.J. Crutzen, A.L. Schmeltekopf, C.J. Howard, D.L. Albritton, E.E. Ferguson, J.A. Davidson and H.I. Schiff, *J. Geophys. Res.*, 81 (1976) 4454.
29 F.C. Fehsenfeld, NOAA Environmental Research Laboratories, Boulder, CO, personal communication, September 1978.
30 R.A. Rasmussen, Oregon Graduate Center, Beaverton, OR, personal communication, July 1980.
31 M.P. Phillips, R.E. Sievers, P.D. Goldan, W.C. Kuster and F.C. Fehsenfeld, *Anal. Chem.*, 51 (1979) 1819.
32 P.D. Goldan, F.C. Fehsenfeld, W.C. Kuster, M.P. Phillips and R.G. Sievers, *Anal. Chem.*, 52 (1980) 1751.

Chapter 6

Wide-range calibration of electron-capture detectors

R.E. KAISER and R.I. RIEDER

CONTENTS

6.1. FUNDAMENTALS OF CALIBRATION

There have been many detailed discussions of the quantitative use of electron-capture detector (ECD) data in gas chromatography and lengthy descriptions of how to use different ECDs in different modes, with different gases, in many analytical applications can be found in the literature. This paper offers an alternative procedure for optimizing and using ECDs for quantitative gas chromatographic (GC) analysis: the "black box" principle and absolute quantitation.

For this purpose it is of no importance which type of detector is used in quantitative gas chromatography: any detector can be understood and treated as a black box for the production of correct, precisely repeatable quantitative analytical data. Therefore, we thought it might be acceptable to describe the quantitative detector behaviour in a way that is valid for a given instrument under constant (although optimized) analytical conditions of temperature, carrier gas composition (including trace contaminants), inlet pressure, outlet pressure, detector cell voltage, and amplifier, integrator and recorder conditions, without including a discussion of the theory of ECDs, as this is available elsewhere in this book.

The principles discussed here are valid for any ECD, under any special conditions of practical use, if the fundamental requirement for quantitative GC analysis holds true, namely that the conditions remain constant throughout the whole working period between calibration and quantitative application.

6.1.1. The two fundamental quantitation equations

The quantitative GC signal can be given as the integral of

(A) voltage (volts) times flow (millilitres/second) → V · ml/s

or

(B) current (amps) times time (seconds) → A·s

Although any GC detector response is based on moles of substance, it is more practical to relate the response to mass given in grams. Calibration allows conversion of all quantitative data to the basis of grams, volts, amps, millilitres and seconds. Therefore, we can write the two fundamental quantitation equations, first for a detector, which is "substance flow dependent":

$$\text{(A) grams of substance } X = \frac{as}{q} \tag{6.1}$$

where q is the "quantity" or the response in amps times seconds, which the detector "produces" per gram of substance X, also known as the "gram-based substance-specific response", and secondly for any detector which is "concentration dependent":

$$\text{(B) grams of substance } X = \frac{v \cdot ml}{q} \tag{6.2}$$

where q is the "quantity" or the response in volts per "concentration unit" given in grams per millilitre.

These two q factors are different for A and B, but both describe quantitatively the function of the detector. Concentration-sensitive detectors produce a signal in volts from the unit concentration, grams/ml. Mass flow-sensitive detectors produce a signal in amps from the unit mass flow, grams/second.

Different ECDs, under certain conditions of carrier gas, flow, pressure, temperature, voltage, etc., show neither pure mass flow- nor pure concentration-dependent behaviour, which fortunately is not a problem in quantitative analysis if all conditions remain constant between the time of calibration and the time of analytical application. This lack of problems depends, however, on the technique of calibration.

The two fundamental quantitation equations both give a result in grams of substance X, i.e., both describe the GC peak area of substance X in terms of grams of the substance.

Both equations are, at first sight, linear, provided that we consider the mass response factor q as a substance-specific constant. In fact, q is by no means a constant and is influenced by many physical and chemical factors, which make quantitation with ECDs difficult. This remark is immediately understandable if one considers the fundamental equations just as the substance-specific calibration line, giving the analytical signal function (see Fig. 6.1). q in Fig. 6.1 is the slope of the analytical signal function and its dependence on the specific nature of the substance and on the amount of substance can be seen; q is constant.

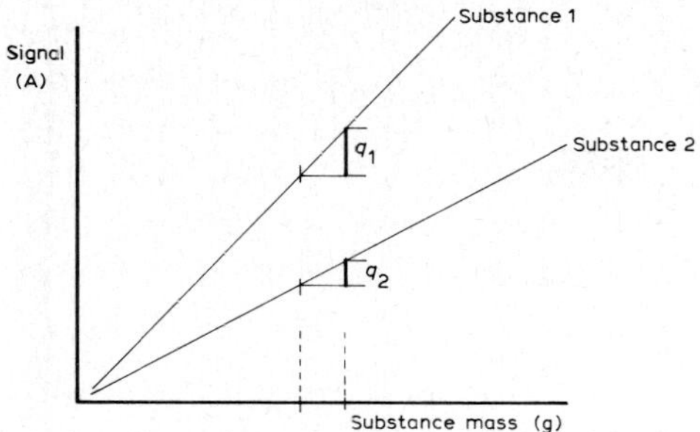

Fig. 6.1. Substance-specific, analytical signal function. $q_{1,2}$ = specific response.

It can be easily understood why the two fundamental quantitation equations A and B are interchangeable or can be used simultaneously: the quantitative GC signal can be expressed as peak area. This area can be approximated to peak height times peak width at half-height. One can measure the peak height in amps and the peak width in seconds, the peak area then being amps times seconds. Alternatively, one also can measure peak height in volts and peak width in millilitres (carrier gas passed during the peak width times period). Whether the signal is measured in amps times seconds (coulombs) or in volts times millilitre it is practically the same. Both measurements are correlated with the mass of the substance in grams.

Although we have discussed quantitation as a "black box" model so far, it would be of some help to the practitioner, using more than one detector, to be able to compare the popular ECD with the equally popular flame ionization detector (FID).

By giving the quantitative content of Fig. 6.1 in a different form, viz., the change of signal function slope with amount of substance in grams, we can better understand the quantitative differences between an FID and an ECD: Fig. 6.2 shows the result one most often finds in practice: the FID has a wider more-or-less linear range, but both detectors need precise calibration as both are never linear enough for accurate quantitative analysis.

Although it is often stated that FIDs are linear over a range of seven orders of magnitude and (linearized) ECDs are considered to be linear over four orders of magnitude, one

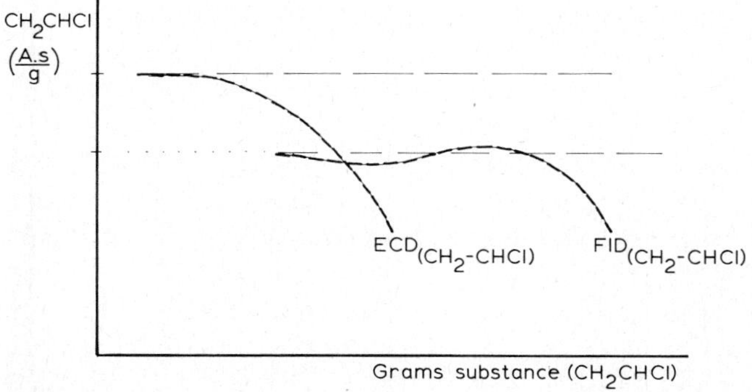

Fig. 6.2. Linearity (or deviation from linearity) can be expressed as the change of q with absolute amount, concentration or flow of analytical substance in the carrier gas, passing through the detector.

has to compare this "linearity" with the analytical quantitative needs: do we want to accept accuracy as less important than repeatability? If not, then the accuracy has to be as correct as repeatability within the statistically possible limitations, and therefore "linearity" becomes fundamental to our discussion.

Linearity is what Fig. 6.2 shows: the change in q over the range of the mass of substance in grams. Many practitioners do not think in such terms and therefore do not measure q data. The quantitative measurement of correct q data is, in fact, the fundamental of calibration.

6.2. CALIBRATION BY CONVENTIONAL LABORATORY TECHNIQUES

A simple calibration technique consists of a few steps and their combination:
(a) Take substance X (substance to be determined) and a standard substance R.
(b) Make a quantitative mixture of both by means of volumetric or mass measurements. Preferably weigh W grams of X ($= W_X$ grams) to W grams R ($= W_R$ grams).
(c) Make a homogeneous mixture of X and R in a suitable solvent, so that in each microlitre the mass ratio of X to R is constant and exactly equal to W_X/W_R.
(d) Separate the substances X and R chromatographically and measure the integrals. Let us call the peak area of substance X A_X and that of substance R A_R.
(e) Compare the mass ratio with the chromatographic peak-area ratio:
Is

$$\frac{W_X}{W_R} = \frac{A_X}{A_R} \ ? \tag{6.3}$$

Possibly not.

We now introduce correction factor, f, to correct the simple calibration equation 6.3. In any case it holds true that

$$\frac{W_X}{W_R} = \frac{A_X \cdot f}{A_R}$$

and therefore the correction factor is

$$f = \frac{W_X \cdot A_R}{W_R \cdot A_X}$$

which too often is considered to be a constant. However, f is never constant, if we consider constancy and linearity as strict quantitative expressions like chromatographic repeatability.

The relationship peak area A_X to the amount of substance X was defined as

$$q_X = \frac{W_X}{A_X}$$

Therefore

$$q_R = \frac{W_R}{A_R}$$

and hence

$${}_R f_X \text{ (for substance } X \text{ relative to } R) = q_X / q_R$$

Thus the simple laboratory calibration correction factor turns out to be the ratio of the two fundamental response data, q, for the two compounds used, X and R.

As this is too simplified for more detailed discussion, we can add important comments. The simple correction factor, f, can never be a constant over a practically useful quantitative range if one of the response factors, q, of which f consists, changes. As there is no detector that is truly linear over the practical ranges in which we want to work quantitatively – considering linearity strictly as quantitative repeatability – the correction factor f is not constant: it depends on the absolute values of the amounts of X and standard substance in grams entering the detector.

In practice we usually overload detectors, but many practitioners simply do not believe this statement, so a quantitative example will be given. Let us use the nearly "unlimited linear FID" for this example. FIDs are called linear over a range of at least 1 to 10^6. Their quantitative working range starts at 10^{-12} g/s giving a measurable signal of about $i = 10^{-14}$ A. Linearity of six orders of magnitude means that the detector will also work at a level of 10^{-6} g/s.

If one uses a packed column, one normally injects between about 1 and 0.1 μl of sample, which is about 1–0.1 mg. Normally a rapidly eluted peak is about 1 s wide at half-height. This means that about 10^{-5} g enter the detector, which is about one order of magnitude above the linear range of the wide-range FID. Let us consider these conditions with an ECD. As its linear range is at least two orders of magnitude less than that of an FID and starts at a two-order smaller range level than that of an FID, one will have overloading with a even four orders of magnitude smaller sample injection. Who really takes care that not more than about 1 ng of substance enters the ECD? Only users of capillaries (sometimes) take care in this respect.

We can repeat, therefore: with the practical range of modern chromatographic quantitation we normally tend to overload our detectors and therefore the correction factor f has to be considered to be non-constant.

On the other hand, for trace amounts of substances, we are confronted with the problem of losses due to chemical sorption. This makes understandable Fig. 6.3, which shows the dependence of the correction factor on the absolute amount of sampled compound X.

A classical rule states that one should choose internal standard compounds so as to be chromatographically as closely related to the substance under analysis as possible. The best would be for them to have identical polarities. Of course, the optimal solution would be to use the substance X itself as the standard. In practice, one tries to have X and R just sufficiently separated from each other, so the peak of X may be partly overlapped by the peak of R. In this case, undesirable side-effects may occur in the detector. As f is composed of two responses, namely q_R and q, we can understand the dependence of the

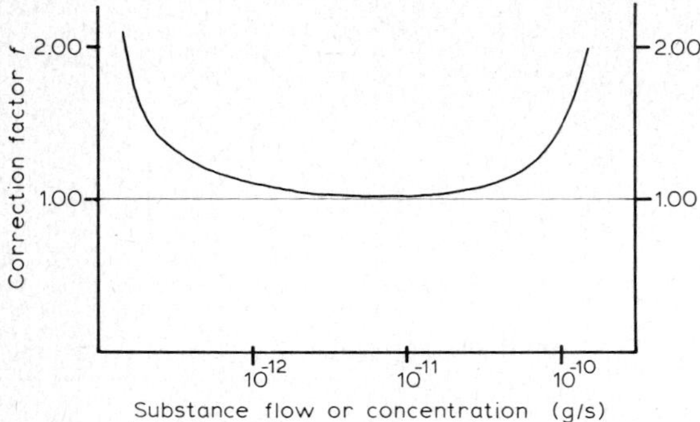

Fig. 6.3. Dependence of the correction factor f on the absolute amount, concentration or flow of analytical substance in the carrier gas, passing through the detector.

simple correction factor of the substance-specific trace behaviour of different compounds as well as of side-effects which can greatly influence the response. Hence a situation arises that is no longer as simple as the process of simple laboratory calibration appears to be. It is not only the relative composition of the calibration mixture, but also the absolute amount of the calibration mixture that is important.

We now have non-linearity plus side-effects. Non-linearity as such leads to problems in quantitative analysis as all quantitation equations – both absolute and relative functions, and internal and external standard techniques – depend on a linear correlation between the signal and the concentration or amount of substance.

There is no doubt about the only way to handle non-linearity: measure repeatability correction factors or response data with many different absolute amounts of substance. Do we have enough time in routine analysis, where quantitatively correct data are important, to follow such time-consuming advice? We do not. Calibration by conventional laboratory techniques cannot handle effectively enough the problems of non-linearity. ECD analysis means non-linear analysis, and here calibration by exponential dilution is of help.

6.3. CALIBRATION BY EXPONENTIAL DILUTION

If one dilutes a full pot of coffee (or any other substance in solution) with a constant flow of water (or any other diluent) and continues carefully to homogenize the dilution (e.g., by stirring), the coffee will be diluted to half its concentration in a constant time. This holds true if the diluted coffee leaves the pot at the same flow-rate as that of the water being added. Let the starting concentration be C_0 (g/ml), the volume of the pot be V (ml), and the flow-rate of diluent be F (ml/s), then the concentration C at each time t (s) is given by the fundamental equation of exponential dilution:

$$C = C_0 \cdot \exp(-Ft/V) \tag{6.4}$$

We built a glass bulb with a volume of about 10 ml and connected to it two capillaries of very narrow bore (less than 0.2 mm I.D.). With a flow-rate of about 10 ml/min of

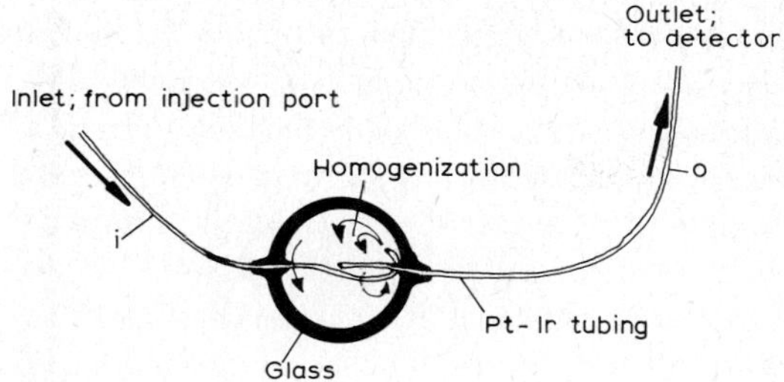

Fig. 6.4. "Quantegg", a glass bulb of several millilitres volume with platinum–iridium tubings (0.1 to 0.15 mm I.D.).

carrier gas through the glass bulb, as shown in Fig. 6.4), the stirring energy pumped into the bulb volume was sufficient to give good constant homogenization of the exponential dilution vessel in the bulb volume. In addition, the inlet capillary introduced tangential stirring. The outlet capillary was precisely centred in the middle of the bulb volume. Capillary i was connected with the (second) chromatograph injection port and capillary o was connected to the detector.

We injected the calibration substance X (pure X diluted in an inert solvent) (preferably no solvent is used, but in practice carrier gas is used). Immediately after injection we obtained a maximum signal from the detector. The change in signal with time approximately followed the exponential dilution curve, but the discrepancy surprised us. We therefore thought at first that the 10-ml exponential dilution vessel was not functioning. To check for systematic errors we split the outlet gas via a dead-volume-free tee-connection into two parallel detector, one an ECD and the other an FID. The FID channel showed a nearly ideal exponential dilution signal with time, but the ECD channel failed at high concentrations, with certain carrier gas compositions, or both. Fig. 6.5 compares the two signals.

Eqn. 6.4 is so simple that with any programmable pocket computer one could calculate the expected time for the "half-signal" value. If one knows the carrier gas flow-rate

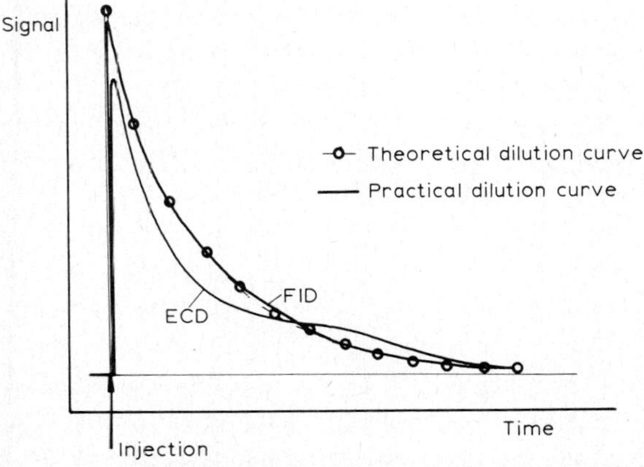

Fig. 6.5. Comparison of theoretical, FID, and ECD experimental exponential dilution function curves.

(which is easy to measure accurately enough) and the volume of the exponential dilution vessel (easy to measure: fill with water and weigh) the equation gives the required data. We therefore calculated the time necessary to measure over 3, 4, 5 and more orders of magnitude. We call this (relative to quantitative calibration) a wide range and hence the procedure is a wide-range calibration technique. We call the whole procedure calibration, as we know the theoretical value and can measure the practical signal; the ratio of the two is the linearity correction factor. We calibrate with substance X and then with substance R and use a laboratory computer to calculate all factors for all concentrations as well as for all absolute amounts of substance. As we no longer depend only on relative data, but on absolute data, we call the technique "direct mode" It is not indirect as with the other techniques.

We never succeeded in making the exponential dilution vessel precisely symmetrical; it looked more like an egg. Nevertheless, one can work quantitatively with it, and therefore we call this tool a "quantegg".

6.4. DIRECT MODE USE OF THE EXPONENTIAL DILUTION TECHNIQUE WITH THE "QUANTEGG"

Based on eqn. 6.4 the following time data are known, if one knows (a) the volume (e.g., 12.00 ml) and (b) the gas flow-rate (e.g., 10.00 ml/min = 0.167 ml/s):

Half-time = 49.9 (s)

1/10-time = 165.8 (s)

One can therefore calculate, independent of the type and amount of sample, the half-concentration time and easily measure this on the chromatogram recorder simply by observing when the signal reaches the half-height level. This can be done n times, if the amplifier sensitivity switch covers a range of $S = 2^n$ positions.

Procedure:

Inject (milligrams, micrograms or nanograms of) any compound with a detector response into the quantegg under constant conditions of pressure, flow and temperature. Switch the detector signal to a low level of sensitivity, so that the maximum signal just can be contained by the recorder on-scale. Switch the sensitivity switch by two units just when the detector signal reaches 50% of full-scale, and measure the time difference from signal jump to signal jump. Do the successive readings seem equal?

Fig. 6.5 shows the power of this first quantitative information with respect to the linearity of the detector response, a basis for simple conditions of accurate quantitative analysis.

One can make further observations. Are the peaks produced by the sensitivity jumps at equal heights? If not, this indicates (a) a bad manual response or, if the starting line for the 2-unit jump is precisely at the 50% full-scale level, this indicates (b) a defective sensitivity switch. Check the shape of the first peak: is it sharp? if yes, the detector is not (heavily) overloaded. Does it look round, as in Fig. 6.6? if so, the detector is overloaded.

Considering precision, in Fig. 6.6 we draw a parallel to the recorder zero-signal line. We measure precisely that the time for each crossover with the signal line. Even if there had been imprecise timing for the 2-unit switching, we have precise signal versus time

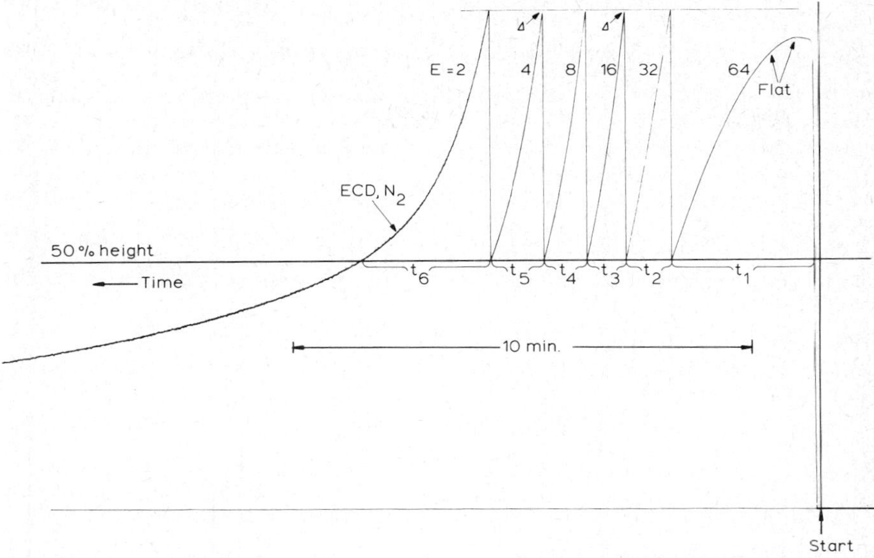

Fig. 6.6. Recorded result of a manual quantegg run. See the flat signal maximum, indicating heavy detector overload. The factors 64 to 2 are the manually adjusted amplifier sensitivity switch positions. The "delta's" indicate systematic (hardware) errors of the sensitivity switch factors. The "t_1 to t_6 time values" have to be equal in the case of a linear detector or a linear detector working range. The 50% height level line offers the time data measuring scale. The time delay between injection (arrow, right side) and the sudden rise of the detector signal demonstrates the possibility of precise flow measurement, if done at the millisecond level, impossible by manual techniques, but simple for on-line computer techniques.

data, provided that the switch factors are precise (some instrument manufacturers guarantee 0.02% accuracy for the sensitivity switch factors).

Now calculate the theoretical signal for the time data one could measure for the baseline. The ratio of the theoretical to the measured signal is the linearity deviation factor. If this is equal to

$$\frac{\text{theoretical signal (for time } t)}{\text{measured signal}} = 1.000$$

then at this concentration level the detector response is linear. If it deviates from unity, we can correct the signal.

Let the signal measurement be done by a computer on-line, and let the calculation be done in real time in the same computer. By this means we can use any non-linear detector quantitatively without any doubt about the accuracy. This is substance-specific auto-linearization, which is not yet available commercially (but ready to be done with a suitable programmable integrator). Of course this necessitates constant conditions between the calibration run and the analysis run, a condition sine qua non for all sequential analysis techniques.

Table 6.1 shows off-line calibration runs with detectors, integrators and computers

TABLE 6.1

EXAMPLE FOR THE NON-AUTOMATIC USE OF THE QUANTEGG TECHNIQUE

We used the chain: gas source – inlet – quantegg – detector – amplifier – integrator – printer/plotter. As integrator we used a Spectra Physics SP 4100, programmed for peak height measurement. The following SP-BASICs program was used and the signal height quantified every five seconds. The SP 4100 printer/plotter printed the peak height in units.
The necessary time data are given as the number of data output times 5.00 in seconds. If an APPLAB interface is connected (IMI Interactive Microware, State College, PA, U.S.A.) to an APPLE 2 home computer, 48 kbyte, and the amplifier signal is fed into the APPLAB interface, then auto-linearization is available. Only a few changes in the BASIC program in Table 6.5 would be necessary to correct the integrator signals automatically.

```
!"  interval in sec = "; input A1:  end
P = PW:  plot 1
run30
return
plot off
gosub 100
A = ((2.5*3peek#C142)/PW) - 1000
!$12. A;
if peek#80C109=#80then 30 else end
Y = 2peek#C234
X = Y+A1
if 2peek#C234=>X then return else 120
gosub 2
```

SPECTRAPHYSICS BASIC program for SP 4100 (R.I. Rieder)

Data print out: (one measurement per five seconds)

Inject time 11:26:54

107424	116902	110375	103451	96374	89243
82521	76378	70657	65456	60518	55931
51654	47772	44144	40776	37643	34742
31992	29502	27170	25042	23042	21223
19543	17993	16553	15229	14002	12879
11859	10914	10037	9222	8499	7806
7186	6611	6083	5600	5149	4740
4362	4014	3697	3403	3137	2889
2661	2450	2262	2083	1921	1771
1632	1507	1392	1286	1188	1094
1016	937	867	805	741	686
638	594	548	513	476	440
409	381	358	335	312	293
276	258	244	225	216	203
191	180	168	157	151	144
138	130	125	117	115	109
103	99	93	90	85	80
77	74	70	70	67	66
66	65	60	59	56	57
53	53	52	47	48	46
45	43	44	43	44	42
40	41	35	37	36	34
35	35	35	34	35	34
32	32	32	33	31	31
28	29	27	27	26	26
24	24	24	24	22	21
17	16	16	13	12	12
10	8	8	5	4	2

available in 1981, for which there were no commercially available instruments for on-line auto-calibration.

6.5. WIDE-RANGE CALIBRATION BY THE "QUANTEGG"

Table 6.2 shows the experimental results. As the detector turned out to be reasonably non-linear, and as most users consider non-linear detectors to be less valuable, we have not revealed the manufacturers of the ECDs tested. In the future it will become much more important for detectors to have a flat optimal response range when the measuring

TABLE 6.2

EXPERIMENTAL CONDITIONS

Quantegg: home made (Institute for Chromatography, Bad Dürkheim, G.F.R.).
Quantegg volume: 14.90 ml (measured by weight of water filling at known temperature)
Carrier gas: nitrogen from direct liquid nitrogen evaporation;
argon–methane (95:5, v/v) from Messer Griesheim, ECD quality.
Temperature: between 25°C and 200°C tested.
Gas flow-rate: between 5 ml/min and 50 ml/min tested. ECD scavenger gas at detector response optimum conditions or through gas by-passing. Pressure: between 0.1 and 3 bar.
Detectors: ECD, different branches, made in 1980.
Amplifier: Carlo-Erba electrometer Model 180. GC amplifiers used in DANI model 3900 or Perkin-Elmer model F22.
Recorder: 0.5 to 200,000 mV f.s.d., 0.3 s recorder Servogor from Goerz/BBC.
Sample: CH_2Cl_2, or $CHCl_3$, or a mixture of both in *n*-heptane (Merck) ranging over 1:100,000 starting with maximum concentrations of 10^{-7} g/ml down to the ECD noise level under normal experimental conditions. Extreme concentration levels until detector saturation have been tested as well.

TABLE 6.3

DEVIATION OF ECD LINEARITY

Normal conditions, clean carrier gas argon–methane (95:5, v/v), not oversampled, just daily analytical conditions. The data show linearity deviations are not constant (S-shaped function for signal deviation from linearity), and remarkably large (–20%!!) for quantitative analysis. Quantegg volume = 14.9 ml; gas flow F = 14.3 ml/min; maximum signal height = 140,000.

Time	Signal height	Corr. factor
38.4	63539.2	1.192
89.4	31769.6	1.055
135	15884.8	1.017
178.8	7942.4	1.009
220.5	3971.2	1.036
262.5	1985.6	1.059
304.2	992.8	1.087
347.4	496.4	1.089
392.1	248.2	1.065
440.1	124.1	0.989

conditions are changed and for the absolute signal levels to remain constant rather than having perfect linearity. However, at present we use primitive integration techniques and primitive quantitation equations, both of which depend on perfect linearity (cf. Table 6.3).

6.6. LIMITATIONS OF THE "QUANTEGG"

The most important condition for quantegg calibrations is a stable and known position of the real signal baseline. Another weak point is that the quantegg signal starts at a maximum, and that at first we may consider this C_0 value as being correct, until we realize that non-linearity applies. From there we have to calculate backwards. If there is no linearity, one has to start the calculation at the lowest signal level. For manual evaluation both are troublesome, but not for on-line or off-line computation, where we can start with any value.

The quantegg function depends on constant conditions of flow-rate and pressure. Therefore, a constant flow-rate is important, but in gas chromatography this can be guaranteed to at least 1%.

The quantegg will give false results if systematic losses by chemisorption on the inner surface occur. However, the surface can be deactivated and heated to the optimal temperature.

The quantegg does not correct systematic calibration errors due to substance losses in the chromatographic column, but one can ascertain their magnitude by using the pneumatic circuit shown in Fig. 6.7.

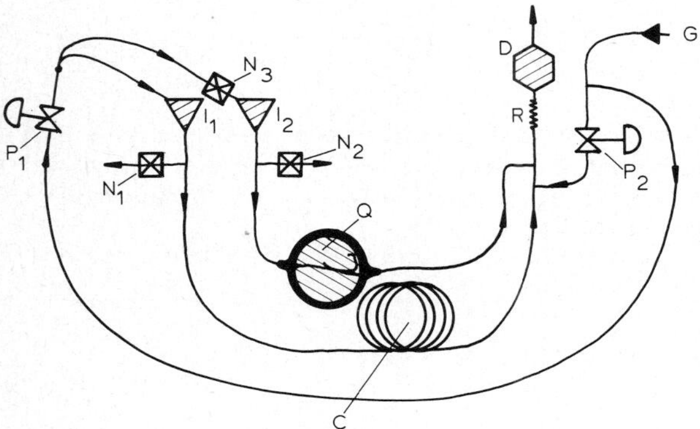

Fig. 6.7. Pneumatic circuit for combined or parallel use of quantegg and a chromatographic (or capillary) column. G = carrier gas source. P_1, P_2 = pressure regulators. I_1, I_2 = sample inlet systems. N_1, N_2, N_3 = needle valves (outlet split for the injection systems; N_3 = carrier gas flow regulator for quantegg. Q = quantegg. C = column or capillary. R = analytical gas flow resistor kept under constant temperature. P_2 allows, together with the gas flows from C and Q, adjustment of the gas flow optimum for detector D. Pressure $P_2 < $ pressure P_1.

The quantegg installation must be gas-tight. The homogenization must be perfect and remain so, but here we have to consider limitations. There is a limiting lowest acceptable flow-rate through the quantegg. Below this value the practical elution curve is steeper than the theoretical shape allows, and homogenization is non-ideal.

The flow-rate through the quantegg must be known and be accurate. This is not easy, but a very precise dead-time measurement can be used for flow metering. As the quantegg itself does (theoretically) not allow any dead time to the signal delay, the flow-rate (F) could be calculated by

$$F = \frac{\text{volume of the quantegg capillary}}{\text{dead time}}$$

It is no problem to measure the quantegg volume exactly, but one has to consider the exact pressure within the bulb for precise time data calculations. It turns out that the quantegg offers simple solutions to the problems encountered in quantitative chromatography, but there are many data to be handled if one wants high accuracy. The data evaluation is not simple. We used an Apple 48 kbyte double disk laboratory computer and used its CP/M version for easy data transfer to and from floppy disks.

TABLE 6.4

TEST OF LINEARITY OF AN ECD

The ECD was run under equal conditions of sample, sample amount temperature, flow, voltage, etc., but with differing carrier gases: A: super clean nitrogen; B: common mixture of argon–methane.

	Time*	Signal height**	Corr. factor***	H-Theor.§
A	164.4	8192	3.003	24602.4
	234.6	4096	2.065	8456.4
	285.9	2048	1.892	3874.9
	338.4	1024	1.703	1743.4
	399.3	512	1.348	690.3
	492.6	256	0.652	167
Quantegg volume = 14.9 ml; gas flow F = 13.6 ml/min; maximum signal height = 300,000.				
B	27.9	31667.2	1.048	33176.7
	73.5	15833.6	1.005	15916.4
	116.4	7916.8	1.007	7975.3
	158.7	3958.4	1.019	4035
	200.1	1979.2	1.047	2071.3
	242.1	989.6	1.064	1053
	284.4	494.8	1.077	532.8
	375	123.7	1.001	123.8
Quantegg volume = 14.9 ml; gas flow F = 14.3 ml/min; maximum signal height = 52,000.				

*Time: since injection of sample into quantegg.
**Signal height: signal height in units (sensitivity times volt times 1000).
***Corr. factor = correction factor to achieve linearity.
§H-Theor. = signal height times corr. factor.

6.7. EVALUATION OF DATA

6.7.1. Data evaluation by manual measurements from a recorder display

Fig. 6.8 compares two records and Table 6.4 contains data from some measuring positions. If one examines both sets of data the whole procedure becomes evident and does not need further comment here, with one exception. Both sets of figures represent wide-range calibrations with CH_2Cl_2, $CHCl_3$ in *n*-heptane* as the test substance. In both examples the same detector and the same amounts of substance were used, the important difference being the carrier gas composition.

The practitioner can imagine how rapidly and simply he can check his own detector conditions with respect to linearity, depending on the working conditions.

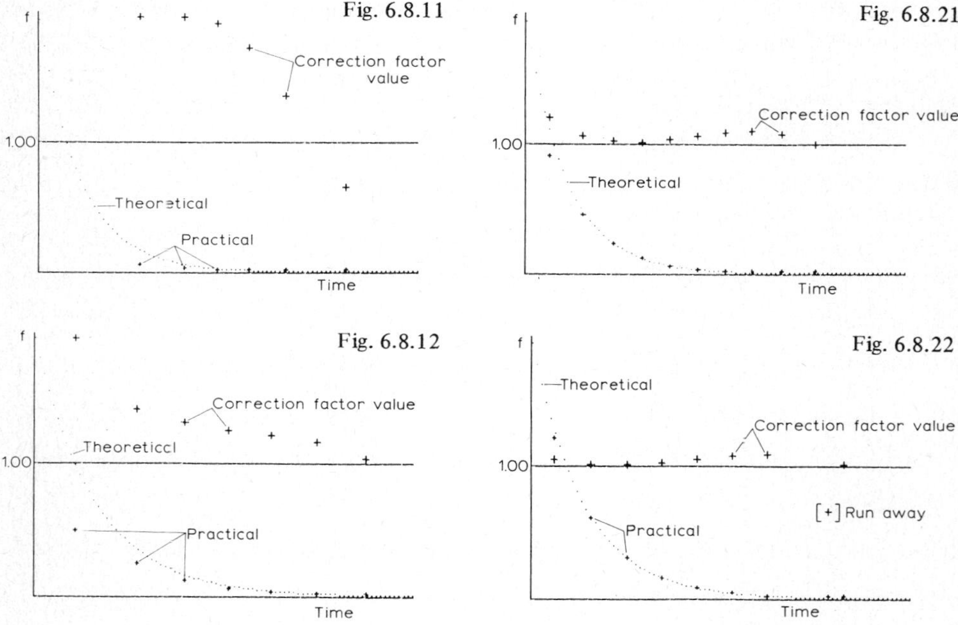

Fig. 6.8. 6.8.11–6.8.22: computer prints of graphical evaluation in on-line or off-line quantegg calibrations. 6.8.11: Heavily overloaded ECD, carrier gas nitrogen. Hopeless to correct anything. 6.8.12: No overloading, but ECD with nitrogen as carrier gas. Detector remains extremely non-linear. 6.8.21: Qualitatively and quantitatively the same conditions as in Fig. 6.8.11, but argon–methane as the carrier gas. 6.8.22: Same as in 6.8.12, but argon–methane as the carrier gas. The measurements are delicate. The gas flow must have an accuracy of better than 99%. The BASIC program (Table 6.5) offers possibilities to check for gas flow and maximum signal errors. The corrections can be checked graphically.

---

*Tests with many other ECD-active compounds run comparably. Results are supported by FID, HCD, in GC and many different detectors in HPLC.

TABLE 6.5

BASIC PROGRAM FOR QUANTEGG

```
]LIST

1  REM                    QUANTEGG
2  CLEAR
9  REM
10  REM ***********************
20  REM
40  REM   CALCULATES LINEARI-
50  REM   TY CORR. FACTORS FOR
60  REM   ALL DETECTORS BY EXP.
70  REM   DILUTION CHAMBER
80  REM   'QUANTEGG' ON-LINE
90  REM   COPYRIGHT IFC DUEW.
100  REM      TEL D-06322 8078
101  REM
102  REM *********************
110  REM
111  POKE 96 * 256,0: POKE 104,96
112  REM
130  TEXT : HOME
140  POKE 34,2: INVERSE : PRINT "  ** QUANTEGG  (CHECKS FOR LIN.RANGE)  "
     : NORMAL : PRINT
150  PRINT : PRINT "-> YOU KNOW PHASE FLOW VALUE ?": HTAB 28: INPUT " Y/N
     ?";FR$
160  IF FR$ = "Y" THEN 200
162  IF FR$ = "N" THEN 170
165  GOTO 112
170  PRINT : PRINT "-> YOU KNOW HALF TIME '0.5-TIME'": HTAB 28: INPUT " Y
     /N ?";FR$
180  IF FR$ = "N" THEN  PRINT : PRINT " ## FIND HALF HEIGHT SIGNAL  ": PRINT
     " -->COORESP.TIME IS '0.5 - TIME'       "
190  PRINT : PRINT "    ENTER THIS VALUE HERE....:": HTAB 28: INPUT " T .
     5 = ";HZ
200  PRINT : PRINT "-> YOU KNOW THE QUANTEGG VOLUME?": HTAB 28: INPUT " Y
     /N ?";FR$
210  IF (FR$ = "Y" AND HZ > 0) THEN 250
220  IF FR$ = "N" THEN 240
230  GOTO 320
240  PRINT : PRINT "YOU MUST KNOW THE VOLUME  ": PRINT "ECACTLY BY  +/- 1
     %  !!": PRINT : PRINT "MEASURE PRECISELY     !! ": END
250  PRINT : PRINT "  ENTER VOLUME IN ML...      ": HTAB 35: INPUT V
260 F = ( - 1) *  LOG (.5) * V / HZ
270 F = 60 *  INT (1000 * F) / 1000
280  PRINT : PRINT "* CALC.VALUE FOR FLOW  :"
290  PRINT "-IF T-0.5 VALUE WAS CORRECT- IS": PRINT "   ";F" ML/MIN
300 F = F / 60
310  FOR P = 1 TO 3500: NEXT P: GOTO 320
320  DIM C(100),H(100),T(100),KF(100)
330  HOME
350  IF V > 0 THEN 400
360  VTAB 14: PRINT "  ENTER VOLUME IN      (ML)
370  PRINT
380  PRINT "                 FLOW    (ML/MIN)"
390  PRINT
400  PRINT "--------------------------------------"
410  PRINT : PRINT "*  SET TIME FOR SIGN.MAXIMUM AS ZERO"
420  PRINT : PRINT "*  ENTER CORRESP. SENSITIVITY FACT."
430  PRINT : PRINT "*  START WITH LARGEST VALUE (SIGNAL CO)"
460  PRINT : PRINT : PRINT
470  IF V > 0 THEN 510
480  INPUT "-->  V (ML)        = ";V
490  IF V = 0 THEN  HOME : PRINT "YOU STOPPED BY <0>, BYE **": END
500  INPUT "     F (ML/MIN)  = ";F:F = F / 60
510 HZ = (( - 1) * V / F) *  LOG (.5):HZ =  INT (10 * HZ + .5) / 10
520  PRINT : PRINT " ** FACTOR  2 DILUTION ALL ";HZ" SEC"
530 ZZ =  INT (10 * (( - 1) * V / F) *  LOG (.1) + .5) / 10
540  PRINT "    FACTOR 10-DILUTION ALL ";ZZ" SEC"
545  PRINT
560  INPUT "PEAKMAX.(IN MM)  = ";CC
570  INPUT "SENSITIVITY FACT.= ";E1:CO = CO * E1
575  PRINT
580  HTAB 8: PRINT "CO NOW IS = ";CO
585  FOR P = 1 TO 2000: NEXT P
590  TEXT : HOME : POKE 34,2: INVERSE : PRINT "* DATA INPUT *     CORRECT
     WITH <-1>  ": NORMAL : PRINT
595  VTAB 2: HTAB 26: PRINT "0.5-T = ";HZ: VTAB 4: PRINT
600  FOR I = 1 TO 100
610  PRINT : HTAB 10: INPUT " SENSITIVITY     E = ";E
620  IF E = 0 THEN 740
630  HTAB 10: INPUT "HEIGHT (MM):   H   = ";H(I):H(I) = H(I) * E: IF H(I)
     < 0 THEN 610
640  HTAB 10: INPUT "CORR.TIME (S)  T   = ";T(I): IF T(I) < 0 THEN 630
645  INPUT "   CORRECT? THEN <RETURN> OTHERW.<-1>";QU$: IF QU$ = "-1" THEN
     640
650  PRINT I
660 C(I) = CO *  EXP (( - 1) * T(I) * F / V)
670  PRINT "--> C(TH) = "; INT (100 * C(I) + .005) / 100;" C(PR) = ";H(I)
```

```
680  IF H(I) = 0 THEN  PRINT "VALUE CANNOT  BE  >>ZERO<<": GOTO 610
690 KF(I) = C(I) / H(I)
700  IF F$ = "D" THEN N = N + 1
710  PRINT : PRINT " * LINEARITY CORRECTION FACTOR = "; INT (100 * KF(I) +
     .05) / 100
720  IF H(I) = 0 THEN 830
730  NEXT
740 N = I - 1
750  TEXT : PRINT : PRINT "CHANCE START SIGN.HEIGHT ? (WAS = ";CO;")": HTAB
     30: INPUT "<Y,N> ? ";QU$
760  IF QU$ = "Y" THEN 780
765  PRINT : PRINT "CHANGE FLOW VALUE? WAS= "; INT (100 * 60 * F) / 100"
     ML/MIN)": HTAB 30: INPUT "<Y,N> ? ";QU$: IF QU$ = "Y" THEN 767
766  GOTO 790
767  PRINT "ENTER NEW VALUE FOR PHASE FLOW": INPUT "  F = ";F:F = F / 60:
      GOSUB 1220
770  GOTO 790
780  PRINT " ENTER SENSITIVITY FACTOR E FOR 'CO'  ": PRINT " VALUE USED T
     IL NOW WAS= ";CO: INPUT " NEW   E*CO -VALUE     = ";CO: GOSUB 1220
790  PRINT "": PRINT : INPUT "*  PRINTOUT VALUES?   <Y> / <N> ";QU$: IF Q
     U$ <  > "Y" THEN 830
800  GOSUB 1160
810  PRINT "**  WANT TO CHANGE STARTING SIGN.HT.?": PRINT "OR WANT TO SEE
      GRAPHICS ?": PRINT "--> CHANGING:   ENTER VALUE-<  >": INPUT "--> G
     RAPHICS? ANTER          <0> HERE  ";W: IF W = 0 THEN 830
820 CO = W: GOTO 790
830  REM
840  IF T(N) =  < 2 * ZZ THEN BE = 2: GOTO 880
850  IF T(N) =  < 3 * ZZ THEN BE = 3: GOTO 880
860  IF T(N) =  < 4 * ZZ THEN BE = 4
870 BE = 4
880 FX = 250 / (BE * ZZ):FY = 159 / (CO * E1)
890  PRINT : PRINT "GRAPHICS RANGES OVER  ";BE: PRINT "ORDERS OF MAGN. OF
      CONCENTRATION   ": PRINT
900  HIMEM: 38400
910  HGR : HCOLOR= 7
920  HPLOT 0,0 TO 0,159: HPLOT  TO 250,159
930  FOR Z = 1 TO 250 STEP 3
940 Y = FY * CO * E1 *  EXP (( - 1) * (Z / FX) * F / V)
950  HPLOT Z,159 - Y
960  NEXT
970  HPLOT 0,79 TO 250,79
980  FOR I = 1 TO N
990 IX = T(I) * FX:YP = FY * H(I) * E1
1000 Y1 = 80 * KF(I)
1010  IF Y1 > 156 THEN Y1 = 156
1020  IF Y1 < 3 THEN Y1 = 3
1030  IF YP > 158 THEN YP = 158
1040  IF YP < 2 THEN YP = 2
1050  IF IX > 250 THEN IX = 250
1060  IF IX < 2 THEN IX = 2
1070  HPLOT IX + 1,159 - YP TO IX - 1,159 - YP
1080  HPLOT IX,159 - YP + 1 TO IX,159 - YP - 1
1090  HPLOT IX,159 - (Y1 + 2) TO IX,159 - (Y1 - 2): HPLOT IX - 2,159 - Y1
      TO IX + 2,159 - Y1
1100  NEXT
1110  PRINT : PRINT " ";BE" ORDERS OF MAGN. RANGING; ---> T"
1120  VTAB 22: INPUT "ADD DATA <D>; END <E>; OTHER 'F;CO'<CO>";F$
1130  IF F$ = "CO" THEN  TEXT : HOME : GOTO 750
1140  IF F$ = "D" THEN  FOR I = N + 1 TO 100: GOTO 610
1150  PRINT " ** BYE **; RESTART  BY 'GOTO 750'": END
1160 D$ =  CHR$ (4): PRINT D$;"PR#1": PRINT "TIME","  SIGN.HT.","CORR.FAC
      T","H-THEOR.": PRINT "----------------------------------------------
1170  FOR I = 1 TO N
1180  PRINT T(I),H(I), INT (KF(I) * 1000 + .5) / 1000, INT (C(I) * 10 + .
      5) / 10
1190  NEXT
1200  PRINT " QUANTEGG VOLUME = ";V;" ML; GAS FLOW F = "; INT (F * 60 * 1
      0) / 10;"ML/MIN; MAX.SIGNAL HEIGHT= ";CO
1210  PRINT : PRINT D$;"PR#0": RETURN
1220  FOR I = 1 TO N:C(I) = CO *  EXP (( - 1) * T(I) * F / V):KF(I) = C(I
      ) / H(I): NEXT : PRINT "TIME             SIGNAL      CORR.FACTOR": PRINT
      "------------------------------------"
1230  FOR I = 1 TO N: PRINT T(I),H(I), INT (100 * KF(I)) / 100: NEXT
1240  PRINT : INPUT " CONTINUE <Y> ";QU$: IF QU$ = "Y" THEN  RETURN
1250  RETURN
1260  REM                  KOESSEN
```

6.7.2. Data evaluation with a laboratory computer

Table 6.5 gives parts of the BASIC program (Applesoft Basic) as well as a program that runs under CP/M conditions, which we wrote to make calibration simple. If one can use an Apple 48 kbyte computer, one disk and a monitor for qualified graphics representation, one will have a screen display as shown in Fig. 6.5; it is much easier to read a picture than numbers in order to be able to make immediate decisions. If line L in Fig. 6.5 (which represents the non-linearity correction factor) deviates by more than 1% from the middleline M, then non-linearity causes systematic analytical errors. We have never seen line L coinciding with line M with any GC detector that we have checked during the past 2 years work with the quanteggs.

LITERATURE

R.E. Kaiser, *Pittsburgh Conference 1980, Atlantic City*, paper No. 111.
R.E. Kaiser and R.I. Rieder, *GIT Fachz. Lab.*, 24 (1980) 633–639.

Chapter 7

Response of the electron-capture detector to compounds with natural electrophores*

JÖRGEN VESSMAN

CONTENTS

7.1. INTRODUCTION

The electron-capture detector (ECD) has been used in gas chromatography (GC) for quantitative purposes for about 15 years. The high sensitivity combined with the selectivity that can be achieved are the reasons why this device has been used so much in the analysis of organic compounds at the trace level in various complex sample types.

This review considers the possibilities of using the ECD in the analysis of non-halogenated compounds that do not contain a nitro group. The introduction of halogenated groups in derivatization reactions will not be considered as such information has been included in other reviews [1–3]. Surveys on the use of electron-capture detection in GC have been published [4–6].

The inventor of the ECD, J.E. Lovelock, ranked various compounds in order of relative absorption coefficients for thermal electrons and proposed the expression "electrophore" for those atoms or structures in a molecule which confer electron-absorbing properties or electron affinity to the molecule [7,8]. Two classes could be distinguished, of which halogen and nitro substituents belong to the simple electrophores. The other class consists of conjugate electrophores, which are found in structures with certain groups which when isolated do not contribute to electron capture, but do so if connected by specific bridges. Typical conjugate electrophores are carbonyl compounds such as biacetyl, fumarate esters, cinnamaldehyde and quinone. Fumaronitrile and azobenzene represent other structures of this kind.

The electron-capture process has been suggested to occur according to two different mechanisms [7–9]. The first is characteristic of several conjugate electrophores and is

*Previously published in *J. Chromatogr.*, 184 (1980) 313.

called the resonance capture or non-dissociative mechanism and results in the formation of a negative molecular ion. The second type is found among halogenated compounds such as pesticides and is called the dissociative mechanism as electron attachment leads immediately to the formation of a negative ion (e.g., a halide ion) and a neutral radical. The negative molecular ion formed upon electron attachment is best detected at low detector temperatures as electron detachment from this ion is favoured by high temperatures. The temperature dependence is related to the overall structure of the molecule and it is of considerable importance for the analyst to be able to determine this effect as the relative response can vary by more than 1000-fold over the useful detector temperature range (100–350°C). Response factors without data on detection temperatures are of limited value [10,11]. It is also important to remember that the temperature of the detection zone is not necessarily the same as that indicated for the detector oven.

The response of the ECD is also influenced by the mode of operation. Modern constant-current ECDs usually collect the electrons via an applied pulsed voltage, which gives better linearity and fewer disturbances than older types operating in the non-pulsed mode. The latter type has been shown in some studies with compounds responding in the non-dissociative mode to have a less pronounced temperature dependence than in the pulsed mode. This was the case for benzophenone and dibutyl phthalate [12]. However, there were also differences between the same type of instrument. From a theoretical point of view the pulsed mode is the one of choice, as pointed out by Wentworth and co-workers [10,13].

Most compounds of interest to the analyst do not capture electrons with thermal energy, which is the reason for the selective response of the detector. Some structures are discussed below that have been quantified by electron-capture detection. They represent mainly the group of conjugate electrophores.

7.2. CONJUGATED CARBONYL COMPOUNDS

Several reports have described the quantitative use of electrophores where a carbonyl group is involved.

Benzophenone (I) is a typical conjugate electrophore and has a pronounced temperature-dependent response. This seems also to be related to the type of detector, as devices with a pulsed voltage exhibit a stronger decrease with increasing temperature than those of the direct-current type [12]. The response for benzophenone is comparable to that of many other well known halogenated electrophores [14] (cf., Table 7.1). It is also of interest that at low temperatures *p*-chlorobenzophenone is less sensitive than benzophenone, which indicates the importance of the conjugated system. At higher temperatures the chloro substituent takes over the electron attachment and *p*-chlorobenzophenone becomes the more sensitive compound.

The relative responses for a series of benzophenones have been reported [14]. Electron-donating, non-electrophore substituents gave an increased response. Some of the benzophenones were used in the quantitation of drugs, which upon oxidation formed the corresponding benzophenones [16].

A similar approach was used for the antidepressant drug amitriptyline, which after

TABLE 7.1

MINIMUM DETECTABLE CONCENTRATION (MDC) OF VARIOUS COMPOUNDS [14, 15]

Compound	MDC (moles/s·10^{15})	Temperature (°C)	Mode
Benzophenone	1.2–3.1	104–163	d.c.
p-Chlorobenzophenone	3.8	104–163	d.c.
2-Benzoylpyridine	4.8	104–163	d.c.
2-(4-Chlorobenzoyl)pyridine	1.2	104–163	d.c.
Benzoylfuran	7.3	170	d.c.
Benzoylthiophene	6.4	170	d.c.
Anthraquinone	4.7	200	d.c.
Dibenzosuberone	5.0	200	d.c.
Di(ethylhexyl) phthalate	52	220	d.c.
8-Methoxypsoralen	2	200	d.c.
Methindione	0.4	215	d.c.
Saccharin	0.3	210	d.c.
Sulfapyridine	0.8	350	Pulsed
N^4-Acetylsulfapyridine	1.2	350	Pulsed

oxidation gave anthraquinone (II). From the temperature dependence, the electron-capture mechanism was stated to be of a dissociative nature [17]. This was also found by Grimsrud and Stebbins [18] in studies involving also halogenated hydrocarbons and anthracene.

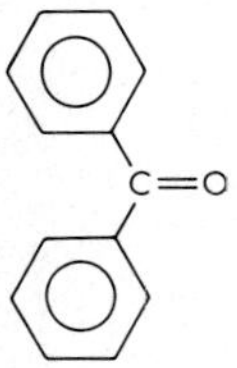

C=O O O C=O N N—H C=O

I 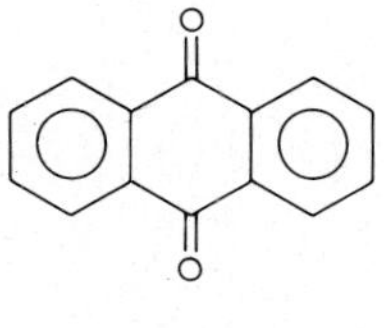 II 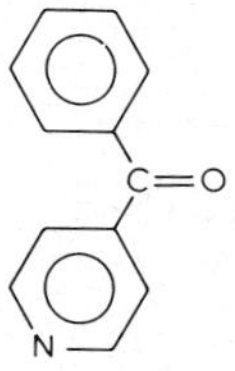 III,1 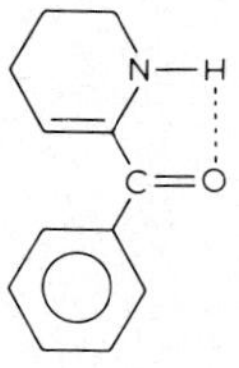 III,2

Benzoylpyridine (III, 1) has about the same response as benzophenone [14]. This was shown to be true also for benzoylfuran and benzoylthiophene, which were obtained by benzoylation of furan and thiophene in order to confer upon them electrophore properties [19]. An unusual oxidation product (6-benzoyl-1,2,3,4-tetrahydropyridine, III, 2) was formed when pipradrol was treated with chromic acid [20]. This conjugated carbonyl compound had an unexpectedly high response, which was postulated to depend on the possibility of forming an intramolecular hydrogen bonding between N–H and the carbonyl group. The corresponding N-methylated compound showed a response which was about 200 times lower. The response of dibenzosuberone (IV) was in the same range as the benzophenone group [17] (Table 7.1). The benzophenone moiety of ketoprofen (V) made possible its determination in biological fluids [21].

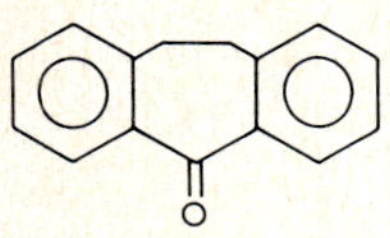

The analysis of phthalates (VI) in various sample types has been of great interest. The conjugate electrophore gives the molecule reasonably good sensitivity, which can be used in trace analysis [22]. In a given situation the selectivity can be good, as was demonstrated in the search for di(ethylhexyl)phthalate (DEHP) in blood plasma [23]. On one occasion a complicated pattern obtained with the ECD was shown to be the same as in gas chromatography–mass spectrometry when the characteristic fragment *m/e* 149 for phthalate was monitored.

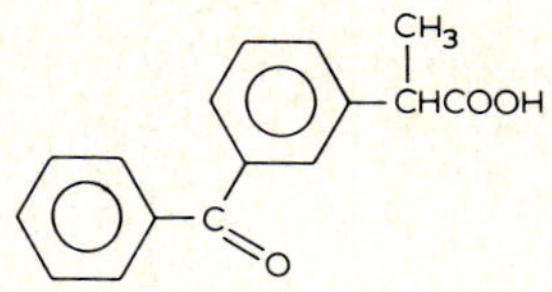

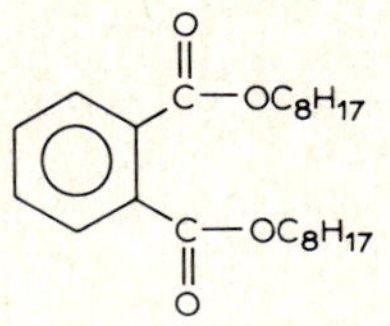

Phthalate esters can make considerable contributions to the contaminant pattern observed with the ECD from organic solvents and plastic devices used in the analytical procedure. If the ECD response for the compound to be analysed exhibits a temperature response that differs from that of the phthalates, i.e., shows an increase with increasing detector temperature, their interference can be diminished. Even for the phthalates the temperature dependence was more pronounced in a detector operating in the pulsed mode than in the d.c. mode [12]. A non-dissociative mechanism was indicated for DEHP from the temperature dependence found in a pulsed detector [24]. A study of the relative response for some phthalates at 230, 290 and 320°C was recently reported [25]. Diallyl phthalate was the only ester which did not lose sensitivity at the higher temperature. The samples included methyl and ethyl benzoates, which as expected were a factor of 100–1000 times less sensitive.

8-Methoxypsoralen (VII) is of importance in the treatment of psoriasis. It has been shown to be a good electrophore, permitting the analysis of nanogram amounts in biological fluids [26]. The temperature-dependent response indicated a non-dissociative mechanism as the detector response at 200°C was about 30 times higher than that at 300°C. It was interesting that the response for the isomer 5-methoxypsoralen was only slightly temperature dependent, which shows that minor structural changes can influence the electron-capture mechanism.

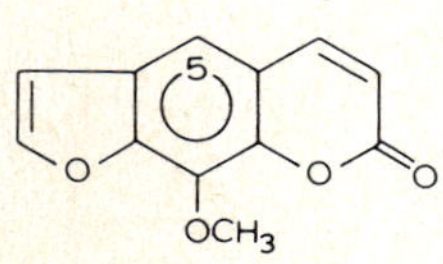

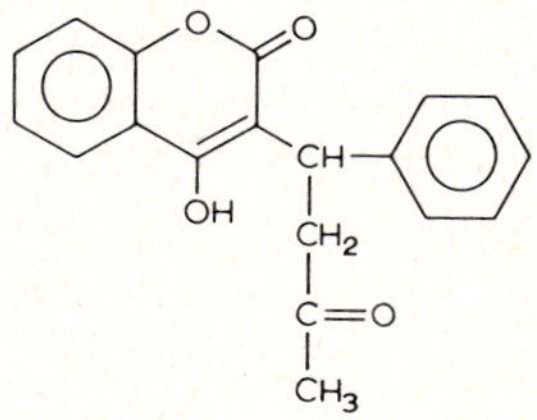

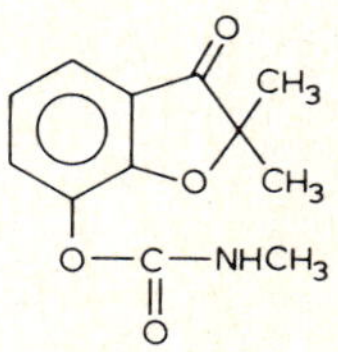

A compound containing a similar structural feature and also of considerable therapeutic interest is warfarin (VIII). This drug can be derivatized with electrophores such as pentafluorobenzyl bromide via the phenolic function, but a technically less complicated procedure makes use of diazomethane, which from an ECD point of view is "transparent", to form the methyl ether. With the detector temperature kept at 300°C the methylation procedure permitted the analysis of 100 ng/g in various tissues [27]. The insecticide carbofuran has one metabolite (IX) with a keto group, which can be quantitated by GC–ECD [28]. The conjugate electrophore in this instance involves only a carbonyl group in an α-position, which as an electrophore could be compared with acetophenone or benzaldehyde. However, the presence of a carbamate group in oxocarbofuran might have an auxiliary function. The advantage of direct analysis instead of derivatizing the carbamate group was stressed. Although a relatively high detection temperature (275°C) was used, the procedure permitted the detection of ca. 100 pg on the column.

Lovelock et al. [29] pointed out the possibility of detecting selectively some steroids with unsaturated ketone structures, such as 4-androstene-3,17-dione (X). The 4-ene-3-one system was indispensible for good detector response. 4-Androstene-3,11,17-trione was the most sensitive compound (X). Melengestrol acetate (XI) has been determined using GC–ECD [30]. The ECD response was comparable to that for the most sensitive haloesters of steroid alcohols. The structural requirements for the high response of this compound type were studied by Koshy [31]. Among the important functional groups were 17α-acetate and substitution with 16-methyl and a 6-keto group. The high ECD response was independent of temperature between 260 and 360°C and permitted 10 pg to be determined.

A similar structural feature can be found in some polyhydroxy steroids, the ecdysones (XII), which act as insect-moulting hormones. Poole and Morgan [32] found that the electrophore was an unsaturated ketone structure (7-ene-6-one), with important contributions from a hydroxy substituent in position 14, either free or preferably silylated. The temperature dependence of the detector response was indicative of a non-dissociative mechanism. For practical reasons a temperature of 300°C was used in the detector, but 5 pg of the trimethylsilyl ethers could be detected.

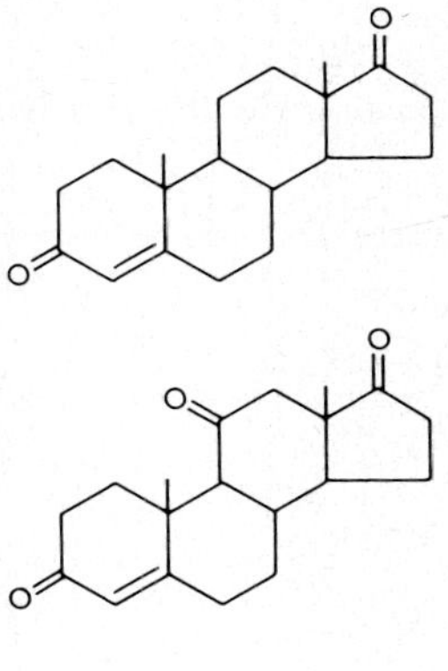

X 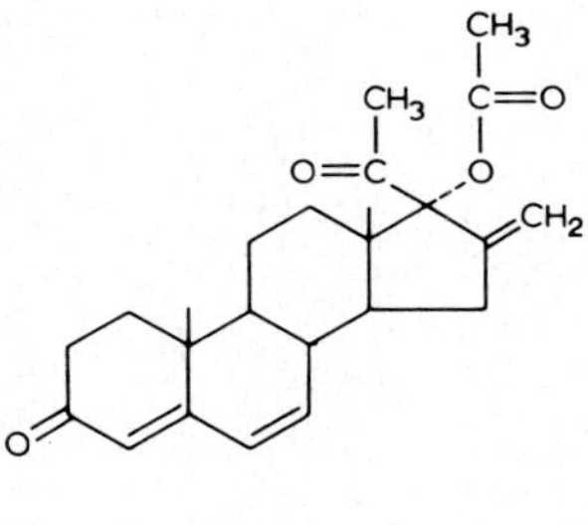 XI 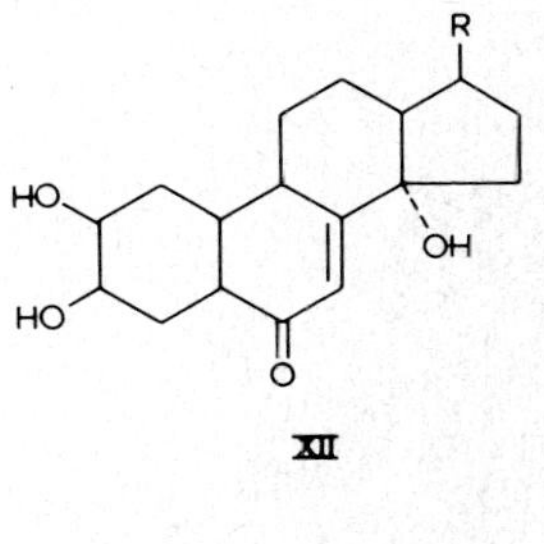 XII

The presence of this structural element, an unsaturated ketone, has made possible the determination, after silylation, of some prostaglandins B (XIII), which are dehydration products of prostaglandins E [33]. The latter did not respond in the ECD.

Of the various electrophores mentioned in Lovelock's early work, the ECD has also been used for the quantitation of biacetyl (XIV, 1) and pentane-2,3-dione in beer [34]. A related compound is pyruvic acid which was quantified in beer as the methyl ester (XIV, 2) [35]. In other sample types quinone derivatives such as vitamin K (XV) have been studied [36]. The compounds required high temperatures for elution, which fixed the detector temperature at a high value (305°C).

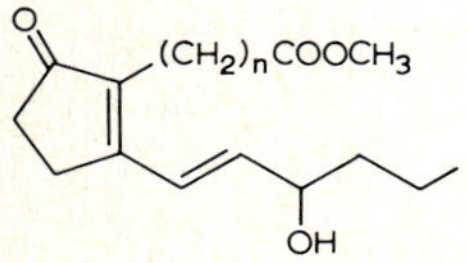

XIII 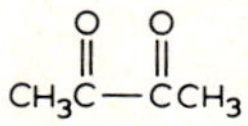 XIV, 1 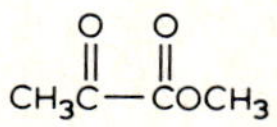 XIV, 2

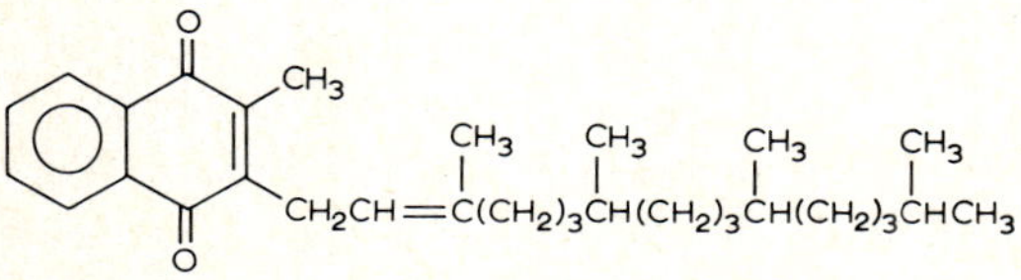

XV 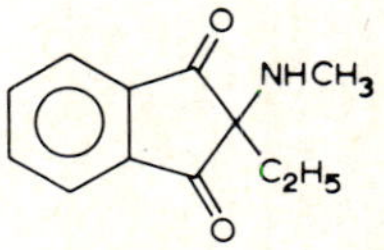 XVI

A structure similar to the phthalate electrophore is present in methindione (XVI), which can be detected at the low picogram level [37]. This compound did not show a temperature-dependent response in a non-pulsed detector over the range 150–215°C, which indicates either a very stable negative molecular ion or a dissociative mechanism. This structure of conjugated β-ketones is present in some other drugs, e.g., phenindione (XVII), which was shown to have comparable sensitivity after masking of the active hydrogen [37]. Another related structure is present in thalidomide (XVIII), which was reported by Zlatkis and Lovelock [8] to have an electron-capture coefficient of ca. 300 (compared to chlorobenzene). The drug was withdrawn from the market before interest in trace analysis by GC–ECD had begun, otherwise it would have been possible to devise a selective and sensitive method.

Saccharin (XIX) can also be placed in the group of conjugated carbonyl electrophores.

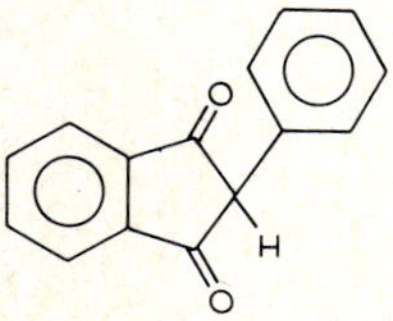

XVII 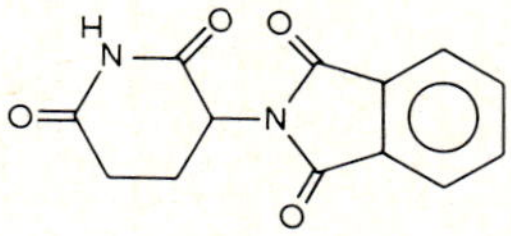 XVIII 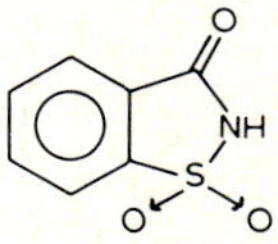 XIX

This compound has been analysed with an ECD in various biological samples [38]. To obtain the best detectability, the acidic sulphonamide function has to be alkylated, preferably in an extractive alkylation reaction [39]. The sensitivity was such that 2 pg could be detected.

The saccharin-like structure has been used in the analysis of a drug candidate (XX, 1), which before GC had to be converted into a less polar structure (XX, 2) to reduce its column adsorptive properties. This change was achieved in a dehydration reaction with trifluoroacetic anhydride without increasing the chromatographic background [40].

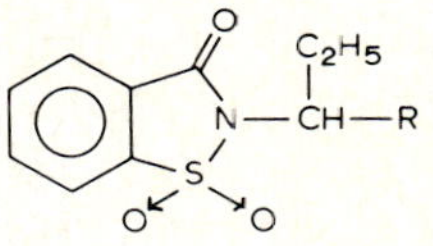

1) R = $CONH_2$
2) R = C≡N

XX

7.3. SULPHONAMIDES

Saccharin can be regarded as a special form of a sulphonamide. It has recently been shown, however, that even conventional sulphonamides of therapeutic interest, e.g., sulfapyridine (XXI, 1) can be gas chromatographed and quantified after either alkylation or acylation [41]. The best sensitivity was obtained after alkylation with pentafluorobenzyl bromide and acylation with heptafluorobutyric anhydride. Of great interest was the observation that both alkylation with a methyl halide (XXI, 2) and acylation with acetic anhydride resulted in derivatives with a sensitivity not much lower than that of the fluorinated derivatives. This means that under certain conditions the sulphonamide group is acting as an electrophore [41]. The conclusion was that the sulphonamide group should be attached to two aromatic rings (delocalized systems) in order to obtain the highest response. The temperature dependence of the detector response has been studied for some model compounds, and for the one most similar to drugs, N-phenylbenzenesulphonamide (XXII), methylation and acetylation gave derivatives that were about 10 times less

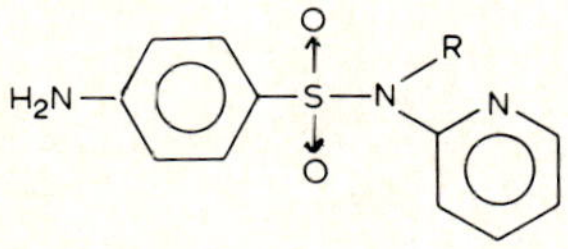

1) R = H
2) R = CH_3

XXI

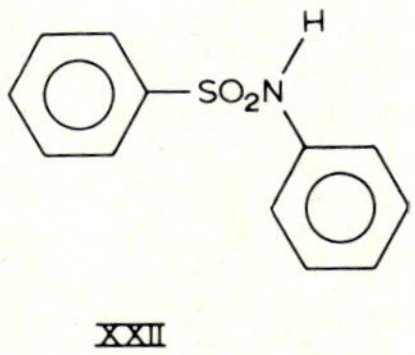

XXII

sensitive at 350°C than at 150°C [42]. The possibility of using reagents that usually give derivatives transparent in the ECD (such as acetic anhydride) or which are volatile (such as the methyl halides) adds considerably to the advantage of using an ECD for a compound with inherent electrophore properties.

7.4. MISCELLANEOUS COMPOUNDS

Very early it was found that hydrocarbons are more or less transparent in the ECD. However, there are exceptions among the polycyclic aromatic hydrocarbons, such as anthracene and benzanthracene [43]. It is also of interest that azulene and cyclo-octatetraene give a strong response. In the latter instance this was explained as being due to the favourable stabilization of the planar negative ion.

A polycyclic heteroaromatic compound having good electrophore properties has been reported for a reaction product (XXIII) between the antihypertensive agent hydralazine and nitrite [44]. Hydralazine, which is labile, was trapped in the biological sample from which the product was isolated and then quantitated in nanogram amounts with an ECD.

Another polycyclic compound, phanquone (XXIV, 1), has also been isolated after derivatization in the aqueous phase and then subjected to electron-capture detection [45]. Phanquone was reacted with methoxylamine to give the bismethoxime derivative (XXIV, 2), which was detected with a pulsed ^{63}Ni ECD at 300°C. A concentration of 15 ng/ml could be quantified in biological fluids.

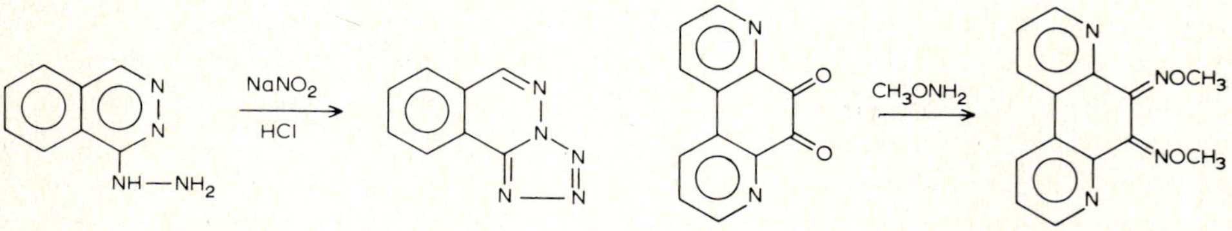

An example of a structure that would not directly be considered as an electrophore is phenylbutazone (XXV), a widely used anti-inflammatory drug. A recent paper nevertheless described the use of an ECD for its selective determination in biological fluids [46]. Concentrations of 10 ng/ml could be quantified accurately and very clean chromatograms were obtained. Sulphinpyrazone has been determined by GC–ECD, the compound undergoing an elimination reaction to give an analogue of phenylbutazone [47].

Although free barbituric acids were reported to have some electron-capture response, no applications have been described until recently. The methylated derivatives of thiopental and thioamytal (XXVI) were studied [48] and at a detector temperature of 300°C, 0.1 and 1 ng, respectively, could be detected.

A structure in part related to the barbiturates is that of pemolinedione (XXVII, 1), produced by acid treatment of pemoline, a drug which is active at the micrograms per millilitre level. Owing to the selective response a very simple assay was developed with

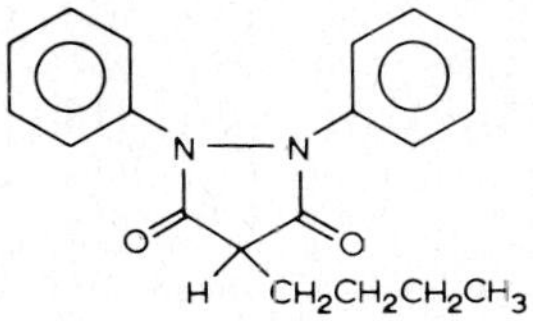

XXV

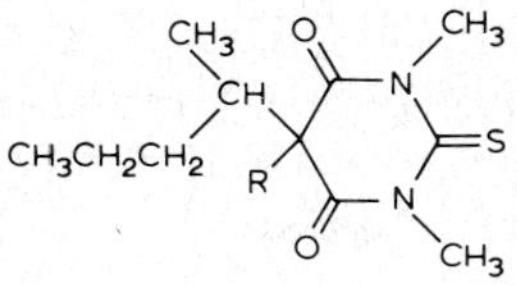

1) R = C_2H_5
2) R = $CH_2CH{=}CH_2$

XXVI

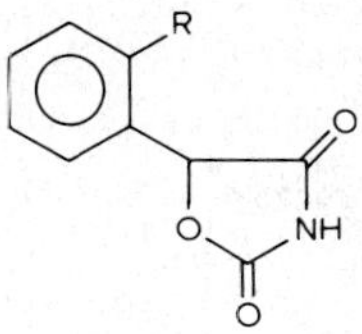

1) R = H
2) R = CH_3

XXVII

XXVII, 2 as internal standard [49]. The response was such that 0.5 ng injected could be detected (detector temperature 230°C, d.c. mode).

A compound that has been reported to have a high electron affinity as the silyl derivative is myo-inosose-2 (XXVIII) [50]. In a pulsed detector at 205°C a few picograms could be detected. Whereas the corresponding inositol was 500 times less sensitive, it was found that one form of fructose was 3–4 times more sensitive. Although the non-conjugated carbonyl group is primarily responsible for the high affinity, the structure of the species that gives the compound the electrophore properties is probably different.

Ethylene bis(isothiocyanate) (XXIX) is another type of structure that has been quantified with the ECD. At a detector temperature of 300°C, 0.02 ppm could be determined [51]. The vicinal position of the isothiocyanate groups might be the reason for the good response in a similar manner to that of bis(1,2-trifluoroacetyl) glycols [52]. A related substituent is found in toluene 2,4-diisocyanate (XXX), which responded well in the ECD [53]. Some compounds with related structures did not give a response.

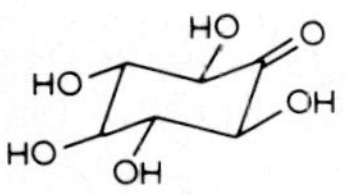

XXVIII

$NCS—CH_2CH_2—SCN$

XXIX

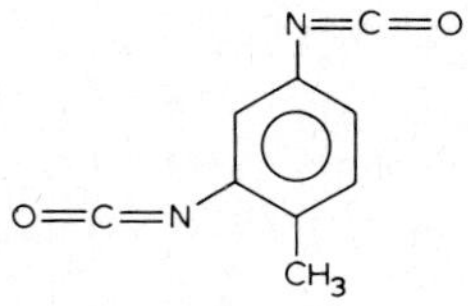

XXX

The correlation between chemical structure and electron-capture response has been studied with some isothiocyanates and isocyanates [54].

Phenylethylmalonamide (XXXI) is a metabolite of primidone. A GC–ECD method has been reported in which the compound was dehydrated with trifluoroacetic anhydride in a similar manner to the reaction described above for the saccharin derivative (XX, 1). Only one of the primary amide groups was converted into a nitrile [55]. This derivative had sufficient response to allow the determination of submicrogram amounts per millilitre. The electrophore component in this compound is not clear.

An unexpected response was shown for one pyrethroid insecticide (XXXII, 1). The electrophore in this compound is not clear [56]. Two related compounds contained a dichlorovinyl substituent instead of a methyl group, but their sensitivities were approximately the same, i.e., around 5 pg injected on-column with a detector temperature of

270°C. This implies that there is an inherently very good electrophore in the structure.

trans-Stilbene (XXXIII) was early pointed out to be a potential electrophore. Hydroxylated metabolites of stilbene were analyzed as chloromethyldimethylsilyl ethers but the responses of the methyl and silyl ethers and the acetates were also investigated [57]. The ranking order was methoxy (0.8), stilbene (1.0), trimethylsilyloxy (1.4), acetoxy (3.2) and chloromethyldimethylsiloxy (40). With the latter derivative 8 ng could be detected.

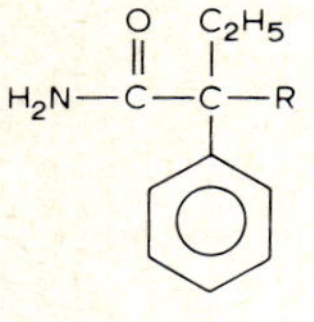

1) R = $CONH_2$
2) R = $C{\equiv}N$

XXXI

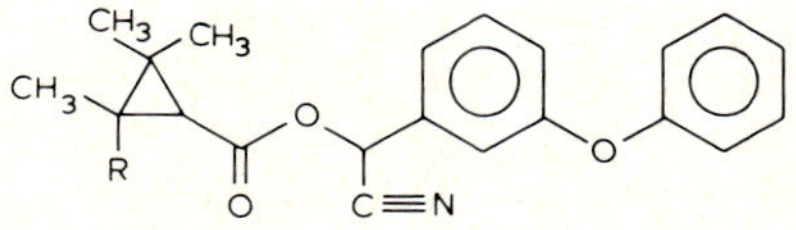

1) R = CH_3
2) R = $CH{=}CCl_2$

XXXII

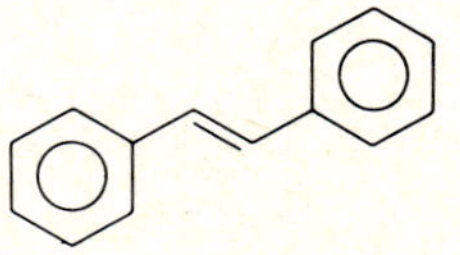

XXXIII

7.5. OTHER ASPECTS

In connection with these various compounds it should be mentioned that, apart from contamination of solvents with phthalates, impurities due to photochemical oxidation of hexane isomers have been reported [58]. The structure of these electrophore contaminants has not been elucidated, but it shows the importance of limiting the presence of unknown electrophores by careful handling and storage of chemicals.

The previous discussion has shown that various structures can be detected and quantified with an ECD. The importance of the conjugated structure in benzophenone and *p*-chlorobenzophenone is best shown by their almost identical responses at low temperatures. There must, certainly, be other structures present in which the halogen content is not the important component in the electrophore structure. This may be the case among many members of the benzodiazepine group, which have a well conjugated system [6]. Other interesting structures from that point of view are indomethacin [59] and griseofulvin [60]. The fact that the dominant electrophore is not always a halogen is still overlooked.

Zlatkis and Lovelock [8] stated in 1965 that the difficulty in predicting the response of conjugate electrophores was related to the fact that the structure one should consider is that of the resultant negative ion, not the parent molecule. In some instances it is now possible to elucidate the structure of negative ions formed under electron-capture conditions. By use of atmospheric pressure ionization mass spectrometry (API-MS), Horning et al. [61] have found the expected structure for, e.g., benzil. One benzodiazepine, flunitrazepam, was shown to undergo a rearrangement [62]. Some conditions in the detector during electron capture might have greater importance than it had earlier been reasonable to believe. The influence of trace amounts of oxygen and water is thus considerable for the detection of some electrophores [18]. Anthracene showed an increased response with very low levels of oxygen, while some halohydrocarbons were affected only at high levels. For anthracene this effect was greatest at high temperature. An exchange reaction between the negative molecular ion and oxygen was proposed. This suggests that control of the carrier gas contamination might be important in order to obtain the best and most reproducible results. In this respect anthraquinone was affected like the halogenated hydrocarbons [18].

In connection with this discussion, it is also of interest that so far no negative ion has been found for benzophenone in the API-MS instrument [63].

The recent availability of negative ion mass spectrometry has opened up interesting possibilities both for the application of electron-capture processes of compounds that form stable negative ions and also for obtaining an insight into the structures formed. The conditions for the formation of electrons with thermal energy are, however, not equivalent in GC–ECD and in electron-capture negative ion GC–MS [64]. In this respect the API instrumentation is more suitable.

An insight into the products formed upon reaction with thermal electrons under conditions prevailing in chemical ionization mass spectrometry operated in the negative ion mode was given in a recent thesis [65].

Molecular anions were found for compounds such as benzophenones, diketones and quinones. Conjugated diacids gave the corresponding cyclic anhydrides. This anion was also found in the case of the phthalate esters.

Of interest was also the observation that sulphonamides gave an abundant ion, most often the base peak, which had a "quinone-like" structure capable of stabilizing the negative charge (XXXIV).

It was also interesting to note that diazepam as a representative for the benzodiazepines, formed an anion with no sign of fragmentation, i.e., no loss of chlorine.

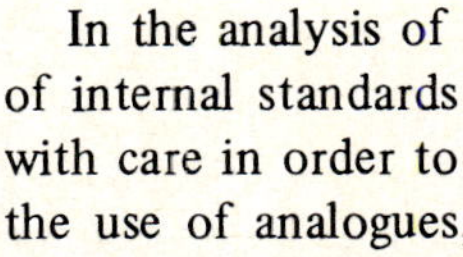

HN= ⟨ ⟩ =S(O)(O)]$^{\dot{-}}$

XXXIV

7.5.1. Quantitation

In the analysis of trace amounts of organic compounds by gas-phase methods, the use of internal standards has found widespread use. The internal standard must be chosen with care in order to optimize the procedure. The best approach is now considered to be the use of analogues, homologues or isomers. However, it is as important to know the properties of the internal standard as those of the compound to be analysed. In connection with electron-capture detection, this means that the mechanism or at least the temperature dependence of the detector response has to be evaluated. Of the compounds covered in this review, examples of differences in this respect can be found among the benzophenones and the alkoxypsoralens. The use of a halogenated compound, i.e., a pesticide that captures thermal electrons only according to a dissociative mechanism, cannot fully compensate for variations in the operating conditions in the same way as an analogue. Examples of ideal internal standards for ECD purposes can be found in papers on the determination of methindione [37], methoxypsoralen [26], di(ethylhexyl) phthalate [23], pemoline [49], phenylethylmalonamide [55], sulfapyridine [41] and a saccharin derivative [40].

The analysis of compounds with inherent electrophore properties can be very simple even if a derivatization reaction has to be used to mask polar functional groups. In comparison with reactions for the introduction of electrophores such as heptafluorobutyryl this makes a large difference, as the latter derivatization products are often accompanied by products from other substrates in the sample extract.

7.6. ACKNOWLEDGEMENTS

Stimulating discussions with Dr. Olle Gyllenhaal and Dr. Colin F. Poole are gratefully acknowledged.

REFERENCES

1 J.D. Nicholson, *Analyst (London)*, 103 (1978) 1, 193.
2 J. Drozd, *J. Chromatogr.*, 113 (1975) 303.
3 K. Blau and G.S. King, in K. Blau and G. King (Editors), *Handbook of Derivatives for Chromatography*, Heyden & Son, London, 1978, pp. 104ff.
4 E.D. Pellizzari, *J. Chromatogr.*, 98 (1974) 323.
5 C.F. Poole, *Chem. Ind. (London)*, (1976) 479.
6 J.A.F. de Silva, in C.E. Pippenger, J.K. Penzy and H. Kutt (Editors), *Antiepileptic Drugs: Quantitative Analysis and Interpretation*, Raven Press, New York, 1978, p. 111.
7 J.E. Lovelock, *Nature (London)*, 189 (1961) 729.
8 A. Zlatkis and J.E. Lovelock, *Clin. Chem.*, 11 (1965) 259.
9 W.E. Wentworth and E. Chen, *J. Gas Chromatogr.*, 5 (1967) 170.
10 E.C.M. Chen and W.E. Wentworth, *J. Chromatogr.*, 68 (1972) 302.
11 C.F. Poole, *J. Chromatogr.*, 118 (1976) 280.
12 J. Vessman and P. Hartvig, *Acta Pharm. Suecica*, 9 (1972) 463.
13 W.E. Wentworth, E. Chen and J.E. Lovelock, *J. Phys. Chem.*, 70 (1966) 445.
14 J. Vessman and P. Hartvig, *Acta Pharm. Suecica*, 8 (1971) 235.
15 O. Gyllenhaal, U. Tjärnlund, H. Ehrsson and P. Hartvig, *J. Chromatogr.*, 156 (1978) 275.
16 J. Vessman, S. Strömberg and G. Rietz, *Acta Pharm. Suecica*, 7 (1970) 363.
17 P. Hartvig, S. Strandberg and B. Näslund, *J. Chromatogr.*, 118 (1976) 65.
18 E.P. Grimsrud and R.G. Stebbins, *J. Chromatogr.*, 155 (1978) 19.
19 H. Brink, O. Gyllenhaal and P. Hartvig, *Acta Pharm. Suecica*, 12 (1975) 235.
20 P. Hartvig, *Acta Pharm. Suecica*, 11 (1974) 109.
21 P. Stenberg, T.-E. Jönsson, B. Nilsson and F. Wollheim, *J. Chromatogr.*, 177 (1979) 145.
22 W. Bunting and E.A. Walker, *Analyst (London)*, 92 (1967) 575.
23 J. Vessman and G. Rietz, *J. Chromatogr.*, 100 (1974) 153.
24 C.F. Poole and D.G. Wibberley, *J. Chromatogr.*, 132 (1977) 511.
25 Y. Hattori, Y. Kuge and S. Asade, *Nippon Kagaku Kaishi*, 8 (1978) 1102.
26 H. Ehrsson, S. Eksborg, I. Wallin, N. Kållberg and G. Swanbeck, *J. Chromatogr.*, 140 (1977) 157.
27 E.M. Odam and M.G. Townsend, *Analyst (London)*, 101 (1976) 478.
28 M.T.H. Ragab, *Anal. Lett.*, 10 (1977) 551.
29 J.E. Lovelock, P.G. Simmonds and W.J.A. VandenHeuvel, *Nature (London)*, 197 (1963) 249.

30 R.A. Devis, K.T. Koshy, M.L. Ogilvie, G.L. Stahl and W.A. Callahan, *J. Ass. Offic. Anal. Chem.*, 55 (1972) 103.
31 K.T. Koshy, *J. Chromatogr.*, 126 (1976) 641.
32 C.F. Poole and E.D. Morgan, *J. Chromatogr.*, 115 (1975) 587.
33 G.H. Jouvenaz, D.H. Nughteren, R.K. Beerthuis and D.H. van Dorp, *Biochim. Biophys. Acta*, 202 (1970) 231.
34 G.A.F. Harrison, W.J. Byrne and E. Collins, *J. Inst. Brew.*, 71 (1965) 336.
35 G.A.F. Harrison and E. Collins, *Am. Soc. Brew. Chem. Proceed.*, (1968) 101.
36 H. Bechtold and E. Jähnchen, *J. Chromatogr.*, 164 (1979) 85.
37 J. Vessman, S. Strömberg and G. Freij, *J. Chromatogr.*, 94 (1974) 239.
38 R.J. Daun, *J. Ass. Offic. Anal. Chem.*, 54 (1971) 1140.
39 P. Hartvig, O. Gyllenhaal and M. Hammarlund, *J. Chromatogr.*, 151 (1978) 232.
40 J. Vessman, in E. Reid (Editor), *Blood Drugs and Other Analytical Challenges*, Ellis Horwood, Chichester, 1978, p. 119.
41 O. Gyllenhaal, B. Näslund and P. Hartvig, *J. Chromatogr.*, 156 (1978) 330.
42 O. Gyllenhaal and P. Hartvig, *J. Chromatogr.*, 189 (1980) 351.
43 J.E. Lovelock, A. Zlatkis and R.S. Becker, *Nature (London)*, 193 (1962) 540.
44 D.B. Jack, S. Brechbühler, P.H. Degen, P. Zbinden and W. Riess, *J. Chromatogr.*, 115 (1975) 87.
45 P.H. Degen, S. Brechbühler, J. Schäublin and W. Riess, *J. Chromatogr.*, 118 (1976) 363.
46 A. Sioufi, F. Candel and F. Marfil, *J. Pharm. Sci.*, 67 (1978) 243.
47 P. Jakobsen and A. Kirstein Pedersen, *J. Chromatogr.*, 163 (1979) 259.
48 R.H. Smith, J.A. MacDonald, D.S. Thompson and W.E. Flacke, *Clin. Chem.*, 23 (1977) 1306.
49 S.-Y. Chu and L.T. Sennello, *J. Chromatogr.*, 137 (1977) 343.
50 W.R. Sherman and S.L. Goodvin, *J. Chromatogr. Sci.*, 7 (1969) 167.
51 W.H. Newsome, *J. Agr. Food Chem.*, 24 (1976) 420.
52 D.D. Clarke, in H. Szymansky (Editor), *Biomedical Applications of Gas Chromatography*, Vol. II, Plenum Press, New York, 1968, p. 137.
53 B.B. Wheals and J. Thomson, *Chem. Ind. (London)*, (1967) 753.
54 M. Satouchi and T. Kojima, *Anal. Lett.*, 5 (1972) 931.
55 J.E. Wallace, H.E. Hamilton, E.L. Shimek, Jr., H.A. Schwertner and K.D. Haegele, *Anal. Chem.*, 49 (1977) 1969.
56 R.A. Chapman and H.S. Simmons, *J. Ass. Offic. Anal. Chem.*, 60 (1977) 977.
57 J.E. Sinsheimer and R.V. Smith, *J. Pharm. Sci.*, 56 (1967) 1280.
58 I.H. Williams, *J. Chromatogr. Sci.*, 11 (1973) 593.
59 A. Arbin, *J. Chromatogr.*, 144 (1977) 85.
60 H.J. Schwarz, B.A. Waldman and V. Madrid, *J. Pharm. Sci.*, 65 (1976) 370.
61 E.C. Horning, D.I. Carrol, I. Dzidic, S.-N. Lin, R.N. Stillwell and J.-P. Thenot, *J. Chromatogr.*, 142 (1977) 481.
62 I. Dzidic, D.I. Carrol, R.N. Stillwell, M.G. Horning and E.C. Horning, *Advan. Mass Spectrom.*, 7A (1978) 359.
63 E.C. Horning, personal communication.
64 E.C. Horning, D.I. Carrol, I. Dzidic, R.N. Stillwell and J.P. Thenot, *J. Ass. Offic. Anal. Chem.*, 61 (1978) 1232.
65 F.W. Crow, Jr., *Applications of Negative Ion Chemical Ionization Mass Spectrometry*, Thesis, University of Virginia, 1978, University Microfilms 79 16 266.

Chapter 8

Sensitive derivatives for the determination of organic compounds by electron-capture gas chromatography

COLIN F. POOLE and ALBERT ZLATKIS

CONTENTS

8.1. INTRODUCTION

Gas chromatography is the technique of choice for the separation of thermally stable volatile organic and inorganic compounds. However, many compounds of biomedical and environmental interest, particularly those containing polar functional groups, are thermally labile at the temperatures required for their separation and cannot be analyzed directly by gas chromatography. To permit these compounds to be separated by gas chromatography the technique of derivatization was developed. The principal reasons for preparing derivatives for gas chromatography can be summarized as

(1) to improve the thermal stability of the compound;

(2) to improve the chromatographic performance of the compound by reducing undesirable column interactions (e.g. irreversible adsorption, skew peak formation);

(3) to change the separation properties of the compound by a purposeful adjustment of its volatility (e.g. to eliminate peak overlaps).

To meet the requirement of high sensitivity for trace analysis, the role of derivatization has been broadened in scope to include its use as a means of introducing a detector oriented tag into a molecule as well as providing the means for maintaining the chromatography column. Detector oriented derivatives enable advantage to be taken of the high sensitivity and discriminatory power of the available selective gas chromatography detectors. Thus they provide the means by which the derivatized compound can be quantitatively identified at trace levels in complex mixtures with a high level of discrimination against the organic background matrix.

The principal selective detectors used in gas chromatography include the electron-capture detector, the flame photometric detector (for S and P), the thermionic detector (for N and P) and the Hall detector (for halogens, N and S). Of these, the electron-capture detector (ECD) has been the most widely studied in conjunction with the technique of derivatization due to the availability of the necessary reagents and the high sensitivity of the detector. The availability of such a sensitive detector as the ECD and the need to derivatize many biologically or environmentally important polar molecules prior to gas chromatography has led to a symbiotic relationship between the detector and the development of detector-oriented reagents to take advantage of its special properties.

8.2. ANATOMY OF A DERIVATIZING REAGENT FOR ELECTRON-CAPTURE DETECTION

Any derivatizing reagent for use in gas chromatography with electron-capture detection must meet certain requirements which are outlined in Fig. 8.1. The derivatizing reagent is constructed from two parts, the organic chain which provides the detector sensitivity and the reactive group which provides the means by which the organic chain is attached to the substrate. The choice of the reactive group controls the range of application of the reagent to different functional groups, the selectivity of the reagent towards certain functional groups in the presence of others and the rate and extent of the reaction of the reagent with the substrate. The organic chain as well as providing the detector-oriented sensitivity must also meet certain chromatographic criteria such as having the necessary volatility, thermal stability and chemical stability. The size or shape of the organic chain can also influence the rate of reaction of the reagent by limiting access of the reactive group to sterically hindered functional groups of the substrate.

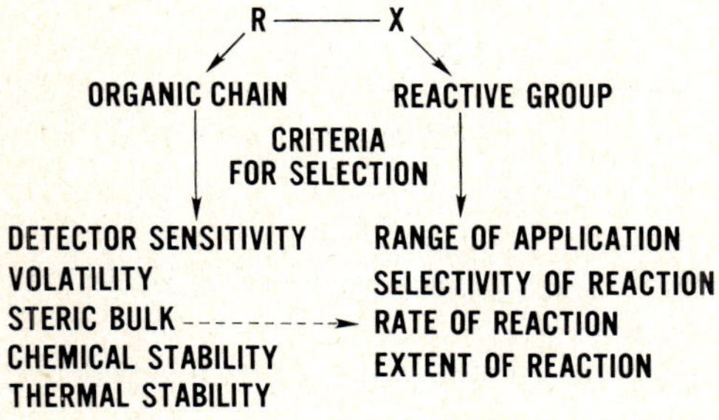

Fig. 8.1. Anatomy of an ECD derivatizing reagent.

8.3. CONSTITUTION OF THE ORGANIC CHAIN FOR A HIGH RESPONSE OF THE ELECTRON-CAPTURE DETECTOR

The response of the ECD to organic molecules is rather ill-defined and covers a wide range (over seven orders of magnitude). The detector has a poor response to alcohols, amines, phenols, aliphatic saturated aldehydes, thioethers, ethers, fatty acid esters, hydrocarbons, aromatics and vinyl type fluorinated hydrocarbons including those containing one chlorine atom (Table 8.1). A high detector response is found among compounds containing electronegative groups such as the halogens or nitro groups as well as some compounds containing unsaturated conjugated systems [1–6].

Clemons and Altshuller [7] have determined the response of the ECD at a fixed detector temperature to a series of fluorocarbons (Table 8.2). The alkyl fluorocarbons have a poor ECD response, only a fraction of that shown by the flame ionization detector. The detector response towards the halogens decreases in the order $I > Br > Cl \gg F$ and increases synergistically with multiple substitution on the same carbon atom. Compounds with a halogen atom attached to a vinyl carbon atom have a lower and compounds with a halogen atom attached to an allyl carbon atom have a higher detector response than the corresponding saturated compounds. The cyclic perfluorocarbon compounds have a higher detector response than their linear saturated analogues of the same carbon number. The above observations are in keeping with the conclusions reached by Devaux and Guiochon [8] for a series of chlorocarbon compounds (Table 8.3), by Castello and D'Amato [9] for some iodoalkanes and by Landowne and Lipsky [10] for a series of isomeric butyl chlorides, bromides and iodides. Although the identity and number of the halogen substituents would seem to be the most important consideration in governing the detector response, subtle variations with small changes in the geometric framework of the alkyl portion have a significant though less dramatic influence on the detector response. An approximate guide to the magnitude of these variations is given in Table 8.4.

The response of the ECD to haloaromatic and nitroaromatic compounds shows similar trends to the alkyl compounds. Zielinski et al. [11] have summarized the response of the ECD to a series of chloronitrobenzenes, anilines and related derivatives (Table 8.5). The chlorobenzene or nitrobenzene compounds have a high detector response increasing with multiple substitution (Cl or NO_2) on the aromatic ring. For the chloronitrobenzene compounds, *meta*-chlorine (relative to NO_2) contributes more to the overall sensitivity of the molecule than *ortho*-chlorine substitution and *para*-chlorine substitution significantly diminishes the response to the detector. The chloroanilines have a much lower detector response than the chloronitrobenzenes. For the chloroanilines, *ortho*-chlorine (relative to NH_2) contributes more to the detector sensitivity than *meta*-chlorine substitution in contrast to the chloronitrobenzene compounds.

As far as the construction of an electrophore to provide the electron-capture detector oriented response to the derivative is concerned, the most generally employed substituents for the organic chain contain alkyl or aryl halogen or nitro groups.

TABLE 8.1

RELATIVE SENSITIVITY (CAPTURE COEFFICIENT) OF VARIOUS ORGANIC COMPOUNDS TO THE ELECTRON-CAPTURE DETECTOR

Chemical classes	K'* scale	Selected examples
Alkanes, alkenes, alkynes, aliphatic ethers, esters and dienes	0.01 – 0.10	Hexane Benzene Cholesterol Benzyl alcohol Naphthalene
Aliphatic alcohols, ketones, aldehydes, amines, nitriles, monofluoro and monochloro compounds	1.0	Vinyl chloride Ethyl acetoacetate Chlorobenzene
Enols, oxalate esters, monobromo, dichloro and hexafluoro compounds	10.0	*cis*-Stilbene *trans*-Stilbene Azobenzene Acetophenone
Trichloro compounds, chlorohydrates, acyl chlorides, anhydrides, barbiturates, thalidomide and alkyl-leads	300	Allyl chloride Benzaldehyde Tetraethyl-lead Benzyl chloride
Monoiodo, dibromo and trichloro compounds, mononitro compounds, lachrymators, cinnamaldehyde, fungistatic compounds and pesticides	1000	Cinnamaldehyde Nitrobenzene Carbon disulfide 1,4-Androstadiene-3,11,17-trione Chloroform
1,2-Diketones, fumarate esters, pyruvate esters, quinones, diiodo, tribromo, polychloro, dinitro compounds and organomercurials	10,000	Dinitrobenzene Diiodobenzene Dimethyl fumarate Carbon tetrachloride

*Values for K' are relative to chlorobenzene, which is arbitrarily given a value of 1.0.

TABLE 8.2

RELATIVE RESPONSE OF THE ELECTRON-CAPTURE DETECTOR TO VARIOUS HALOGEN-CONTAINING COMPOUNDS

All measurements were made with a 300 mC tritium d.c. detector at 100°C.

Compound	Relative response	Compound	Relative response
CF_4	1.0	$CF_3CF{=}CFCF_3$	$1.3 \cdot 10^5$
$CF_3CF_2CF_3$	1.0	$ClF_2C\text{-}CFCl_2$	$1.6 \cdot 10^5$
CF_3Cl	3.3	$(CF_3)_2\text{-}CF_2CF_2CF_2CF_2$ (ring)	$3.0 \cdot 10^5$
CHF_2Cl	10.0	$(CF_3)_2C{=}CF_2$	$3.0 \cdot 10^5$
CH_3CHF_2	30.0	$F_3C\text{-}CHClBr$	$4.0 \cdot 10^5$
$CF_2{=}CFCl$	$1.0 \cdot 10^2$	$CF_3CF_2CF_2I$	$6.0 \cdot 10^5$
CF_3CF_2Cl	$1.7 \cdot 10^2$	$BrF_2C\text{-}CF_2Br$	$7.7 \cdot 10^5$
$CF_2{=}CCl_2$	$6.7 \cdot 10^2$	CH_2Br_2	$1.0 \cdot 10^6$
$ClF_2C\text{-}CF_2Cl$	$6.7 \cdot 10^3$	$CFCl_3$	$1.2 \cdot 10^6$
CF_2Cl_2	$3.0 \cdot 10^4$	CF_2Br_2	$1.7 \cdot 10^6$
$CHCl_3$	$3.3 \cdot 10^4$	CCl_4	$2.2 \cdot 10^6$
$ClHC{=}CCl_2$	$6.7 \cdot 10^4$	C_6F_6	$3.3 \cdot 10^6$
CF_3Br	$8.7 \cdot 10^4$	$IF_2C\text{-}CF_2I$	$4 \cdot 10^6$
$CF_2CF_2CF_2CF_2$ (ring)	$1.2 \cdot 10^5$		

TABLE 8.3

RELATIVE RESPONSE FACTOR FOR SOME HALOCARBON COMPOUNDS DETERMINED WITH A 150 mC TRITIUM PULSED ELECTRON-CAPTURE DETECTOR OPERATED AT 190°C

Compound	Relative response factor	Compound	Relative response factor
Benzene	0.06	1,4-Dichlorobutane	15.00
Toluene	0.20	Chlorobenzene	75.00
Acetone	0.50	*cis*-1,2-Dichloroethylene	90.00
2-Fluorotoluene	0.55	1,1-Dichlorobutane	110.00
4-Fluorotoluene	0.55	1,2-Dichloroethane	190.00
Di-*n*-butyl ether	0.60	1-Bromopropane	255.00
Methylbutyrate	0.90	1-Bromobutane	280.00
1-Butanol	1.00	Bromocyclopentane	280.00
1-Chlorobutane	1.00	*trans*-1,2-Dichloroethylene	370.00
1-Chloropentane	1.00	Bromobenzene	450.00
1-Chlorohexane	1.10	*n*-Propylpentafluoropropionate	450.00
1-Chloroheptane	1.50	1-Bromoprop-2-ene	$4 \cdot 10^3$
1-Chlorooctane	1.60	2,3-Butanedione	$5 \cdot 10^4$
1-Chloro-2-methylpropane	1.70	Chloroform	$6 \cdot 10^4$
2-Chlorobutane	2.00	1-Iodobutane	$9 \cdot 10^4$
n-Heptyltrifluoroacetate	4.50	1,1-Dibromoethane	$1.1 \cdot 10^5$
2-Chloro-2-methylpropane	12.00	Carbon tetrachloride	$4.0 \cdot 10^5$

TABLE 8.4

RELATIONSHIP BETWEEN MOLECULAR STRUCTURE AND RELATIVE CAPTURE COEFFICIENT (SENSITIVITY)

Parameter	K' *
Halogen series	
I	$9 \cdot 10^4$
Br	$3 \cdot 10^2$
Cl	1
F	$\ll 1$
Substitution on carbon atom	
Tertiary	10
Secondary	2
Primary	1
Frequency on carbon atom	
Tetra-	$4 \cdot 10^5$
Tri-	$6 \cdot 10^4$
Di-	10^2
Mono-	1
Positional isomer (di-, tri-, etc.)	
Alpha-	10
Beta-	5
Delta-	1
Geometrical isomer	
trans-	4
cis-	1

*Capture coefficients are relative to the lowest value of series, which is arbitrarily given a value of 1.0.

8.3.1. The influence of substitution in the reagent organic chain on derivative volatility

In terms of response by electron-capture detection, the order among the halogens is $I > Br > Cl \gg F$ which is the reverse order of the volatility of their compounds to gas chromatography. Although compounds containing iodine and bromine are more sensitive to the ECD than fluorine, their greater mass produces less volatile derivatives which may be inconvenient for the analysis of high molecular weight compounds. Closely bound fluorine atoms in an alkyl or aryl compound have a remarkable property, in that they show very little increase in boiling point compared to hydrocarbons containing the same number of carbon atoms in spite of their much higher molecular weight (the increase in molecular weight being offset by a decrease in intermolecular bonding forces in the fluorocarbon) [12]. The high volatility of fluorine-containing compounds enables multiple substitution of fluorine into the organic chain without a large change in volatility (Table 8.6) and also favors an increase in ECD sensitivity, especially if the captured electron can be stabilized by delocalization. The heptafluorobutyryl ester/amide derivatives owe much of their popularity for use in gas chromatography to the volatility conferred by the perfluorocarbon group and the presence of the conjugated carbonyl

TABLE 8.5

RELATIVE RESPONSE FACTORS FOR SOME HALOAROMATIC AND NITROAROMATIC COMPOUNDS DETERMINED WITH A 250 mC TRITIUM ELECTRON-CAPTURE DETECTOR

Detector temperature unknown.

Compound	Relative response
Chloronitrobenzenes	
4-Chloronitrobenzene	1.00
2,4-Dichloronitrobenzene	1.11
3,4-Dichloronitrobenzene	1.13
2-Chloronitrobenzene	1.66
2,4,5-Trichloronitrobenzene	2.10
2,5-Dichloronitrobenzene	2.12
3-Chloronitrobenzene	2.21
2,3-Dichloronitrobenzene	2.34
Fluoronitrobenzenes	
2-Fluoronitrobenzene	0.74
4-Fluoronitrobenzene	0.69
3-Fluoronitrobenzene	0.21
Dinitrobenzenes	
1,3-Dinitrobenzene	1.63
1,2-Dinitrobenzene	1.29
Chloroanilines	
2,4,6-Trichloroaniline	0.0164
2,3-Dichloroaniline	0.0048
2,4-Dichloroaniline	0.0039
2,5-Dichloroaniline	0.0013
2-Chloroaniline	0.0001
3-Chloroaniline	0.00003
Miscellaneous	
1,4-Dichlorobenzene	1.88
2-Nitroaniline	1.63
2-Bromonitrobenzene	1.22
3-Nitroaniline	0.26
2-Nitro-4-chloroaniline	0.21
2-Bromoaniline	0.0017
Fluorobenzene	0.000107
2-Fluoroaniline	0.0000007

group which helps to stabilize the captured electron.

Nitroaromatic compounds in general, often show poor volatility characteristics accompanied by poor peak shape especially when separated on non-polar stationary phases. The use of polar stationary phases improves the peak shape often at the expense of a large increase in retention time. Derivatizing reagents containing aromatic nitro groups are rarely used for the analysis of high molecular weight compounds by electron-capture gas chromatography.

TABLE 8.6

RELATIVE VOLATILITY OF A SERIES OF $RR_1(CH_3)$Si-CHOLESTEROL ETHERS

Determined on a 1.0 m × 2.0 mm I.D. nickel column of 1% OV-101 on Gas-Chrom Q (100–120 mesh), temperature 250°C, nitrogen flow-rate 75 ml/min.

R	R_1	Relative retention time
CH_3	CH_3	1.00
$CF_3(CH_2)_2$	CH_3	1.26
$CF_3(CF_2)_2(CH_2)_2$	CH_3	1.37
$ClCH_2$	CH_3	2.10
C_6F_5	CH_3	3.14
C_6F_5	$CH(CH_3)_2$	4.57
CH_2Br	CH_3	5.13
C_6F_5	CH_2Cl	6.26
C_6F_5	$C(CH_3)_3$	6.30
CH_2I	CH_3	12.82

8.4. TEMPERATURE DEPENDENCE AND MECHANISMS OF THE RESPONSE OF THE ELECTRON-CAPTURE DETECTOR

The response of the ECD is markedly temperature dependent. For an oversimplified picture, one can imagine that an electron absorbing compound (AB) can attach a thermal electron by one of four basic mechanisms

$$AB + e^- \rightleftharpoons AB^- \quad \text{(I)}$$

$$AB + e^- \longrightarrow A^{\cdot} + B^- \quad \text{(II)}$$

$$AB + e^- \longrightarrow AB^- \begin{cases} \longrightarrow A^{\cdot} + B^- & \text{(III)} \\ \longrightarrow A^- + B^{\cdot} & \text{(IV)} \end{cases}$$

Mechanism I represents non-dissociative attachment to produce a stable negative molecular ion. A potential energy curve for such an ion lies below that of the neutral molecule. Thus, on electron attachment this energy difference must be liberated by either radiation or through collisions with other molecules, by energy exchange. An increase in temperature would increase the population of higher vibrational levels, thus decreasing the probability for attachment. Mechanism II represents dissociative attachment to produce a negative ion and a radical in a single step. For this process the potential energy curve would cross that of the neutral molecule, at a level corresponding to a vibrationally excited state. This process would have an energy of activation, which would be that energy necessary to populate those states where the dissociative curve crosses. Thus increasing the temperature increases the probability of attachment, since it serves to increase the population of the excited levels. Mechanisms III and IV also describe a dissociative pathway, wherein an unstable negative ion is first formed which can then dissociate by either of the two paths shown. For a more detailed discussion of the relationship between detector temperature and mechanism the reader is referred to Chapter 3.

To provide an adequate theory of the operation and to predict the sensitivity to different compounds of the ECD, the principal problem has always been a lack of knowledge of the products formed in the detector by electron attachment. These products are expected to include neutral molecules, radicals and positive and negative ions. Two relatively new techniques, atmospheric pressure ionization mass spectrometry (Chapter 14) and plasma chromatography (Chapter 15) have been used to study the product ions formed in an ECD. In related studies, the neutral volatile products formed in an ECD have been rechromatographed on a second gas chromatography column and identified by their retention data [13–15]. For example the neutral products identified from pentachloronitrobenzene and tetrachloronitrobenzene were pentachlorobenzene and tetrachlorobenzene respectively. The products formed by exposing certain compounds to electrons of near thermal energy generated in a corona discharge have been identified by gas chromatography–mass spectrometry [16]. In this case, the nature of the neutral products was used to confirm the mechanism of the electron attachment process. Thus for non-dissociative capture by the molecule AB in the presence of a large excess of AB, the expected products are as follows

$$AB^- + AB \longrightarrow \begin{cases} ABAB + e^- \\ ABA + B^- \\ ABB + A^- \end{cases}$$

Only in the case of associative attachment will the molecular structure of the original molecule be retained and a predominance of products of the form ABAB, ABA, and/or ABA would indicate that non-dissociative electron attachment had occurred. For dissociative attachment, the negative species A^- and B^- can transfer their electrons to another AB by the following reactions

$$A^- + AB \longrightarrow AB^- + A^\bullet$$

$$B^- + AB \longrightarrow AB^- + B^\bullet$$

The radicals $A^\bullet$ and $B^\bullet$ can react with AB as follows

$$A^\bullet + AB \longrightarrow A_2 + B^\bullet$$

$$B^\bullet + AB \longrightarrow B_2 + A^\bullet$$

Therefore, the predominance of products of the type A_2 and B_2 indicates that dissociative electron attachment was the initial step. The reaction products identified from benzaldehyde, diethyl maleate, *cis*-stilbene, chlorobenzene and benzyl chloride are summarized in Table 8.7.

For experimental purposes, a plot of ln $KT^{3/2}$ vs. $1/T$ (K = capture coefficient, T = detector temperature in $^\circ$K) can be used to provide some insight into the detection mechanism and to indicate the optimal detector temperature for maximum response [17]. Idealized plots of ln $KT^{3/2}$ vs. $1/T$ for the four detection mechanisms I–IV are given in Fig. 8.2. Compounds capturing electrons by mechanism I have their highest response at low detector temperatures and are easily differentiated from the other three mechanisms. Compounds capturing electrons by mechanisms II and III are not differ-

TABLE 8.7

PRINCIPAL PRODUCTS FROM THE REACTION OF ELECTRONS WITH NEAR THERMAL ENERGIES AND ELECTRON ABSORBING COMPOUNDS

Electron absorbing compound	Principal products	Mechanism of electron attachment
Benzaldehyde C_6H_5CHO	Biphenyl $(C_6H_5)_2$ Benzophenone $(C_6H_5)_2CO$ Benzil $(C_6H_5CO)_2$	Non-dissociative
Diethyl maleate $C_2H_5OOC\text{-}CH{=}CH\text{-}COOC_2H_5$	Diethyl succinate	Non-dissociative
cis-Stilbene $C_6H_5CH{=}CHC_6H_5$	Bibenzyl $(C_6H_5CH_2)_2$ Deoxybenzoin $(C_6H_5COCH_2C_6H_5)$ 1,2-Diphenylethanol $(C_6H_5CHOHCH_2C_6H_5)$	Non-dissociative
Chlorobenzene C_6H_5Cl	Biphenyl $(C_6H_5)_2$ *m*-Chlorobiphenyl *o*-Chlorobiphenyl *p*-Chlorobiphenyl	Dissociative
Benzyl chloride $C_6H_5CH_2Cl$	Benzyl alcohol Phenol Bibenzyl *cis*- + *trans*-Stilbene	Dissociative

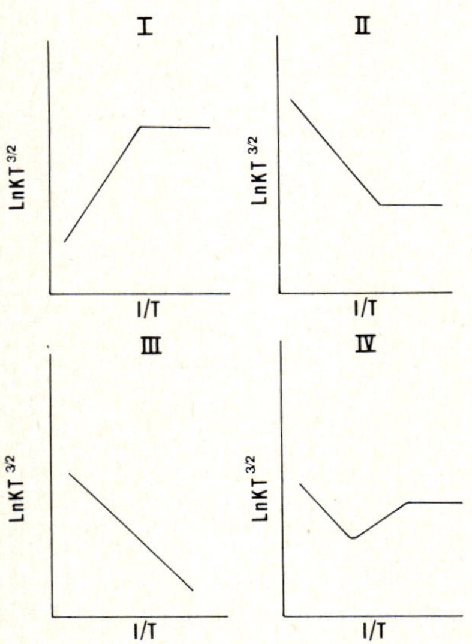

Fig. 8.2. Idealized plots of ln $KT^{3/2}$ vs. $1/T$ for the four basic mechanisms of electron-capture.

entiated from each other by plots of this type as they show similar temperature dependence. They are, however, easily differentiated from compounds which capture electrons by mechanism I. The presence of a region of positive slope for mechanism IV (similar to mechanism I) is taken to be evidence for the formation of a negative molecular ion intermediate. Thus at intermediate and low detector temperatures, the electron-capture process is related to the electron affinity of the molecule and at high temperatures (negative slope) the electron-capture process can be related to the activation energy for dissociation.

For most practical purposes, it is not necessary to calculate K as it is proportional to the peak area of the compound recorded on the chromatogram. Thus the mechanism for electron-capture can also be deduced from a plot of ln $AT^{3/2}$ vs. $1/T$, where A = peak area for a fixed mass of compound injected at the different detector temperatures T (°K). The form of the plots ln $KT^{3/2}$ and ln $AT^{3/2}$ vs. $1/T$ take the same shape and are interpreted in the same way.

Graphically, the effect of detector temperature on the detector response can be more clearly seen by plotting the peak area for a fixed mass of compound against detector temperature. The change in response with detector temperature for the 3,5-bis(trifluoromethyl)benzeneboronate derivative of pinacol [18] and for some heptafluorobutyramide and heptafluorobutyrate derivatives of amines and alcohols [19] are shown in Figs. 8.3 and 8.4 respectively. In the latter study, the change in detector response was studied over

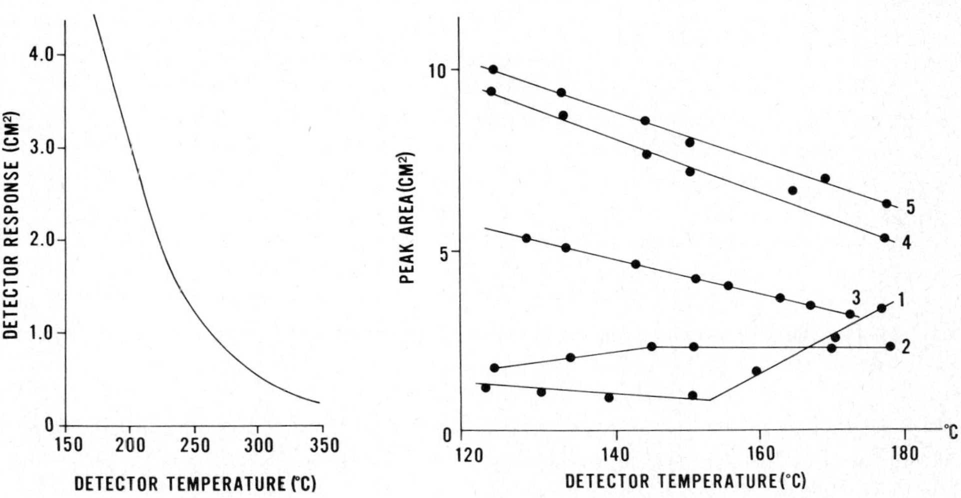

Fig. 8.3. The effect of detector temperature on detector response for the 3,5-bis(trifluoromethyl)-benzeneboronate derivative of pinacol.

Fig. 8.4. Relationship between ECD response and detector temperature for (1) *n*-octanol heptafluorobutyrate, (2) methoxyphenamine heptafluorobutyramide, (3) methamphetamine heptafluorobutyramide, (4) 1-bromonaphthalene, (5) phenmetrazine heptafluorobutyramide.

a narrow detector temperature range and illustrates the importance of maintaining the detector temperature accurately at a fixed value to obtain good precision in analysis. Wentworth and Chen [17] have stated that the temperature of the detector cell should be capable of being set at a specific temperature and controlled to ±0.3°C to obtain a precision of 1.0%.

The effect of detector temperature on detector response to a particular compound is of considerable analytical significance. Although often overlooked by many workers, optimization of detector temperature is very important prior to making a comparison of the detector response to different compounds or derivatives. For the normal range of detector temperatures employed, the change in response of the detector with temperature can vary from zero to over one thousand fold for a given compound (Table 8.8). Failure to optimize the detector temperature provides misleading information as the detection limits measured may be woefully in error [20, 21]. It is also a common misconception that a derivative of one compound establishes the mechanism and optimal temperature for the determination of that derivative of all other compounds. This may be true as a rule of thumb for a single derivative of a homologous series of compounds but is not necessarily true for the same derivative of different compounds.

TABLE 8.8

THE INFLUENCE OF DETECTOR TEMPERATURE ON THE RESPONSE OF THE ELECTRON-CAPTURE DETECTOR TO SOME HALOCARBON COMPOUNDS

Compound	Detection limit ($\times 10^{-9}$ g)		
	Detector temperature (°C)		
	80	227	350
CCl_4	0.01	0.01	0.01
$CHCl_3$	1.0	0.1	0.05
CH_2Cl_2	1000	40	8
CH_2ClCH_2Cl	1000	20	1

8.4.1. Experimentally verified mechanisms of electron-capture for some common derivatives

In a comprehensive study, Zlatkis and co-workers [22, 23] have determined the mechanism of electron-capture for seven electron-capturing derivatives of *n*-hexanol (Fig. 8.5), cyclohexanol (Fig. 8.6), phenol (Fig. 8.7), *n*-hexylamine (Fig. 8.8), cyclohexylamine (Fig. 8.9), and aniline (Fig. 8.10). For *n*-hexanol and cyclohexanol, the various derivatives behave identically with both alcohols as is evidenced by the ln $KT^{3/2}$ vs. $1/T$ plots. The pentafluoropropionates, heptafluorobutyrates, chloroacetates and the chlorodifluoroacetates capture by a dissociative mechanism. For the chloroacetate and the chlorodifluoroacetate derivatives, dissociative capture is most probably occurring by the breaking of the carbon–chlorine bond (electron-capture by mechanism II). Due to

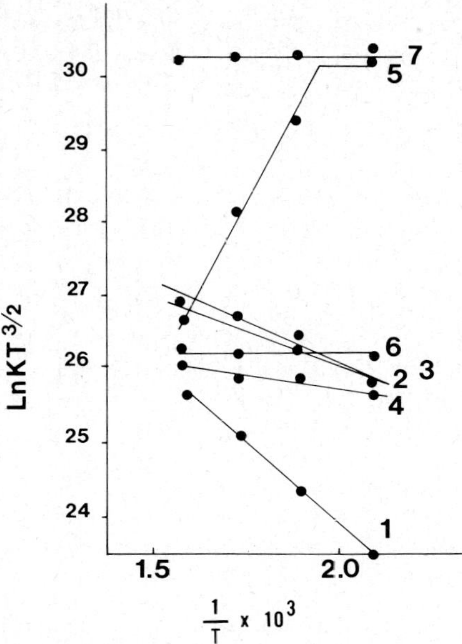

Fig. 8.5. Plots of ln $KT^{3/2}$ vs. $1/T$ for the (1) pentafluoropropionate, (3) heptafluorobutyrate, (3) chloroacetate, (4) chlorodifluoroacetate, (5) pentafluorobenzoate, (6) pentafluorophenacetate, (7) pentafluorophenoxyacetate of n-hexanol.

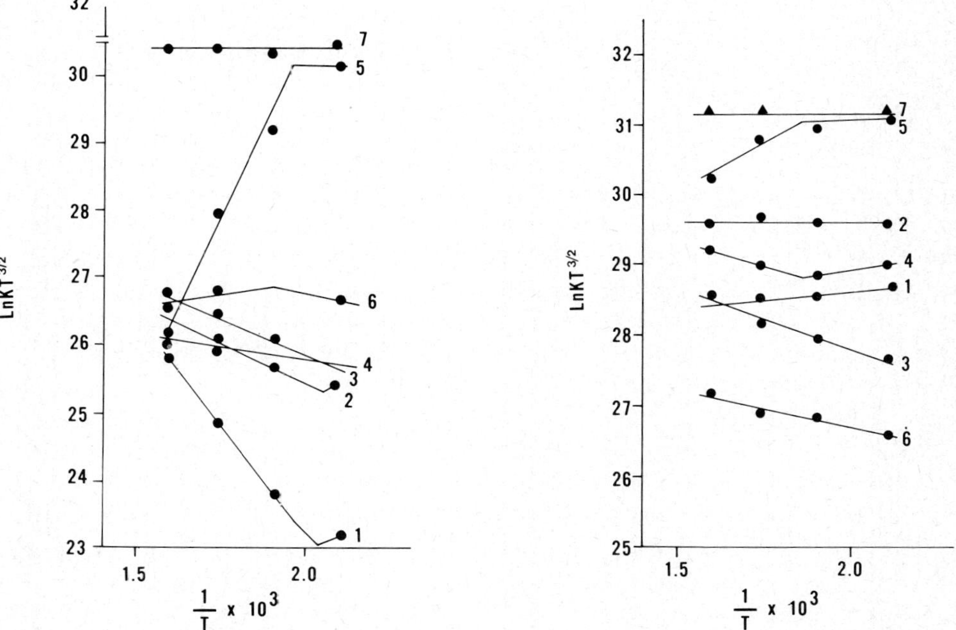

Fig. 8.6. Plots of ln $KT^{3/2}$ vs. $1/T$ for some acetyl derivatives of cyclohexanol. For the identification of the derivatives see Fig. 8.5.

Fig. 8.7. Plots of ln $KT^{3/2}$ vs. $1/T$ for some acetyl derivatives of phenol. For the identification of the derivatives see Fig. 8.5.

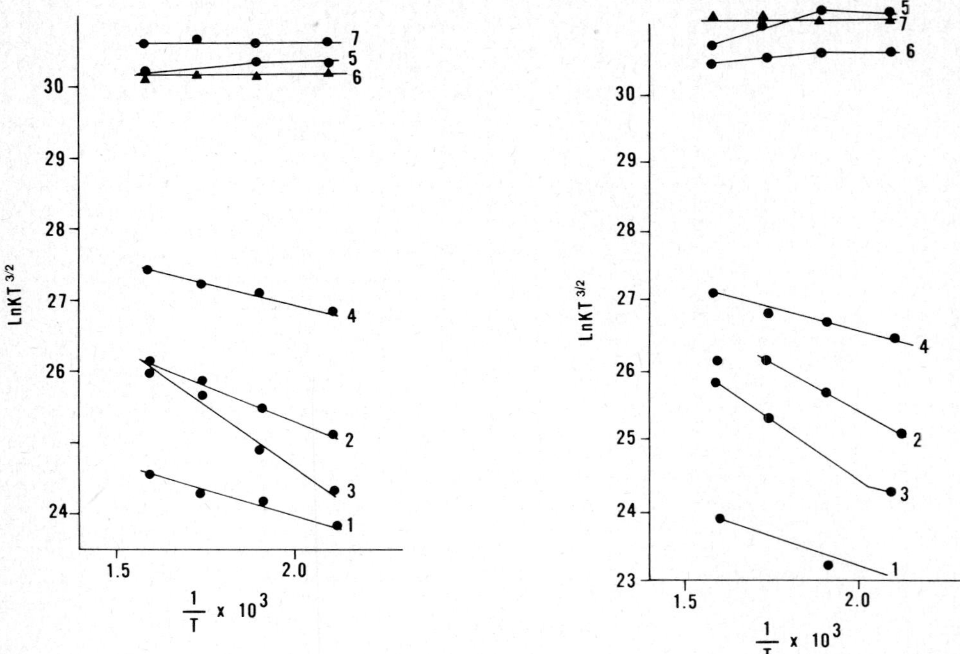

Fig. 8.8. Plots of ln $KT^{3/2}$ vs. $1/T$ for some acetamide derivatives of *n*-hexylamine. For the identification of the derivatives see Fig. 8.5

Fig. 8.9. Plots of ln $KT^{3/2}$ vs. $1/T$ for some acetamide derivatives of cyclohexylamine. For the identification of the derivatives see Fig. 8.5.

the high energy of the carbon–fluorine bond, dissociation in the pentafluoropropionates and heptafluorobutyrates most probably occurs to give the propionate and butyrate radicals respectively (electron-capturing by mechanism IV). The pentafluorobenzoate derivative shows simple non-dissociative electron-capture (mechanism I). The pentafluorophenacetyl and pentafluorophenoxyacetyl derivatives have a slope that is virtually zero. For the pentafluorophenoxyacetates, the value of the electron-capture coefficient increases with decreasing detector temperature suggesting a non-dissociative pathway for electron attachment. In contrast, the pentafluorophenacetates appear to be genuinely capturing by a dissociative pathway. Such behavior could be explained if dissociation occurred by mechanism IV to give the pentafluorophenacetate radical. The correct choice of mechanism would have to be based upon a knowledge of the electron affinities of the resulting radicals and upon the dissociation energies of the bonds involved in their formation. The pentafluoropropionate and heptafluorobutyrate derivatives of phenol capture electrons by a non-dissociative pathway (mechanism I) and the pentafluorophenacetyl derivative by a dissociative mechanism. Otherwise the phenol derivatives investigated show a similar mechanism of electron-capture to the cyclohexanol derivatives. As for the alcohols, mechanistically the alkyl amine derivatives show a similar temperature dependence for both amines. The heptafluorobutyramides and pentafluoropropionamides capture electrons by a dissociative mechanism most probably by formation of the pentafluoropropionamido and heptafluorobutyramido radicals. The chloroacetyl and chlorodifluoroacetyl derivatives are both dissociative capturers. Since the carbon–nitrogen bond is weaker than the carbon–oxygen bond in the alcohol derivatives, it is possible that dissociation occurs in this case to give the corresponding amido radical rather than by breaking of the carbon–chlorine bond. The mechanism of electron-capture for the pentafluorobenzyl and pentafluorophenoxyacetyl derivatives is non-dissociative (mechanism I) as was found for the alcohol derivatives, the only difference being that the magnitude of the temperature variation for the pentafluorobenzylamine derivatives is very much reduced compared to the alcohol derivatives. As with phenol, the aniline derivatives are characterised by clear changes in mechanism compared to the alkyl compounds and exhibit an increase in detector response. With the exception of the chloroacetamide and the chlorodifluoroacetamide derivatives (dissociative, mechanism II), the other derivatives show a non-dissociative electron-capture mechanism (mechanism I).

Gyllenhaal and Hartvig [24] have determined the temperature dependence of the ECD response to some alkylated and acylated N-ethylbenzenesulfonamide derivatives (Fig. 8.11).

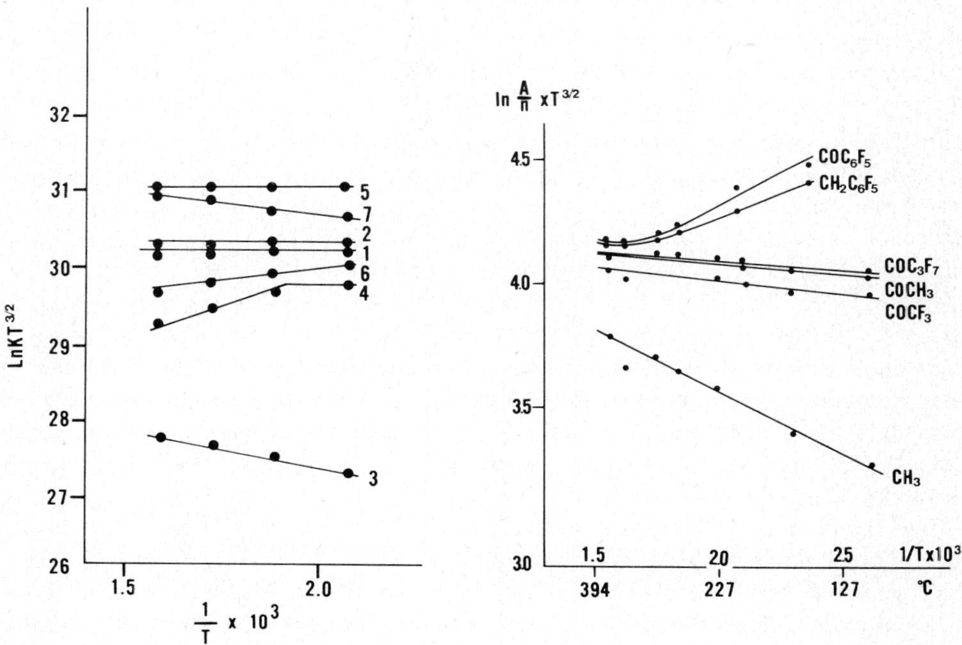

Fig. 8.10. Plots of ln $KT^{3/2}$ vs. $1/T$ for some acetamide derivatives of aniline. For the identification of the derivatives see Fig. 8.5.

Fig. 8.11. Temperature dependence of the ECD response for N-ethylbenzenesulfonamide derivatives.

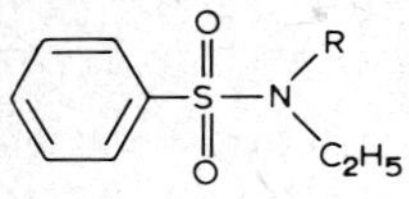

$C_6H_5SO_2N(R)C_2H_5$

When $R = CH_3$, $COCF_3$, $COCH_3$, or COC_3F_7 the sulfonamide derivatives capture electrons by a dissociative mechanism. For the derivatives in which $R = CH_2C_6H_5$ or COC_6F_5, the electron-capture mechanism was non-dissociative (mechanism I).

The flophemesyl, ISP-flophemesyl and CM-flophemesyl derivatives of *n*-octanol show a dissociative mechanism of electron-capture detection (Fig. 8.12) [25]. The pentafluorophenyl group forms a stable anion and is a good leaving group in nucleophilic reactions in solution. Also, the silicon–oxygen bond is much stronger than the silicon–carbon bond. As the pentafluorophenyl group can accommodate a negative charge much more readily than CH_3 or OR, the electron-capture mechanism has been represented diagramatically as shown below [26].

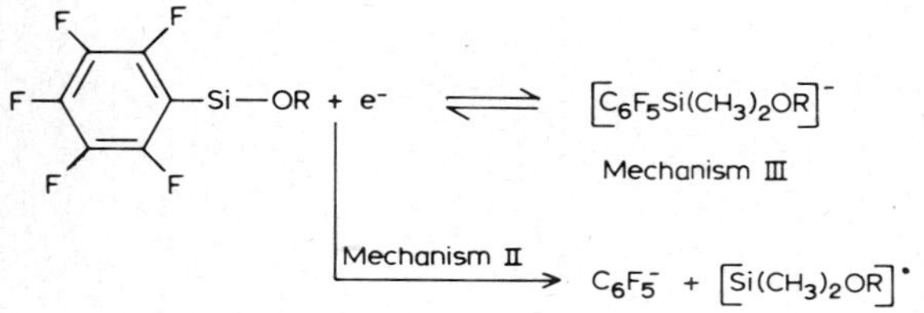

$$C_6F_5\text{–Si–OR} + e^- \rightleftharpoons [C_6F_5Si(CH_3)_2OR]^- \quad \text{Mechanism III}$$

$$\xrightarrow{\text{Mechanism II}} C_6F_5^- + [Si(CH_3)_2OR]^\bullet$$

The CM-flophemesyl derivative shows a similar mechanism of electron-capture to the flophemesyl and ISP-flophemesyl derivatives but in this case two centers for electron attachment are available, the pentafluorophenyl group and the chloromethyl group. Without a knowledge of the products formed, it is impossible to dinstinguish between these two centers and it is likely that an electron can be captured to different extents by both centers. The *tert.*-buflophemesyl derivative of octanol (Fig. 8.12) shows regions of both dissociative and non-dissociative electron-capture and is typical of a type IV mechanism involving the formation of an unstable negative molecular ion which undergoes dissociation to produce a radical and a negative ion at high detector temperatures. Compared to the flophemesyl derivative, the high electron affinity of the *tert.*-butyl radical may favor dissociation of the silicon–*tert.*-butyl bond as opposed to the silicon–pentafluorophenyl bond.

Boronic acids with electron-capturing properties have been prepared for the selective and sensitive analysis by gas chromatography of bifunctional compounds [27]. The mechanism of their respoinse to the electron-capture detector has been studied for a series of pinacol derivatives (Figs. 8.13 and 8.14) [18, 28]. The 4-bromobenzeneboronate (Fig. 8.13A), 2,4-dichlorobenzeneboronate (Figs. 8.13C, 8.14C), 3,5-dichlorobenzeneboronate (Figs. 8.13D, 8,14A), 2,6-dichlorobenzeneboronate (Fig. 8.14B) and 2,4,6-trichlorobenzeneboronate (Fig. 8.14D) capture electrons by a dissociative mechanism as would be predicted for compounds of this type. Of note is the moderate electron-capturing ability of the benzeneboronate derivative by a non-dissociative mechanism (Fig. 8.13E) and the poor electron-capturing properties of naphthaleneboronate by a

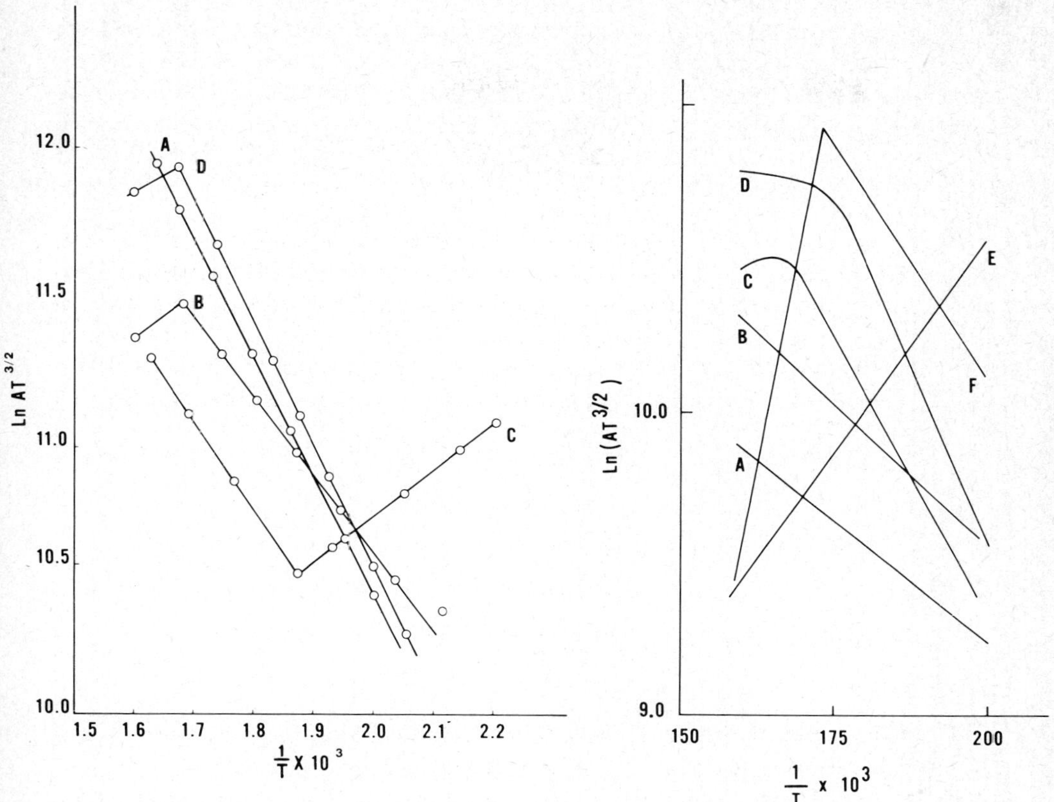

Fig. 8.12. Temperature dependence of the response of the electron-capture detector towards the (A) flophemesyl, (B) ISP-flophemesyl, (C) *tert.*-buflophemesyl, (D) CM-flophemesyl derivatives of *n*-octanol.

Fig. 8.13. Plots of ln $AT^{3/2}$ vs. $1/T$ for (A) 4-bromobenzeneboronate, (B) naphthaleneboronate, (C) 2,4-dichlorobenzeneboronate, (D) 3,5-dichlorobenzeneboronate, (E) benzeneboronate and (F) 3-nitrobenzeneboronate derivatives of pinacol.

dissociative mechanism (Fig. 8.13B). This should be contrasted with the fact that benzene is virtually transparent to thermal electrons and that naphthalene captures electrons weakly in a non-dissociative manner. Thus, the attachment of the boronate group to the aromatic ring introduces electron-capturing properties into the molecule which can override the expected mechanism of electron-capture if the boronate group has no influence on the detector response. The 3,5-bis(trifluoromethyl)benzeneboronate derivative (Fig. 8.14E) captures electrons by a non-dissociative type mechanism. The 3-nitrobenzeneboronate derivative (Fig. 8.13F) shows regions of both dissociative and non-dissociative capture with a narrow detector temperature plateau for maximum response. However, this mechanism is not of the type IV variety as it shows non-dissociative behavior at high detector temperatures and dissociative capture at low detector temperatures.

Obviously at the molecular level the interplay among the electronic energy levels available to the attached electron is very complex and would require further work to elucidate the mechanism of electron-capture in this case. The 4-iodobutaneboronate and 4-chlorobutaneboronate derivatives of pinacol have also been investigated and found to be electron-capturing by a dissociative mechanism. The 2,4-dichlorobenzeneboronate derivative of guaifenesin was shown to be electron-capturing by a dissociative mechanism and the 3,5-bis(trifluoromethyl)benzeneboronate derivative by a non-dissociative mechanism [30].

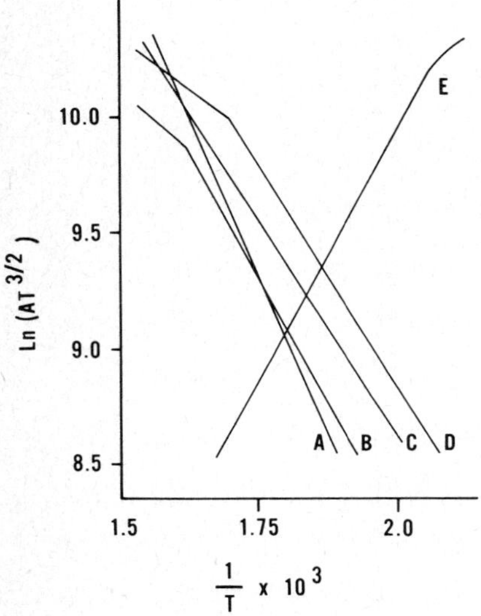

Fig. 8.14. Plots of ln $AT^{3/2}$ vs. $1/T$ for the pinacol derivatives of (A) 3,5-dichlorobenzeneboronic acid, (B) 2,6-dichlorobenzeneboronic acid, (C) 2,4-dichlorobenzeneboronic acid, (D) 2,4,6-trichlorobenzeneboronic acid and (E) 3,5-bis(trifluoromethyl)benzeneboronic acid.

8.4.2. Speculative mechanisms of electron-capture for some common derivatives

Several interpretations have been put forward to explain the response of the electron-capture detector to certain groups of derivatives. Generally, only the response of the derivative to the detector has been measured so that these explanations must be considered to be speculative as they lack any supporting evidence from temperature dependence studies or other physical measurements.

The order of response to the electron-capture detector for a series of cholesterol haloacetate derivatives was found to be chloroacetate > dichloroacetate > bromoacetate > trichloroacetate > trifluoroacetate (Table 8.9) [31]. This is not the order of response that would be expected if electron attachment occurred at the halogen carbon atom and led the authors to speculate that electron attachment must be occurring at the carbonyl carbon of the acetate group. Although the unsubstituted acetate does not exhibit any

TABLE 8.9

ELECTRON-CAPTURE DETECTOR RESPONSE TO SOME CHOLESTEROL HALOACETATE DERIVATIVES

Tritium d.c. detector operated at 200°C.

Derivative	Sensitivity (mole$\cdot$s^{-1})
Cholesterol chloroacetate	$1.25\cdot10^{-15}$
Cholesterol dichloroacetate	$8.45\cdot10^{-15}$
Cholesterol bromoacetate	$1.11\cdot10^{-14}$
Cholesterol trichloroacetate	$2.26\cdot10^{-14}$
Cholesterol trifluoroacetate	$1.35\cdot10^{-13}$

significant electron affinity under the experimental conditions employed, the α-halogenated derivatives are capable of promoting electron attachment by virtue of the influence of the halogen atom on the carbonyl double bond which tends to increase its polarity. However, this effect cannot be due to an electron shift from the α-carbon atom toward the carbonyl group along their common bond in the usual manner, for this would yield the reverse effect. The observed phenomena can therefore only come about by the direct influence of the electron-attracting halogen on the carbonyl double bond across intramolecular space. This occurs when the halogen atom is in the *cis*-position with respect to the carbonyl oxygen as noted in structure I.

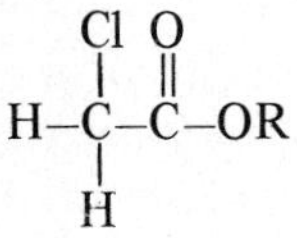

```
  Cl O              H  O
  |  ||             |  ||
H-C--C-OR         H-C--C-OR
  |                 |
  H                 Cl

  (I)               (II)
```

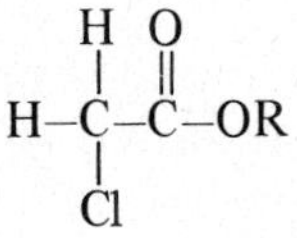

Although it should be emphasized that free rotation is occurring in the gas phase, of the two rotational isomers (I and II) the *cis*-configuration could be present long enough for interaction to occur and permit electron attachment at the carbonyl carbon (III). Once this has happened, the more thermodynamically favored rotational isomer (IV) could stabilize the resulting radical-ion by a similar interspatial effect.

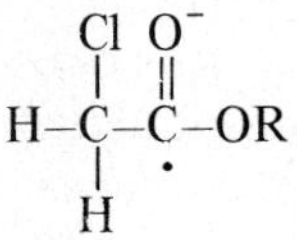

```
  Cl O⁻             H  O⁻
  |  ||             |  |
H-C--C-OR         H-C--C-OR
  |  •              |  •
  H                 Cl

  (III)             IV)
```

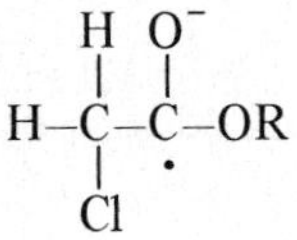

For the bromoacetate, the interspatial effect on the carbonyl group is comparatively less than the chloroacetate because bromine has a lower attraction than chlorine for electrons and also the bulkier bromine atom tends to limit the formation of the *cis*-configuration.

For the di- and trichloroacetate derivatives, increasing the number of chlorine atoms should inherently enhance the chances of having a chlorine atom in the favorable *cis*-position at any given time. It would appear that an opposing force is also operational in this case involving the hindrance to electron attachment resulting from a chlorine atom in a gauche configuration to the carbonyl oxygen atom. For the trichloroacetate derivative, no rotational isomerism can exist and it is more probable that electron attachment occurs at the α-carbon atom, but the overall electron affinity is markedly reduced because of the opposing electron attachment of the carbonyl group. The dichloroacetate derivative occupies the usual intermediate picture between the chloroacetate and trichloroacetate derivatives. The polarity of the carbonyl group can be increased by the chlorine atom in the *cis*-position (V and VI) but not to the extent noted with the chloroacetate since the presence of a second chlorine atom in an opposing position (VI and VII) may possibly produce a partial reversal of this effect.

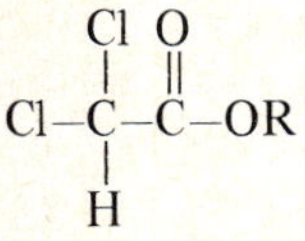

```
    Cl O              Cl O              H  O
    |  ||             |  ||             |  ||
 Cl-C--C-OR        H--C--C-OR       Cl--C--C-OR
    |                 |                 |
    H                 Cl                Cl
```

(V) favorable 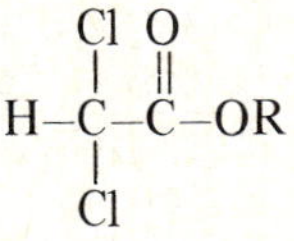 (VI) semifavorable 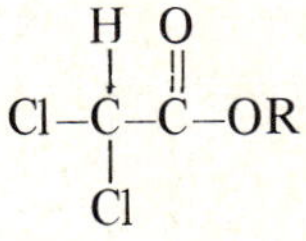 (VII) unfavorable

Based on the above assumptions, the authors predicted that the monofluoroacetate derivative should evoke the highest overall response from the ECD since its polarizing effect on the carbonyl group would be greater than that of chlorine. This was not found to be the case with the monofluoroacetate derivative of thymol which was much less sensitive than the monochloroacetate derivative [32]. In reviewing the sensitivity of the ECD to a series of haloacetate derivatives of testosterone, Dehennin et al. [33] concluded that the balance of probability favored electron-capture at the halogen carbon atom rather than the carbonyl carbon as discussed above for the cholesterol haloacetates.

Martin and Rowland [34] have compared the response of the electron-capture detector to various derivatives of pharmaceutically important amines. The response of the ECD to eight derivatives of amphetamine is summarized in Table 8.10. Reduction of the carbonyl group of amphetamine heptafluorobutyramide to the corresponding N-heptafluorobutyl derivative resulted in a markedly lower detector response, while formation of the Schiff base derivative, which contains the polarizable C=N group, produced little loss of sensitivity. Similar trends were observed for the pentafluorobenzyl derivatives. These observations tend to support the hypothesis that electron-capture occurs primarily at the carbonyl group rather than the side chain. The pentafluorobenzamides always exhibited a greater response (60–200 fold) compared to the corresponding heptafluorobutyramides when derived from primary amines. While the initial electron attachment at the carbonyl group remains the same in the two derivatives, the highly electronegative pentafluorophenyl ring can resonate with the carbonyl group to provide a coplanar highly electron-delocalized system. Also, once attached, an electron can be stabilized in the pentafluorobenzene ring or by the molecule as a whole in the excited state through a non-dissociative electron-capture process. Such a stabilization is far less likely in the heptafluorobutyl group. To explain the relatively poor sensitivity of some chloroacetamide

TABLE 8.10

COMPARISON OF THE RESPONSE OF THE ELECTRON-CAPTURE DETECTOR TO VARIOUS DERIVATIVES OF AMPHETAMINE

Tritium detector, detector temperature 180–195°C.

Derivative	Structure	Response (Coulomb·10^3·$mole^{-1}$)
N-Pentafluorobenzamide	$C_6F_5\text{-}C(=O)\text{-}NHR$	225
N-Pentafluorobenzylidine	$C_6H_5CH{=}NR$	43
N-*p*-Nitrobenzamide	$O_2NC_6H_4C(=O)\text{-}NHR$	11
N-Pentafluorobenzylamine	$C_6F_5CH_2NHR$	3.7
N-2,4-Dinitroaniline	$(O_2N)_2C_6H_3NHR$	2.8
N-Heptafluorobutyramide	$F_3CCF_2CF_2\text{-}C(=O)\text{-}NHR$	2.2
N-Heptafluorobutyrylidine	$F_3CCF_2CF_2\text{-}CH{=}NR$	2.1
N-Heptafluorobutylamine	$F_3CCF_2CF_2CH_2NHR$	0.41

derivatives, Clarke et al. [35] have assumed that electron attachment occurs at the carbonyl group but is destabilized by a contribution to the resonance form by the loan pair of electrons on nitrogen. This contention is further supported by the increased sensitivity of aromatic amines in which the electron density on the amino nitrogen is lowered by the electron-withdrawing ability of the benzene ring. Compared to the chloroacetamide and the trifluoroacetamide derivatives a significantly higher detector response was observed for the N-pentafluoropropionamide and N-heptafluorobutyramide derivatives. It was assumed that the increased detector sensitivity in the perfluorocarbon acetamide derivatives was due to electron-capture at carbon atoms number 2, 3, or 4 in the perfluorocarbon chain which are far removed from the effect of the loan pair of electrons on nitrogen.

8.5. A COMPARISON OF ELECTRON-CAPTURING DERIVATIVES IN TERMS OF DETECTOR SENSITIVITY

The comparison of derivatives in terms of response to the ECD is a noble task and one important aspect of the optimization of an analytical scheme for the determination of a particular compound. However, the examples selected for illustration in this section should be considered to be advisory only, as rarely were all derivatives determined under optimal detector conditions. It has become common practice to make all measurements under the same set of operating conditions. This ignores the fact that the detector response is very much dependent on the selected operating parameters and can vary over several orders of magnitude for the same derivative determined under different conditions. The importance of optimizing the detector temperature in this respect was commented on earlier. The criterion for selection of a particular derivative in terms of sensitivity must be a comparison of the various derivatives under optimized detector conditions; this criterion is rarely met in the majority of studies published in the current scientific literature.

Exley and Dutton [36] have compared the response of the ECD to a wide range of derivatives of estradiol, estratriol, and estrone (Table 8.11). The greatest response was observed with the iodomethyldimethylsilyl and 2-(iodomethyldimethylsiloxy)propyl

TABLE 8.11

THE ELECTRON-CAPTURE DETECTOR RESPONSE TO VARIOUS DERIVATIVES OF 17β-ESTRADIOL AND ESTRONE

Pulsed tritium detector at 215°C.

Compound	Detection limit ($\times 10^{-12}$ g)
Estrone derivatives	
3-(perfluoro-3,5,7,8-tetrachlorooctanoic) ester*	10
3-(perfluorooctanoic) ester	10
3-(iodomethyldimethylsilyl) ester	50
3-methyl ester 17-pentafluorophenylhydrazone	150
3-(hexa-2:2,3:3,4:4-fluoro-5-oxo-6-thianonate)**	150
3-nonochloroacetate	300
3-dichloroacetate	500
Estradiol derivatives	
3-[2-(iodomethyldimethylsiloxy) propyl ether] 17-iodomethyldimethylsilyl ether	4
3-[1-(iodomethyldimethylsiloxy) ethyl ether] 17-iodomethyldimethylsilyl ether	4
3,17-di(iodomethyldimethylsilyl) ether	6
3-(*p*-iodobenzenesulfonyl) ester*** 17-heptafluorobutyric ester	10
3,17-di(heptafluorobutyric) ester	12
3-(*p*-iodobenzenesulphonyl) ester***	30
3-methyl ether, 17-iodomethyldimethylsilyl ester	50
3,17-di(hexafluoroglutaryl) ester	150
3-pentafluorophenyl ester	200
3-(3,5-diiodobenzoyl) ester***	200
3-(3:4:5-triiodobenzyl) ester***	300
3,17-bis(monochloroacetate)	300
3,17-bis(dichloroacetate)	500
Estriol derivatives	
3,16,17-triiodomethyldimethylsilyl ether	1
3,16,17-triheptafluorobutyric ester	10

```
     F  Cl F  Cl F  Cl F  O
     |  |  |  |  |  |  |  ||
*Cl–C––C––C––C––C––C––C––C–O–
     |  |  |  |  |  |  |  |
     F  F  F  F  F  F  F
```

```
                          O                  O
                          ||                 ||
**CH3–CH2–CH2–S–C–CF2–CF2–CF2–C–O–
```

***Pulsed ^{63}Ni detector at 320°C.

ether derivatives, the latter being preferred for analysis due to its greater hydrolytic stability to thin-layer chromatography. The iodinated benzenesulfonyl and benzoyl esters of 17β-estradiol were all less sensitive than the diheptafluorobutyrate. Unexpectedly, it was found that increasing the number of iodine atoms on the benzene ring lowers the response to the ECD. Most of the other derivatives tested had a lower response than either the heptafluorobutyrate or iodomethyldimethylsilyl ether derivatives.

Dehennin et al. [33] have prepared an extensive range of haloacetate derivatives of testosterone and compared their response to the ECD relative to that of the bis(heptafluorobutyrate) derivative (Table 8.12). As can be seen, under the selected conditions, the response to the detector covers about six orders of magnitude. The acetate and trifluoroacetate derivatives being the least sensitive, the replacement of one fluorine atom by chlorine in the chlorodifluoroacetate derivative increases the detector sensitivity by about

TABLE 8.12

RELATIVE MOLAR ELECTRON-CAPTURE DETECTOR SENSITIVITY OF SOME HALOACETATE DERIVATIVES OF TESTOSTERONE

Tritium (200 mC) detector, d.c., detector temperature 215°C.

Position of the derivative		Molar electron-capture detector response
3	17	
Acetate	Acetate	$2 \cdot 10^{-4}$
Trifluoroacetate	Acetate	$2 \cdot 10^{-4}$
Trifluoroacetate	Trifluoroacetate	3
Acetate	Heptafluorobutyrate	5
Chlorodifluoroacetate	Acetate	15
Trifluoroacetate	Heptafluorobutyrate	19
Pentafluoropropionate	Acetate	31
Chlorodifluoroacetate	Heptafluorobutyrate	34
Heptafluorobutyrate	Acetate	44
Heptafluorobutyrate	Chloroacetate	46
Heptafluorobutyrate	Trifluoroacetate	46
Chlorodifluoroacetate	Chlorodifluoroacetate	55
Pentafluoropropionate	Pentafluoropropionate	60
Heptafluorobutyrate	Heptafluorobutyrate	66
Heptafluorobutyrate	Pentafluoropropionate	93
Heptafluorobutyrate	Heptafluorobutyrate	100
Heptafluorobutyrate	Chlorodifluoroacetate	106
Heptafluorobutyrate	Pentadecafluorooctanoate	115
Heptafluorobutyrate	Pentafluorobenzoate	127
	Acetate	$1 \cdot 10^{-4}$
	Trifluoroacetate	$4 \cdot 10^{-4}$
	Chloroacetate	4
	Pentafluoropropionate	5
	Heptafluorobutyrate	19
	Chlorodifluoroacetate	34
	Pentafluorobenzoate	50
	Pentadecafluorooctanoate	60

five orders of magnitude which in turn is about an order of magnitude more sensitive than the chloroacetate derivative. The order of sensitivity for the perfluoroacetates was trifluoroacetate < pentafluoropropionate < heptafluorobutyrate < pentadecafluoro-octanoate < pentafluorobenzoate. Similar, although by no means quantitatively identical data were obtained by Nakagawa et al. [37] (Table 8.13) and by Wilkinson [38] (Table 8.14) for some haloacetate derivatives of testosterone and β-phenylethylamine respectively. Increasing the length of the alkyl perfluorocarbon chain from the trifluoroacetate to the pentadecafluorooctanoate increases the derivative response by a little over two orders of magnitude. For the perfluorocarbon acetamide derivatives, there was an increase in response of about 125 fold between the trifluoroacetamide and the pentafluoropropionamide derivatives, a five fold increase between the pentafluoropropionamide and heptafluorobutyramide derivatives and a further three fold increase between the heptafluorobutyramide and pentadecafluorooctamide derivatives. The effect of increasing the length of the perfluorocarbon acyl chain on detector response is clearly shown for some steroid

TABLE 8.13

RELATIVE RESPONSE OF THE ELECTRON-CAPTURE DETECTOR TO SOME HALOACETATE DERIVATIVES OF TESTOSTERONE

Tritium detector, pulsed modulated, detector temperature 205°C.

Testosterone derivative	Relative detector response	
	FID	ECD
Monochloroacetate	0.99	0.20
Monochlorodifluoroacetate	1.02	1.41
Trifluoroacetate	1.00	0.01
Pentafluoropropionate	1.04	0.28
Heptafluorobutyrate	1.00*	1.00**
Pentadecafluorooctanoate	1.08	2.10
Bis(heptafluorobutyrate)	1.11	2.30

*2.7 Coulomb·mole^{-1}.
**1.18·10^{3} Coulomb·mole^{-1}.

TABLE 8.14

RELATIVE RESPONSE OF THE ELECTRON-CAPTURE DETECTOR TO A SERIES OF HALO-ACETAMIDE DERIVATIVES OF β-PHENYLETHYLAMINE

Tritium detector, detector temperature 210°C.

Derivative	Response ECD/FID (Coulomb · mole^{-1})
Trifluoroacetamide	2.8
Pentafluoropropionamide	350
Heptafluorobutyramide	1.57·10^{3}
Pentadecafluorooctanamide	4.82·10^{3}
Pentafluorobenzamide	8.30·10^{4}

derivatives in Table 8.15 [39]. Increasing the number of fluorine atoms beyond five leads to a more gradual increase in detector response than that observed between the trifluoroacetate and the pentafluoropropionate derivatives. Thus the hexadecafluorononanoate and eicosafluoroundecanoate derivatives show a similar detector response which is only 2–3 times greater than that of the heptafluorobutyrate. It is very noticeable with the derivatives of estradiol that virtually all the electron affinity of the molecule is associated with derivatization of the phenolic group as derivatization of the alcohol group produces a poor detector response.

The responses of a variety of amide and Schiff base derivatives of β-phenylethylamine have been compared under identical detector operating conditions. The results are summarized in Table 8.16 [40]. The non-halogen substituted amide and Schiff base deriva-

TABLE 8.15

RELATIVE RESPONSE OF THE ELECTRON-CAPTURE DETECTOR TO SOME PERFLUOROACYL STEROID DERIVATIVES

Tritium detector operated at 225°C.

Derivative	Testosterone (17-mono)	Estradiol (17-mono)	Estradiol (3,17-di)
Monochloroacetate	0.22		
Heptafluorobutyrate	1.00	0.03	2.65
Pentadecafluorooctanoate	1.82	0.05	2.74
Hexadecafluorononanoate	2.30	0.08	3.34
Eicosafluoroundecanoate	2.55	0.07	3.00

TABLE 8.16

RELATIVE RESPONSE OF THE ELECTRON-CAPTURE DETECTOR TO A SERIES OF ACETAMIDE AND SCHIFF BASE DERIVATIVES OF β-PHENYLETHYLAMINE

Detector, ^{63}Ni (10 mC), pulsed, detector temperature 300°C.

Derivative	Sensitivity (mole • s^{-1})	
	FID	ECD
Free base	$4.2 \cdot 10^{-13}$	$2.1 \cdot 10^{-12}$
Acetamide	$4.2 \cdot 10^{-13}$	$1.1 \cdot 10^{-11}$
Trifluoroacetamide	$4.7 \cdot 10^{-13}$	$1.1 \cdot 10^{-14}$
Pentafluoropropionamide	$2.6 \cdot 10^{-13}$	$1.2 \cdot 10^{-15}$
Heptafluorobutyramide	$3.4 \cdot 10^{-13}$	$2.2 \cdot 10^{-16}$
N-(2,4-Dinitrophenyl) amine	$7.1 \cdot 10^{-13}$	$4.3 \cdot 10^{-16}$
Acetone Schiff base	$4.0 \cdot 10^{-13}$	$7.1 \cdot 10^{-13}$
Benzaldehyde Schiff base	$2.4 \cdot 10^{-13}$	$3.1 \cdot 10^{-12}$
Trifluoroacetone Schiff base	$3.5 \cdot 10^{-13}$	$1.6 \cdot 10^{-13}$
Heptafluorobutyraldehyde Schiff base	$6.4 \cdot 10^{-13}$	$3.1 \cdot 10^{-16}$
Perfluorooctanaldehyde Schiff base	$5.2 \cdot 10^{-13}$	$6.0 \cdot 10^{-17}$
Pentafluorobenzaldehyde Schiff base	$2.4 \cdot 10^{-13}$	$9.1 \cdot 10^{-17}$
p-Chlorobenzaldehyde Schiff base	$4.5 \cdot 10^{-13}$	$6.3 \cdot 10^{-15}$
p-Nitrobenzaldehyde Schiff base	$4.1 \cdot 10^{-13}$	$5.3 \cdot 10^{-15}$

tives have a poor ECD response. The trifluoroacetamide, pentafluoropropionamide and heptafluorobutyramide derivatives were more sensitive to detection than the parent amine by factors of approximately $2 \cdot 10^2$, $2 \cdot 10^3$, and 10^4 respectively. For the Schiff base derivatives, introduction of a chlorine or nitro group into benzaldehyde improves the detector sensitivity by approximately 500 and 600 fold respectively. The pentafluorobenzaldehyde was the second most sensitive of the derivatives examined with a response slightly less than the perfluorooctanaldehyde Schiff base and about three fold greater than the heptafluorobutyraldehyde Schiff base.

The responses to the ECD of some pentafluorophenyl-containing derivatives of β-phenylethylamine and N-methyl-β-phenylethylamine have been compared in order to elucidate the structural requirements of the derivatives for a high detector response [41]. The results are summarized in Table 8.17. The most electron-capturing of the derivatives

TABLE 8.17

RELATIVE RESPONSE OF THE ELECTRON-CAPTURE DETECTOR TO SOME PENTAFLUOROPHENYL CONTAINING AMINE DERIVATIVES

Detector ^{63}Ni (10 mC), pulsed, detector temperature 300°C.

No.	Structure	Sensitivity*
I	$C_6H_5CH_2CH_2$–N=CHC_6F_5	100
II	$C_6H_5CH_2CH_2$–NHCOC_6F_5	78
III	$C_6H_5CH_2CH_2$–NHCO$CH_2C_6F_5$	11
IV	$C_6H_5CH_2CH_2$–NH$CH_2C_6F_5$	4.1
V	$C_6H_5CH_2CH_2$–N($CH_2C_6F_5$)	34
VI	$C_6H_5CH_2CH_2$—N(CH_3)— ring (F, F, F, F, CHO)	47
VII	$C_6H_5CH_2CH_2$—N(CH_3)— ring (F, F, F, F) —CHO	24
VIII	$C_6H_5CH_2CH_2$—NH— ring (F, F, F, F, $COOCH_3$)	53
IX	$C_6H_5CH_2CH_2$—NH— ring (F, F, F, F) —$COOCH_3$	22

*Relative to $C_6H_5CH_2CH_2$-N=CHC_6F_5 on a Coulomb · mole^{-1} basis.

tested was the pentafluorobenzaldehyde Schiff base (I) followed by the pentafluorobenzoyl derivative (II). Separation of the double bond from the ring by a methylene group, as in pentafluorophenylacetylphenylethylamine (III) causes a reduction in the electron-capturing properties by a factor of 7 (relative to II). When no double bond is present in the molecule (IV and V), the sensitivity declines by 25 to 35 fold compared to the Schiff base derivative. Even when two pentafluorophenyl groups are present in the same derivative (V), although its response is increased by a factor of eight compared to the monosubstituted amine, the sensitivity is still less than that of those compounds with a double bond on the carbon atom adjacent to the pentafluorophenyl group. Thus, while the highly electronegative pentafluorophenyl group does confer some electron-capturing properties to the molecule, the presence of a polarizable carbon (C=O, C≡N), adjacent to this aromatic ring, enhances the effect even further. Derivatives VI and IX still possess a polarizable carbon atom adjacent to the highly electron-withdrawing tetrafluorophenyl group and therefore still retain considerable electron-capture sensitivity. The reduced sensitivity of these compounds (relative to I) may be accounted for by the lone pair of electrons on the nitrogen atom feeding into the ring, thereby lowering the electron density and electron-capturing potential of the system.

McCallum and Armstrong [32] have compared the relative response of the electron-capture detector to a series of ester and ether derivatives of thymol (2-hydroxy-1-isopropyl-4-methylbenzene) (Table 8.18). With the detector conditions employed, the pentafluoropropionate was slightly more sensitive than the heptafluorobutyrate. The monofluoroacetate showed only a fraction of the sensitivity of the monochloroacetate derivative which in turn was less sensitive than the perfluorocarbon acyl esters. The greatest response was obtained with the pentafluorobenzyl ether and the pentafluorobenzoate ester.

The ECD response to a series of halocarbondimethylsilyl ether derivatives of cholesterol at a fixed detector temperature has been determined (Table 8.19) [42]. The derivatives showed the expected order of response to the ECD with the halomethyldimethylsilyl ethers following the order $CH_2I > CH_2Br > CH_2Cl$ and the silyl ethers containing alkyl fluorocarbon groups showing very little response to the detector. The re-

TABLE 8.18

RELATIVE RESPONSE OF THE ELECTRON-CAPTURE DETECTOR TO SOME HALOACETATE AND ETHER DERIVATIVES OF THYMOL

Detector temperature 200°C.

Derivative	Relative detector response
Heptafluorobutyrate	1.0
Pentafluoropropionate	1.3
Monochloroacetate	0.3
Monofluoroacetate	$7 \cdot 10^{-3}$
Pentafluorobenzyl ether	5.9
Pentafluorobenzoate	6.9
2,4-Dinitrophenyl ether	0.3

sponse of the pentafluorophenyldimethylsilyl ether (flophemesyl ether) is 50 times more sensitive to the ECD than the 2′-pentafluorophenylethyldimethylsilyl ether, illustrating the importance of charge delocalization between the *d*-orbitals of silicon and the π-orbitals of the pentafluorophenyl ring in influencing the response of these derivatives to the detector. Changing the nature of the alkyl substituent in the pentafluorophenyldimethylsilyl ether derivatives has little influence on their response to the ECD (Table 8.20) [25]. Only the chloromethyl group of the substituents R_1 has any significant electron-capturing properties in its own right, but even its influence on the overall detector response is small.

A comparison has been made of the ECD response to a series of electron-capturing boronic acids as their pinacol derivatives (Table 8.21) [18]. From the response of the chlorobenzeneboronic acid derivatives it can be seen that both the number and position of the chlorine atoms relative to the boronate group have a significant influence on the response of the derivative to the detector. A greater response is shown by the 2,4-dichlorobenzeneboronate derivative (*o*- and *p*-chlorine substitution) compared to the 3,5-dichlorobenzeneboronate (*m*-chlorine substitution). The detector response is diminished by the presence of two *o*-chlorine substituents as in the 2,6-dichlorobenzene-

TABLE 8.19

RESPONSE OF THE ELECTRON-CAPTURE DETECTOR TO A SERIES OF HALOCARBONDIMETHYLSILYL ETHER DERIVATIVES OF CHOLESTEROL

Detector, ^{63}Ni (10 mC), pulsed, detector temperature 300°C.

Halocarbon chain	Minimum detectable quantity ($\times 10^{-9}$ g)
$CF_3(CH_2)_2-$	1500
$CF_3(CF_2)_2(CH_2)_2-$	115
$ClCH_2-$	75
C_6F_5-	4
$C_6F_5(CH_2)_2-$	200
CH_2Br-	0.5
CH_2I-	0.005

TABLE 8.20

RESPONSE OF THE ELECTRON-CAPTURE DETECTOR TO A SERIES OF $RR_1(CH_3)Si$-OCTANOL DERIVATIVES

Detector, ^{63}Ni (8 mC), constant-current, optimized detector temperature.

R	R_1	Abbreviated name	Minimum detectable quantity ($\times 10^{-12}$ g)	Optimal detector temperature (°C)
C_6F_5	CH_3	Flophemesyl	4.0	350
C_6F_5	$(CH_3)_2CH$	ISP-Flophemesyl	5.0	320
C_6F_5	$(CH_3)_3C$	*tert.*-Buflophemesyl	6.0	350
C_6F_5	CH_2Cl	CM-Flophemesyl	0.9	320

TABLE 8.21

A COMPARISON OF THE VOLATILITY AND ECD SENSITIVITY OF THE ELECTRON-CAPTURING BORONIC ACIDS

Boronic ester	Relative volatility*	Minimum detectable quantity ($\times 10^{-12}$ g) pinacol	Optimal detector temperature (°C)
3,5-Bis(trifluoromethyl)-benzeneboronates	0.3 ± 0.05	3.0	180
Benzeneboronates	1.0	200.0	150
4-Iodobutaneboronates	1.8 ± 0.5		
4-Bromobenzeneboronates	3.9 ± 0.8	3.0	350
2,6-Dichlorobenzeneboronates	4.3 ± 2.0	18.0	380
2,4-Dichlorobenzeneboronates	4.7 ± 1.7	4.0	380
3,5-Dichlorobenzeneboronates	5.0 ± 1.1	11.0	380
2,4,6-Trichlorobenzeneboronates	6.9 ± 1.8	4.0	380
3-Nitrobenzeneboronates	11.7 ± 3.4	4.0	300
Naphthaleneboronates	18.5 ± 4.6	2550.0	350

*Based on the comparison of the retention times of a series of bifunctional compounds compared to the benzeneboronate derivative [30].

boronate and 2,4,6-trichlorobenzeneboronate derivatives. The 4-bromobenzeneboronate and 3-nitrobenzeneboronate derivatives show similar detector sensitivity to that of the 2,4-dichlorobenzeneboronate derivative. The benzeneboronate and in particular the naphthaleneboronate derivative are too insensitive to the ECD for use in trace organic analysis. The moderately good response of the ECD to the benzeneboronate derivative was unexpected and enabled this derivative to be detected at lower levels than was possible with a flame ionization detector. The use of benzeneboronate derivatives in conjunction with an ECD could be useful for the selective analysis of bifunctional compounds were the discriminatory power of the detector is a more important parameter than the need to obtain very low detection limits.

A few studies have been made of the suitability of some derivatizing reagents for the determination of a wide range of organic compounds. Martin and Rowland [34] have compared the response of the ECD to a series of heptafluorobutyramide and pentafluorobenzamide derivatives of some pharmaceutically important primary and secondary amines (Table 8.22). For the pentafluorobenzamide derivatives, the detector response varies by about two orders of magnitude, although they all contain the same derivative group. The derivatizing reagent is responsible for introducing an electrophore into the molecule. The response of the detector is, however, dependent on the overall structure of the molecule after derivatization. As such, changes in the response of the detector are observed when subtle variations are made to the molecular framework. The detector response is influenced by such factors as the conjugation of weakly electron-capturing groups with the derivative portion of the molecule, electronic inductive effects modifying the site of electron attachment and stereochemical interactions preventing the formation of a favored (resonance) stabilized center for electron attachment. Thus the derivatizing reagent confers the initial electron-capturing properties on the molecule, but the magni-

TABLE 8.22

ELECTRON-CAPTURE RESPONSE OF VARIOUS HEPTAFLUOROBUTYRAMIDE AND PENTAFLUOROBENZAMIDE DERIVATIVES OF PRIMARY AND SECONDARY AMINES

Amine	Minimum detectable quantity (Coulomb $\cdot 10^3 \cdot$ mole^{-1})	
	Heptafluorobutyramide	Pentafluorobenzamide
β-Phenethylamine	2.0	300
Amphetamine	2.2	225
α-Methylbenzylamine	2.5	180
p-Methoxyphenethylamine	1.9	135
Mescaline	3.7	155
Phentermine	1.1	29
Phenmetrazine	8.8	28
Mephentermine	2.2	9.2
Methamphetamine	4.0	6.1
Methoxyphenamine	1.6	3.2

TABLE 8.23

THE RESPONSE OF THE ELECTRON-CAPTURE DETECTOR TO SOME HEPTAFLUOROBUTYRAMIDE AND HEPTAFLUOROBUTYRATE DERIVATIVES

Detector temperature 165°C.

Compound	Minimum detectable quantity	
	mole $\cdot$ s$^{-1} \cdot 10^{-15}$	picograms
Amines		
Aniline	0.3	0.3
Benzylamine	4.7	5.8
β-Phenylethylamine	11	15
Naphthylamine	0.6	0.8
Amphetamine	9.2	12
Phentermine	14	23
Chlorphentermine	2.2	4.6
N-Methyl-β-phenylethylamine	5.4	8.2
Methamphetamine	5.5	9.2
Cyclopentamine	12	19
Methoxyphenamine	24	48
Phenmetrazine	1.3	2.5
Desipramine	0.8	2.3
Alcohols		
Benzyl alcohol	0.9	1.1
β-Phenylethanol	2.3	3.2
γ-Phenylpropanol	4.7	7.2
1,1′-Biphenylmethanol	0.4	0.8
Tropine	5.1	8.1
Octanol	5.4	7.9

TABLE 8.24

RELATIVE RESPONSE OF THE ELECTRON-CAPTURE DETECTOR TO SOME PENTAFLUOROBENZYL, 2,6-DINITRO-4-TRIFLUOROMETHYLPHENYL ETHER, AND 2,4-DINITROPHENYL ETHER DERIVATIVES OF PHENOL

Tritium detector at 210°C.

Phenol	Derivative*		
	PFB	DNT	DNP
Phenol	0.18	0.14	0.23
p-Chlorophenol	0.29	0.19	0.31
3,4,5-Trimethylphenol	0.36	0.16	0.20
Carbofuran phenol	0.30	0.13	0.25
p-Nitrophenol	0.29	0.10	0.06
1-Naphthol	0.39	0.20	0.19

*Response relative to aldrin (1.00).

TABLE 8.25

RELATIVE RESPONSE OF THE ELECTRON-CAPTURE DETECTOR TO SOME DERIVATIVES OF FOUR AGRICULTURALLY IMPORTANT CHEMICALS

^{63}Ni pulsed detector operated at 250°C.

Compound	Derivative	Electron-capture detector response*
Diethylstilbestrol (structure: HO–C₆H₄–C(C₂H₅)=C(C₂H₅)–C₆H₄–OH)	Trifluoroacetate	1.5
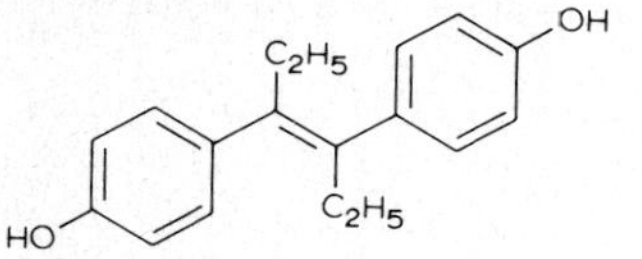	Pentafluoropropionate	15
	Heptafluorobutyrate	23
	Pentadecafluoroundecanoate	21
Clopidol (structure: OH, Cl, Cl, CH₃, N, CH₃)	Trifluoroacetate	24
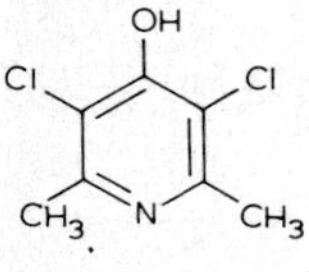	Pentafluoropropionate	32
	Heptafluorobutyrate	45
	Pentadecafluoroundecanoate	34
Linuron (structure: Cl, Cl, NH–C(=O)–N(CH₃)(CH₃))	Trifluoroacetamide	2.1
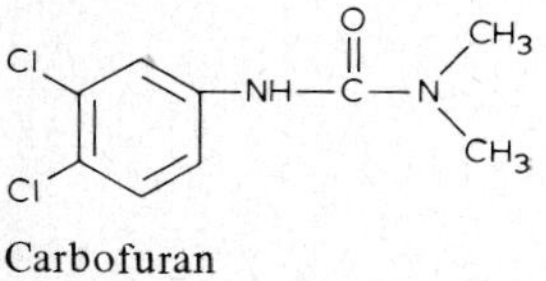	Pentafluoropropionamide	4.8
	Heptafluorobutyramide	7.7
	Pentadecafluoroundecanonamide	13.3
Carbofuran (structure: CH₃, CH₃, O, O–C(=O)–NH–CH₃)	Trifluoroacetamide	0.9
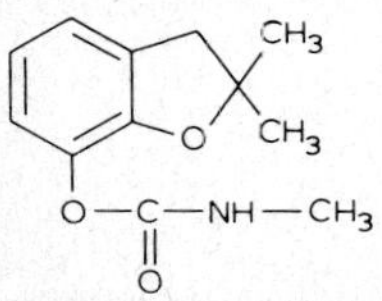	Pentafluoropropionamide	2.3
	Heptafluorobutyramide	4.4
	Pentadecafluoroundecanonamide	11.0

*Relative peak area per nanogram of parent compound.

tude of the detector response is a function of the molecule as a whole. Similar trends, although of a smaller magnitude, were observed for some heptafluorobutyramide and heptafluorobutyrate derivatives of some amines and alcohols (Table 8.23) [19].

A number of phenols have been determined as their pentafluorobenzyl, 2,6-dinitro-4-trifluoromethylphenyl ether and 2,4-dinitrophenyl ether derivatives (Table 8.24) [43]. The variation in response factors is relatively small considering that some of the phenols derivatized contain additional electron-capturing groups. The presence of the trifluoromethyl group in the dinitrophenyl reagents does not significantly change the electron-capture response in the 2,6-dinitro-4-trifluoromethylphenyl ether derivatives but does confer greater volatility compared to the 2,4-dinitrophenyl derivatives.

Ryan and Lawrence [44] have compared the response of the ECD to four agricultural chemicals which require derivatization prior to analysis (Table 8.25). Although a general trend of increase in detector response with an increase in the number of CF_2 groups can be discerned for the compounds containing amino groups, for the phenols, the pentadecafluoroundecanoate derivative had a lower response than the heptafluorobutyrate under the conditions used for the determination.

8.6. THE PREPARATION OF ELECTRON-CAPTURING DERIVATIVES

There are several general reviews and books pertaining to the methods and techniques for the analytical preparations of derivatives [6,27,45–52]. For experimental requirements and for a more comprehensive literature review of particular derivatives of individual compounds, the reader is referred to these sources. In the following sections, some comments of a general nature will be given to aid the reader in making an initial selection of a particular derivative reaction.

8.6.1. Halocarbonsilyl ethers and esters

The trimethylsilyl reagents are the most versatile and widely used derivatizing reagents in gas chromatography [53,54]. Nearly all functional groups which can present a problem in gas chromatography can be converted to trimethylsilyl derivatives (Fig. 8.15). The trimethylsilyl group shows no particular electron-capturing properties "per se" but the necessary detector-oriented response can be conferred on the trimethylsilyl group by

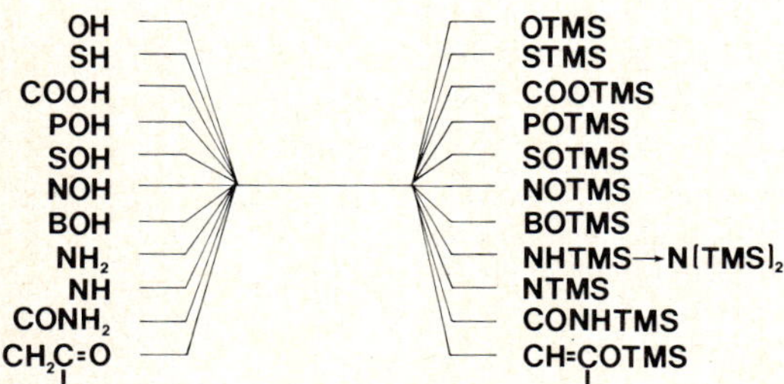

Fig. 8.15. Organic functional groups forming trimethylsilyl derivatives.

introducing a halogen atom (Cl, Br, I) into one of the methyl groups or by replacing a methyl group with a pentafluorophenyl ring (flophemesyl reagents).

The pentafluorophenyldimethylsilyl reagents, contracted to flophemesyl for convenience, offer the best compromise between derivative volatility and detector sensitivity. They have been used to detect trace levels of alcohols, phenols, carboxylic acids, amines, steroids, and drugs [25,26,42,55–57]. Three reagents, flophemesyl chloride, flophemesylamine, and flophemesyldiethylamine are commercially available (Lancaster Synthesis, St. Leonardgate, Great Britain; Alfa Products Division, Ventron Corporation, Danvers, MA, U.S.A.) and in combination they enable a wide range of sterically crowded functional groups to be quantitatively derivatized. Flophemesylamine is a particularly useful reagent that can be used to react selectively with unhindered primary and secondary hydroxyl groups in the presence of tertiary hydroxyl or unprotected ketone groups. The flophemesyl derivatives have similar hydrolytic stability to the trimethylsilyl ethers and if a more robust derivative is required due to the need for extensive sample clean-up prior to analysis, then the much more hydrolytically stable *tert.*-buflophemesyl and ISP-flophemesyl reagents can be used [25, 28].

The halomethyldimethylsilyl ethers are usually prepared under mild conditions to minimize the formation of by-products. The range of application of the derivatives is limited by the silyl donor power of the available reagents. Derivatives are usually prepared with a mixture of 1,3-bis-(halomethyl)-1,1,3,3-tetramethyldisilazane and the halomethyldimethylchlorosilane with diethylamine in hexane [59–62]. The iodomethyldimethylsilyl reagents are unstable and cannot be conveniently stored. The iodomethyldimethylsilyl derivatives are prepared in situ after first preparing the chloromethyldimethylsilyl or the bromomethyldimethylsilyl ether and introducing iodine by halide ion-exchange with a saturated solution of sodium iodide in acetone [61, 62]. When vigorous reaction conditions are used to prepare the halomethyldimethylsilyl derivatives, displacement of the halogen from the halomethyl group or expulsion of the whole halomethyl group with the formation of a non electron-capturing derivative or dimer can occur [50, 54]. Short-chain aliphatic acids, when derivatized with bromomethyldimethylchlorosilane and diethylamine, formed bromomethyltetramethylmethyldisiloxane esters rather than the expected product [63].

$$\mathrm{R{-}\underset{\underset{O}{\|}}{C}{-}OH + BrCH_2{-}\underset{\underset{CH_3}{|}}{\overset{\overset{CH_3}{|}}{Si}}{-}Cl \xrightarrow[\text{diethylamine}]{\Delta H} R{-}\underset{\underset{O}{\|}}{C}{-}OCH_2{-}\underset{\underset{CH_3}{|}}{\overset{\overset{CH_3}{|}}{Si}}{-}O{-}\underset{\underset{CH_3}{|}}{\overset{\overset{CH_3}{|}}{Si}}{-}CH_2Br}$$

Under carefully controlled conditions, quantitative yields of the halomethyldimethylsilyl derivatives of alcohols, steroids, bile acids, phenols, prostaglandins, and carbohydrates can be prepared [54]. As an approximate guide to the detector response for the analysis of steroids, the chloromethyldimethylsilyl ethers have similar sensitivity towards the ECD as to the flame ionization detector; the bromomethyldimethylsilyl ethers are several-fold more sensitive with detection limits below the nanogram level; the iodomethyldimethylsilyl ethers are the most sensitive of all and can be determined at the low picogram level. In terms of volatility, the iodomethyl- and bromomethyldimethylsilyl derivatives have relatively long retention times (see Table 8.6).

8.6.2. Haloacyl esters

The haloacyl anhydrides are probably the most studied reagents for the introduction of an electrophore into compounds with a protonic functionality (except carboxylic acids) (Fig. 8.16). The use of the appropriate anhydride, sometimes in the presence of an acid acceptor, such as pyridine, leads to a rapid quantitative reaction of all but the most sterically hindered groups. The use of the anhydride can lead to undesirable side reactions (dehydration, enolization, etc.) with sensitive molecules due to the strong acid conditions prevailing in the reaction medium. For these compounds, acylation can be performed using the acylimidazole reagents in which the by-product of the reaction is the weakly amphoteric imidazole. The perfluoroacyl derivatives have good gas chromatographic properties and high volatility. The chloroacetyl and bromoacetyl derivatives have long retention times by comparison with the perfluoroacyl derivatives often accompanied by poor peak shape and poor thermal stability. In general terms, the monochloroacetyl and chlorodifluoroacetyl derivatives are more sensitive than the trifluoroacetyl derivatives. Increasing the fluorocarbon chain length of the fluorocarbonacyl derivatives increases their ECD response without inconveniently increasing their retention times. The heptafluorobutyryl derivatives are considered to be the best compromise between detector sensitivity and volatility for most applications.

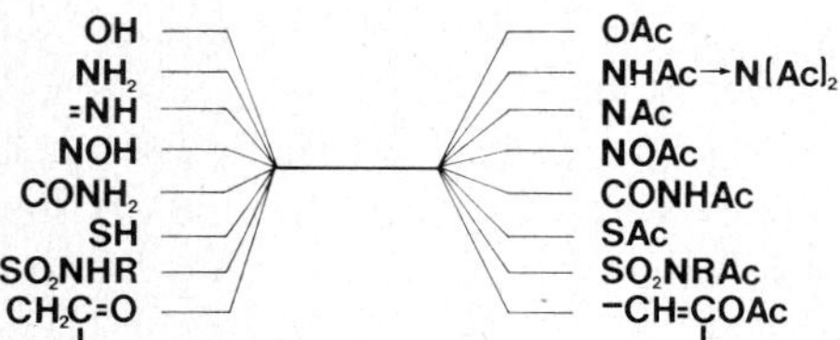

Fig. 8.16. Organic functional groups forming haloacyl derivatives.

8.6.3. Organic reagents containing a pentafluorophenyl group

Reagents for the preparation of pentafluorophenyl-containing derivatives are summarized in Table 8.26. The pentafluorophenyl-containing reagents can be used to derivatize a broad spectrum of organic compounds including tertiary amines as well as offering a high degree of specificity in some cases, for example the determination of primary amines with pentafluorobenzaldehyde and carboxylic acids with pentafluorobenzyl alcohol [6]. The pentafluorophenyl-containing derivatives are generally easy to prepare, are volatile with good gas chromatographic properties and provide a high response to the ECD. Their reactions proceed smoothly and few unexpected by-products have been identified. With secondary amines, pentafluorobenzaldehyde can react to form mixtures of ring-substituted products involving hydrogen fluoride elimination (Table 8.27, structures VI and VII) and with α-hydroxyamines cyclic derivatives can be formed [41]. Pentafluorobenzyl bromide reacts with clonidine to form a cyclic derivative again involving amine attack at the ring with fluoride elimination [64].

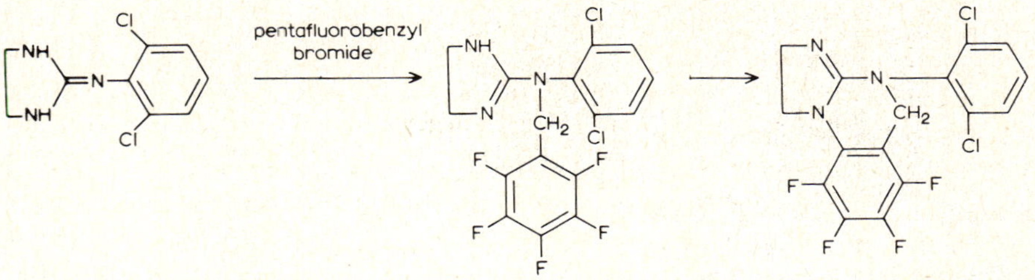

Pentafluorobenzyl bromide is also a potent lachrymatory agent and is unstable under some conditions used for extractive alkylation [65]. Pentafluorophenylhydrazine and pentafluorobenzyl hydroxylamine derivatives of ketone compounds can give two peaks separated by gas chromatography due to the formation of *syn*- and *anti*-geometrical isomers [42,66–68].

TABLE 8.26

REAGENTS FOR THE INTRODUCTION OF THE PENTAFLUOROPHENYL GROUP INTO ORGANIC MOLECULES

Reagent	Functional group type
Pentafluorobenzoyl chloride	Amines, phenols, alcohols
Pentafluorobenzyl bromide	Carboxylic acids, phenols, mercaptans, sulfonamides
Pentafluorobenzyl alcohol	Carboxylic acids
Pentafluorobenzaldehyde	Primary amines
Pentafluorobenzyl chloroformate	Tertiary amines
Pentafluorophenacetyl chloride	Alcohols, phenols, amines
Pentafluorophenoxyacetyl chloride	Alcohols, phenols, amines
Pentafluorophenylhydrazine	Ketones
Pentafluorobenzylhydroxylamine	Ketones

8.6.4. Reagents for the selective derivatization of bifunctional compounds

Bifunctional compounds are characterized by having at least two functional groups in close proximity that can be reacted with certain specific reagents to form cyclic derivatives [27]. Many physiologically active compounds among the steroids, prostaglandins, lipids, catecholamines and nucleosides for example are bifunctional and their selective analysis as cyclic derivatives with electron-capture detection is potentially a powerful method for their determination at trace levels. Cyclizing reagents for bifunctional compounds with electron-capturing properties are few in number and limited almost entirely to the cyclic derivatives formed with halogen-substituted ketones, anhydrides and boronic acids. Some typical reactions are shown in Fig. 8.17. Of these reagents, only the boronic acids have a broad range of application being suitable for the derivatization of alkyl 1,2-diols, 1,3-diols, 1,4-diols, 1,2-enediols, 1,3-hydroxyamines, 1,3-diamines, and aromatic compounds with phenol or amine groups in the *ortho* positions. Of the electron-

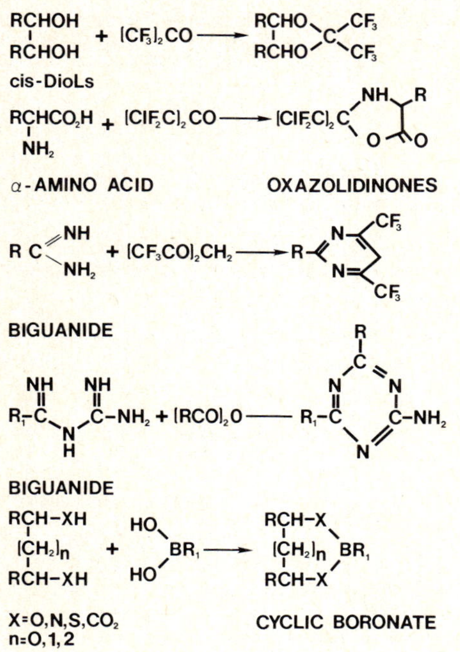

RCHOH / RCHOH + $(CF_3)_2CO$ ⟶ RCHO / RCHO $>C<$ CF_3 / CF_3

cis-DioLs

$RCH(NH_2)CO_2H$ + $(ClF_2C)_2CO$ ⟶ $(ClF_2C)_2C$ (NH, R, O, O)

α-AMINO ACID OXAZOLIDINONES

$RC(=NH)NH_2$ + $(CF_3CO)_2CH_2$ ⟶ R–(N, N, CF_3, CF_3)

BIGUANIDE

R_1–C(=NH)–NH–C(=NH)–NH_2 + $(RCO)_2O$ ⟶ R_1–C, N, C–R, N, C–NH_2, N

BIGUANIDE

RCH–XH / $(CH_2)_n$ / RCH–XH + HO / HO $>BR_1$ ⟶ RCH–X / $(CH_2)_n$ BR_1 / RCH–X

X=O,N,S,CO_2
n=0,1,2

CYCLIC BORONATE

Fig. 8.17. Reagents forming cyclic electron-capturing derivatives of bifunctional compounds.

capturing boronic acids evaluated, the 3,5-bis(trifluoromethyl)benzeneboronic acid, 2,4-dichlorobenzeneboronic acid, 4-bromobenzeneboronic acid, and 4-iodobutaneboronic acid were found to be the most generally useful reagents [18]. The derivatives are volatile with good chromatographic properties and have electron-capture detection limits at the low picogram level.

The reaction between halogen-substituted analogs of acetylacetone and biguanides is very selective and can be used for their derivatization in biological fluids without preliminary isolation. The cyclic derivatives formed with hexafluoroacetylacetone are the most volatile and hydrolytically stable of the derivatives evaluated and have good electron-capturing properties [69, 70]. The reaction between biguanides and organic acid anhydrides results in the formation of cyclic 2,4-disubstituted 2,6-amino-1,3,5-*s*-triazines which have good thermal and gas chromatographic properties [71, 72]. The anhydrides tested include chlorofluoroacetic, dichlorofluoroacetic, heptafluorobutyric, pentafluoropropionic, and trifluoroacetic anhydride. Using chlorodifluoroacetic anhydride as cyclizing reagent, phenformin could be determined with the ECD at below the 1.0 ng/ml level in plasma [71].

Substitution of hydrogen with electronegative groups such as chlorine or fluorine in acetone enhances the acidic character of the carbonyl group and promotes the formation of a series of stable monofunctional adducts and cyclic derivatives not observed with aliphatic ketones. The condensation of hexafluoroacetone and 1,3-dichlorotetrafluoroacetone with α-substituted carboxylic acids leads to the formation of a series of stable five-membered ring derivatives (for example, oxazolidinones) which after silylation or acylation of unprotected functional groups have good gas chromatographic properties and electron-capture detection limits at the picogram level [73–76].

8.6.5. Miscellaneous electron-capturing derivatizing reagents

Nitrophenyl-containing reagents can be used to prepare derivatives of amines, phenols, mercaptans, and carbonyl compounds [77]. The reagents employed include 1-fluoro-2,4-dinitrobenzene, 2,4-dinitrobenzenesulfonic acid, 2,4,6-trinitrobenzenesulfonic acid, 2,4-dinitrophenylhydrazine, 2,6-dinitro-4-trifluoromethylchlorobenzene and 2-nitro-4-trifluoromethylchlorobenzene. The 2,6-dinitro-4-trifluoromethyl and 2-nitro-4-trifluoromethyl derivatives of primary amines have shorter retention times on gas chromatography than the corresponding dinitrophenyl derivatives, and, additionally, these derivatives have a somewhat greater sensitivity than the dinitrophenyl derivatives towards electron-capture detection [78]. 2,4-Dinitrobenzenesulfonic acid has a much higher specificity for amines than 1-fluoro-2,4-dinitrobenzene which also reacts with mercaptans, hydroxyl compounds and the imidazole nitrogen. 2,4-Dinitrobenzenesulfonic acid is water soluble, aiding its removal by solvent extraction prior to chromatographic analysis [79, 80]. Strong dipolar interactions between the aromatic nitro group and the separation medium can result in broad peaks with poor shape unless some care is taken in deactivating the support and in the selection of the stationary phase for the separation. The derivatives have good electron-capturing properties with detection limits below the nanogram level [80–82].

Carboxylic acids can be esterified with 2-chloroethanol [83–85], 2,2,2-trichloroethanol [86–88], 1-chloro-1,1,3,3,3-pentafluoro-2-propanol [89, 90], 2,2,3,3,3-pentafluoropropanol [91–95] and hexafluoroisopropanol [96, 97]. The 2-chloroethyl esters have poor ECD properties but the detection limits for the trichloroethyl esters are below the nanogram level. The fluorine-containing alcohols are usually used in combination with other ECD derivatizing reagents to improve their sensitivity to the detector. A fluorocarbon acyl anhydride can be employed both as a catalyst for the esterification reaction and to simultaneously derivatize other functional groups present in the molecule. The simultaneous introduction of the fluorocarbon ester and fluorocarbon acyl or acetamide groups produces derivatives with good chromatographic and electron-capturing properties.

Trichloroethyl chloroformate reacts with tertiary amines to form the trichloroethyl carbamate derivatives [98, 99]. The trichloroethyl carbamate derivatives have a slightly lower ECD response than the corresponding pentafluorobenzyl derivative but gave fewer interferences and less pronounced solvent fronts.

8.7. CONCLUSIONS

The marriage between the electron-capture detector and detector oriented derivatization techniques has been a long and happy one. A combination of the reaction selectivity of the derivatizing reagents coupled with the high sensitivity and selectivity of the detector has proven to be the tool by which an immense amount of information on the levels and effects of a wide range of environmentally and biomedically important compounds has been gathered.

Although the electron-capture detector has been in use for about twenty years, certain misunderstandings concerning its use remain. With respect to derivatization techniques, detection limits should be quoted at the optimal detector temperature for their determination. It has become common practice when comparing the ECD response to a series of derivatives of a compound, to make all measurements at a fixed detector temperature. It is meaningless to make such comparisons if some derivatives have a much larger detector response at a different detector temperature. Also, detection limits measured under optimal conditions are a property of both the reagent and the compound derivatized. As such they are compound dependent. Therefore, for a range of compounds with the same functional group different detector responses can be expected after derivatization due to the influence of the original compound structure on the electron affinity of the electrophore. Failure to appreciate the above points has often led to disappointment in making interlaboratory comparisons of electron-capture data.

REFERENCES

1 J.E. Lovelock, *Nature (London)*, 189 (1961) 729.
2 J.E. Lovelock, P.G. Simmonds, W.J.A. VandenHeuvel, *Nature (London)*, 197 (1963) 249.
3 A. Zlatkis and J.E. Lovelock, *Clin. Chem.*, 11 (1965) 259.
4 J.E. Lovelock, in L. Augenstein, R. Mason and B. Rosenberg (Editors), *Physical Processes in Radiation Biology*, Academic Press, New York, 1964, p. 183.
5 E.D. Pellizzari, *J. Chromatogr.*, 98 (1974) 323.
6 C.F. Poole, *Chem. Ind.*, (1976) 479.
7 C.A. Clemons and A.P. Altshuller, *Anal. Chem.*, 38 (1966) 133.
8 P. Devaux and G. Guiochon, *J. Gas Chromatogr.*, 5 (1967) 341.
9 G. Castello and G.D'Amato, *J. Chromatogr.*, 54 (1971) 157.
10 R.A. Landowne and S.R. Lipsky, *Anal. Chem.*, 34 (1962) 726.
11 W.L. Zielinski, L. Fishbein and R.O. Thomas, *J. Chromatogr.*, 30 (1967) 77.
12 R.D. Chambers, *Fluorine in Organic Chemistry*, Wiley, New York, 1973, pp. 140 and 274.
13 S. Kapila and W.A. Aue, *J. Chromatogr.*, 108 (1975) 13.
14 C.R. Hastings, T.R. Ryan and W.A. Aue, *Anal. Chem.*, 47 (1975) 1169.
15 S. Kapila and W.A. Aue, *J. Chromatogr.*, 148 (1978) 343.
16 D.E. Durbin, W.E. Wentworth and A. Zlatkis, *J. Amer. Chem. Soc.*, 92 (1970) 5131.
17 W.E. Wentworth and E. Chen, *J. Gas Chromatogr.*, 5 (1967) 170.
18 C.F. Poole, S. Singhawangcha and A. Zlatkis, *J. Chromatogr.*, 186 (1979) 307.
19 T. Walle and E. Ehrsson, *Acta Pharm. Suecica*, 7 (1970) 389.
20 E.C.M. Chen and W.E. Wentworth, *J. Chromatogr.*, 68 (1972) 302.
21 C.F. Poole, *J. Chromatogr.*, 118 (1976) 280.
22 B.C. Pettitt, P.G. Simmonds and A. Zlatkis, *J. Chromatogr. Sci.*, 7 (1969) 645.
23 A. Zlatkis and B.C. Pettitt, *Chromatographia*, 2 (1969) 484.
24 O. Gyllenhaal and P. Hartvig, *J. Chromatogr.*, 189 (1980) 351.
25 C.F. Poole, W.-F. Sye, S. Singhawangcha, F. Hsu, A. Zlatkis, A. Arfwidsson and J. Vessman, *J. Chromatogr.*, 199 (1980) 123.
26 P.M. Burkinshaw, E.D. Morgan and C.F. Poole, *J. Chromatogr.*, 132 (1977) 548.
27 C.F. Poole and A. Zlatkis, *J. Chromatogr.*, 184 (1980) 99.
28 C.F. Poole, S. Singhawangcha and A. Zlatkis, *J. Chromatogr.*, 158 (1978) 33.

29 C.F. Poole, S. Singhawangcha and A. Zlatkis, *Analyst (London)*, 104 (1979) 82.
30 S. Singhawangcha, C.F. Poole and A. Zlatkis, *J. Chromatogr.*, 183 (1980) 433.
31 R.A. Landowne and S.R. Lipsky, *Anal. Chem.*, 35 (1963) 632.
32 N.K. McCallum and R.J. Armstrong, *J. Chromatogr.*, 78 (1973) 303.
33 L. Dehennin, A. Reiffsteck and R. Scholler, *J. Chromatogr. Sci.*, 10 (1972) 224.
34 S.B. Martin and M. Rowland, *J. Pharm. Sci.*, 61 (1972) 1235.
35 D.D. Clarke, S. Wilk and S.E. Gitlow, *J. Gas Chromatogr.*, 4 (1966) 310.
36 D. Exley and A. Dutton, *Steroids*, 14 (1969) 575.
37 K. Nakagawa, N.L. McNiven, E. Forchielli, A. Vermeulen and R.I. Dorfman, *Steroids*, 7 (1966) 329.
38 G.R. Wilkinson, *Anal. Lett.*, 3 (1970) 289
39 M.A. Kirschner and J.P. Taylor, *Anal. Biochem.*, 30 (1969) 346.
40 A.C. Moffat and E.C. Horning, *Anal. Lett.*, 3 (1970) 205.
41 A.C. Moffat, E.C. Horning, S.B. Martin and M. Rowland, *J. Chromatogr.*, 66 (1972) 255.
42 C.F. Poole, A. Zlatkis, W.-F. Sye, S. Singhawangcha and E.D. Morgan, *Lipids*, 15 (1980) 734.
43 J.N. Seiber, D.G. Crosby, H. Fouda and C.J. Soderquist, *J. Chromatogr.*, 73 (1972) 89.
44 J.J. Ryan and J.F. Lawrence, *J. Chromatogr.*, 135 (1977) 117.
45 K. Blau and G.S. King (Editors), *Handbook of Derivatives for Chromatography*, Heyden, London, 1977.
46 D.R. Knapp, *Handbook of Analytical Derivatization Reactions*, Wiley-Interscience, New York, 1979.
47 J. Drozd, *J. Chromatogr.*, 113 (1975) 303.
48 W.P. Cochrane, *J. Chromatogr. Sci.*, 17 (1979) 124.
49 S. Ahuja, *J. Pharm. Sci.*, 65 (1976) 163.
50 L.M. Cummins, in I.I. Domsky and J.A. Perry (Editors), *Recent Advances in Gas Chromatography*, Marcel Dekker, New York, 1971, p. 313.
51 J.D. Nicholson, *Analyst (London)*, 103 (1978) 1, 193.
52 C.F. Poole and A. Zlatkis, *Anal. Chem.*, 52 (1980) 1002A.
53 C.F. Poole, in K. Blau and G.S. King (Editors), *Handbook of Derivatives for Chromatography*, Heyden, London, 1977, p. 152.
54 C.F. Poole and A. Zlatkis, *J. Chromatogr. Sci.*, 17 (1979) 115.
55 E.D. Morgan and C.F. Poole, *J. Chromatogr.*, 89 (1974) 225.
56 E.D. Morgan and C.F. Poole, *J. Chromatogr.*, 104 (1975) 351.
57 A.J. Francis, E.D. Morgan and C.F. Poole, *J. Chromatogr.*, 161 (1978) 111.
58 C.F. Poole, S. Singhawangcha, L.-E. Chen Hu, W.-F. Sye, R. Brazell and A. Zlatkis, *J. Chromatogr.*, 187 (1980) 331.
59 C. Eaborn, D.R.M. Walton and B.S. Thomas, *Chem. Ind.*, (1976) 827.
60 C. Eaborn, C.A. Holder, D.R.M. Walton and B.S. Thomas, *J. Chem. Soc. C*, (1969) 2502.
61 B.S. Thomas, *J. Chromatogr.*, 56 (1971) 37.
62 E.K. Symes and B.S. Thomas, *J. Chromatogr.*, 116 (1976) 163.
63 J.B. Brooks, J.A. Liddle and C.C. Alley, *Anal. Chem.*, 47 (1975) 1960.
64 P.O. Edlund, *J. Chromatogr.*, 187 (1980) 161.
65 O. Gyllenhaal, *J. Chromatogr.*, 153 (1978) 517.
66 K. Kobayashi, M. Tanaka and S. Kawai, *J. Chromatogr.*, 187 (1980) 413.
67 Y. Hoshika and G. Muto, *J. Chromatogr.*, 152 (1978) 224.
68 K.T. Koshy, D.G. Kaiser and A.L. Van Der Silk, *J. Chromatogr. Sci.*, 13 (1975) 97.
69 T. Kawabata, H. Oshima, T. Ishibashi, M. Matsui and T. Kitsuwa, *J. Chromatogr.*, 140 (1977) 47.
70 M.S. Lennard, J.H. Silas, A.J. Smith and G.T. Tucker, *J. Chromatogr.*, 133 (1977) 161.
71 S.B. Matin, J.H. Karam and P.H. Forsham, *Anal. Chem.*, 47 (1975) 545.
72 D. Alkalay, J. Volk and M.F. Bartlett, *J. Pharm. Sci.*, 65 (1976) 525.
73 P. Husek, V. Felt and M. Matucha, *J. Chromatogr.*, 180 (1979) 53.
74 P. Husek and V. Felt, *J. Chromatogr.*, 152 (1978) 363.
75 P. Husek and V. Felt, *J. Chromatogr.*, 152 (1978) 546.

76 P. Husek, *J. Chromatogr.*, 91 (1974) 483.
77 D.J. Edwards, in K. Blau and G.S. King (Editors), *Handbook of Derivatives for Chromatography*, Heyden, London, 1977, p. 391.
78 D.G. Crosby and J.B. Bowers, *J. Agric. Food Chem.*, 16 (1968) 839.
79 P.S. Doshi and D.J. Edwards, *J. Chromatogr.*, 176 (1979) 359.
80 D.J. Edwards and K. Blau, *Anal. Biochem.*, 45 (1972) 387.
81 I.C. Cohen and B.B. Wheals, *J. Chromatogr.*, 43 (1969) 233.
82 I.C. Cohen, J. Norcup, J.H.A. Ruzicka and B.B. Wheals, *J. Chromatogr.*, 49 (1970) 215.
83 W.H. Gutenmann and D.J. Lisk, *J. Assoc. Office Anal. Chem.*, 47 (1964) 353.
84 D.W. Woodham, *J. Agric. Food Chem.*, 19 (1971) 186.
85 A.S.Y. Chau and K. Terry, *J. Ass. Offic. Anal. Chem.*, 58 (1975) 1294.
86 S. Mierzwa and S. Witek, *J. Chromatogr.*, 136 (1977) 105.
87 R.V. Smith and S.L. Tsai, *J. Chromatogr.*, 61 (1971) 29.
88 C.C. Alley, J.B. Brooks and G. Choudhary, *Anal. Chem.*, 48 (1976) 387.
89 E. Watson, B. Travis and S. Wilk, *Life Sci.*, 15 (1974) 2167.
90 E. Watson and S. Wilk, *Psychopharmacologia*, 42 (1975) 57.
91 D.D. Godse, J.J. Warsh and H.C. Stancer, *Anal. Chem.*, 49 (1977) 915.
92 E. Watson, S. Wilk and J. Roboz, *Anal. Biochem.*, 59 (1974) 441.
93 C.-G. Fri, F.-A. Wiesel and G. Sedvall, *Life Sci.*, 14 (1974) 2469.
94 S. Wilk and M. Orlowski, *Anal. Biochem.*, 69 (1975) 100.
95 C.Y. Lee and T.E. Acree, *J. Agric. Food Chem.*, 19 (1971) 184.
96 B.A. Davis, D.A. Durden, P. Pun-Li and A.A. Boulton, *J. Chromatogr.*, 142 (1977) 517.
97 S.W. Dziedzic and S.E. Gitlow, *J. Neurochem.*, 22 (1974) 333.
98 P. Hartvig and B. Nashland, *J. Chromatogr.*, 133 (1977) 367.
99 P. Hartvig, K.-E. Karlsson, L. Johansson and C. Lindberg, *J. Chromatogr.*, 121 (1976) 235.

Chapter 9

The detection of inorganic and organometallic compounds by electron-capture gas chromatography

COLIN F. POOLE and ALBERT ZLATKIS

CONTENTS

9.1. INTRODUCTION

The application of the electron-capture detector to the analysis of inorganic and organometallic compounds is limited primarily by the poor chromatographic properties of many of these substances. Most metals and their salts are either ionic or too involatile to be separated by gas chromatography. A further problem is the high chemical reactivity of many organometallic compounds which results in their loss at low levels due to irreversible interactions with the chromatographic system. However, when these problems can be overcome, excellent sensitivity can be obtained with the electron-capture detector for appropriately substituted inorganic compounds.

9.2. THE DETERMINATION OF METALS AS THEIR CHELATE DERIVATIVES

The general area of the separation of metals as their neutral chelate derivatives by gas chromatography has been extensively reviewed [1–8]. A limit to the success of this approach has been the paucity of suitable reagents which can confer the necessary volatility, thermal stability and chemical inertness (with respect to the components of the chromatographic system) on the metal ion. Numerous examples exist of metal chelate derivatives which can not be quantitatively eluted at low levels by gas chromatography. For trace metal analysis using the electron-capture detector, the most frequently employed complexing ligands are 1,1,1-trifluoropentane-2,4-dione (HTFA) and 1,1,1,5,5,5-hexafluoropentane-2,4-dione (HHFA). Chromatographically stable complexes are formed with ions such as Be(II), Al(III) and Cr(III) which form chelate derivatives having a preferred coordination number that is twice the oxidation number of the metal.

By contrast, metal ions such as Ni(II), Co(II), Fe(II), and lanthanide(III) which readily adduct additional ligands to assume a coordination number greater than twice their oxidation number are often non-quantitatively eluted from the gas chromatographic column. The HTFA chelate derivatives formed with trivalent hexacoordinate metal ions exist in two isomeric forms (*cis* and *trans*) which may be resolved into two peaks depending on the chromatographic conditions selected for the analysis. Only in the case of the HTFA derivatives of beryllium, aluminium and chromium have extensive environ-

TABLE 9.1

DETECTION LIMITS FOR SOME METAL CHELATES DETERMINED BY GAS CHROMATOGRAPHY WITH ELECTRON-CAPTURE DETECTION

Metal	Ligand*	Detector operating parameters	Detection limit of the metal (g)	Ref.
Be(II)	HTFA	^{3}H, d.c.	$4 \cdot 10^{-13}$	9
	HTFA	^{3}H, d.c., 180°C	$8 \cdot 10^{-14}$	10
	HTFA	^{3}H, pulsed, 176°C	$4 \cdot 10^{-14}$	11
	HTFA	^{3}H, d.c., 200°C	$1 \cdot 10^{-12}$	12
	HTFA	^{63}Ni, d.c.	10^{-11} g/g	16
Cr(III)	AcAc	^{3}H, d.c.	$6.5 \cdot 10^{-14}$	27
	HTFA	^{3}H, d.c.	$2.0 \cdot 10^{-16}$	27
	AcAc	^{3}H, d.c.	$8.8 \cdot 10^{-8}$	28
	HTFA	^{3}H, d.c.	$9 \cdot 10^{-11}$	28
	HHFA	^{3}H, d.c.	$3.3 \cdot 10^{-11}$	28
	HTFA	^{3}H, 190°C	5 ng/ml	13
	HTFA	^{63}Ni, pulsed, 200°C	$3 \cdot 10^{-14}$	14
	HTFA	^{63}Ni, d.c. 245°C	$2.5 \cdot 10^{-13}$	15
	HTFA	^{63}Ni, pulsed	$1 \cdot 10^{-13}$	17
Al(III)	HTFA	^{3}H, d.c.	$2 \cdot 10^{-15}$	27
	AcAc	^{3}H, d.c.	$5.1 \cdot 10^{-5}$	28
	HTFA	^{3}H, d.c.	$2.7 \cdot 10^{-9}$	28
	HHFA	^{3}H, d.c.	$4.8 \cdot 10^{-10}$	28
Cu(II)	HTFA	^{3}H, d.c.	$1.7 \cdot 10^{-11}$	27
	HATFP	^{63}Ni, pulsed, 200°C	$1 \cdot 10^{-8}$	29
	H_2bnTFA_2	^{3}H, d.c.	$1 \cdot 10^{-13}$	29
Ni(II)	HATFP	^{63}Ni, pulsed	$8 \cdot 10^{-12}$	29
	H_2enTFA_2	^{3}H, d.c.	$7 \cdot 10^{-13}$	29
Rh(II)	HTFA	^{3}H, d.c.	$2.2 \cdot 10^{-12}$	30
Pb(II)	DEDTC	^{63}Ni, d.c., 270°C	$6 \cdot 10^{-6}$	31
Zn(II)	DEDTC	^{63}Ni, d.c., 270°C	$3 \cdot 10^{-6}$	31
Ni(II)	DEDTC	^{63}Ni, d.c., 270°C	$2 \cdot 10^{-6}$	31
Hg(II)	DEDTC	^{63}Ni, d.c., 270°C	$3 \cdot 10^{-6}$	31
Cd(II)	DEDTC	^{63}Ni, d.c., 270°C	$4 \cdot 10^{-6}$	31
Co	HFOD	–	$4.4 \cdot 10^{-11}$	32

*HTFA = 1,1,1-trifluoroacetylacetonate, AcAc = acetylacetonate, HHFA = 1,1,1,5,5,5-hexafluoroacetylacetonate, HATFP = 4-amino-1,1,1-trifluoropent-3-en-2-one, H_2bnTFA_2 = N,N′-butylenebis-(trifluoroacetylacetoneimine), [bis-(trifluoroacetylacetone)butylenediimine], H_2enTFA_2 = N,N′-ethylenebis-(trifluoroacetylacetoneimine), [bis-(trifluoroacetylacetone)ethylenediimine], DEDTC = diethyldithiocarbamate, HFOD = heptafluorodimethyloctanedione.

mental and biomedical applications of the chelate derivatization technique with electron-capture detection been employed [9–26].

There is no consensus in the literature concerning the optimum detector operating conditions to obtain maximum sensitivity for the chelate derivatives. The quoted detection limits for the metal chelate derivatives by electron-capture detection are summarized in Table 9.1. Most measurements have been made with the tritium electron-capture detector which has precluded a study of the influence of detector temperature on detector response due to the limited upper operating temperature of the tritium source. In spite of this, the quoted detection limits illustrate the excellent detector sensitivity of the electron-capture detector for the HTFA derivatives of chromium and beryllium. It is believed that the detector response to these derivatives is due to both the metal ion and the trifluoroacetylacetonate ligand. The evidence offered for this is the general observation that the fluorine-containing ligands have a much greater detector response than their hydrocarbon analogues.

9.3. THE DETERMINATION OF ORGANOARSENIC COMPOUNDS

Cacodylic acid (hydroxydimethylarsine oxide) reacts rapidly and quantitatively with hydriodic acid (provided that sample water is limited to less than 20% v/v of the reagent HI) to form dimethylarsine iodide which can be detected down to the 0.05 ppm level with a ^3H electron-capture detector at 200°C [33].

$$CH_3{-}\overset{\overset{\displaystyle O}{\|}}{\underset{\underset{\displaystyle CH_3}{|}}{As}}{-}OH \xrightarrow{HI} CH_3{-}\underset{\underset{\displaystyle CH_3}{|}}{As}{-}I$$

Cacodylic acid Dimethylarsine iodide

Inorganic and methylated arsenic compounds can be determined as their diethyldithiocarbamate complexes (Table 9.2) [34]. For these compounds, the observed trend in detector response is in the opposite direction to that predicted based on the number of diethyldithiocarbamate moieties introduced. This either reflects a genuine unexpected

TABLE 9.2

ELECTRON-CAPTURE DETECTOR RESPONSE TO SOME ARSENIC DIETHYLDITHIOCARBAMATE COMPLEXES

Compound*	Electron-capture detection limit (ng/ml)**
$As(DEDTC)_3$	73
$CH_3As(DEDTC)_2$	40
$(CH_3)_2As(DEDTC)$	15

*DEDTC = diethyldithiocarbamate complex.
**Detector temperature 109°C.

trend in the electron-capturing ability of these compounds or represents increasing adsorption and/or decomposition of the diethyldithiocarbamate derivatives on the gas chromatographic column.

9.4. THE DETERMINATION OF ORGANOMERCURY COMPOUNDS

The recognition of organomercury compounds as the causative agent of the outbreak of the tragic neurological diseases inflicted upon the inhabitants of Minamata and Niigata in Japan and the discovery of toxic levels of these compounds in the inland waters of Scandinavia and the Northern United States of America has prompted numerous studies

TABLE 9.3

REAGENTS FOR THE FORMATION OF VOLATILE DERIVATIVES OF INORGANIC MERCURY

Reagent	Reaction products	Detection limit (Hg)	Electron-capture detector operating conditions	Ref.
Tetramethyltin	$CH_3HgCl + (CH_3)_3SnCl$ yield 85–90%, reaction time 5.0 min			46
	$RHgCl \xrightarrow{CuBr_2} CH_3HgBr$	1.36 ppb	^3H, d.c., 150°C	
	C_2H_5HgBr	1.30 ppb	150°C	
	C_6H_5HgBr	1.12 ppb	185°C	
Tetraalkyltin	$RHg^+ + R_3SnCl$ $R = CH_3, C_2H_5, C_3H_7$ yield 10–20%			47
$[Co(III)(CN)_5R]^{3-}$	RHg $R = CH_3, C_2H_5, C_3H_7$	10 ng/ml	^3H, d.c., 210°C	47
$[ClCH_2Cr(III)(H_2O_5]^{2+}$	$ClCH_2Hg^+$ derivative is thermally unstable			
$R_2Tl(III)CO_2R'$	RHg^+ $R = CH_3, C_2H_5$ $R' = CH_3, C_3H_7$ non-quantitative reaction			47
Tetraphenylborate	$C_6H_5Hg^+$ reagent unstable in acid solution			47
$C_6H_5SO_2H$	$C_6H_5HgCl + SO_2 + HCl$	$2 \cdot 10^{-10}$ g	^{63}Ni, pulsed, 280°C	48
Sodium 2,2′-dimethyl-2-silapentane-5-sulfonate	$CH_3HgCl \xrightarrow{KI} CH_3HgI$	<2.5 ppb	^{63}Ni, pulsed, 250–275°C	49
$C_6F_5SO_2H$	$(C_6F_5)_2Hg$	20 ppb	^3H, d.c., 210°C	50
$NaB(C_6H_5)_4$	$2Hg(C_6H_5)_2 + 3HCl + NaCl + B(OH)_3$	45 pg	^3H	51
$HgBr_2 + KBr$	$R_2Hg \rightarrow RHgBr$ conversion of dialkylmercurials			
	$R = CH_3$	10 pg	^3H, d.c., 165°C	52
	$R = C_2H_5$	25 pg		
	$R = C_3H_7$	50 pg		
	$R = C_4H_9$	25 pg		
	$R = C_6H_5$	200 pg		

on the distribution and levels of these compounds in the environment. The organomercuric halides and organomercurials have favorable gas chromatographic properties and this coupled with their high response to the electron-capture detector makes gas chromatography with electron-capture detection the method of choice for their analysis at trace levels in biological samples. The gas chromatographic properties of organomercurial compounds have been reviewed by Fishbein [35] and Mushak [36]. Of note with respect to the electron-capture detection of organomercurials in environmental samples are the early studies of Sumino et al. [37,38], Yamaguchi and Matsumoto [39] and Nishi and Horimoto [40,41] carried out in Japan and the subsequent development of improved extraction and sample cleanup techniques developed by Westoo [42,43], Uthe et al. [44] and Von Burg et al. [45] in particular. Although there is no consensus of opinion in the available literature as to the optimum operating parameters for the electron-capture detector, detection limits for alkylmercurial compounds at the low picogram level and at the tens of picogram level for arylmercurial compounds have been obtained. The use of the electron-capture detector in the study of the concentration, distribution, speciation, impact, and fate of organomercurials in the environment is now too extensive to be reviewed in this section. Gas chromatography with electron-capture detection remains the principal analytical tool for the study of organomercurial compounds in the environment.

The facile alkylation of mercury in the environment coupled with the high sensitivity of alkylmercurial compounds to electron-capture detection has prompted studies on the use of alkylating reagents as a means of derivatizing inorganic mercury for gas chromatographic analysis with electron-capture detection. The alkylating reagents evaluated for the derivatization reaction are summarized in Table 9.3. Among the more common reagents are the arylsulfinates (Peters' reaction) [48,50], $[Co(III)(CN)_5CH_3]^{3}$; [47], sodium tetraphenylborate in basic or weakly acidic solution [47,51] and tetramethyltin [46,47]. Several of the reactions given in Table 9.3 are rapid, quantitative and can be used directly with sample digests. The electron-capture detector responds to parts per billion concentrations of the derivatives being competitive in sensitivity terms with other instrumental techniques for the determination of inorganic mercury. Thus gas chromatography with electron-capture detection can be used to determine the total mercury content of a sample as well as to study the speciation of the organomercurials in that sample.

9.5. THE DETERMINATION OF SELENIUM AS PIAZSELENOLS

Selenium(IV) reacts with substituted 1,2-diaminobenzene or 2,3-diaminonaphthalene at low pH in aqueous solution to form a stable cyclic piazselenol derivative which can be extracted into an organic solvent. When the 1,2-diaminobenzene reagent is used in large excess, the reaction is quantitative and the piazselenol derivative can be determined at trace levels by gas chromatography with electron-capture detection. The structure and nomenclature for naming the piazselenol derivatives is illustrated below.

NH_2 NH_2 X —Se(IV)→ X, 1 N, Se, N 3, 4, 5, 6, 7

piazselenol

The optimum pH for piazselenol formation in aqueous solution is midway between the acid dissociation constant for the diprotonated diamine and the first acid dissociation constant of selenous acid, the reactive species in the derivative formation reaction being identified as the monoprotonated diamine and the undissociated selenous acid.

Shimoishi [53] has compared the properties of thirteen substituted 1,2-diaminobenzene reagents for the determination of selenium by formation of piazselenol derivatives. The reagent 1,2-diamino-3,5-dibromobenzene was selected as the reagent of choice based on its gas chromatographic properties, electron-capture sensitiv-

TABLE 9.4

ANALYTICAL PROPERTIES OF SOME PIAZSELENOL DERIVATIVES

Shimadzu Model GC-5A, ^{63}Ni electron-capture detector, and detector temperature 280°C.

Piazselenol	Retention index*	Distribution ratio**	Relative electron-capture detector sensitivity
Piazselenol	1307	172	1.0
5-Methylpiazselenol	1421	379	1.4
5-Ethylpiazselenol	1515	4010	1.2
5-Methoxypiazselenol	1541	297	1.7
5-Ethoxypiazselenol	1611	1661	1.6
5-Chloropiazselenol	1454	1405	17
5-Bromopiazselenol	1547	2576	30
5-Nitropiazselenol	1653	250	128
5,6-Dichloropiazselenol	1632	N.B.	102
4,6-Dibromopiazselenol	1812	N.B.	363
4,7-Dibromopiazselenol	1864	N.B.	172
4-Bromo-6-nitropiazselenol	1908	946	255
Naphthylpiazselenol	1800	N.B.	25

*1 m × 3 mm I.D. column of 15% SE-30 on Chromosorb W (60–80 mesh), 180°C.
**Calculated from the amount of piazselenol back-extracted from toluene by 1.0 *M* hydrochloric acid.
N.B. = not back-extracted.

TABLE 9.5

OPTIMUM CONDITIONS FOR THE DETERMINATION OF PIAZSELENOLS

Pye Unicamb GCV, ^{63}Ni constant-current detector, detector current 10 nA, detector temperature 350°C.

Piazselenol	Column temperature (°C)	Gas chromatography* retention time (min)	Electron-capture detector response Se ($\times 10^{-12}$ g)
Piazselenol	180	1.59	90
Naphthylpiazselenol	230	4.69	20
5,6-Dichloropiazselenol	220	1.78	1.0
5-Nitropiazselenol	220	3.94	1.0

*1.5-ft. column of 7% OV-225 on Gas-Chrom Q (100–120 mesh), nitrogen flow-rate 35 ml/min.

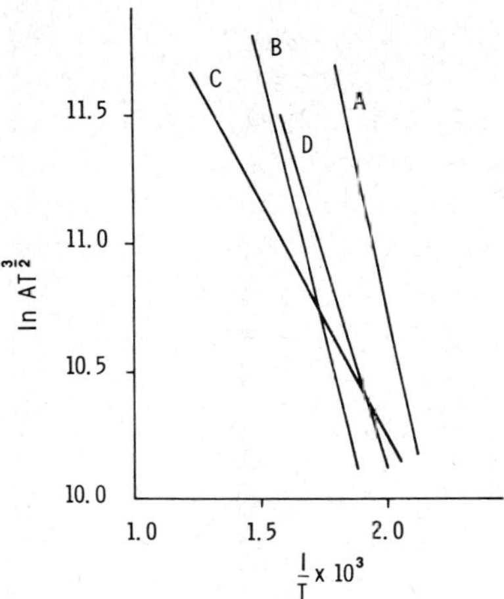

Fig. 9.1. Plots of ln $AT^{3/2}$ vs. $1/T$ for (A) piazselenol, (B) naphthylpiazselenol, (C) 5-nitropiazselenol, and (D) 5,6-dichloropiazselenol. The negative slope of the lines indicates a dissociative mechanism of electron capture.

ity and distribution ratio between aqueous solution and organic solvents. The results are summarized in Table 9.4. The response of the electron-capture detector to the different substituents in the piazselenol structure was $H < CH_3$ or $C_2H_5 < CH_3O$ or $C_2H_5O < Cl < Br < NO_2$. The highest detector response was observed for 4,6-dibromopiazselenol. The response to the 4,6-dibromopiazselenol derivative was about three times greater than the response observed for 5-nitropiazselenol. Poole et al. [54] have determined the response of the electron-capture detector to four piazselenol derivatives under optimum operating conditions (Table 9.5). The mechanism of electron-capture for the four piazselenols was shown to be dissociative (Fig. 9.1) and the maximum detector response was obtained at high detector temperatures. Without some knowledge of the products generated by capture of the thermal electrons, it was not possible to propose a molecular basis for the electron-capture process for these compounds.

Chronologically, Nakashima and Toei [55] were the first to describe the use of the electron-capture detector to determine selenium as its 5-chloropiazselenol derivative. Using a ^3H detector at 200°C, the detection limit for selenium was found to be 0.04 μg. To improve the detection limit of the selenium derivative, Shimoishi and Toei prepared the 5-nitropiazselenol derivative [56]. The detection limit under optimum conditions for selenium as the 5-nitropiazselenol derivative was found to be $1.0 \cdot 10^{-12}$ g [54]. The 5-nitropiazselenol derivative has been used to determine selenium in pure sulfuric acid [56], pure tellurium [57], sea water [58], plant material [59], copper and its salts [60,61], milk, milk products and albumin [62], arsenic and arsenic oxide [63] and in human blood, hair, urine and placenta, bovine liver and orchard leaves [54]. Young and Christian [64] used the naphthylpiazselenol derivative to determine selenium in human blood and urine. The detection limit for selenium was found to be 0.5 ng in this study and 0.2 ng by Poole et al. [54]. The 5,6-dichloropiazselenol derivative was used to determine selenium in food products such as vegetables, fruits, mushrooms, soil and whole egg powder [65]. Using a 1.0-g sample and a final solution volume of 20 ml, a detection limit of 0.01 ppm of selenium was found. Under optimum conditions the detection limit for selenium as its 5,6-dichloropiazselenol derivative is 1.0 kg [54]. The 4,6-dibromopiazselenol derivative was used to determine selenium in bovine liver [53] and in human blood [66]. The detection limit for selenium as its 4,6-dibromopiazselenol was stated to be 2.5 times better than obtained with the 5-nitropiazselenol derivative.

The reaction between selenous acid and the monoprotonated form of the diamine reagent is very selective and virtually free from interferences. None of the cations Ag^+, Cu^{2+}, Sn^{2+}, Pb^{2+}, Hg^{2+}, Cd^{2+}, Mg^{2+}, Mn^{2+}, Ni^{2+}, Zn^{2+}, Ga^{3+}, In^{3+}, As^{3+}, Te^{3+}, Sb^{3+}, Bi^{3+}, Al^{3+}, Cr^{3+}, Fe^{3+}, Ge^{4+}, V^{5+}, Mo^{6+} present in over a 1000-fold excess concentration compared to selenium interfere in the formation of the piazselenol derivative [63]. The reaction is selective for Se(IV) and Se(0) and Se(VI) are not detected by this method without prior conversion to Se(IV). Biological samples are usually digested under strong oxidizing conditions prior to piazselenol formation and this results in the oxidation of Se(0) to Se(IV). Shimoishi and Toei [56] used a bromine–bromide redox buffer to oxidize Se(0) to Se(IV) in samples of pure concentrated sulfuric acid. Se(VI) can be easily reduced to Se(IV) by heating on a water bath with concentrated hydrochloric acid for a few minutes [54,62,66]. By analyzing a sample with or without hydrochloric acid reduction enables a value for total selenium and Se(IV) to be determined and by difference Se(VI) to be calculated. As far as biological samples are concerned, the concentration of Se(0) is considered to be vanishingly small.

Although chemical interferences in the formation of the piazselenol derivative are not a problem, chromatographic interference by co-extracted electron-capturing compounds from biological samples can be. To some extent this can be eliminated by careful choice of the gas chromatographic column used for the separation. In a comparative study of three sample digestion techniques for the analysis of trace selenium levels in human placenta, it was found that very clean chromatograms were obtained when a nitric acid–magnesium nitrate digestion technique was employed (Fig. 9.2) [54]. The sample is

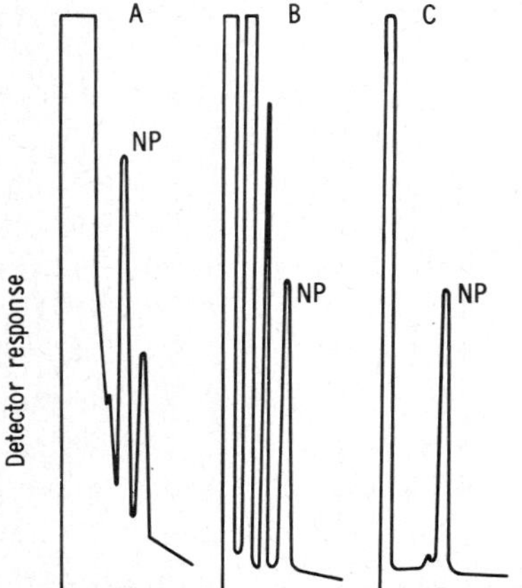

Fig. 9.2. Determination of selenium as its 5-nitropiazselenol derivative (NP) in 0.1 g of standard reference placental material after digestion with (A) nitric acid, (B) nitric acid–perchloric acid–sulfuric acid–hydrogen peroxide, and (C) nitric acid–magnesium nitrate.

TABLE 9.6

COMPARISON OF THE RESPONSE OF THE ELECTRON-CAPTURE DETECTOR TO SOME ALKYLSELENIUM COMPOUNDS

Compound	Detector response ratio ECD/FID
Dimethyl selenide	0.01
Diethyl selenide	0.002
Dipropyl selenide	0.001
Dimethyl diselenide	130
Diethyl diselenide	135
Dipropyl diselenide	150
Ethyl selenocyanate	320

initially dissolved completely at a low temperature ($<100^\circ C$), excess acid removed and the nitrates decomposed at a higher temperature. Finally, the sample digestion is completed by heating in a muffle furnace at 500°C for 0.5 h. Under these conditions the organic matrix is essentially eliminated entirely from the chromatogram. This can be seen by comparing the chromatograms A, B, and C in Fig. 9.2.

The electron-capture detector has also been used to determine some alkylselenium compounds [67]. The operating conditions for the electron-capture detector were not given, but the response of the detector compared to the flame ionization detector indicated that the alkyl diselenides and selenocyanates had good electron-capture properties while the alkyl selenides showed only a poor detector response (Table 9.6).

9.6. THE DETERMINATION OF INORGANIC ANIONS

The cyanide anion reacts with chloramine-T (sodium *p*-toluene sulfonchloramide) at ice-bath temperatures to form cyanogen chloride [68].

$$CN^- + \text{chloramine-T} \rightarrow ClCN$$

Reaction times are 5–6 min and the detection limit for the cyanogen chloride derivative was 0.025 μg/ml using an electron-capture detector. This detection limit corresponds to 12.5 pg of cyanide anion as cyanogen chloride injected on-column. The cyanide and thiocyanate anions in aqueous solution can also be determined as cyanogen bromide after reaction with bromine [69].

$$CN^- + Br_2 \rightarrow BrCN + Br^-$$

$$SCN^- + 4Br_2 + 4H_2O \rightarrow BrCN + SO_4^{2-} + 7Br^- + 8H^+$$

The thiocyanate anion can be quantitatively determined in the presence of cyanide by adding an excess of formaldehyde solution to the sample which converts the cyanide ion to the unreactive cyanohydrin. The detection limit for the cyanide and thiocyanate anions was less than 0.01 ppm using a ^3H detector at 150°C.

An organic solution of tetraheptylammonium carbonate can be used for the extraction of halide anions from aqueous solution. Injection of the organic phase into the gas chromatograph results in a thermal decomposition of the tetraheptylammonium salts to form the 1-haloheptanes of the chloride, bromide, and iodide anions [70]. The fluoride salt pyrolyzes to form heptene and hydrogen fluoride.

$$[R_4N]^+ X^- \xrightarrow{\Delta H} R_3N + RX, \text{ where } R = C_7H_{15} \text{ and } X = Cl, Br, I$$

A ^3H electron-capture detector was used to detect the haloheptanes, but the method was rendered unusable in practice due to the large amount of involatile triheptylamine formed in the pyrolysis reaction which condensed in the detector and completely masked its response. The fluoride anion reacts specifically and quantitatively with triethylsilanol and other alkylsilanol compounds to form alkylfluorosilane derivatives which are stable to gas chromatography [71].

$$(C_2H_5)_3SiOH + F^- \rightarrow (C_2H_5)_3SiF$$

However, the detection limit for the triethylfluorosilane derivative was much higher by electron-capture detection than that measured with the flame ionization detector. In order to improve the detection limit for the fluoride anion determination and to increase the selectivity of the derivatization reaction the use of the pentafluorophenyldimethylchlorosilane (flophemesyl chloride) reagent has been suggested [72]. Iodine in acid solution reacts with acetone to form monoiodoacetone which can be detected at high sensitivity with an electron-capture detector [73–75]. Strong acid conditions are required to inhibit the formation of the diiodoacetone derivative. The reaction is specific for iodine and iodide is determined after oxidation with potassium iodate. The other halide anions are not oxidized by iodate to the reactive molecular form and do not interfere in the derivatization reaction. Using a ^{63}Ni pulsed (500 μs) electron-capture detector at 150°C, the detection limit for total iodine (iodine + iodide) in milk was established as 0.004 ppm [75]. In two reactions, the concentration of free iodine and after iodate oxidation the concentration of iodine + iodide can be determined for the same sample.

The nitrate anion can be determined in aqueous solution after conversion to nitrobenzene by reaction with benzene in the presence of sulfuric acid [76, 77]. The yield of nitrobenzene was 90 ± 8% for nitrate anion concentrations in the range 0.12–62 ppm. The derivative was prepared by vigorously shaking the unhomogeneous mixture of benzene and acidified aqueous sample at 75°C for approximately 5 min. The nitrite anion can be determined after oxidation to nitrate with potassium permanganate. The detection limit for the nitrate anion was found to be less than 0.1 ppm using a ^{63}Ni pulsed (150 μs) electron-capture detector at 275°C.

9.7. THE DETERMINATION OF MISCELLANEOUS INORGANIC COMPOUNDS

The electron-capture detector has been used to determine trace quantities of nitrogen dioxide in the atmosphere [78, 79]. Using a ^3H detector at 200°C a detection limit of less than 3 ppm was established. Chlorine trifluoride can be detected at less than 0.1 ppm in

TABLE 9.7

ELECTRON-CAPTURE DETECTOR RESPONSE TO SOME SULFUR GASES

Compound	Detector temperature (°C)	Detection limit (ppb, v/v)*
H_2S	225	9
	250	12
COS	250	0.5**
CH_3SH	250	5
$(CH_3)_2S$	250	50
CS_2	250	1
	350	0.2
$(CH_3)_2S_2$	250	12
	350	3
C_4H_4S	250	500

*Pye 104, ^{63}Ni 10-mCi, 150-μs pulse.
**Highest response found at a pulse frequency of 500 μs.

the atmosphere by electron-capture detection [80]. Phosgene was detected in air at less than 1 ppb with a d.c. ^{3}H electron-capture detector [81]. The electron-capture detector has been used to detect sulfur gases in carbon dioxide at the ppb level (Table 9.7) [82]. The detector response to dimethyldisulfide was about five times greater at a detector temperature of 350°C compared to that at 250°C, in keeping with a dissociative mechanism of electron-capture. For the other sulfur gases, the highest response occurred at a detector temperature between 225 and 275°C. By comparison with the sulfur specific flame photometric detector, it was found that the electron-capture detector was as sensitive to disulfides, less sensitive to mercaptans and much less sensitive to thioethers than the flame photometric detector.

The inorganic fluorine compounds F_2, MoF_6, UF_6, SbF_3, and SbF_5 can be separated by gas chromatography and detected with an electron-capture detector [83, 84]. No detection limits for the fluorides were given. The fluorides are very corrosive and special precautions were required in the handling and separation of these compounds. Lead tetra-alkyls used as anti-knock additives in gasoline can be determined with an electron-capture detector [85]. Detection limits were not quoted in this study but the relative electron absorption coefficient of tetraethyllead was found to be 115 compared to a value of 1.0 for chlorobenzene. Triphenyltin, diphenyltin and phenyltin complexes extracted from aqueous solution as their chlorides are readily reduced to the equivalent hydride with lithium aluminium hydride [86]. The hydrides could be detected down to the 0.2 ng level with a ^{3}H electron-capture detector at 210°C. The organotin hydrides are strong reducing agents and any contact with oxidizable substances has to be avoided. Elemental sulfur when gas chromatographed produced three identifiable peaks [87]. The largest peak accounts for 98.8% of the total molar mass and could be determined with a ^{3}H d.c. electron-capture detector down to the 0.1–0.4 ng level.

REFERENCES

1 R.W. Moshier and R.E. Sievers, *Gas Chromatography of Metal Chelates*, Pergamon, Oxford, 1965.
2 D.N. Sokolov, *Zh. Anal. Khim.*, 27 (1972) 993.
3 T. Kuwamoto, *Bunseki Kagaku*, 21 (1972) 445.
4 G. Guiochon and C. Pommier, *Gas Chromatography in Inorganics and Organometallics*, Ann Arbor Sci. Publ., Ann Arbor, MI, 1973.
5 J.A. Rodriguez-Vasquez, *Anal. Chim. Acta*, 73 (1974) 1.
6 C.A. Burgett, *Sep. Purif. Methods*, 5 (1976) 1.
7 P. Mushak, in K. Blau and G.S. King (Editors), *Handbook of Derivatives for Chromatography*, Heyden, London, 1977, p. 433.
8 P.C. Uden and D.E. Henderson, *Analyst (London)*, 102 (1977) 889.
9 W.D. Ross and R.E. Sievers, *Talanta*, 15 (1968) 87.
10 M.L. Taylor and E.L. Arnold, *Anal. Chem.*, 43 (1971) 1328.
11 K.J. Eisentrant, D.J. Griest and R.E. Sievers, *Anal. Chem.*, 43 (1971) 2003.
12 M.H. Nowier and J. Cholak, *Environ. Sci. Technol.*, 3 (1969) 927.
13 L.C. Hansen, W.G. Schribner, T.W. Gilbert and R.E. Sievers, *Anal. Chem.*, 43 (1971) 349.
14 J. Savory, P. Mushak, F.W. Sunderman, R.H. Estes and N.O. Roszel, *Anal. Chem.*, 42 (1970) 294.
15 G.H. Booth and W.J. Darty, *Anal. Chem.*, 43 (1971) 831.
16 G. Kaiser, E. Grallath, P. Tschopel and G. Tolg, *Z. Anal. Chem.*, 259 (1972) 257.
17 T.R. Ryan and C.R.H. Vogt, *J. Chromatogr.*, 130 (1977) 346.
18 W.D. Ross and R.E. Sievers, *Anal. Chem.*, 41 (1969) 1109.
19 M. Miyazaki and H. Kaneko, *Chem. Pharm. Bull.*, 18 (1970) 1933.
20 J.K. Forman, T.A. Gough and E.A. Walker, *Analyst (London)*, 95 (1970) 797.
21 C. Genty, C. Houin, P. Molherbe and R. Schott, *Anal. Chem.*, 43 (1971) 235.
22 W.D. Ross and R.E. Sievers, *Environ. Sci. Technol.*, 6 (1972) 155.
23 M.L. Lee and D.C. Burrell, *Anal. Chim. Acta*, 66 (1973) 245.
24 T.A. Gasnik, *Anal. Chem.*, 47 (1975) 165.
25 R.J. Lovett and G.F. Lee, *Environ. Sci. Technol.*, 10 (1976) 67.
26 Th. Stiefel, K. Schulze, H. Zorn and G. Tölg, *Arch. Toxicol.*, 45 (1980) 81.
27 D.K. Albert, *Anal. Chem.*, 36 (1964) 2034.
28 W.D. Ross, *Anal. Chem.*, 35 (1963) 1596.
29 P.C. Uden, D.E. Henderson and A. Kamalizad, *J. Chromatogr. Sci.*, 12 (1974) 591.
30 W.D. Ross, R.E. Sievers and G. Wheeler, *Anal. Chem.*, 37 (1965) 598.
31 M. Ahmad and A. Aziz, *J. Chromatogr.*, 152 (1978) 542.
32 W.D. Ross and R.E. Sievers, *Develop. Appl. Spectroscop.*, 8 (1970) 181.
33 C.J. Soderquist, D.G. Crosby and J.B. Bowers, *Anal. Chem.*, 46 (1974) 155.
34 E.H. Daughtrey, A.W. Fitchett and P. Mushak, *Anal. Chim. Acta*, 79 (1975) 199.
35 L. Fishbein, *Chromatogr. Rev.*, 13 (1970) 133.
36 P. Mushak, *Environ. Health Perspect.*, 4 (1973) 55.
37 S. Kitamura, T. Tsukamoto, K. Hayakawa, K. Sumino and T. Shibata, *Med. Biol.*, 72 (1966) 274.
38 K. Sumino, *Kobe J. Med. Sci.*, 14 (1968) 115.
39 S. Yamaguchi and H. Matsumoto, *Kurume Med. J.*, 16 (1969) 33.
40 S. Nishi and Y. Horimoto, *Bunseki Kagaku*, 17 (1968) 1247.
41 S. Nishi and Y. Horimoto, *Bunseki Kagaku*, 17 (1968) 75.
42 G. Westoo, *Acta Chem. Scand.*, 20 (1966) 2131.
43 G. Westoo, *Acta Chem. Scand.*, 22 (1968) 2277.
44 J.F. Uthe, J. Solomon and B. Grift, *J. Ass. Offic. Anal. Chem.*, 55 (1972) 583.
45 R. von Burg, F. Faris and J.C. Smith, *J. Chromatogr.*, 97 (1974) 65.
46 C.J. Cappon and J.C. Smith, *Anal. Chem.*, 49 (1977) 365.
47 P. Zarnegar and P. Mushak, *Anal. Chim. Acta*, 69 (1974) 389.
48 P. Jones and G. Nickless, *J. Chromatogr.*, 76 (1973) 285.

49 P. Jones and G. Nickless, *J. Chromatogr.*, 89 (1974) 201.
50 P. Mushak, F.E. Tibbetts, P. Zarnegar and G.B. Fisher, *J. Chromatogr.*, 87 (1973) 215.
51 V. Luckow and H.R. Russel, *J. Chromatogr.*, 150 (1978) 187.
52 M. Shariat, *J. Chromatogr. Sci.*, 17 (1979) 527.
53 Y. Shimoishi, *J. Chromatogr.*, 136 (1977) 85.
54 C.F. Poole, N.J. Evans and D.G. Wibberley, *J. Chromatogr.*, 136 (1977) 73.
55 S. Nakashima and K. Toei, *Talanta*, 15 (1968) 1475.
56 Y. Shimoishi and K. Toei, *Talanta*, 17 (1970) 165.
57 Y. Shimoishi, *Bull. Chem. Soc. Jap.*, 44 (1971) 3370.
58 Y. Shimoishi, *Anal. Chim. Acta*, 64 (1973) 465.
59 Y. Shimoishi, *Bull. Chem. Soc. Jap.*, 47 (1974) 997.
60 Y. Shimoishi, *Bull. Chem. Soc. Jap.*, 48 (1975) 2797.
61 A. Meyer, E. Grallath, G. Kaiser and G. Tolg, *Z. Anal. Chem.*, 281 (1976) 201.
62 Y. Shimoishi, *Analyst (London)*, 101 (1976) 298.
63 M. Akiba, Y. Shimoishi and K. Toei, *Analyst (London)*, 101 (1976) 644.
64 J.W. Young and C.D. Christian, *Anal. Chim. Acta*, 65 (1973) 127.
65 T. Stijve and E. Cardinale, *J. Chromatogr.*, 109 (1975) 239.
66 K. Kurahashi, S. Inoue, S. Yonekura, Y. Shimoishi and K. Toei, *Analyst (London)*, 105 (1980) 690.
67 C.S. Evans and C.M. Johnson, *J. Chromatogr.*, 21 (1966) 202.
68 J.C. Valentour, V. Aggarwal and I. Sunshine, *Anal. Chem.*, 46 (1974) 924.
69 G. Nota and R. Palombari, *J. Chromatogr.*, 84 (1973) 37.
70 D.R. Matthews, W.D. Schults and J.A. Dean, *Anal. Lett.*, 6 (1973) 513.
71 R. Bock and H.-J. Semmler, *Z. Anal. Chem.*, 230 (1967) 161.
72 H. Ehrson, Stockholm, personal communication.
73 R.A. Hasty, *Mikrochim. Acta*, (1971) 348.
74 R.A. Hasty, *Mikrochim. Acta*, (1973) 621.
75 St. Grays, *J. Chromatogr.*, 100 (1974) 43.
76 D.J. Glover and J.C. Hoffsommer, *J. Chromatogr.*, 94 (1974) 334.
77 J.C. Hoffsommer, D.J. Glover and D. Krubose, *J. Chromatogr.*, 103 (1975) 182.
78 M.E. Morrison, R.G. Rinker and W.H. Corcoran, *Anal. Chem.*, 36 (1964) 2256.
79 M.E. Morrison and W.H. Corcoran, *Anal. Chem.*, 39 (1967) 255.
80 S. Zolty and M.J. Prager, *J. Gas Chromatogr.*, 5 (1967) 533.
81 L.J. Priestley, F.E. Critchfield, N.H. Ketcham and J.D. Cavender, *Anal. Chem.*, 37 (1965) 70.
82 M.E. Pick, *J. Chromatogr.*, 171 (1979) 305.
83 O. Pitak, *Chromatographia*, 2 (1969) 304.
84 O. Pitak, *Chromatographia*, 3 (1970) 29.
85 J.E. Lovelock and A. Zlatkis, *Anal. Chem.*, 33 (1961) 1958.
86 C.J. Soderquist and D.G. Crosby, *Anal. Chem.*, 50 (1978) 1435.
87 D.L. Struble, *J. Chromatogr. Sci.*, 10 (1972) 57.

Chapter 10

Environmental applications of the electron-capture detector – pesticides

W.P. COCHRANE and R.B. MAYBURY

CONTENTS

10.1. INTRODUCTION

Since its inception in 1960 [1] the electron-capture detector (ECD) has played a major role as a gas chromatographic (GC) detection technique in the analysis of pesticides, polychlorinated biphenyls and other organic pollutants of environmental interest. During the intervening 20 years the ECD has undergone continuous development. Extensive data now exist on detection mechanisms, design, operation, response characteristics and optimization of the various types and these topics have been comprehensively reviewed up to 1974 by Pellizzari [2] and more recently by Holland and Greenhalgh [3].

The various types of ECDs that have been commercially available for pesticide residue work differ in ionizing source used, design of the detector cell and the different modes of electrical stimulation and measurement, with each of these parameters having a profound effect on the performance of the ECD. Historically, there have been four types of ECDs commonly used in pesticide analysis – three utilizing radioactive sources (tritiated titanium, ^{3}HTi; tritiated scandium, ^{3}HSc; and nickel-63, ^{63}Ni) and one non-radioactive source (helium glow discharge).

Foils with ^{3}H imbedded in titanium were the only type available in the early 1960s and, therefore, these have been the traditional source of β-particles which ionize the carrier gas, generally argon, helium or nitrogen, to produce electrons [4]. The ^{3}HTi sources are characterized by high specific activities (e.g. >200 mCi) which resulted in a large standing current and increased sensitivity and linear range when compared to sources employing other radioisotopes. Since ^{3}H is easily lost from these foils the upper operational temperature limit is 225°C which results in the foil being extremely susceptible to contamination from GC column eluates. The d.c. mode of operation was also subject to instability and anomalous responses such that the d.c. mode characteristically exhibited drifting baselines, sensitivity variations and limited dynamic range (typically $10–10^{2}$) [5]. While the ^{3}HTi ECD is extremely sensitive to halogens, NO_2, S, other electronegative elements and some organometallic compounds, it is relatively unresponsive to gross quantities of hydrocarbons, alcohols, ketones, etc. In the case of DDT in hexane, for example, the response ratio is 10^{6}:1 when DDT is present in quantities as low as 1 pg. It should be noted that although the ^{3}HTi pin-cup ECD has been a valuable and reliable tool in pesticide laboratories for 20 years it is no longer commercially available.

It was quickly realized that a stable ionisation source which would not volatilize under normal operating conditions was needed for high-temperature GC in the pesticide area. In 1973, Hartmann [6] published a detailed evaluation of scandium tritide as a high-temperature ionization source. ^{3}HSc sources have a maximum temperature limit of 325°C and source activities were available that approach 1 C/cm^{2}. This provided a higher baseline current (min. $4 \cdot 10^{-8}$ A) than any other electron-capture β-source. When operated in the constant current/variable frequency pulsed mode, the linear range of the detector is significantly increased, i.e. $10^{2}–10^{4}$ [7]. Although this radiation source has lost favour with commercial detector manufacturers, Brechbühler et al. [8] reported the detection of halogenated organic volatiles in water, polychlorinated biphenyls (PCBs) and pesticide residues in foods as well as bile and amino acids in body fluids using a micro-ECD (cell volume <100 mm^{3}) coupled with high-resolution glass capillary columns. Operating in the preferred constant current mode linearity was 10^{4} with a minimum detection limit of <50 fg lindane at a signal-to-noise ratio of 5:1.

To still further increase the temperature range of ECDs, a number of alternative radiation sources have been reported. A tritium source based on a rare-earth matrix has been used at temperatures as high as 400°C [9] and a ^{147}Pm source on a gold foil gave sensitivities of 0.1 ppb for pesticide analysis at 400°C [10]. However, it is the ^{63}Ni ECD that has become the widely accepted ECD due to its stability at higher operating temperatures (up to 400°C) [11]. Originally, since the ^{63}Ni source has a very much lower specific activity (max. 15 mCi), its sensitivity and linear dynamic range were much less than of ^{3}H sources. However, by operating the ^{63}Ni detector in the constant current mode [7,12] the linear range is extended by a factor of 200 to about $2 \cdot 10^{4}$. However, some substances of interest are restricted by their chemical properties, to a limited linear dynamic range and these non-linearities have been reported on by Sullivan and Burgett [13].

During the 1960s, Beckman Instruments marketed a non-radioactive source ECD [14] which was used in pesticide residue work. The unit was basically an ionization detector utilizing a helium glow discharge. There were, therefore, two forms of the detector developed – an electron-capture form and a helium ionization form. These were similar in configuration except that in the helium ionization form, the orientation of the discharge electrodes was 180° from that of the ECD form. The maximum operating temperature was 400°C instead of the 225°C attainable with the ^3H ECDs. An average discharge potential of approx. −100 V through the helium stream produced, among other species, low-energy electrons and photons. Entry of electron-absorbing sample components reduced the electron population and therefore the background current. The magnitude of current reduction was a non-linear function of the concentration of the components of interest; therefore, a calibration curve must be used for quantitative results. This detector also employed the addition of 1–3 ml/min CO_2 diluent to the helium carrier gas prior to the ECD. Since the diluent molecules collide inelastically with the electrons this resulted in a decrease in their random movement and made the electrons more susceptible to collection. The principal advantage of CO_2 over other diluents was its tendency to increase detector selectivity by absorbing photons which might otherwise ionize other sample components, e.g., hydrocarbons, and possibly detract from the signal of the electron-capturing component. The limit of detection for lindane was approx. 2 pg. Subsequently, a modified helium "photoionization" detector was described by Villalobos and co-workers [15,16] and Freeman and Wentworth [17] capable of detecting inert gases as well as organic compounds. The interaction of photons with ionizable molecules in the carrier gas produced electrons which were collected, amplified and detected. This detector was designed for use in process gas chromatographs. Although these Beckman detectors are no longer commercially available, reports on the analysis of polybrominated biphenyls in plants [18] and soils [19] show they are still being used.

A decade later, in 1975, an ECD was constructed which uses photoionization produced with resonance lamps [20]. Although microwave excitation of both argon and hydrogen (25% hydrogen in helium) were investigated, it was concluded that the Lyman-α-resonance line of hydrogen, with triethylamine added to the gas stream (as the electron source compound), produced a viable photo-ECD operating in the d.c. mode. The lower limit of detection of CCl_4 and $CHCl_3$ was 50 pg and 200 pg respectively. In general, the photo-ECD was 10–150 times less sensitive than a ^3H ECD operated in the pulsed mode. A photoionization detector employing interchangeable sealed UV lamp sources has been marketed commercially and shown to rival the flame ionization detector in noise and background but has a 30-fold increase in sensitivity for carbon compounds [21].

Employing the ECDs described above for the routine analysis of pesticide residues and other chlorinated organics in biological and environmental substrates usually involves quite a few trade-offs between speed, reliability and sensitivity. Various contaminants including bleed from the septum and liquid phase, oxygen from the atmosphere or carrier gas and excessive amounts of co-extractives markedly affect linearity and detection limits. Also, the choice of GC column and amount of preceding clean-up of samples can drastically influence the lifetime (and intervals between rejuvenation) of the foil. Similarly, the choice of derivatization reaction in pesticide residue analysis is crucial since it is utilized for a variety of reasons, but primarily to enhance detection, allow determination of active hydrogen-containing (e.g., OH, NH, etc.) pesticides or their metabolites and for the confirmation of identity. These factors are discussed in more detail in the appropriate section.

10.2. ORGANOCHLORINE PESTICIDES

The application of the ECD to the residue analysis of organochlorine (OC) pesticides is an enormous field. This covers a period of over 20 years from the initial application of ECD to detect and quantitate lindane, five cyclodiene insecticides and DDT in various crop extracts [22,23]. The major members of the OC class of compounds are given in Table 10.1 and normally these compounds are highly lipid soluble and hence are not readily attacked by bacteria or leached from the soil. Such characteristics render them

TABLE 10.1

LISTING OF ORGANOCHLORINE INSECTICIDES WITH RETENTION TIME AND ELECTRON-CAPTURE RESPONSE DATA

Organochlorine insecticide	Relative retention time (aldrin = 1.00)*	Electron-capture response** (ng)
Hexachlorobenzene	0.42	0.4
α-BHC	0.46	0.4
Chlordene	0.56	0.6
γ-BHC (lindane)	0.58	0.5
δ-BHC	0.60	1.8
Heptachlor	0.81	0.7
Chlordene epoxide	0.92	0.6
Aldrin	1.00	0.8
Octachlor epoxide (oxychlordane)	1.38	1.0
Heptachlor epoxide	1.47	1.0
trans-Chlordane	1.57	1.0
cis-Chlordane	1.73	1.0
Photodieldrin B	1.86	2.0
p,p'-DDE	1.88	1.5
Endosulfan I (thiodan I)	1.89	1.5
Perthane	2.02	200.0
Dieldrin	2.22	1.5
Endrin	2.55	2.0
Chlordecone (kepone)	2.67	5.0
Endosulfan II (thiodan II)	2.92	2.0
p,p'-DDT	3.28	4.0
8-Monohydromirex (mirex photoproduct)	3.52	5.0
p,p'-Methoxychlor	4.8	9.0
Mirex	5.1	9.0
Endosulfan sulphate (thiodan sulfate)	5.4	5.0
Endrin ketone (delta keto 153)	6.0	5.0
Photodieldrin A	7.5	6.0

*Using 10% DC-200–15% QF-1 (1:1) on Chromosorb W HP.

**Nanograms required for 50% full-scale recorder deflection [27].

exceptionally persistent thus giving rise to the major problem of pollution of the global environment [24,25]. The partitioning characteristic of the OC pesticides between aqueous and lipid phases in successive organisms leads to bioaccumulation of the compounds as they move up the food chain. This phenomenon was responsible for the disastrous effects observed on certain species of wildlife in the 1960s. Subsequently in the early 1970s severe restrictions were placed on the permitted use of many OC insecticides and maximum residue limits were introduced for many foods by certain countries especially the U.S.A. To monitor residues in all of these areas, the GC–ECD system of choice has been used extensively and applications have covered all substrates imaginable – water, blood, urine, soil, sediments, plants, vegetables, air, animal and fish species and all food commodities. As a result of the restricted use of DDT, aldrin, dieldrin and heptachlor in the 1970–1973 era, a gradual fall in residue levels has been observed in many Western countries in recent years.

10.2.1. Foods and humans

Prior to the establishment of residue tolerances, an adequate method of analysis must be available. It was established early that the only reasonable approach to monitoring the food supply was to simplify analyses by developing methods which measure more than one chemical at a time. In 1970 [26] there existed multi-residue methodology that would tentatively identify and quantitate about 55 organochlorine and about 15 organophosphorus pesticide chemicals in 450 different food products. The end determination method was via electron-capture, alkali flame ionization or flame photometric detection. Constant expansion and improvement of the multi-residue methods by the U.S. Food and Drug Administration [27] and agencies such as the Association of Official Analytical Chemists [28] resulted in method behaviour data being available on 183 pesticides in non-fatty foods and 142 pesticide chemicals in fatty foods by 1978. Theoretically any of hundreds of pesticides, their metabolites and degradation products and other environmental contaminants could be present in any sample. To help overcome the enormous problem of quick tentative identification, a protocol of relative retention time data was compiled on 5 to 8 different packed GC columns varying in phase type and loading [28,29]. This has been compiled in computer format, 46 pages in 1978 [29], for cross-reference of retention time, detector sensitivity, method recovery data, etc. (see Table 10.2). ECDs used to measure OC residues with this multi-residue approach have traditionally been set to produce 1/2 full-scale recorder deflection for 1 ng heptachlor epoxide. Aliquots of the final cleaned-up sample extract equivalent to 20–30 mg non-fatty sample or 3 mg fat are injected into the chromatogram for initial examination. As example, this has resulted in limits of quantitation for heptachlor epoxide being 10 ppb in non-fatty foods and 70 ppb in fat samples. Different methods and GC columns using electron-capture quantitation have been collaboratively studied in other countries, such as Canada [30], G.F.R. [31], etc. Using such multi-residue methodology, the Total Diet studies of the U.S. Food and Drug Administration [32] provide exhaustive information on levels and exact dietary sources of many persistent OC insecticides. For example, it has been shown that dieldrin residue levels in environmental substrates, which are indicative of aldrin or dieldrin use, have decreased significantly since the mid 1960s, the peak

TABLE 10.2

FIRST PAGE OF COMPUTER PRINTOUT FROM FDA PESTICIDE ANALYTICAL MANUAL SHOWING ELECTRON-CAPTURE RESPONSE CHARACTERISTICS AND METHODOLOGY STATUS [27]

05/23/79 PESTICIDES BY FDA NO. PAGE 1

STD NO.	RRT/A. DC-200	RRT/A. MIXED	EC/GLC SENS. NG/FSD/2	TID/GLC SENS. NG/FSD/2	RRT/P. DC-200	RRT/P. MIXED	PREFERRED NAME	MOLECULAR FORMULA	THRU FAT METHOD	THRU NONFAT METHOD	MIXED ETHER	METHYLENE CHLORIDE	THRU CHAR
2	1.00	1.00	0.8	-	1.04	.53	ALDRIN	C12H8CL6	C	C	6	1	-
5	1.96	2.54	9999.9	-	2.03	1.35	ARAMITE *	C15H23CLO4S	NR	P	15	-	-
5	2.10	2.74	9999.9	-	2.18	1.46	ARAMITE *	C15H23CLO4S	NR	P	15	-	-
6	0.41	0.54	200.	58.	.43	.29	ATRAZINE	C8H14CLN5	-	P	50	0	-
8	-	0.68	1.5	-	-	.36	BHC (TECHNICAL)*	C6H6CL6	-	-	-	-	-
8	0.38	0.46	1.5	-	.39	.24	BHC (TECHNICAL)*	C6H6CL6	-	-	-	-	-
8	0.47	0.58	1.5	-	.49	.31	BHC (TECHNICAL)*	C6H6CL6	-	-	-	-	-
10	3.32	5.4	3.	-	3.44	2.9	BULAN	C16H15CL2NO2	P	P	15	2	-
12	0.19	0.26	2.	-	.20	.14	DICAMBA METHYL ESTER (DATA FOR CHEMICAL OTHER THAN THE REFERENCE STD)	C8H6CL2O3	-	-	-	-	-
15	9.5	25.	100.	50.	9.9	13.	COUMAPHOS	C14H16CLO5PS	-	NR	0	3	C
16	-	1.82	65.	-	-	.97	CHLORTHION*	C8H9CLNO5PS	C	C	15	-	-
16	-	2.27	65.	-	-	1.21	CHLORTHION*	C8H9CLNO5PS	C	C	15	-	-
16	1.05	1.35	65.	-	1.09	.72	CHLORTHION*	C8H9CLNO5PS	C	C	15	-	-
17	1.17	2.10	1.5	-	1.21	1.12	CAPTAN	C9H8CL3NO2S	-	P	50	3	-
18	0.29	0.40	7.	-	.30	.21	CHLORANIL	C6CL4O2	-	-	-	-	-
19	1.36	1.51	2.	-	1.41	.80	CHLORBENSIDE	C13H10CL2S	P	P	6	1	-
20	0.44	0.53	12.	-	.46	.28	CHLORDANE (TECHNICAL)*	C10H6CL8	C	C	6	1	-
20	0.62	0.82	12.	-	.64	.44	CHLORDANE (TECHNICAL)*	C10H6CL8	C	C	6	1	-
20	0.72	1.24	12.	-	.75	.66	CHLORDANE (TECHNICAL)*	C10H6CL8	C	C	6	1	-
20	0.79	1.34	12.	-	.82	.71	CHLORDANE (TECHNICAL)*	C10H6CL8	C	C	6	1	-
20	0.95	1.45	12.	-	.99	.77	CHLORDANE (TECHNICAL)*	C10H6CL8	C	C	6	1	-
20	1.14	1.58	12.	-	1.18	.84	CHLORDANE (TECHNICAL)*	C10H6CL8	C	C	6	1	-
20	1.42	1.82	12.	-	1.47	.97	CHLORDANE (TECHNICAL)*	C10H6CL8	C	C	6	1	-
20	1.59	1.92	12.	-	1.65	1.02	CHLORDANE (TECHNICAL)*	C10H6CL8	C	C	6	1	-
20	2.56	2.36	12.	-	2.65	1.26	CHLORDANE (TECHNICAL)*	C10H6CL8	C	C	6	1	-
21	2.26	2.73	75.	-	2.34	1.45	CHLOROBENZILATE (ELUTION VARIABLE)	C16H14CL2O3	P	C	1550	3	-
22	0.31	0.41	2000.	-	.32	.22	CHLORPROPHAM (CIPC)	C10H12CLNO2	C	C	15	2	-
24	0.33	1.03	-	4.	.34	.55	DICROTOPHOS (BIDRIN)	C8H16NO5P	-	NR	-	-	C
26	1.34	2.59	-	11.	1.39	1.38	CROTOXYPHOS (CIODRIN)	C14H19O6P	-	NR	-	-	C
27	0.33	0.13	-	6.	.34	.07	NALED	C4H7BR2CL2O4P	-	NR	0	0	P
28	1.81	1.88	1.5	-	1.88	1.00	DDE, P, P'-	C14H8CL4	C	C	6	1	-
31	0.36	0.39	5.	-	.37	.21	DDT (TECHNICAL)*	C14H9CL5	-	-	-	-	-
31	2.44	2.48	5.	-	2.53	1.32	DDT (TECHNICAL)*	C14H9CL5	-	-	-	-	-
31	3.03	3.28	5.	-	3.14	1.75	DDT (TECHNICAL)*	C14H9CL5	-	-	-	-	-
32	0.54	0.85	3.	-	.56	.45	DICHLONE (ELUTION VARIABLE)	C10H4CL2O2	P	P	0	23	NR
34	1.83	2.22	1.5	-	1.90	1.18	DIELDRIN	C12H8CL6O	C	C	15	2	-
35	-	5.4	7.	-	-	2.9	DILAN*	C15.5H14CL2NO2	P	P	15	-	-
35	2.28	3.6	7.	-	2.36	1.9	DILAN*	C15.5H14CL2NO2	P	P	15	-	-
35	2.75	4.4	7.	-	2.85	2.3	DILAN*	C15.5H14CL2NO2	P	P	15	-	-
35	3.32	4.7	7.	-	3.44	2.5	DILAN*	C15.5H14CL2NO2	P	P	15	-	-

usage years. Aldrin and dieldrin use were banned in the U.S.A. in 1974. However, human milk and tissue studies do not show a corresponding decline; between 1963 and 1976 the average dieldrin levels maintained by the body remained in the range 0.15–0.22 ppm. Thus it was concluded that the concentration of dieldrin in the human body has reached a constant level at which the amount ingested and absorbed equals the amount metabolized and excreted [33].

Ever since Laug et al. [34] first reported pesticide residues in human milk in 1951, and particularly after the introduction of multi-residue screening by GC–ECD techniques, numerous studies have been reported on the sources, bioaccumulation metabolism and current levels of OCs and PCBs in humans [35]. In India, GC–ECD analyses of butter samples showed that DDT residues ranged from 0.42 to 11.36 ppm and exceeded the FAO/WHO practical residue limit of 1.25 ppm in 90% of the samples [36]. Also infant formula contained DDT residues above the practical residue limit. Most DDT residues were in the form of *p,p'*-TDE (DDD), with smaller amounts of *p,p'*-DDT and *p,p'*-DDE, for both commodities. Also, a 1975 U.S.A. market basket survey of fresh whole milk consumed by infants showed that the major OCs were DDE (0.001–0.005 ppm, whole product basis), dieldrin (0.001–0.002 ppm), heptachlor epoxide, HCB and lindane [37]. In a comparison of a high pesticide use area (Mississippi delta) and a low pesticide use area, the residue levels found in human milk samples were 19.17 ppm and 2.36 ppm (fat basis), respectively [35] for total DDT and metabolites. While the 19.17 ppm ΣDDT level is the highest recorded, β-BHC, heptachlor epoxide, oxychlordane and dieldrin also occurred as significant residues. These figures only demonstrated that DDT is stored in greater amounts than are other OC pesticides and it is found in almost every adipose tissue sample analyzed [38]. Again, like dieldrin, since the ban on the U.S.A. domestic use of DDT in 1972 the adipose tissue residue levels of ΣDDT have decreased slightly but the frequency of finding DDT or its metabolites remains high [39].

10.2.2. Soils and aqueous media

OC residues occur in soils from the direct incorporation of the parent OC pesticide to kill soil-borne pests and also from fallout from crop application, rainfall, drift, etc. Since the OC insecticides persist longer in soil than on foliage or in water, many reports on OC residues in the various types of field and orchard soils have been published over the years. The field was reviewed by Edwards in 1972 [40]. The methodology used in such studies has also been the subject of intense study by a number of workers, all utilizing the ECD as the end determinative method of choice for parent OC, metabolites and degradation products. Chiba and Morley [41] tested 21 solvent extraction systems for aldrin and dieldrin from various soils while Pionke et al. [42] observed that the highest recovery of OC insecticides was obtained by Soxhlet extraction using hexane–acetone (41:59). Williams [43] further increased recoveries by adding water to the soil prior to Soxhlet extraction and in a collaborative test of 12 laboratories and their procedures [44] it was noted that water addition appeared to give better recoveries and that hexane–acetone (1:1) was the solvent system of choice. Saha [45] also concluded that the moisture content of the soil was most important. He found that the addition of ca. 20% water to air-dried soils gave 92–98% recovery of OC insecticides extracted with hexane–

2-propanol, hexane–acetone or benzene–methanol, all as 1:1 mixtures. It was further stated that the ability of water to absorb dieldrin from soil is so great that all other factors of extraction are of little consequence. In the official AOAC method for determining OC insecticides in soil, the sample is premoistened with 0.2 *M* ammonium chloride solution rather than water; the salt causes the soil colloids to expand for better contact with the solvent [46]. However, Soxhlet refluxing usually fares best in comparison with shaking or blending in terms of efficiency of pesticide recovery [44]. This approach has been used for many years and in the EPA Manual for Environmental Analysis [47], OC pesticides together with other lipid-soluble substances are extracted from homogenised air-dried soil samples by continuous Soxhlet reflux with acetone–hexane (1:1). After solvent stripping, alumina and Florisil column clean-up steps and determination by ECD, average recoveries have been >83% for lindane (85.25%), heptachlor epoxide (87.3%), *p,p'*-DDE (83.08%), dieldrin (88.25%), *p,p'*-DDD (91.17%) and *p,p'*-DDT (94.17%).

In addition to the above application of the ECD to the methodology for OC residues in soil, it has also had extensive use in soil persistence, metabolism, degradation and leaching studies, as well as the translocation of OC residues from soil to plant and soil–pesticide interactions [40,48].

One area that EC-GC analysis has had a tremendous impact on is the estimation of the soil persistence of various OCs and the identification of metabolites and degradation products. Table 10.3 shows the relative persistence of some OC insecticides in soil as reported in various publications. As expected large variations occur depending on application rate, soil type and length of monitoring. However, each OC soil study must be viewed individually since OC residues have been detected in field soils 15 years after a single application of DDT, aldrin and lindane [52], 15 years after BHC, DDT and chlordane [53] and 20 years after an OC pesticide application [54]. In data collected in fiscal year 1970 from 1506 sites located in 35 States, the U.S. National Soils Monitoring Program found dieldrin, ΣDDT (average 0.3 ppm) aldrin, chlordane and heptachlor

TABLE 10.3

PERSISTENCE OF SOME ORGANOCHLORINE INSECTICIDES IN SOILS

Organochlorine insecticide	Edwards, 1966 [49]		Nash and Woolson, 1967 [50]		Menzies, 1972 [51]
	Dosage (lb./acre)	Years for 95% disappearance*	% organochlorines remaining	Years	Half-life (years)
DDT	1–2.5	4–30 (10)	39	17	3–10
Dieldrin	1–3	5–25 (8)	31	15	1–7
Endrin	–	–	41	1	4–8
Toxaphene	–	–	45	14	10
Chlordane	1–2	3–5 (4)	40	14	2–4
Heptachlor	1–3	3–5 (3.5)	16	14	7–12
Aldrin	1–3	1–6 (3)	28	14	1–4
Lindane	1–2.5	2–7 (6.5)	10	14	2

*Average given in parentheses.

epoxide residues most frequently in soils [55], highest residue levels being found in the corn and cotton cropping regions; lowest levels were in the hay, general farming and small-grain soils. The list of OC residues increased when urban soils were monitored; here ΣDDT, aldrin, dieldrin, photodieldrin, chlordane, heptachlor, heptachlor epoxide, endrin, endrin ketone and endosulfan sulfate were found [56]. Five U.S.A. metropolitan areas were included in the urban soil program and the highest detected levels found were aldrin, 0.74 ppm; dieldrin, 6.02 ppm; ΣDDT, 17.56 ppm; chlordane, 140.69 ppm; toxaphene, 4.95 ppm; and heptachlor epoxide, 1.95 ppm. In agricultural soils aldrin gradually declined during 1968–1973 while its primary oxidation product, dieldrin, peaked about 1969 and then started to decline. Similar findings have also been reported in Japanese field soils [57]. In a comparison of ΣDDT and chlordane residues in U.S.A. urban and agricultural soils sampled in 1973, urban areas are more heavily contaminated with the OCs than agricultural soils in the same locality. Only in the southern U.S.A., where DDT was used heavily in cotton cultivation, do the agricultural ΣDDT residues exceed urban concentrations [58]. DDT readily degrades in anaerobic soil conditions to DDD (TDE) and slowly in aerobic soils to DDE. This also leads to marked differences in ΣDDT persistence in such soils. This was clearly demonstrated in an Arizona study which suggested the ΣDDT half-life in irrigated soils was ca. 7 years and only 2.5 years in desert or non-irrigated soils [59], with the bulk of the ΣDDT residues steadily shifting with time, to higher proportions of DDE. Cliath and Spencer [60] suggested that the disappearance of DDT from well-aerated soil was due to conversion to DDE and subsequent dissipation of DDE into the atmosphere.

To estimate the aerial transportation of DDT, one indicator has been the results obtained from the GC–ECD analysis of rain water samples. Table 10.4 shows the levels of DDT residues found in various studies for the years 1965–1974. The two major metabolites found were *p,p'*-DDD and *p,p'*-DDE with the latter being present in relatively smaller amounts. Atmospheric transportation and deposition were also believed to be the mechanisms responsible for the wide distribution of lindane and α-BHC in Western Canada [67]. In this 1971–1977 study, approximately 1400 samples were analysed for 13 OC residues at the ppt level (Table 10.5). Concentrations of both lindane and α-BHC at times exceed 0.01 μg/l. The only other OCs detected were heptachlor, aldrin and endosulfan II. Also, in this instance, it was speculated that isomerization of lindane to the

TABLE 10.4

DDT LEVELS FOUND IN RAIN WATER, 1965–1974

Year	Region	DDT, μg/kg (ppb)	Reference
1965	Great Britain	0.141–0.246	61
1965	Ohio, U.S.A.	0.075–0.370	62
1966–1967	Antarctica	0.040*	63
1967–1968	Eastern Canada	ND–1.33	64
1970–1971	Hawaii	0.001–0.13*	65
1974	New York, U.S.A.	ND–0.002	66

*No DDD or DDE residues were detected.

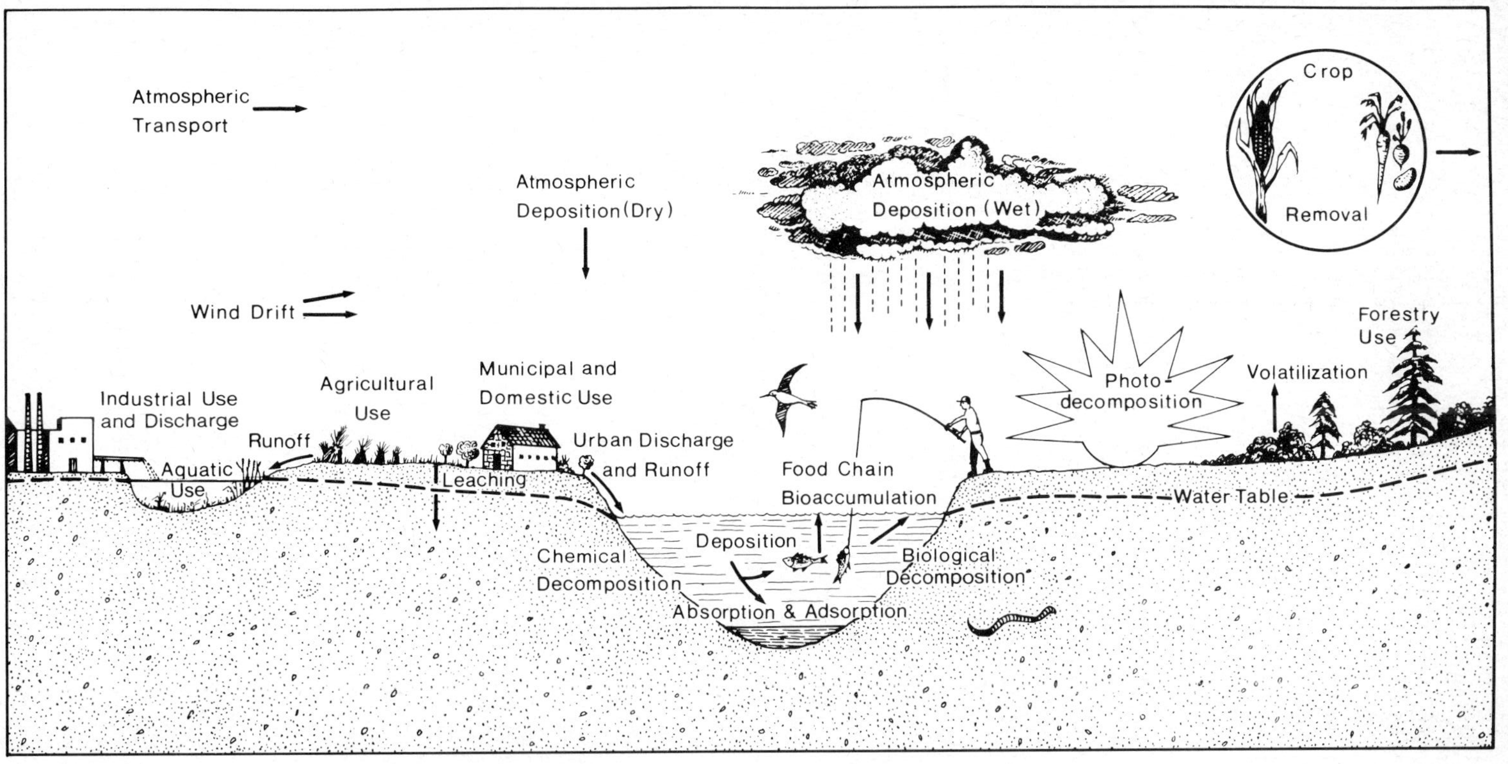

Fig. 10.1. Pesticide uses and environmental pathways. (From ref. 67.)

α-BHC isomer accounted for the abundance of α-BHC found. Many similar studies have been carried out in other areas of the world. For example, the BHC and DDT contents in Czechoslovakian rivers ranged from 0.01–0.80 μg/l, in the Danube from 0.01–0.60 μg/l and in ground waters from 0.02–0.03 μg/l [70] and these results agreed well with the actual agricultural usage of OCs in the area studied. Similarly, levels ranged from 0.007–0.604 ppb (μg/l) for total BHC isomers (α-, β-, γ- and δ-isomers) in Japanese tap and river surface waters [71].

OC compounds can also enter surface waters in association with soil particles, and it has been reported that OC residues in particulate matter can be at levels 10,000–100,000 times greater than measured in the filtrate [72]. In a study of waters in Hawaii a comparison of the ratio of OC residues in canal waters to residues in algae, sediment and fish from the same canals found 1:4,000; 1:9,000 and 1:32,000, respectively [73]. Thus suspended sediment and bottom material has also been extensively studied as part of the ecological chain (Fig. 10.1).

TABLE 10.5

DETECTION LIMITS OF ORGANOCHLORINE PESTICIDES IN VARIOUS TYPES OF WATER

Insecticide	Detection limit in surface and waste water [68], μg/l (ppb)*	Detection limit in river and drinking water [69], pg***
α-BHC	0.001**	1.2
γ-BHC (lindane)	0.001**	1.5
Heptachlor	0.001**	1.4
Heptachlor epoxide	0.001**	1.8
Aldrin	0.001**	1.6
Dieldrin	0.001**	2.4
p,p'-TDE (*p,p'*-DDD)	0.001**	6.8
p,p'-DDT	0.001**	9.3
p,p'-DDE	0.001**	2.4
o,p'-DDT	0.001	6.2
Endrin	0.01**	3.2
p,p-Methoxychlor	0.05**	–
Endosulfan I	0.01**	2.4
Endosulfan II	0.01**	3.9
cis-Chlordane	0.005	2.2
trans-Chlordane	0.005	2.0
Mirex	0.001	–
Aroclor 1242	0.02	–
Aroclor 1254	0.02	–
Aroclor 1260	0.02	–
Total PCBs	0.02	–

*Based on extraction of 1-l water sample and ^{63}Ni electron-capture detection using various packed columns.

**Organochlorines monitored in surface waters in Western Canada [67].

***From standards and ^{3}HSc electron-capture detection using 60-m SE-30 capillary column (split ratio 1:4). Method based on extraction of 6-l water sample.

10.2.3. Fish and wildlife

From Fig. 10.1 it can be seen that substantial amounts of OC residues reach the aquatic environment. Therefore, the residue analysis of fish has been carried out to establish accumulation, action and effects of pesticides on fish and their ecosystems. Much of the literature on residues, persistence, elimination and hazards due to residues was reviewed by Johnson in 1973 [74]. It has been well demonstrated that OC insecticides will accumulate in lipid rich tissues and it has been further suggested that the extent of contamination may be directly related to the lipid content of a particular organism. This may be illustrated by examining ΣDDT residues reported for cod livers (average lipid content 30%) and cod flesh (average lipid content 0.2%) [75]. In cod collected from sampling sites off the Atlantic Coast of the Maritime Provinces of Canada during the summer of 1971, 12 samples had mean ΣDDT levels of 5.9 ppm and 0.04 ppm in their livers and flesh, respectively. The average ΣDDT residue content of 34 cod livers expressed on a lipid basis was more than 20 ppm. The average ΣDDT level found in commercial (refined) cod liver oils was only 1.9 ppm showing a 90% reduction in DDT residues during processing. The major metabolites were *p,p'*-DDE and *p,p'*-DDD and these varied in relative contributions in samples taken from different sampling areas, e.g. Bay of Fundy and northern coast of New Brunswick.

More recently, mirex and photomirex have been identified as major contaminants in Lake Ontario biota [76]. This surprising discovery sparked a considerable effort to monitor mirex concentrations in sediment, herring gull eggs and various species of fish from the lake [77]. Mirex and photomirex ranked as major OC residues (0.04–0.23 ppm) along with DDE (0.41–0.97 ppm) and PCBs (2.31–5.77 ppm) in coho salmon in Lake Ontario. Muscle is the major storage tissue for OC residues in salmon by virtue of size (2.4 kg) and lipid content (approx. 8.2%). Since the concentration of an OC contaminant in salmon and herring gull eggs will be directly proportional to the food of the adults, in this case alewives and smelts, the ratio of the two levels can legitimately be termed as a bioconcentration factor. The mean bioconcentration factor was 2.9 for accumulation of PCBs, DDE, mirex and photomirex in coho salmon muscle from alewives and smelts. Similarly the bioconcentration factor was 50 ± 10 for PCBs with more than 5 chlorines, DDE, mirex and photomirex in herring gull eggs. Further, assuming an apparent bioconcentration factor from water of 500,000 for accumulation of these OC residues in smelts and alewives, the corresponding values for coho salmon and herring gull eggs were 1,500,000 and 25,000,000. Since the PCBs interfered in the GC–ECD analysis of mirex and its various related photo-products [78] a nitration method was developed which removed the PCBs, as nitro compounds, and allowed the detection and confirmation of much smaller amounts of mirex and its degradation products than was previously possible [79]. In carp samples, mirex ranging from 0.061–0.088 ppm and photomirex at the 0.018–0.028 ppm level were easily detected in the presence of 0.5–166 ppm PCBs. Again the literature is extensive on OC residues in fish in North America, e.g. OC residues in Lake Michigan fish, 1929–1966 [80] as well as studies conducted in other parts of the world as illustrated by chlordane residues in fish from Tokyo Bay [81] and in aquatic environments in Iran [82].

In the wildlife area, by far the most interesting and well publicized environmental application of the ECD has been its role in correlating the egg-shell thinning/reproductive failure phenomenon in certain raptorial and fish-eating birds to the presence of DDT and its metabolites. In Britain, a study conducted in 1961–1963 revealed that the reproductive status of the peregrine falcon had declined to 56% of the recorded pre-1940 level. This was linked to the high incidence of broken eggs in the nests which had increased considerably since 1950. Early evidence implicated toxic chemical residues as a possible cause and as early as 1961 the GC–ECD analysis of eggs revealed the presence of dieldrin, DDE, heptachlor epoxide and lindane residues. After Radcliffe [83] reported the correlation between egg-shell thinning, subsequent status of the British peregrine falcon breeding population and their exposure to persistent OCs in 1967, it was soon established that the peregrine falcon in the United States had undergone egg-shell thinning to the same degree (approx. 20%) dating from exactly the same time (1946–1947, i.e. commercial introduction of DDT after the second World War) [84]. DDE has been shown to be the principal agent inducing egg-shell thinning in several species including the peregrine falcon and brown pelican. The higher the level of DDE in the eggs the greater the degree of shell thinning. This dose–response relationship is logarithmic [85] such that the percent change in egg-shell thickness was greater at the lower concentration levels of DDE. In a study of the North American peregrine, Pearce et al. [86] tentatively concluded that 15–20 ppm DDE (wet weight basis) was the critical level in the egg for causing hatching failure. In the British peregrine an average level of 13.6 ppm DDE in eggs corresponded to an average 21% thinning. During a 1968–1973 study, DDE levels in Alaskan peregrine eggs were well above this level, at 20–70 ppm, while samples of Aleutian Peale's falcon contained 3–5 ppm DDE corresponding to a 8% thinning which was regarded as insufficient to cause population decline. It was eventually demonstrated that post-war egg-shell thinning had occurred in no less than 13 species of North American raptors and 15 species of other bird groups. The brown pelican, for example, is unusually susceptible to shell thinning, a 15% decrease in thickness being associated with DDE residues of 4–5 ppm [85,87]. The egg-shell thinning response induced by DDE in the brown pelican was similar to that found in the prairie falcon. Also double-crested cormorants showed severe egg-shell thinning and population declines at 7–11 ppm DDE and 10–12 ppm PCBs [88]. In contrast, herring gulls on Lake Huron and Lake Superior reproduce normally despite residues of 15–20 ppm DDE and 50–60 ppm PCBs in their eggs [86].

Monitoring of this egg-shell thinning/population decline phenomenon continued throughout the 1970s. King et al. [89] found significant decreases in egg-shell thickness in 15 of 22 species of aquatic birds in Texas in 1970. The highest decreases (9–15%) were observed in white pelicans, brown pelicans and great blue herons and a correlation between ΣDDT residues and shell thickness was obtained for the brown pelican, great blue heron, white-faced ibis, gull-billed tern and reddish egret in which the mean residues ranged from 0.41–5.55 ppm. Also a correlation with PCBs was obtained for the brown pelican (9.73 ppm) and reddish egret (1.50 ppm). These differences emphasize the difficulties in drawing conclusions about the meaning of residues in eggs of various bird species. As a further example, DDE (range 1–21.4 ppm) was the only contaminant correlated with 2.3% decrease in egg-shell thickness in the eggs of Western grebe [90]. Even though thinning had occurred, there appeared to be no detectable effect on reproduction.

With the banning of DDT, dieldrin, etc., in the early 1970s various residue monitoring programmes have shown a gradual decline in the level of residues found in various substrates. There has also been an overall trend to decreasing DDE levels in the eggs of the British peregrine falcon and this had led to a 1979 national population level of 75% compared to the pre-1940 level [91]. Similar reproductive successes have been recorded for the status of the brown pelican in South Carolina for the years 1973–1977 [92] which occurred simultaneously with declining OC residues (except PCBs). Declining DDE contamination in anchovies (from 3.24 ppm to 0.12 ppm), the major food source of the brown pelican of the southern California coast, has paralleled their significantly improved productivity since 1971 [93].

Other OC residues, particularly dieldrin, have been implicated in bird population declines. Nest success in the brown pelican was related to residues of DDE and dieldrin in egg samples [87]. Deaths of some peregrine falcons may have been due to the use of dieldrin as a seed-dressing which was eventually banned in Britain in 1975. Moderate egg residues of dieldrin, 1–3 ppm, continued to occur in 1977–1978 in Southern Scotland and Northern England [91]. A mean level of 0.86 ppm in Scottish golden eagle eggs was associated with lowered reproductive success and reproduction improved markedly when the average dieldrin level dropped to 0.34 ppm [94]. More recently, data on bald eagles from 29 U.S.A. States for the years 1975–1977 showed that the brains of five eagles contained possible lethal levels of dieldrin (6.8–11 ppm) and two possibly died of endrin (0.71–1.2 ppm) poisoning [95].

The above examples, drawn from the areas of foods, soil, water, fish and wildlife is but a brief summary intended to show the vast field in which GC–ECD analysis has played an essential role. However, the interpretation of ECD results in the OC area is also fraught with difficulties. This is well illustrated in the finding of "apparent" OC residues in 32 of 34 samples of 1910 soils during GC–ECD analysis performed in 1970 [96]. Here peaks corresponding to heptachlor epoxide on a QF-1–OV-17 column and to aldrin on a QF-1–DC-200 column provided the greatest interference. These interferences arose from indigenous compounds in the soils and by judicious use of two or more GC columns, these "discrepancies" could be resolved satisfactorily. A similar situation also exists in the ECD analysis of certain root crops, primarily onions and carrots, in which a GC peak was found to have an identical retention time to aldrin on most non-polar columns. One interference was eventually identified as sulfur and a number of reagents used to specifically remove the interference [97]. However, one of the most complicating factors in OC residue analysis in both identification and quantitation has been the ubiquitous presence of the PCBs. This has led to the development of many specialized techniques for the separation and estimation of OCs in the presence of PCBs and these are discussed in Section 10.3.

10.3. POLYCHLORINATED BIPHENYLS

10.3.1. Occurrence

PCBs have been available as commercial products with a variety of applications for over fifty years. They were marketed under the trade name of Aroclors by Monsanto in

North America and in other parts of the world under various names, such as Clophen (G.F.R.), Phenoclor (France) and Kennechlor (Japan). These PCB formulations, since their development in 1929, have been utilized chiefly as electrical coolant and insulating fluids in transformers and capacitors, but also in heat exchange and hydraulic systems, as plasticizers in paints, adhesives, caulking compounds and dye-carriers in carbonless copy paper. Their escape from these systems or their disposal in wastes has led to worldwide environmental contamination with PCBs. The presence of PCBs in the environment became apparent only after the ECD had been extensively applied to the monitoring of organochlorine pesticide residues, during the early 1960s. Chromatograms of wildlife sample extracts were found to exhibit many electron-capturing responses which could not be attributed to organochlorine pesticide residues. These were first identified as PCBs by Jensen [98] who confirmed their presence with mass spectrometry.

PCBs are produced by the chlorination of biphenyl to various extents, resulting in such products as Aroclor 1242 and Aroclor 1260, which consists of a mixture of chlorinated biphenyls with an average chlorine content of 42% and 60% respectively. Fig. 10.2 illustrates the chromatograms obtained from some representative formulations of Aroclor and a chromatogram of a typical environmental sample extract.

Because of their resistance to breakdown and their other physical and chemical properties which are similar to those of the organochlorine pesticides, particularly DDT and its metabolites, the PCBs tend to bioaccumulate towards the top of the food web, namely in the fatty tissues of predatory fish, mammals and birds. PCBs have been found in all parts of the environment. Their presence and distribution have been reviewed by Risebrough et al. [99], Peakall [100], Jensen [101] and Nisbet and Sarofin [102].

Pal et al. [103] have reviewed the fate of PCBs in soil–plant systems as well as in other environmental substrates. Fishbein [104] has reviewed some of the chromatographic and biological properties of PCBs. Since the chemical structures of PCBs and DDT residues are related and their GC–ECD retention times and detector responses show much similarity, undoubtedly some cases of misidentification of peaks in environmental samples have occurred in the past. The overlapping chromatograms of the two groups of

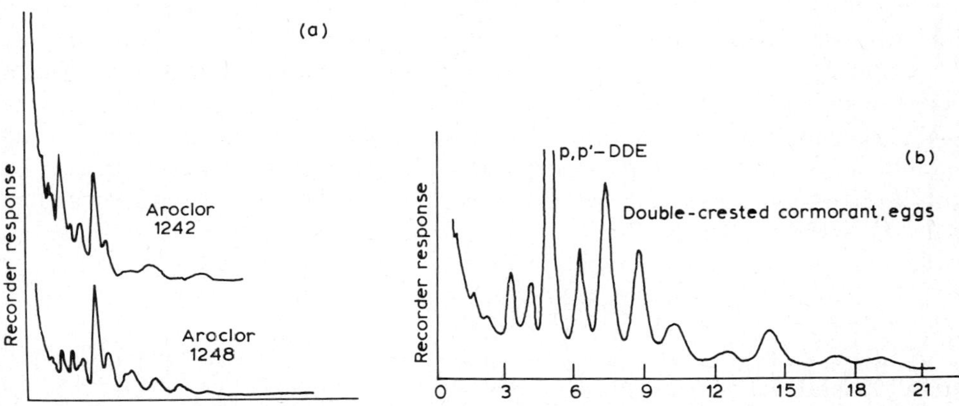

Fig. 10.2. (a). GC–ECD tracings of two Aroclor preparations on 4% SE-30 at 200°C. (b) Typical environmental extract showing combination of organochlorine and PCB peaks. (From ref. 109.)

compounds give rise to the need for prior separation by column chromatography. A number of methods have been reported, including those of Reynolds [105] on Florisil, Armour and Burke [106] on silicic acid–Celite and Berg et al. [107] on a charcoal column. Griffin and Morin [108] optimized chromatographic conditions for the separation of *p,p'*-DDE from Aroclor 1254 on silica, using azulene as indicator.

10.3.2. Quantitation

Various methods have been proposed for the quantitation of PCBs in environmental and other substrates. One approach by Berg et al. [107] involved quantitation of the total PCBs present by measuring the amount of decachlorobiphenyl produced from exhaustive chlorination of the residues with antimony pentachloride. This method, however, does not give any information on the relative contribution of PCB components with different chlorine content and has other disadvantages which have been discussed by Duinker et al. [110]. Most commonly, for the purpose of quantitation, the pattern of peaks obtained by GC–ECD has been matched to the closest corresponding technical formulation of Aroclor. Where direct contamination with a formulation has occurred, a reasonable match can be obtained in this way, but metabolized material presents a changed pattern of GC peaks which may not correspond to any one particular Aroclor. Beezhold and Stoud [111] proposed that Aroclor standards, mixed if necessary, should reflect a best attempt to quantitatively duplicate the peak response of the residue. This method was evaluated in an AOAC collaborative study by Sawyer [112], who recommended the use of either single or mixed Aroclor standards, utilizing all peaks that could be attributed to chlorobiphenyls.

The relative electron-capture responses of the various PCB congeners have an influence on the quantitation of mixtures. In an early OECD study [113] this fact was noted and a scheme for calibration of individual electron-capture responses was sought. Risebrough [114] in quantitating levels in marine birds and fish assumed that the average PCB response was similar to that given by the same weight of *p,p'*-DDE. Sissons and Welti [115] stated the need to think in terms of individual PCBs rather than complex mixtures, both from the analytical and toxicological point of view. They synthesized forty individual PCBs and predicted the structures of the components of Aroclor 1254, 1260 and 1242, utilizing GC retention indices on a surface-coated open-tubular (SCOT) column. Rote and Murphy [116] determined the average electron-capture response of the biphenyls substituted with from one to seven chlorine atoms and established a semi-logarithmic relationship between detector response and average chlorine content. From the response curve (Fig. 10.3), the content of the various PCBs in each Aroclor mixture was established and this was then employed to calculate the total weight of PCB components in environmental samples. The authors recommended this method be used by all workers until a more accurate quantitation could be achieved using synthesized PCB components.

Zitko et al. [117] reported on the retention times and electron-capture responses of some 23 individual synthesized chlorobiphenyls. They found that the detector response increases strongly with the number of chlorine atoms in the molecule, the response of decachlorobiphenyl being 500 times that of 4-chlorobiphenyl. Most of this increase occurs in the mono- and trichlorobiphenyl range as seen in Table 10.6. Chlorine substitution in positions 2 and 6 decreases and vicinal substitution, particularly in positions 3 and 4, increases the electron-capture response. The response for the tetra-, penta-. hexa- and heptachlorobiphenyls does not vary by more than a factor of 2 and since these

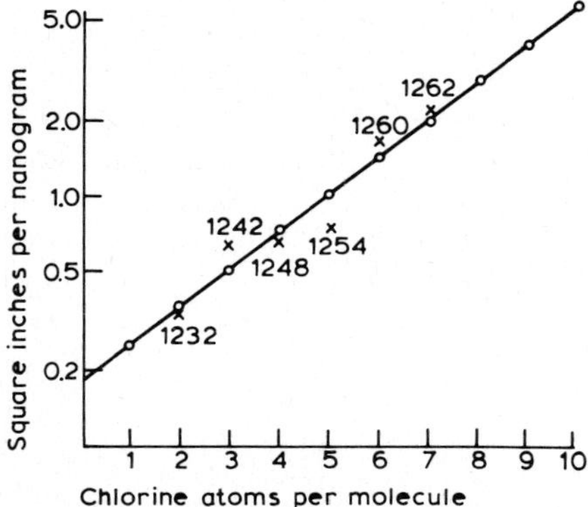

Fig. 10.3. ECD response curve for mono- to decachlorobiphenyls with observed (X) and theoretical (○) values for the chlorine content for some Aroclors. (From ref. 116.)

TABLE 10.6

RETENTION TIME AND ELECTRON-CAPTURE DETECTOR RESPONSE OF CHLOROBIPHENYLS [117]

Compound	Relative retention time (*p,p'*-DDE = 1.00)	Relative response (ng) (*p,p'*-DDE = 1.00)
4-Chlorobiphenyl	0.17	0.0033
2-Chlorobiphenyl	0.11	0.0030
3-Chlorobiphenyl	0.14	0.0006
4,4'-Dichlorobiphenyl	0.30	0.0152
3,3'-Dichlorobiphenyl	0.25	0.0155
2,2'-Dichlorobiphenyl	0.15	0.0131
3,4-Dichlorobiphenyl	0.28	0.0388
2,4-Dichlorobiphenyl	0.19	0.0450
2,6-Dichlorobiphenyl	0.16	0.0815
2,4,4'-Trichlorobiphenyl	0.39	0.293
2,4,6-Trichlorobiphenyl	0.24	0.276
2,2',4,4'-Tetrachlorobiphenyl	0.51	0.206
3,3',4,4'-Tetrachlorobiphenyl	0.96	0.770
2,2',6,6'-Tetrachlorobiphenyl	0.33	0.0403
3,3',5,5'-Tetrachlorobiphenyl	0.70	0.625
2,3,4,5-Tetrachlorobiphenyl	0.69	0.715
2,3,5,6-Tetrachlorobiphenyl	0.51	0.505
2,3,4,5,6-Pentachlorobiphenyl	1.00	1.30
2,2',4,4',6,6'-Hexachlorobiphenyl	0.78	0.545
3,3',4,4',5,5'-Hexachlorobiphenyl	2.98	1.15
2,2',3,3',4,4',6,6'-Octachlorobiphenyl	2.62	1.58
2,2',3,3',5,5',6,6'-Octachlorobiphenyl	2.45	1.53
Decachlorobiphenyl	8.20	1.61

compounds are the most common constituents of environmental PCBs, they may be used as a fairly reliable basis for quantitation.

Webb and McCall [118] established the weight of PCB represented by each GC–ECD peak of a given Aroclor by using a combination of elemental analysis, GC with electrolytic conductivity detection and GC–mass spectrometry to obtain the empirical formula of the compound and the chlorine content. Using this information, the response factors for each peak were calculated for the ECD and a table was prepared for each Aroclor showing the weight percent of each GC–ECD peak in the mixture. Fig. 10.4

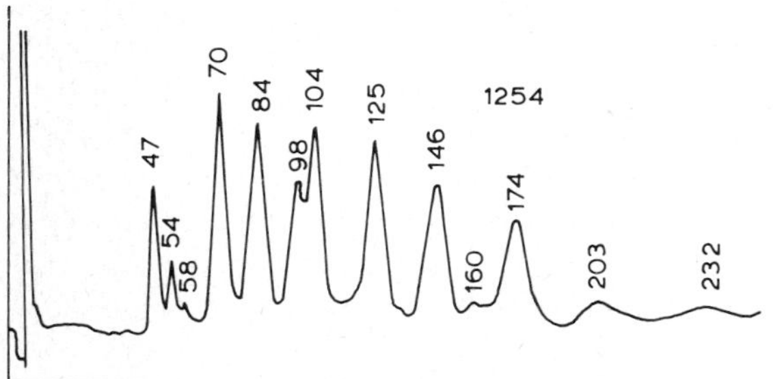

Fig. 10.4. ^{63}Ni ECD chromatogram of Aroclor 1254 on 3% SE-30 at 200°C. (From ref. 118.)

TABLE 10.7

COMPOSITION OF AROCLOR 1254 [118]

Relative retention time*	Mean weight percent	Relative std. dev.**	Number of chlorines***
47	6.2	3.7	4
54	2.9	2.6	4
58	1.4	2.8	4
70	13.2	2.7	4 } 25%
			5 } 75%
84	17.3	1.9	5
98	7.5	5.3	5
104	13.6	3.8	5
125	15.0	2.4	5 } 70%
			6 } 30%
146	10.4	2.7	5 } 30%
			6 } 70%
160	1.3	8.4	6
174	8.4	5.5	6
203	1.8	18.6	6
232	1.0	26.1	7
Total	100.0		

*Retention time relative to *p,p'*-DDE = 100. Measured from first appearance of solvent.
**Standard deviation of six results as a percentage of the mean of the results.
***From GC–mass spectrometric data. Peaks containing mixtures of isomers are bracketed.

shows the electron-capture chromatogram of Aroclor 1254 and Table 10.7 gives the composition of Aroclor 1254 obtained by the above procedure.

Chau and Sampson [119] carried out a survey of GC–ECD methods used by laboratories for PCB analysis. They applied a number of these procedures to the quantitation of several mixtures of Aroclors. They found that the method of Webb and McCall [118] gave the best precision and accuracy since it was more generally applicable than other methods to analysis of PCB residues from a wide variety of sources and it eliminates the need for mixed standards. The method does, however, depend on the availability of standards of known composition and it is limited to one set of chromatographic conditions. Chau and Sampson recommended the Webb-McCall approach for incorporation into procedures employed by laboratories analyzing for PCBs.

Sawyer [120] used the Webb–McCall method to recalculate the results obtained in his early study [112] when a total response comparison had been used. He found that the individual peak method of Webb and McCall demonstrated greater precision. In chicken fat residues of Aroclor 1242 he found that in the total peak comparison, the response is dominated by higher chlorobiphenyls which cause a disproportionately high response on the ECD. He concurred with the recommendation of Chau and Sampson [119] that the individual peak method is preferable. Several lots of Aroclor mixtures were analysed by the Hall electrolytic conductivity detector for chlorine measurement and mass spectrometry to establish molecular weights of individual PCB peaks. A collaborative study was then conducted by Sawyer [121] which compared the original AOAC total response quantitation [112] with the approach of Webb and McCall [118] and an individual peak comparison using a mixture of synthetically prepared PCB standards. From the point of view of both precision and accuracy, the individual peak method was found to be an improvement over the existing AOAC methods.

10.3.3. Applications of high-resolution gas chromatography with electron-capture detection to PCB analysis

Recently there has been an increased availability and application of high-resolution capillary GC for the more or less complete resolution of PCB congeners and toxicologists have shown interest in those congeners which exhibit the greatest biological activity or toxicity. Poland and Glover [122] examined sixteen halogenated biphenyls for their ability to induce hepatic aryl hydroxylase activity in the chicken embryo. Of these, 3,3′,4,4′-tetrachlorobiphenyl (TCB) and 2,2′,4,4′,5,5′-hexachlorobiphenyl (HCB) were found to be active, resulting in an intoxication syndrome similar to that produced by 2,3,7,8-tetrachlorodibenzo-*p*-dioxin (TCDD). This high toxicity appears to be associated with the more planar character of compounds not substituted with chlorine at the *o,o′*-positions. Huckins et al. [123] found that high-resolution capillary GC with ECD failed to resolve 3,4,4′-trichlorobiphenyl and 3,3′,4,4′-tetrachlorobiphenyl from components in Aroclor 1254 having *o,o′*-chlorine disubstitution. These workers utilized columns of activated carbon on shredded polyurethane foam to separate non-*o,o′*-chlorine disubstituted PCBs from milligram quantities of Aroclor 1016, 1242, 1248, 1254 and 1260. Aroclor 1248 was found to contain the highest levels of 3,3′,4,4′-TCB. Analysis of fish, utilizing carbon-foam chromatography and high-resolution capillary GC resulted in the detection of 3,3′,4,4′-TCB at levels of 0.11 ppm. From these findings, they suggested that analysis of PCB residues should include determination of levels of non-*o,o*′-chlorine disubstituted PCBs. Kamops et al. [124] found that Florisil column chromatography would separate 3,3′,4,4′-TCB and 3,3′,4,4′,5,5′-HCB from most of the other PCBs in technical Aroclors.

High-resolution glass capillary chromatography with ECD was used by Zell et al. [125] who identified 83 PCBs in marine fish samples. They found four octachlorobiphenyls in technical PCB mixtures whose levels and relative proportions changed very little in the environment. This made it possible, in some cases, to identify the type of PCB formulation that was originally responsible for the contamination.

Onuska and Comba [126] reviewed the quantitative procedures used by workers for PCBs and described the chromatography of Aroclor mixtures using wall-coated open-tubular (WCOT) glass capillary columns and an ECD. They presented relative retention indices and response factors for individual congeners using the method of Sissons and Welti [115] and found that retention indices could be predicted with considerable accuracy by this means. The identity of the predicted compound was confirmed by mass spectrometry.

Tuinstra and Traag [127] applied glass capillary GC–ECD to the routine analysis of PCBs and organochlorine pesticide residues in agricultural products. They demonstrated the feasibility of automated analysis of milk fat and feeds for PCBs using as standards some 21 individual chlorinated biphenyls.

The trend in quantitative GC–ECD analysis of PCBs is towards the resolution of individual isomers with capillary GC and the use of reference standards of synthesized chlorinated biphenyls. In both environmental and other samples, there should be particular emphasis on the identification and quantitation of the planar PCB congeners because of their toxicological implications.

10.4. ORGANOPHOSPHATES

Early studies by Burke and Giuffrida in 1964 [128] on multiresidue methodology utilizing the ECD included in the analytical scheme some organophosphorus (OP) pesticides such as ronnel, parathion, chlorthion which respond well to the ECD. The response of OP compounds containing halogens, nitro and other electron-capturing groups is greater on the ECD than to any other detector as illustrated by the response data listed in Table 10.8. In 1966 Brody and Chaney [129] introduced the flame photometric detector which, while not as sensitive as the ECD for many OP compounds, had the advantage of selectivity and a reduced need for sample clean-up thereby making it the detector of choice for OP residue analysis.

Nevertheless, the ECD has still been utilized in some circumstances, where the type of substrate or electron-capture sensitivity of the compounds justify it. Table 10.9 shows a number of recent applications where the ECD was preferred for the analysis of OP residues without derivatization.

Addition of an electron-capturing group to increase sensitivity, to improve GC characteristics or for confirmatory purposes, has been another approach for the intact OPs. Coburn and Chau [137] hydrolyzed ronnel, cruformate, methyl parathion, fenithrothion and parathion, then derivatized the corresponding phenols with pentafluorobenzyl bromide to produce the pentafluorobenzyl ethers. The derivatization of OP compounds, possessing a NH or NH_2 moiety, for confirmation purposes has been carried out in the case of monocrotophos residues in strawberries using trifluoroacetic anhydride and GC–ECD analysis [138].

The analysis of OP metabolites frequently involves derivatization to a compound capable of being detected by electron capture. Cook and Moore [139] reported a method

TABLE 10.8

RELATIVE RETENTION TIMES TO PARATHION AND RESPONSES TO GIVE 1/2 FULL-SCALE DEFLECTION ON THE ELECTRON-CAPTURE AND THERMIONIC DETECTORS FOR SOME COMMON ORGANOPHOSPHORUS PESTICIDES AND METABOLITES [27]

Common name	Relative retention time on DC-200	Electron-capture sensitivity (ng)	Thermionic detector sensitivity (ng)
Thiometon	–	142.0	16.0
Dichlorvos	0.07	6.0	0.5
Mevinphos (phosdrin)	0.13	18.0	1.5
Phorate	0.36	25.0	1.7
Dimethoate	0.41	4.5	1.8
Diazinon oxygen analogue	0.45	300.0	2.5
Diazinon	0.52	25.0	2.0
Parathion-methyl oxygen analogue	0.56	11.0	5.0
Parathion-methyl	0.71	3.0	2.5
Parathion oxygen analogue	0.79	10.0	3.5
Fenchlorphos (ronnel)	0.81	10.0	3.0
Malathion	0.90	25.0	5.0
Parathion	1.00	5.0	2.0
Chlorpyrifos	1.02	1.4	3.0
Bromophos	1.13	1.3	3.0
Pirimiphos-ethyl	1.16	150.0	4.0
α-Chlorfenvinphos	1.19	3.5	4.0
Phenthoate	1.28	5.0	3.0
β-Chlorfenviphos	1.29	2.5	4.0
Tetrachlorovinphos (gardona)	1.57	3.0	8.0
Methyl trithion oxygen analogue	1.58	10.0	21.0
Methyl trithion	2.20	5.0	8.0
Fensulfothion oxygen analogue	2.53	–	10.0
Ethion	2.55	12.0	7.0
Carbophenothion (trithion)	2.93	4.0	15.0
Phosmet (imidan)	4.0	1.5	15.0
EPN	4.4	5.0	16.0
Azinphos-methyl (guthion)	5.2	50.0	25.0
Phosalone	5.5	7.0	30.0
Leptophos (phosvel)	5.8	10.0	20.0
Azinphos-ethyl (ethyl-guthion)	6.4	45.0	40.0

TABLE 10.9

SOME RECENT APPLICATIONS OF THE ELECTRON-CAPTURE DETECTOR FOR ORGANO-PHOSPHORUS PESTICIDE RESIDUE ANALYSIS

Compound	Sample type	Reference
Dichlorvos	Tissues and crops	Elgar et al., 1970 [130]
Morestan	Fruits and vegetables	Tjan and Konter, 1971 [131]
Chlorfenvinphos and 2,2′,4′-trichloroacetophenone	Eggs and poultry tissues	Sherman and Kerrick, 1972 [132]
Coumaphos	Milk	Eschle et al., 1973 [133]
Tetrachlorovinphos	Grain and grain products	Hall et al., 1973 [134]
Cyolane	Alfalfa pellets and rumen contents	Tadjer and Egyed, 1974 [135]
Chlorpyrifos	Tobacco	Toet and Toet, 1976 [136]

for the determination of malathion residues in fish, oyster and shrimp tissues involving esterification of the acid metabolites with diazomethane and analysis by GC–ECD. The GC–ECD analysis of the hydrolysis product of etrimfos in corn and alfalfa was achieved after conversion to its pentafluorobenzoate derivative [140]. The phenolic metabolites of chlorpyrifos and leptophos were analysed in vegetable tissue by GC–ECD following conversion to the trimethylsilyl derivative [141], while *p*-nitrophenol, a metabolite of parathion, was detected by GC–ECD following derivatization with diazomethane [142].

It can be concluded therefore that, despite the high sensitivity of the ECD to many OP compounds, in all but some special situations, the flame photometric and the alkali flame ionization detectors are preferred because of their greater selectivity.

10.5. N-METHYLCARBAMATES

The sensitivity of most N-methylcarbamates to the ECD is poor and there have been few direct GC–ECD methods for their residue analysis [143]. In addition, many tend to undergo thermal dissociation on the GC column. Ragab [144] used a direct electron-capture method for the analysis of 3-oxocarbofuran in water at the 1-ppb level but he reported no response for the parent carbofuran and very little for the major metabolite, 3-hydroxycarbofuran. Carbofuran and other phenol-producing carbamates have more commonly been converted to an electron-capturing derivative prior to GC–ECD analysis. Ralls and Cortes [145] brominated carbaryl in a sealed tube and the resulting electron-capturing derivative enabled carbaryl to be detected in green beans at the 1-ppm level. Gutenman and Lisk [146] developed a method for carbaryl which involved hydrolysis to produce 1-naphthol which was then brominated and acetylated to give a detection limit of 0.5 ng by ECD. Butler and McDonough [147] developed a sensitive method applicable to carbamates in general, which involved alkaline hydrolysis followed by trichloroacetylation of the phenol to form the electron-capturing trichloroacetate. Matacil, baygon and carbofuran were detected at the 0.01-ng level and carbaryl, MC-A-600 and mesurol at the 0.02-ng level. Lau and Marxmiller [148] acetylated the intact carbamate moiety of landrin, using trifluoroacetic anhydride to achieve an electron-capture detection limit of

0.1 ng. Holden et al. [149] reacted the amine moiety of carbamates, produced by hydrolysis under basic conditions, with 1-fluoro-2,4-dinitrobenzene at pH 7 to form the dinitroaniline derivative which was detected by GC–ECD. Cohen et al. [150] used the same reagent at pH 11 to form the phenol derivative of some carbamates. As most insecticidal carbamates produce the same amine on hydrolysis, the method of Holden et al. was not specific for any particular carbamate. However, derivatization of the phenol portion of the carbamate produces an unique product in most cases.

Caro et al. [151] described a method for carbofuran analysis in soils with ammonium acetate extraction and GC–ECD determination of the dinitrophenyl ether of carbofuran at pesticide concentrations as low as 0.1 ppm.

Holden [152] developed a general procedure for determining N-methylcarbamates in plant material as the dinitrophenyl ethers, with a detection limit of 0.01 ppm, utilizing either an ECD or nitrogen-selective detector. The method underwent an AOAC collaborative study in 1975 [153] for residues of propoxur, carbofuran, carbamolate and carbaryl in various plant materials and was adopted as an official, first action.

Published methods for the GC of N-methylcarbamates in recent years have tended to favour the more selective nitrogen detectors over the ECD.

10.6. HERBICIDES

Herbicides are applied directly to plants, soils or water to achieve various types of weed control. They, in fact, comprise a large group of organic and inorganic (e.g. the arsenicals) compounds. Those herbicides that can be analysed by GC have been roughly categorized into 10 major and minor chemical groups (see Table 10.10). A survey of the herbicide literature reveals a voluminous amount of data and studies on the efficacy of individual compounds in various environments, e.g. different crops, soils, weed types, climatic conditions, etc. Also the use of GC–ECD analysis varies according to the chemical nature or polarity of the herbicide of concern. In the case of the *s*-triazine group the intact herbicide has been studied for its persistence in soil [154], leaching characteristics, translocation into corn, etc. With the phenoxy and benzoic acid herbicides, the residue methodology entails derivatization to a compound (a) more amenable to GC analysis and (b) more responsive to electron-capture detection, e.g. 4-chloro-2-methylphenoxyacetic acid (MCPA), which only has one chlorine atom, is less sensitive to electron-capture detection than 2,4-D, 2,4,5-T, etc.

The major applications of the ECD in herbicide analyses have been, of course, in soil persistence [48,155], plant uptake, leaching and appearance of residues in run-off water [67], degradation studies (both chemical and biochemical) [156], the fate and accumulation of residues in animals from ingesting treated crops and volatilization/drift problems associated with different formulation types and/or methods of application (aerial versus ground-rigs). Cochrane and Purkayastha [157] extensively reviewed the use of the ECD and other detectors in herbicide residue analysis up to 1971. In addition, the use of chemical derivatization techniques for both quantitative and confirmatory analysis in herbicide work has been updated in a number of reviews by Cochrane (ref. 97 and references cited therein). Table 10.10 outlines some of the more recent applications of the ECD. Of the 10 classes of herbicides listed, the more important major groups are the triazines,

TABLE 10.10

MAJOR GROUPS OF HERBICIDES AND SOME RECENT APPLICATIONS UTILIZING ELECTRON-CAPTURE DETECTION

Herbicide group	Individual examples	Applications		Reference
		Substrate	Detection limit	
Triazines	Atrazine, simazine	Soil	0.1 ppm	Young and Chau, 1973 [158]
	Propazine			Albright et al., 1974 [159]
	Simazine	Fish and mud	0.05 ppm	Hesselberg and Johnson, 1972 [160]
	Cyanazine and metabolites	Wheat and potatoes	0.01 ppm	Beynon et al., 1972 [161]
	Metribuzin and metabolites	Crops and soil	0.01 ppm	Hyzak and Zimdahl, 1974 [162]
				Thornton and Stanley, 1977 [163]
Carbamates	Triallate	Soils	0.01 ppm	Head and McKercher, 1971 [164]
				Fryer and Kirkland, 1970 [165]
		Oats and barley	0.05 ng	Thiele and Zimdahl, 1976 [166]
	Triallate/ trifluralin, etc.	Soils	0.05 ppm	Smith, 1974 [167]
Phenylureas	Linuron	Carrots	0.005 ppm	Løkke, 1974 [168]
		Soil	0.1 ppm	Khan et al., 1975 [169]
Amides	Kerb®*	Alfalfa	0.01 ppm	Adler et al., 1972 [170]
	CDAA**	Milk	0.01 ppm	St. John, Jr. and Lisk, 1974 [171]
	Propachlor	Soil	0.2 ppm	Molnau et al., 1973 [172]
	Diphenamid	Soil	0.15 ppm	Payne, Jr. et al., 1974 [173]
		Water	0.5 ppb	
2,6-Dinitro-anilines	Trifluralin	Soil	10 ppb	Menges and Tamex, 1974 [174]
				Smith, 1972 [175]
	Nitralin/benefin	Soils	40 ppb	Miller et al., 1975, 1976 [176,177]
	Benfluralin	Soils/carrots	2 ppb	Businelli et al., 1975 [178]
	Oryzalin	Crops and soil	0.01 ppm	Sieck et al., 1976 [179]
Phenoxy acid	2,4-D	Total diet samples	0.02 ppm	Corneliussen, 1972 [180]
		Wheat and soil	0.01 ppm	Cochrane and Russell, 1975 [181]
		Air	15 ng/m^3	Que Que et al., 1975 [182]
	2,4,5-T	Soil	0.005 ppm	McKone and Hance, 1972 [183]
	2,4,5-TP (fenoprop)	Apples	0.05 ppm	Leidy et al., 1975 [184]
			0.01 ppm	Cochrane et al., 1976 [185]
	MCPA	Water, mud, rice	0.01 ppm	Soderquist and Crosby, 1975 [186]

TABLE 10.10 (*continued*)

Herbicide group	Individual examples	Applications		References
		Substrate	Detection limit	
	2,4-D/MCPA	Water	0.096 ppb/ 0.06 ppb respectively	Mierzwa and Witek, 1977 [187]
	2,4-D/2,4,5-T/MCPA	Soil	0.03 ppm	Purkayastha, 1974 [188]
	2,4-DB/2,4-DP/2,4,5-T	Soil	0.01 ppm	Smith, 1976 [189]
	2,4-D/2,4,5-T/fenoprop	Milk and cream	0.05 ppm	Bjerke et al., 1972 [190]
Uracils	Bromacil	Orchard soil	0.01 ppm	Leistra et al., 1975 [191]
	Terbacil	Orchard soil	0.01 ppm	Marriage et al., 1977 [192]
		Blueberries	1 ppb	Cessna, 1977 [193]
Benzoic acids	Dicamba	Soil	0.02 ppm	Alton and Stritzke, 1973 [194]
		Wheat and barley	0.05 ppm	Khan, 1975 [195]
Benzonitriles	Bromoxynil	Soil	0.05 ppm	Crouch and Pallin, 1974 [196]
	Dichlobenil	Water	2 ppb	Rice et al., 1974 [197]
		Gooseberries	5 ppb	McKone et al., 1971 [198]
Miscellaneous	TOK®***	Rice and wheat	0.01 ppm	Adler and Wargo, Jr., 1974 [199]
	Glyphosphate	Water	2.5 ppb	Comes et al., 1976 [200]
	Picloram	Soils	1 ppb	Lutz et al., 1973 [201]
		Surface water	1 ppb	Bovey et al., 1974 [202]

*N-(1,1-Dimethylpropynyl)-3,5-dichlorobenzamide.
**2-Chloro-N,N-diallylacetamide.
***2,4-Dichloro-1-(4-nitrophenoxy)benzene.

phenoxy acids, 2,6-dinitroanilines, phenylurea and carbamates. These individual groups are briefly discussed below.

10.6.1. Triazines

The majority of the triazines can be analysed intact by GC with good response being obtained via electron-capture detection. However, nitrogen-selective detectors are generally preferred because less interference from co-extractives is encountered and they have a wider application across the triazine group [203]. A comparison of the ECD and Coulson conductivity detector (CCD) showed the ECD to be much more sensitive to the chloro-*s*-triazines such as atrazine, simazine, metribuzin, etc., with half-full scale responses (1/2 f.s.d.) being in the range 1–3 ng using a 3% OV-1 packed column at 195°C and ^{63}Ni detection [204]. No response was observed for prometone, a methoxy-substituted *s*-triazine, on electron capture while a 1/2 f.s.d. was obtained with 7 ng using the CCD. In a comparison of three detection systems for atrazine, ametryne and terbutryne in water the alkali flame ionization detector could detect 1-ppb levels while UV or polarographic detection limited the lower detection level to 10 ppb [205]. Similarly, quantitative determination of atrazine was possible with the CCD for water and soil samples with or without clean-up. Analysis could not be performed on soil samples using ECD, even with clean-up steps included, due to large interfering co-extractive peaks [204]. The use of chemical derivatization in the triazine group has been primarily in the confirmation of identity of the parent compound or to render the polar dehalogenated (e.g. hydroxyatrazine) or dealkylated metabolites more amenable to GC analysis. Here most derivatizations involve reaction of the NH or OH moieties with various alkylation, silylation, chlorination, methoxylation or dinitrophenylation reagents, followed by selective nitrogen detection [97]. In an attempt to increase detector sensitivity, the heptafluorobutyryl (HFB) derivatives of ten triazines were prepared using heptafluorobutyrylacetic acid in benzene with pyridine as catalyst [206]. The mono- and di-HFB derivatives produced were 300 times more ECD-sensitive and 5–10 times more CCD-sensitive than the parent triazines, thereby allowing the detection of 0.13–0.86 ppm residue levels in peas, tomatoes and potatoes.

10.6.2. Phenoxy acids

Traditionally, the chlorophenoxyacetic acid herbicides have been esterified to their methyl esters prior to analysis by ECD [157]. The methyl esters were preferred because of their ease of preparation and reasonably short GC retention times on packed columns. However, efforts have been made to further increase ECD sensitivity by use of esterification reagents such as 2-chloroethanol, 2,2,2-trifluoroethanol and pentafluorobenzyl bromide [97], the major drawback being the high background noise levels produced even after column clean-up to remove excess reagents. Also, in certain soil types, although the 2-chloroethyl ester of 2,4-D was found to have increased sensitivity and gave a longer GC retention time than the 2,4-D methyl ester, it was still subject to interferences [207]. Mierzwa and Witek [187] found the trichloroethyl ester to have the greatest increase in sensitivity when applied to 2,4-D and MCPA analysis in water. Using a ^{63}Ni ECD the detection limit for MCPA was 150 pg while 240 pg was required for 2,4-D when using a 15% QF-1–10% DC-200 column at 195°C. This allowed a lower limit of 0.06 ppb MCPA and 0.096 ppb 2,4-D to be detected in 1-l water samples. Using the pentafluorobenzyl esters, concentration of 0.1 μg/l MCPA and 0.2 μg/l 4-chloro-2-methylphenoxybutyric acid (MCPB) could also be determined in natural waters using ^{63}Ni electron-capture detection [208].

Another approach has been to prepare esters, e.g. propyl, butyl, etc., that exhibit longer retention times than the methyl esters in order to eliminate or curtail interference problems [183]. The technique of transbutylation of the methyl ester using boron trifluorides–*n*-butanol has also been employed in the confirmation of identity. This procedure has the advantage of using the same GC–ECD system as for the methyl ester and since yields are normally 85–90% cross-checking of quantitative results can also be carried out. Farwell et al. [209] compared the detection limits of chlorophenoxy ester levels in air using electron-capture detection and GC–mass spectrometry. With mass spectrometry–single ion monitoring the detection limit for the 2,4-D butyl ester was 1 pg while 50 pg was required with ^{63}Ni electron-capture detection. In the analysis of actual air samples the ^{63}Ni ECD and GC–mass spectrometry quantitative techniques exhibited good agreement while a ^{3}H ECD gave consistently high results for which no explanation was given.

10.6.3. 2,6-Dinitroanilines

Since this class of herbicides possesses 2 nitro groups, they have excellent ECD and nitrogen-selective detector characteristics. The 1/2 f.s.d. responses of trifluralin and benefin are 700 pg and 400 pg respectively using ^{63}Ni ECD while their corresponding CCD response is 25 ng [204]. Normally, ECD has been used in the analysis of their residues in crops and soils and this has resulted in very low detection limits (Table 10.10). However, the two isomeric herbicides, trifluralin and benefin display identical retention times on some commonly used GC packed columns, e.g. OV-17 [204]. In one instance [210], separation was achieved using a Durapak column at 175°C. Even with trifluralin appearing at ca. 49 min and benefin at ca. 52 min baseline separation was not achieved. Another alternative would be 3% OV-1 at 195°C (or lower) which gave relative retention times (parathion = 1.00) of 0.31 and 0.41 for trifluralin and benefin respectively [204].

Again utilizing the nitro groups, chromous chloride is a very facile reagent for use in their confirmation. This conversion of the NO_2 to NH_2 in essence destroys the electron-capturing ability of the herbicides, therefore the N-mode CCD or alkali flame ionization detector is best used in these instances [97].

10.6.4. Ureas and carbamates

Of the ten commonly used substituted phenylurea herbicides, eight possess (a) halogenated atom(s) attached to one of the benzene rings, thereby making them amenable to electron-capture detection [157]. However, like the carbamates, on-column decomposition has been a major problem. In the case of linuron this was overcome by the use of on-column injection, short GC columns (1.5 m 1.5% XE-60 or 3% OV-101) and temperatures below 165°C (both column and injector) [169]. Typically, 1/2 f.s.d. ^{63}Ni electron-capture responses are obtained with 0.4–5.0 ng of the halogenated ureas [204] while 100–200 ng are required for similar CCD responses. The phenylureas, similar to the other herbicides, have been examined for persistence and degradation in water, soils and crops [156,157] (Table 10.10). Buchert and Løkke [211] showed that the thermal stability of the mono-, di- and trisubstituted ureas was related to the presence of an amide-H group. Therefore, since tetrasubstituted phenylureas are thermally stable, much methodology activity has been directed towards derivatization reactions involving replacement of this active H, utilizing reagents which would also increase the electron-capture response. These derivatization reactions including silylation, trifluoroacetylation (and similar reactions involving other halogenated reagents) and alkylation have been discussed in detail by Cochrane [97].

The application of the ECD to herbicidal carbamates is limited to those possessing electron-capturing moieties. These are typically diallate, triallate (Table 10.10), chlorpropham [157], carbyne and Swep. In most instances nitrogen-selective detectors or high-performance liquid chromatography procedures have been preferred where possible, to eliminate excessive clean-up or GC on-column problems as outlined in Section 10.5.

Where derivatization has been used a variety of bromo-, chloro-, fluoro- and nitro-containing reagents have been extensively studied in combination with electron-capture detection [97].

10.7. MISCELLANEOUS

10.7.1. Fungicides

These compounds fall into a variety of chemical classes, only a few containing chlorine or other electron-capturing groups. The world fungicide market is dominated by the dithiocarbamates, followed by the phthalimides, copper compounds, mercurials and others [212]. Del Re et al. [213] have reviewed analytical methods for formulation and residue analysis of fungicides.

The phthalimide group, the most prominent members of which are the fungicides captan, captafol and folpet, are readily amenable to GC–ECD. Burke and Holswade [214] detected amounts of 4 ng captan or folpet by ECD compared to 3 μg captan and 8 μg folpet by microcoulometric detection. Captafol can be detected by GC–ECD to a limit of 0.01 ppm in fruits [215]. Other workers have achieved similar sensitivities in other substrates. In the authors' laboratory for many samples a nitrogen- or sulfur-selective detector has been found to give comparable detection limits and may be preferable to the ECD since less sample clean-up is required.

The 3,5-dichloroaniline fungicides have been analyzed by various workers using GC–ECD for residues in vines, grapes and wine [213].

Quintozene (pentachloronitrobenzene) has been used as a fungicide on a variety of vegetable crops. Dejonckheere et al. [215] in Belgium found that quintozene applied to lettuce crops persisted in the soil until the following growing season. These workers also detected HCB residues in the same soil and lettuce, the presence of which may be explained by the fact that technical grade quintozene contained 2.28–2.36% of HCB. In addition, residues of the quintozene metabolites pentachloronitroaniline and pentachloronitroanisole were detected. HCB is used as a fungicide against bunt in cereal seeds and improper use of treated seed as animal feed may give rise to residues in milk, eggs and other products [216]. HCB is found widely as an environmental contaminant in samples analyzed by GC–ECD but its widespread presence is a result of its utilization not as a fungicide but as an important intermediate in many industrial processes [217]. Rourke et al. [218] has reported HCB as a contaminant of laboratory plastic bottles, use of which could give rise to misleading GC–ECD results in residue analysis.

Pentachlorophenol (PCP) has a variety of uses, such as a herbicide on rice and a defoliant for cotton and soybeans, but the largest quantities are utilized as a fungicide for wood preservation. Firestone [219] has reviewed the chemistry and analysis of PCP and its contaminants. PCP and related phenols have entered the food chain from their uses on farm buildings and equipment, from contamination of animal feeds and as a result of using treated wood chips as animal bedding. Routes of entry into the environment are from wood-preserving plants and on-site wood treatment. PCP has been used in the leather and tanning industries, contaminated wastes from which are sometimes utilized as ingredients in animal feeds [220].

In the case of contaminated poultry litter, chlorophenols are bacterially converted to chloroanisoles, which are absorbed by the birds, producing musty flavours in the flesh. Curtis et al. [221] employed GC–ECD to detect the presence of pentachloroanisole and 2,3,4,6-tetrachloroanisole in chicken meat. PCP is normally derivatized to the methyl or ethyl ether before GC analysis [222], although Chau and Coburn [223] used acetic anhydride to form the PCP acetate from residues in natural and waste water and Erney [224] used the same method for PCP in milk. It is possible to avoid the derivatization step in some circumstances and Borsetti [225] has reported the direct GC–ECD of PCP from milk and blood of dairy cattle, utilizing a column of 2% SP-1000–1% orthophosphoric acid on 80–100 mesh Chromosorb W AW.

10.7.2. Synthetic pyrethroids

The natural pyrethroids, components of an extract from pyrethrum flowers (*Chrysanthemum cinerafolium*), have been used as an insecticide in various forms for many years, being introduced into Europe from Eastern Asia in about 1820. Chemically speaking, they are esters of various alcohols with chrysanthemum mono- and dicarboxylic acids. Their poor stability, especially in the presence of sunlight, is a serious limitation to their agricultural applications. Their lack of electron-capturing properties means that their residue analysis is dependent on colorimetric methods or relatively insensitive GC procedures with a flame ionization detector. The first synthetic pyrethroid, allethrin, was produced in 1949 and did nothing to overcome the disadvantages of the natural product. In 1967 resmetrhrin was produced in Great Britain by Elliot. This was followed by permethrin, a chlorinated compound which is comparatively stable to UV light. Other compounds were synthesized elsewhere including some with cyano groups (fenpropanate, fenvalerate) and others with both cyano and chlorine substituents (cypermethrin). These latter are highly electron-capturing compounds, with quantitation possible at the 20-pg level using a ^{63}Ni ECD [226]. Residues down to the level of 0.01 ppm have been reported in asparagus, carrots, tomatoes, lettuce and onions [227]. George et al. [228] published a residue study on permethrin and bioethanomethrin involving saponification followed by derivatization of the alcohol moiety with trichloroacetyl chloride and of the acid with trichloroethanol. GC–ECD of these derivatives and the original pyrethroids gave comparable results and a detection limit of less than 0.04 ppm. In France, Mestres [229] has reported on residue methodology for decamethrin in a variety of crops with a detection limit using electron-capture detection down to 0.001 ppm. It was found that direct GC of decamethrin was difficult since a GC column temperature of 245°C was necessary and the operating temperature limit of the ^{3}H ECD was 220°C. A dibromo derivative of decamethrin produced by transesterification could be chromatographed at a column temperature below 220°C.

Several of the synthetic pyrethroids are very toxic to some fish and anthropods [230] and because of their high persistance relative to natural pyrethroids, methods have been developed for residue monitoring in soil [231,232] and other environmental substrates.

10.8. CONCLUSION

The drawbacks of the early ECDs for environmental analysis, which have already been indicated were (1) the inability of the tritiated radioactive sources to be operated at a temperature much above 200°C, leading to easy contamination by sample co-extractives, and (2) limited linearity of the detector, resulting in difficulties and frequent errors in quantitation. These disadvantages have now been overcome to a great extent by the development of the more temperature-stable radioactive sources, such as ^{63}Ni which can be used up to 400°C and also through the introduction of linearized detectors in recent years. The extended linear range facilitates the use of these detectors in automated analysis [233]. While the ^{63}Ni detector is basically less sensitive than the tritium designs, improvements in detector configuration have generally overcome this difference, with the minimum detectable amount of a typical organochlorine compound being in the 1– 10 pg range (see Table 10.11). In comparison with other detectors commonly used in pesticide work, the ECD is generally the most sensitive for low-level analysis of environmental samples, e.g. chlorinated compounds in water, chlorodioxins in herring gull eggs, etc. However, with improvements in the sensitivity of the electrolytic conductivity (Hall) detector, coupled with its greater selectivity, it is possible that in certain circumstances, e.g. water analysis, similar concentrations of organochlorine compounds in the ppb range can be analyzed with minimal or no clean-up. This is certainly the case with monochlorinated compounds.

The current trend in GC–ECD is towards the development and use of capillary columns coupled with micro-volume detectors to provide better peak resolution and greater sensitivity.

TABLE 10.11

PERFORMANCE CHARACTERISTICS OF SOME GAS CHROMATOGRAPHY DETECTORS COMMONLY USED IN PESTICIDE ANALYSIS

Detector	Application	Selectivity*	Linearity	Detection limit (pg)
Electron-capture detector	Cl, Br, I, NO_2, etc.	10^5–10^6	$2 \cdot 10^4$	1**
Coulson conductivity detector (Hall)	Cl, Br	$>10^6$	10^5	10
	N	10^5	10^3	10–20
	S	10^4	–	100
Flame photometric detector	P	10^6	$5 \cdot 10^4$	4***
	S	10^4	$5 \cdot 10^2$ §	70
Alkali flame ionization detector	P	10^5	$2 \cdot 10^4$	1***
	N	10^4	$2 \cdot 10^4$	10
	Others, e.g. As	10^2	–	200

*With respect to carbon.
**Using aldrin.
***Using parathion.
§Quadratic response.

REFERENCES

1 J.E. Lovelock and S.R. Lipsky, *J. Amer. Chem. Soc.*, 82 (1960) 431.
2 E.D. Pellizzari, *J. Chromatogr.*, 98 (1974) 323–361.
3 P.T. Holland and R. Greenhalgh, in H.A. Moye (Editor), *Analysis of Pesticide Residues*, Wiley, New York, 1980, Ch. 2, pp. 51–136.
4 E.R. Adlard, *CRC Critical Reviews in Analytical Chemistry*, CRC Press, Cleveland, OH, May 1975, pp. 27–32.
5 J.C. Steelhammer and W.E. Wentworth, *J. Chem. Phys.*, 51 (1969) 1802.
6 C.H. Hartmann, *Anal. Chem.*, 45 (1973) 733–739.
7 R.J. Maggs, P.L. Joynes, A.J. Davies and J.E. Lovelock, *Anal. Chem.*, 43 (1971) 1966–1971.
8 B. Brechbühler, L. Gay and H. Jaeger, *Chromatographia*, 10 (1977) 478–486.
9 D. Fenimore, P.R. Lay and A. Zlatkis, *Anal. Chem.*, 43 (1971) 1972–1975.
10 J.A. Lubkowitz and W.C. Parker, *J. Chromatogr.*, 62 (1971) 53–61.
11 P.G. Symonds, D.C. Fenimore, B.C. Petit, J.E. Lovelock and A. Zlatkis, *Anal. Chem.*, 39 (1967) 1428–1433.
12 J.E. Lovelock, *J. Chromatogr.*, 99 (1974) 3–12.
13 J.J. Sullivan and C.A. Burgett, *Chromatographia*, 8 (1975) 176–180.
14 Beckman Instruments Inc., *Instruction Manual 1337, Electron Capture Detector and Power Supply for GC-4 Gas Chromatographs*, June 1965.
15 R. Villalobos, *ISA Trans.*, 7 (1968) 38–46.
16 R. Villalobos, L.E. Braun and D. Richards, Paper presented at *16th National I.S.A. Symposium*, Pittsburgh, PA, May 25–27, 1970.
17 R.R. Freeman and W.E. Wentworth, *Anal. Chem.*, 43 (1971) 1987–1991.
18 L.W. Jacobs, S.F. Chou and J. Tiedie, *J. Agr. Food Chem.*, 24 (1976) 1198–1203.
19 A.B. Filonow, L.W. Jacobs and M.M. Mortland, *J. Agr. Food Chem.*, 24 (1976) 1203–1204.
20 W.E. Wentworth, A. Tishbee, C.F. Batten and A. Zlatkis, *J. Chromatogr.*, 112 (1975) 229–246.
21 J.N. Driscoll, *J. Chromatogr.*, 134 (1977) 49–55.
22 E.S. Goodwin, R. Goulden, R. Richardson and J.G. Reynolds, *Chem. Ind. (London)*, (1960) 1220–1230.
23 E.S. Goodwin, R. Goulden and R. Reynolds, *Analyst (London)*, 86 (1961) 697–701.
24 C.A. Edwards, *Persistant Pesticides in the Environment*, CRC Press, Cleveland, OH, 1970.
25 G.T. Brooks, *Chlorinated Insecticides*, Vol. II, CRC Press, Cleveland, OH, 1971.
26 J.A. Burke, *Res. Rev.*, 34 (1971) 59–90.
27 *Pesticide Analytical Manual*, `Vols. I, II, III, U.S. Department of Health, Education and Welfare, Food and Drug Administration, Washington, DC, 1978.
28 J.F. Thompson, J.B. Mann, A.O. Apodaca and E.J. Kantor, *J. Ass. Offic. Anal. Chem.*, 58 (1975) 1037–1050.
29 B. McMahon and J.A. Burke, *J. Ass. Offic. Anal. Chem.*, 61 (1978) 640–652.
30 *Canadian Manual on Analytical Methods for Pesticide Residues in Foods, Cat. No. H44-2869-REV.*, Information Canada, Ottawa, 1973.
31 *Methodensammlung zur Rückstandsanalytik von Pflanzenschutzmitteln, 4. Lieferung*, Verlag Chemie, Weinheim/Bergstrasse, 1976.
32 Food and Drug Administration, *Program Evaluation, PY 75. Total Diet Studies, Adult (7302.00)*, U.S. Department of Health, Education and Welfare, 1979, 20 pp. and attachments.
33 L.B. Ackerman, *Pestic. Monit. J.*, 14 (1980) 64–69.
34 E.P. Laug, F.M. Kunzl and C.S. Preckett, *Arch. Ind. Hyg. Occup. Med.*, 3 (1951) 245–246.
35 R.W. Barrett, A.J. D'Ercole, J.D. Cain and R.D. Arthur, *Pestic. Monit. J.*, 13 (1979) 47–51 and references cited therein.
36 G.S. Dhaliwal and R.L. Kalra, *Pestic. Monit. J.*, 51 (1978) 91–93.
37 R.D. Johnston, D.D. Manske, D.H. New and D.S. Podrebrac, *Pestic. Monit. J.*, 13 (1979) 87–98.
38 F.W. Kuntz, A.R. Yobs and S.C. Strassman, *Bull. Soc. Pharmacol. Environ. Pathol.*, 4 (1976) 17–19.
39 F.W. Katz, A.R. Yobs, S.C. Strassman and J.F. Viar, Jr., *Pestic. Monit. J.*, 11 (1977) 61–63.

40 C.A. Edwards, In C.A.I. Goring and J.W. Hamaker (Editors), *Organic Chemicals in the Soil Environment*, Vol. 2, Marcel Dekker, New York, 1972, Ch. 8, p. 513.
41 M. Chiba and H.V. Morley, *J. Agr. Food Chem.*, 16 (1968) 916–922.
42 H.B. Pionke, G. Chesters and D.E. Armstrong, *Agron. J.*, 60 (1968) 289–292.
43 I.H. Williams, *J. Ass. Offic. Anal. Chem.*, 51 (1968) 715–717.
44 E.A. Woolson and P.C. Kearney, *J. Ass. Offic. Anal. Chem.*, 52 (1969) 1202–1206.
45 J.G. Saha, *J. Ass. Offic. Anal. Chem.*, 54 (1971) 170–174.
46 E.A. Woolson, *J. Ass. Offic. Anal. Chem.*, 57 (1974) 604–609.
47 *Manual of Analytical Methods for the Analysis of Pesticides in Humans and Environmental Samples, EPA-600/8-80-03*, U.S. Environmental Protection Agency, Health Effects Research Laboratory, Research Triangle Park, NC, June 1980.
48 S.U. Khan, *Pesticides in the Soil Environment*, Elsevier, Amsterdam, 1980.
49 C.A. Edwards, *Res. Rev.*, 13 (1966) 83–132.
50 R.G. Nash and E.A. Woolson, *Science*, 157 (1967) 924–925.
51 C.M. Menzie, *Ann. Rev. Entomol.*, 17 (1972) 199–222.
52 E.P. Lichtenstein, T.W. Fuhremann and K.R. Schulz, *J. Agr. Food Chem.*, 19 (1971) 718–721.
53 D.K.R. Stewart and D. Chisholm, *Can. J. Soil Sci.*, 61 (1971) 379–383.
54 R.G. Nash, W.G. Harris, P.D. Ensor and E.A. Woolson, *J. Ass. Offic. Anal. Chem.*, 56 (1973) 728–732.
55 A.B. Crockett, G.B. Wiersma, H. Tai, W.G. Mitchell, P.F. Sand and A.E. Carey, *Pestic. Monit. J.*, 8 (1974) 69–97.
56 A.E. Carey, P. Douglas, H. Tai, W.G. Mitchell and G.B. Wiersma, *Pestic. Monit. J.*, 13 (1979) 17–22.
57 M. Suzuki, Y. Yamato and T. Watanabe, *Pestic. Monit. J.*, 11 (1977) 88–93.
58 A.E. Carey, *Pestic. Monit. J.*, 13 (1979) 23–27.
59 G.W. Ware, B.J. Estesen, N.A. Buck and W.P. Cahill, *Pestic. Monit. J.*, 12 (1978) 1–3.
60 M.M. Cliath and W.F. Spencer, *Environ. Sci. Technol.*, 6 (1972) 910–914.
61 D.C. Abbot, R.B. Harrison, J.O. Tatton and J. Thompson, *Nature (London)*, 211 (1966) 259–261.
62 J.M. Cohen and C. Pinkerton, *Advan. Chem. Ser.*, 60 (1966) 163–176.
63 T.J. Peterle, *Nature (London)*, 224 (1969) 620.
64 P.A. Pearce, L.M. Reynolds and D.B. Peakall, *Pestic. Monit. J.*, 11 (1978) 199–204.
65 A. Bevenue, J.N. Ogata and J.W. Hylin, *Bull. Environ. Contam. Toxicol.*, 8 (1972) 238–241.
66 D.B. Peakall, *Atmos. Environ.*, 10 (1976) 899–900.
67 W.D. Gummer, *Water Quality Interpretative Report No. 4, Cat. No. EN-36-518/4*, Supply and Services Canada, Ottawa, 1979.
68 *Environment Canada (1979) Analytical Methods Manual*, Inland Waters Directorate, Water Quality Branch, Ottawa, 1979.
69 N.V. Brodtmann, Jr. and W.E. Koffskey, *J. Chromatogr. Sci.*, 17 (1979) 97–110.
70 J. Uhnák, M. Sackmauerová, A. Szokolay and O. Pal'ušová, *J. Chromatogr.*, 91 (1974) 545–547.
71 Y. Yamato, M. Suzaki and T. Watanabe, *J. Ass. Offic. Anal. Chem.*, 61 (1978) 1135–1139.
72 J.O. Keith and E.G. Hunt, *N. Amer. Wildlife Natur. Resources Conf. Trans.*, 31 (1966) 150–177.
73 A. Bevenue, J.W. Hylin, Y. Kawano and T.W. Kelly, *Pestic. Monit. J.*, 6 (1972) 56–64.
74 D.W. Johnson, in C.A. Edwards (Editor), *Environmental Pollution by Pesticides*, Plenum Press, London, 1973, Ch. 5, pp. 181–212.
75 G.G. Sims, C.E. Cosham, J.R. Campbell and M.C. Murray, *Bull. Environ. Contam. Toxicol.*, 14 (1975) 505–512.
76 D.J. Hallet, R.J. Norstrom, F.I. Onuska, M.E. Comba and R. Sampson, *J. Agr. Food Chem.*, 24 (1976) 1189–1193.
77 R.J. Norstrom, D.J. Hallett and R.A. Sonstegard, *J. Fish. Res. Bd. Can.*, 35 (1978) 1401–1409.
78 R.J. Norstrom, D.J. Hallett, F.I. Onuska and M.E. Comba, *Environ. Sci. Technol.*, 14 (1980) 860–866.
79 R.J. Norstrom, H.T. Won, M.V.H. Holdrenet, P.G. Calway and C.D. Naftel, *J. Ass. Offic. Anal. Chem.*, 63 (1980) 37–42.

80 W.J. Neidermyer and J.J. Hickey, *Pestic. Monit. J.*, 10 (1976) 92–95.
81 T. Miyazaki, K. Akiyama, S. Kaneko, S. Horii, T. Yanagishi and K. Kodera, *Bull. Environ. Contam. Toxicol.*, 25 (1980) 305–312.
82 A. Sodegren, R. Djirsarai, M. Gharibzodeh and A. Moinpour, *Pestic. Monit. J.*, 12 (1978) 81–86.
83 D.A. Radcliffe, *Nature (London)*, 215 (1967) 208–210.
84 J.J. Hickey and D.W. Anderston, *Science*, 162 (1968) 271–273.
85 L.J. Blus, C.D. Gish, A.A. Belisle and R.M. Prouty, *Nature (London)*, 235 (1972) 376–377.
86 P.A. Pearce, D.A. Peakall and L.M. Reynolds, *Pestic. Monit. J.*, 13 (1979) 61–68.
87 L.J. Blus, B.S. Neely, Jr., A.A. Belisle and R.M. Prouty, *Environ. Pollut.*, 7 (1974) 81–91.
88 S. Postupalsky, *Pesticide Section Report No. 20*, Canadian Wildlife Service, Ottawa, 1971.
89 K.A. King, E.L. Flickinger and H.H. Heldebrand, *Pestic. Monit. J.*, 12 (1978) 16–21.
90 M.L. Lindnall and B.V. Low, *Pestic. Monit. J.*, 14 (1980) 108–111.
91 D.A. Ratcliffe, *The Peregrine Falcon*, T. & A.D. Poyser, Calton, 1980, Ch. 13, pp. 304–332.
92 L.J. Blus, T.G. Lamont and B.S. Neely, Jr., *Pestic. Monit. J.*, 12 (1979) 172–184.
93 D.W. Anderson, J.R. Jehl, Jr., R.W. Risebrough, L.A. Woods, Jr., L.R. Deweese and W.G. Edgecomb, *Science*, 180 (1975) 806–808.
94 J.D. Lockie, D.A. Ratcliffe and R. Balharry, *J. Appl. Ecol.*, 6 (1969) 381–389.
95 T.E. Kaiser, W.L. Reichel, L.N. Locke, E. Cromartie, A.J. Krynitsky, T.G. Lamont, B.M. Hulhem, R.M. Prouty, C.J. Strafford and D.M. Swineford, *Pestic. Monit. J.*, 13 (1980) 145–149.
96 B.E. Frazier, G. Chesters and G.B. Lee, *Pestic. Monit. J.*, 4 (1970) 67–70.
97 W.P. Cochrane, in J.F. Lawrence and R. Frei (Editors), *Chemical Derivatisation in Analytical Chemistry, Vol. 1, Chromatography*, Plenum Press, New York, 1981.
98 S. Jensen, *New Scientist*, 32 (1966) 612.
99 P.W. Risebrough, P. Riecha, S.G. Herman, D.B. Peakall and O.N. Kirven, *Nature (London)*, 220 (1968) 1098–1102.
100 D.B. Peakall, *Res. Rev.*, 44 (1972) 1–21.
101 S. Jensen, *Ambio*, 1 (1972) 123.
102 I.C.T. Nisbet and A.F. Sarofin, *Environ. Health Perspect.*, 1 (1972) 21.
103 D. Pal, J.B. Weber and M.R. Overcash, *Res. Rev.*, 4 (1980) 45–98.
104 L. Fisbein, *J. Chromatogr.*, 68 (1972) 345–426.
105 L.M. Reynolds, *Res. Rev.*, 34 (1971) 27–57.
106 J.A. Armour and J.A. Burke, *J. Ass. Offic. Anal. Chem.*, 53 (1970) 761–768.
107 O.W. Berg, P.L. Diosady and G.A. Rees, *Bull. Environ. Contam. Toxicol.*, 7 (1972) 338–347.
108 D.A. Griffin and A.B. Morin, *J. Ass. Offic. Anal. Chem.*, 63 (1980) 959–964.
109 O. Hutzinger, S. Safe and V. Zitko, *Analabs Research Notes*, 12 (1972) 1–11.
110 J.C. Duinker, M.T.J. Hillebrand, K.H. Palmork and S. Whilhelmsen, *Bull. Environ. Contam. Toxicol.*, 25 (1980) 956–964.
111 F.L. Beezhold and V.F. Stout, *Bull. Environ. Contam. Toxicol.*, 10 (1973) 10–15.
112 L.D. Sawyer, *J. Ass. Offic. Anal. Chem.*, 56 (1973) 1015–1023.
113 S. Jensen, B. Nucci and G. Widmark, *OECD Study of Analysis of PCB*, Report from the Institute of Analytical Chemistry, Stockholm, 1968.
114 R.W. Risebrough, *Chemical Fallout*, Charles C. Thomas, Springfield, IL, 1969, pp. 5–30.
115 D. Sissons and D. Welti, *J. Chromatogr.*, 60 (1971) 15–32.
116 J.W. Rote and R.G. Murphy, *Bull. Environ. Contam. Toxicol.*, 6 (1971) 377–384.
117 V. Zitko, O. Hutzinger and S. Safe, *Bull. Environ. Contam. Toxicol.*, 6 (1971) 160–163.
118 R.G. Webb and A.C. McCall, *J. Chromatogr. Sci.*, 11 (1973) 366–373.
119 A.S.Y. Chau and R.C.J. Sampson, *Environ. Lett.*, 8 (1975) 89–101.
120 L.D. Sawyer, *J. Ass. Offic. Anal. Chem.*, 61 (1978) 272–281.
121 L.D. Sawyer, *J. Ass. Offic. Anal. Chem.*, 61 (1978) 282–292.
122 A. Poland and E. Glover, *Mol. Pharmacol.*, 13 (1977) 924–938.
123 J.N. Huckins, D.S. Stalling and J.D. Petty, *J. Ass. Offic. Anal. Chem.*, 63 (1980) 750–755.
124 L.R. Kamops, W.J. Trotter, S.J. Young, A.C. Smith, J.A.G. Roach and S.W. Page, *Bull. Environ. Contam. Toxicol.*, 23 (1979) 51–56.
125 M. Zell, H.J. Neu and K. Ballschmiter, *Z. Anal. Chem.*, 292 (1978) 97–107.

126 F.I. Onuska and M. Comba, *Environ. Sci. Rev.*, 16 (1980) 285–302.
127 L.G.M.Th. Tuinstra and W.A. Traag, *Meded. Fac. Landbouww. Rijksuniv.*, 44/2 (1979) 886–893.
128 J. Burke and L. Giuffrida, *J. Ass. Offic. Anal. Chem.*, 47 (1964) 326–342.
129 S.S. Brody and J.E. Chaney, *J. Gas Chromatogr.*, 4 (1966) 42–46.
130 K.E. Elgar, R.G. Marlow and B.L. Mathews, *Analyst (London)*, 95 (1970) 875–878.
131 G.H. Tjan and Th. Konter, *J. Ass. Offic. Anal. Chem.*, 54 (1971) 1122–1123.
132 M. Sherman and R.B. Herrick, *J. Agr. Food Chem.*, 20 (1972) 985–990.
133 J.L. Eschle, D.D. Oehlar, B.F. Hogan and M.C. Ivey, *J. Econ. Entomol.*, 66 (1973) 290–291.
134 R.C. Hall, D.L. Ballee, G.W. Bennett and J.E. Fahey, *J. Econ. Entomol.*, 66 (1973) 315–318.
135 G.S. Tadjer and M.N. Egyed, *Bull. Environ. Contam. Toxicol.*, 12 (1974) 173–176.
136 M.J. Toet and L. Toet, *Beitr. Tabakforsch.*, 8 (1976) 334–338.
137 J.A. Coburn and A.S.Y. Chau, *J. Ass. Offic. Anal. Chem.*, 57 (1974) 1272–1278.
138 J.F. Lawrence and H.A. McLeod, *J. Ass. Offic. Anal. Chem.*, 59 (1976) 637–640.
139 G.H. Cook and J.C. Moore, *J. Agr. Food Chem.*, 24 (1976) 631–634.
140 M.C. Bowman, C.L. Holder and L.G. Rushing, *J. Agr. Food Chem.*, 26 (1978) 35–42.
141 H.E. Braun, *J. Ass. Offic. Anal. Chem.*, 57 (1974) 182–188.
142 J.E. Woodrow, J.N. Seiber, D.G. Crosby, K.W. Moilanen, C.J. Sonderquist and C. Mouter, *Arch. Environ. Contam. Toxicol.*, 6 (1977) 175–191.
143 I.H. Williams, *Res. Rev.*, 38 (1971) 1–20.
144 M.T.H. Ragab, *Anal. Lett.*, 10 (1977) 551–564.
145 J.W. Ralls and A. Cortes, *J. Gas Chromatogr.*, 2 (1964) 132–135.
146 W.H. Gutenman and D.J. Lisk, *J. Agr. Food Chem.*, 13 (1965) 48–50.
147 L.I. Butler and L.M. McDonough, *J. Agr. Food Chem.*, 16 (1968) 403.
148 S.C. Lau and R.L. Marxmiller, *J. Agr. Food Chem.*, 18 (1970) 413–415.
149 E.R. Holden, W.M. Jones and M. Beroza, *J. Agr. Food Chem.*, 17 (1969) 56–59.
150 I.C. Cohen, J. Norcup, J.H.A. Ruzicka and B.B. Wheals, *J. Chromatogr.*, 49 (1970) 215–221.
151 J.H. Caro, D.E. Glotfelty, H.P. Freeman and A.W. Taylor, *J. Ass. Offic. Anal. Chem.*, 56 (1973) 1319–1323.
152 E.R. Holden, *J. Ass. Offic. Anal. Chem.*, 56 (1973) 713–717.
153 E.R. Holden, *J. Ass. Offic. Anal. Chem.*, 58 (1975) 562–565.
154 K.E. Savage, *Weed Sci.*, 25 (1977) 55–59.
155 R.P. Upchurch, in C.A.I. Goring and J.W. Hamaker (Editors), *Organic Chemicals in the Soil Environment*, Vol. 2, Marcel Dekker, New York, 1972, Ch. 7, p. 443.
156 P.C. Kearney and D.D. Kaufman, *Degradation of Herbicides*, Marcel Dekker, New York, 1969.
157 W.P. Cochrane and R. Purkayastha, *Toxicol. Environ. Chem. Rev.*, 1 (1971) 137–268.
158 H.Y. Young and A. Chu, *J. Agr. Food Chem.*, 21 (1973) 711–713.
159 R. Albright, N. Johnson, T.W. Sanderson, R.M. Fark, R. Melton, L. Fisher, G.R. Wells, W.R. Parsons, V.C. Scott, J.L. Speake, J.R. Stallworth, B.G. Moore and A.W. Hayes, *Bull. Environ. Contam. Toxicol.*, 12 (1974) 378–384.
160 R.J. Hesselberg and J.L. Johnson, *Bull. Environ. Contam. Toxicol.*, 7 (1972) 115–120.
161 K.I. Beynon, G. Stoydin and A.N. Wright, *Pestic. Sci.*, 3 (1972) 379–387.
162 D.L. Hyzak and R.L. Zimdahl, *Weed Sci.*, 22 (1974) 75–79.
163 J.S. Thornton and C.W. Stanley, *J. Agr. Food Chem.*, 25 (1977) 380–386.
164 W.K. Head and R.B. McKercher, *Can. J. Soil Sci.*, 51 (1971) 423–430.
165 J.D. Fryer and K. Kirkland, *Weed Res.*, 10 (1970) 133–158.
166 G.H. Thiele and R.L. Zimdahl, *Weed Sci.*, 24 (1976) 183–185.
167 A.E. Smith, *J. Chromatogr.*, 97 (1974) 103–106.
168 H. Løkke, *Pestic. Sci.*, 5 (1974) 749–757.
169 S.V. Khan, R. Greenhalgh and W.P. Cochrane, *Bull. Environ. Contam. Toxicol.*, 13 (1975) 602–610.
170 I.L. Adler, C.F. Gordon, L.D. Haines and J.P. Wargo, Jr., *J. Ass. Offic. Anal. Chem.*, 55 (1972) 802–805.
171 L.E. St. John, Jr. and D.J. Lisk, *Bull. Environ. Contam. Toxicol.*, 11 (1974) 529–531.

172 M.P. Molnau, W.G. Lovely and H.P. Johnston, *Weed Sci.*, 21 (1973) 185–188.
173 W.R. Payne, Jr., J.D. Pope, Jr. and J.E. Benner, *J. Agr. Food Chem.*, 22 (1974) 79–82.
174 R.M. Menges and S. Tamez, *Weed Sci.*, 22 (1974) 67–71.
175 A.E. Smith, *J. Agr. Food Chem.*, 20 (1972) 829–831.
176 J.H. Miller, P.E. Keeley, C.H. Carter and R.J. Thallen, *Weed Sci.*, 23 (1975) 211–214.
177 C.H. Miller, T.J. Manaco and T.J. Sheets, *Weed Sci.*, 24 (1976) 288–291.
178 M. Busonelli, F. Tafuri, L. Searponi and C. Maracchini, *Pestic. Sci.*, 6 (1975) 475–480.
179 R.F. Sieck, W.S. Johnson, A.F. Cockerill, D.N.B. Mallen, D.J. Osborne and S.J. Barton, *J. Agr. Food Chem.*, 24 (1976) 617–620.
180 P.E. Corneliussen, *Pestic. Monit. J.*, 5 (1972) 313–330.
181 W.P. Cochrane and J.B. Russell, *Can. J. Plant Sci.*, 55 (1975) 323–325.
182 S.S. Que Que, R.G. Sutherland and M. Velter, *Environ. Sci. Technol.*, 9 (1975) 62–66.
183 C.E. McKone and R.J. Hance, *J. Chromatogr.*, 69 (1972) 204–206.
184 R.B. Leidy, M.D. Jackson, W.A. Skrotch and T.J. Sheets, *Bull. Environ. Contam. Toxicol.*, 13 (1975) 338–341.
185 W.P. Cochrane, R. Greenhalgh and N.E. Looney, *Can. J. Plant Sci.*, 56 (1976) 207–210.
186 C.J. Soderquist and D.G. Crosby, *Pestic. Sci.*, 6 (1976) 17–33.
187 S. Mierzwa and S. Witek, *J. Chromatogr.*, 136 (1977) 105–111.
188 R. Purkayastha, *J. Agr. Food Chem.*, 22 (1974) 453–458.
189 A.E. Smith, *Weed Res.*, 16 (1976) 19–22.
190 E.L. Bjerke, J.L. Herman, P.W. Miller and J.H. Wetlers, *J. Agr. Food Chem.*, 20 (1972) 963–967.
191 M. Leistra, J.H. Smelt and R. Zandvoort, *Weed Res.*, 15 (1975) 243–247.
192 P.B. Marriage, S.V. Khan and W.J. Saidak, *Weed Res.*, 17 (1977) 219–225.
193 A.J. Cessna, *J. Agr. Food Chem.*, 25 (1977) 432–433.
194 J.D. Altom and J.F. Stritzke, *Weed Sci.*, 21 (1973) 556–560.
195 S.V. Khan, *J. Ass. Offic. Anal. Chem.*, 58 (1975) 1027–1031.
196 R.V. Crouch and E.M. Pallin, *Pestic. Sci.*, 5 (1974) 281–285.
197 C.P. Rise, H.C. Sikka and R.S. Lynch, *J. Agr. Food Chem.*, 22 (1974) 533–534.
198 C.E. McKone, R.J. Hance and D.J. Burchill, *Weed Res.*, 11 (1971) 283–291.
199 I.L. Adler and J.P. Wargo, Jr., *J. Ass. Offic. Anal. Chem.*, 58 (1975) 551–553.
200 R.D. Comes, V.F. Bruns and A.D. Kelley, *Weed Sci.*, 24 (1976) 47–50.
201 J.F. Lutz, G.C. Byers and T.J. Sheets, *J. Environ. Qual.*, 2 (1973) 485–488.
202 R.W. Bovey, E. Burnett, C. Richardson, M.G. Merkle, J.R. Baur and W.F. Knisel, *J. Environ. Qual.*, 3 (1974) 61–64.
203 R. Purkayastha and W.P. Cochrane, *J. Agr. Food Chem.*, 21 (1973) 93–98.
204 W.P. Cochrane and B.P. Wilson, *J. Chromatogr.*, 63 (1971) 364–369.
205 C.E. McKone, T.H. Bryant and R.J. Hance, *Analyst (London)*, 97 (1977) 653–655.
206 R. Bailey, G. LeBel and J.F. Lawrence, *J. Chromatogr.*, 161 (1978) 251–257.
207 D.W. Woodham, W.G. Mitchell, C.D. Loftis and C.W. Collier, *J. Agr. Food Chem.*, 19 (1971) 186–188.
208 H. Agemain and A.S.Y. Chau, *Analyst (London)*, 101 (1976) 732–737.
209 S.O. Farwell, E.W. Bowes and D.F. Adams, *Anal. Chem.*, 48 (1976) 420–425.
210 J.R. Koons and E.W. Day, Jr., *J. Chromatogr. Sci.*, 10 (1972) 176–178.
211 A. Büchert and H. Løkke, *J. Chromatogr.*, 115 (1975) 682–686.
212 A.J.A. Pearson, *Int. Pest Contr.*, 20, March/April (1978) 19–22.
213 A. Del Re, P. Fontana, G.F. Marchini, G.P. Molinari and E. Rossi, *Res. Rev.*, 74 (1980) 99–131.
214 J.A. Burke and W. Holswade, *J. Ass. Offic. Anal. Chem.*, 49 (1966) 374–385.
215 W. Dejonckheere, W. Steurbaut and R.H. Kips, *Bull. Environ. Contam. Toxicol.*, 13 (1975) 720–729.
216 I.S. Taylor and F.P. Keenan, *J. Ass. Offic. Anal. Chem.*, 53 (1970) 1293–1295.
217 *Assessing Potential Ocean Pollutants*, National Academy of Sciences, Washington, DC, 1975, pp. 188–288.
218 D.R. Rourke, W.F. Mueller and R.S.H. Yang, *J. Ass. Offic. Anal. Chem.*, 60 (1977) 233–235.

219 D. Firestone, *FDA By-Lines*, 2 (1977) 57–89.
220 L.D. Metcalfe, *J. Ass. Offic. Anal. Chem.*, 55 (1972) 542–546.
221 R.F. Curtis, C. Dennis, J.M. Gee, M.C. Gee, N.M. Griffiths, D.E. Land, J.P. Peel and D. Robinson *J. Sci. Food Agr.*, 25 (1974) 811–828.
222 G. Yip, *J. Ass. Offic. Anal. Chem.*, 54 (1971) 966–969.
223 A.S.Y. Chau and J.A. Coburn, *J. Ass. Offic. Anal. Chem.*, 57 (1974) 389–393.
224 D.R. Erney, *J. Ass. Offic. Anal. Chem.*, 61 (1978) 214–216.
225 A. Borsetti, *J. Agr. Food Chem.*, 28 (1980) 710–714.
226 R.A. Chapman and H.S. Simmons, *J. Ass. Offic. Anal. Chem.*, 60 (1977) 977–978.
227 R.A. Chapman and C.R. Harris, *J. Chromatogr.*, 166 (1978) 513–518.
228 D.A. George, J.E. Halfhill and L.M. McDonough, *Synthetic Pyrethroids*, ACS Symposium Series 42 (1972) 201–210.
229 R. Mestres, *Trav. Soc. Pharm. Montpellier*, 38 (1978) 181–191.
230 J. Miyamoto, *Environ. Health Perspect.*, 14 (1976) 15–28.
231 G.H. Fujie and O.H. Fullmer, *J. Agr. Food Chem.*, 26 (1978) 395–398.
232 I.H. Williams and M.J. Brown, *J. Agr. Food Chem.*, 27 (1979) 130–132.
233 R. Greenhalgh and W.P. Cochrane, *J. Chromatogr.*, 188 (1980) 305–313.

Chapter 11

Environmental applications of the electron-capture detector – dioxins

F. BRUNER

CONTENTS

11.1. INTRODUCTION: THE DIOXIN PROBLEM

If a book reviewing the theory and practice of the electron-capture detector (ECD) had been published 5 years ago, there is little doubt that a chapter devoted to dioxins would not have been sought by the Editors. The enormous interest in polychlorodibenzodioxin (PCDD) and in particular in one of the 22 isomers of the tetrachloro compound (TCDD), 2,3,7,8-TCDD, followed the explosion of a safety valve in an industrial plant producing trichlorophenol at Seveso in northern Italy in July 1976. As the explosion was the result of overheating of the plant reactor, 2,3,7,8-TCDD, the most toxic [1] compound of the series, was produced in much larger amounts than in the normal process, where its concentration is restricted to low parts per million levels [2]. A large, highly populated area was severely polluted. This disaster resulted in many families losing their homes and the economic activity of the entire region was completely upset.

Many children have been affected by chloracne, a skin disease caused by TCDD. In addition, the long-term consequences of the absorption of TCDD on future generations are unknown, as it has been demonstrated that this type of compound has teratogenic, mutagenic and carcinogenic properties.

The Seveso event made public authorities more aware of the dangers of PCDDs [3] and considerable research activity has been undertaken. It was immediately recognized that, although the 2,3,7,8-TCDD isomer seems to be the most toxic, most of the other dioxins have similar properties but these vary to a large extent depending on the structure [4]. These compounds are produced in a wide range of industrial processes and may be found as undesirable combustion products [5–7].

Thus, the analytical problems connected with PCDDs, which began with the determination of 2,3,7,8-TCDD only, nowadays involve the analysis of all dioxins, including the determination of all isomers (75 compounds with similar chemical properties). Another class of toxic compounds, polychlorodibenzofurans (PCDFs), is found in samples of environmental interest and their determination interferes with that of dioxins. Thus, an important problem is the preparation of the sample for the final gas chromatographic (GC) analysis from the original matrix. Some of the most common matrices in which PCDDs are to be determined are as follows: (1) soils; (2) waters; (3) particulate matter; (4) fumes and fly ashes; (5) sludges; (6) human and animal fluids (blood, urine, milk); (7) human and animal tissues; (8) industrial products.

The variety of matrices from which PCDDs have to be isolated and determined implies the need for purification processes that must be extremely accurate in order to achieve the following aims:

(a) Complete or at least a known and reproducible recovery of PCDDs from the original sample.

(b) Pre-separation of other organic compounds that may interfere with the determination and that, having similar physical and chemical properties, follow the dioxins in the gross extraction process.

Of course, these requirements are closely linked and have to be met in the overall "sample purification" process. The main characteristic that makes the determination of dioxins one of the most difficult and challenging is that these compounds are fortunately present at very small concentrations in most instances. However, because of their extremely high activity, they must be determined down to the lowest level possible with the technology available. It is interesting that in about 5 years the determination level was lowered by a factor of 10^6, from a few nanograms in 1976 to a few femtograms in 1980 [8,9].

In conclusion, the final step of this extremely difficult analytical problem implies the use of the most sophisticated techniques that should ensure the highest specificity and sensitivity. Nowadays the use of capillary columns in combination with high-resolution mass spectrometry seems to meet these requirements.

The ECD played a very important role in early work and it is the author's opinion that it may still have great importance in routine analysis, especially in some particular problems. However, the successful use of the ECD is dependent on a high degree of purity of the sample and extremely efficient GC columns.

11.2. INTERFERENCES IN THE DETERMINATION OF POLYCHLORODIBENZODIOXINS WITH ELECTRON-CAPTURE DETECTION AND PURIFICATION PROCEDURES

The major source of PCDDs is in the massive production and use of pentachlorophenol (PCP) and other chlorophenols that are widely used as herbicides, insecticides and wood preservatives. These compounds contain a variety of chlorine-containing contaminants and by-products that have different structures but a behaviour that is similar to that of PCDDs, and also their molecular weights lie in the same range. The names and formulae of the most common interfering classes of substances are listed in Table 11.1.

TABLE 11.1

POLYCHLORINATED COMPOUNDS: GENERAL FORMULAE

Name	Abbreviation	General structure
Polychlorodibenzo-*p*-dioxins	PCDDs	Cl_m Cl_n
Polychlorobenzofurans	PCBFs	Cl_m Cl_n
Polychlorodiphenyl ethers	PCDPEs	Cl_m Cl_n
Polychlorobiphenyls	PCBs	Cl_m Cl_n
Polychloromethoxybiphenyls	PCMBs	OCH_3 Cl_m Cl_n
Polychlorobenzenes	PCBzs	Cl_m
Chloropesticides	CPs	No general formula

The ECD shows a very high response toward all these classes of compounds and some of them occur in biological and environmental samples at concentration levels that are often considerably higher than those of the dioxins to be determined. Even very highly efficient capillary columns are unable to ensure a satisfactory separation of dioxins from the other polychlorinated compounds.

Other contaminants may easily come from the solvents used, especially when large volumes must be concentrated or taken to dryness, which applies in most instances [10].

It has been shown that dioxins are always accompanied by at least some of the classes of organochlorine compounds mentioned and that very careful clean-up procedures must be followed to obtain samples in which dioxins, and especially 2,3,7,8-TCDD, can be determined with acceptable accuracy. The clean-up procedures should also remove from the dioxin sample most organic compounds which, even though they do not contain electron-capture detectable groups, are present in such high concentrations in the original sample that they affect the detector background considerably.

The sample preparation procedure may differ greatly according to the nature of the original matrix, and we shall not describe the initial steps. Solvent extraction is the first sample treatment stage and usually *n*-hexane (see, for example, ref. 11) or a medium-polarity solvent mixture, such as *n*-hexane–acetone [12], is used. If phenolic compounds are suspected to be present, an alkaline reflux with alcoholic potassium hydroxide is suggested, followed by solvent extraction.

Most matrices of dioxin samples contain lipids, especially human and animal tissues or soil and vegetable materials. Treatment with concentrated sulphuric acid is advisable in order to remove the organic compounds by carbonization. This can be done either by adding to the solution a few drops of concentrated sulphuric acid or by passing the solution through a Celite–sulphuric acid column. The former method is more effective. PCDDs and PCDFs can be now separated from electron-capturing substances such as PCBs, chlorinated pesticides and PCDPEs by column chromatography using Florisil [13] or alumina [13–15] and eluting with different solvent mixtures. The solution must be dried with granular sodium sulphate prior to this operation. Several solvent mixtures can be used, but the commonest is 2% dichloromethane in *n*-hexane–50% dichloromethane in *n*-hexane. After elution, the 2% dichloromethane in *n*-hexane fraction contains PCDDs and PCBFs and the other fraction contains PCBs and chlorinated pesticides.

The final solution can then be evaporated to eliminate dichloromethane, which is undesirable for ECD operation, and the residue is dissolved in *n*-hexane. In recent work [16,17], multi-step columns have been devised with which the various clean-up procedures using chromatographic principles can be carried out in sequence with consequent time savings.

In a recent paper, Kooke et al. [18] have shown that the best procedure for the extraction of PCDDs and PCDFs from fly ash consists of Soxhlet extraction with benzene or toluene. The results are further improved by treatment with 1 *N* hydrochloric acid prior to extraction.

Phillipson and Puma [19] have shown recently that polychlorinated methoxybiphenyls (PCMBs) are present as contaminants in fish samples and that they can be separated from PCDDs and PCDFs by elution on a silica gel column with *n*-hexane. The interfering compounds are retained on the silica gel column.

Fig. 11.1 shows the extraction and clean-up procedure most widely used (see, for example, ref. 20). It is important to note that such a procedure is not compulsory or without alternatives. Depending on the nature of the sample matrix, the abundance of dioxins in the sample and other circumstances, some steps may be changed or added. For this reason, the clean-up procedure described here is only a general guide.

Very recently, an important source of contamination was found and eliminated by Nestrick and Lamparski [10]. Often during the extraction and clean-up procedures the sample needs to be evaporated in order to remove the solvent; with small volumes of solvent and when the sample must be taken to dryness, this is usually done by passing a stream of clean gas over the sample. Contamination occurs, even when using very pure gases, from unknown substances that are detected by the ECD. The authors invented a device, called Femtogas, to make the gas stream extra pure. The Femtogas packing is

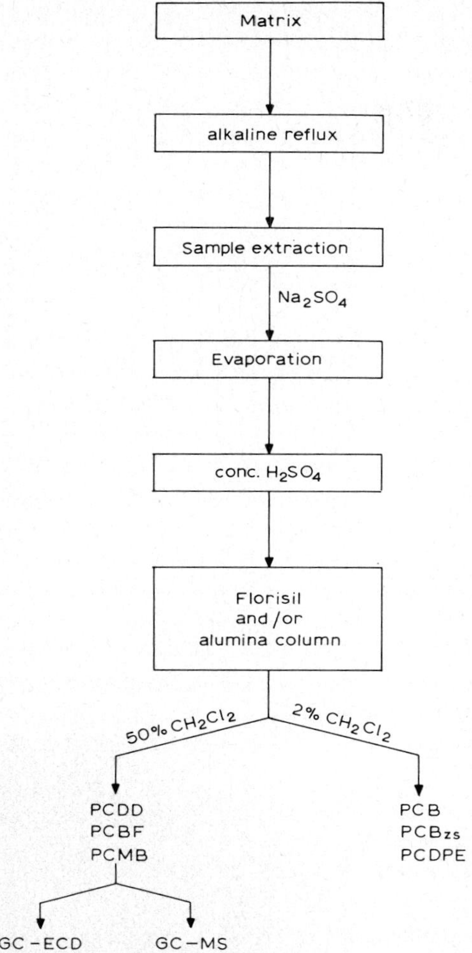

Fig. 11.1. Flow chart of sample preparation for GC analysis of PCDDs.

prepared with equal parts by weight (3 g) of Carbopack B (Supelco) and active carbon (Amoco), which are dissolved in chloroform and the slurry is placed in an ultrasonic bath; 3 g of Apiezon L dissolved in Chromosorb W are added and the resulting slurry is again placed in an ultrasonic bath.

The mixture is mechanically mixed with 2 g of Chromosorb W AW. The resulting mixture is placed in appropriate metal traps that are set on the gas line. Further trapping systems are placed in line, such as silica-supported sulphuric acid and Carbosieve S. According to the authors, the purification device has a long life (2 years) and ensures very clean chromatograms, as shown in Fig. 11.2. A comparison of chromatograms a and b shows that in practice no impurities contaminate the TCDD if the purification device is used. The amount of impurities when an unpurified gas used is very high, as can be seen from chromatogram b.

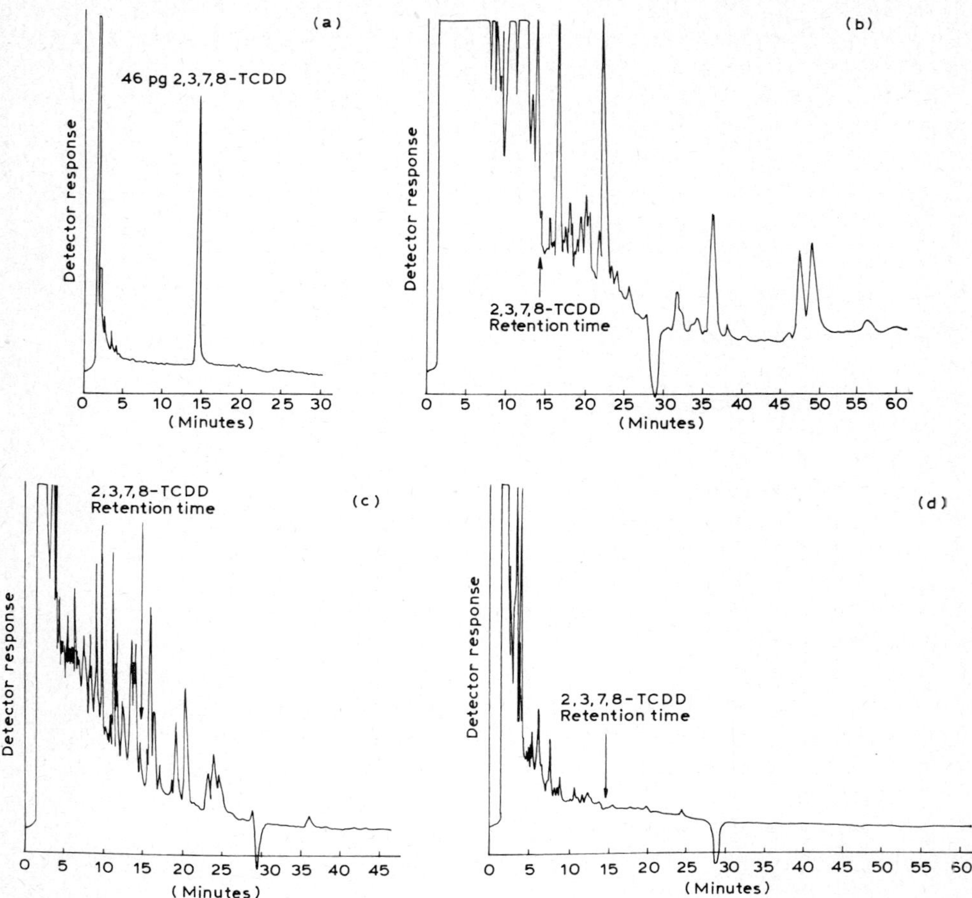

Fig. 11.2. Capillary column GC–ECD chromatograms for residue contamination tests: (a) background for isooctane residue solvent containing 46 pg of 2,3,7,8-TCDD; (b) initial system; (c) molecular sieves and activated charcoal system; (d) Femtogas purification device. (From ref. 10.)

11.3. THE ANALYSIS OF POLYCHLORODIBENZODIOXINS BY GAS CHROMATOGRAPHY WITH ELECTRON-CAPTURE DETECTION

It has been shown above that, in spite of using very careful sample preparation and clean-up procedures, the 50% dichloromethane–*n*-hexane fraction contains all the PCDDs and PCDFs possibly present in the original sample. It has also been shown [2,21] that these compounds may easily be formed at the same time in the pyrolysis of polychlorobenzophenols with other chlorinated compounds. To date, no GC column has been prepared that is able to separate all PCDDs and PCDFs. However, it can be estimated that of the order of 10^6 theoretical plates are needed in order to separate all isomers of the

PCDD series, the most difficult being the 22 isomers of the tetrachloro compounds and the 10 isomers of the hexachloro compounds. Such efficient columns cannot be made at present, so that some immediate conclusions have to be made:

(1) Electron-capture detection does not indicate if both PCDDs and PCDFs are present in the sample.

(2) Electron-capture detection is not feasible if one has to identify a particular TCDD isomer in a sample when the presence of most of its isomers is suspected.

However, the use of electron-capture detection seems to be feasible if the following circumstances occur:

(a) The sample contains only one or a few PCDD isomers and, in the latter instance, the type and number of isomers are known.

(b) No PCDFs are present in the sample or their number is small and known.

(c) The identity of the compound to be analysed can be checked by a more specific detection system.

For example, in the Seveso accident it was immediately recognized that the only TCDD present was the 2,3,7,8-isomer, accompanied by a small amount of the 1,3,7,8-isomer [12,22], which can easily be separated.

11.4. DETECTION OF POLYCHLORODIBENZODIOXINS WITH AN ELECTRON-CAPTURE DETECTOR

ECD calibration graphs for PCDDs are not available in the literature. The data reported in Figs. 11.3 and 11.4 were obtained by A. Cecinato [23] of the Air Pollution Institute of the Italian National Research Council in Rome.

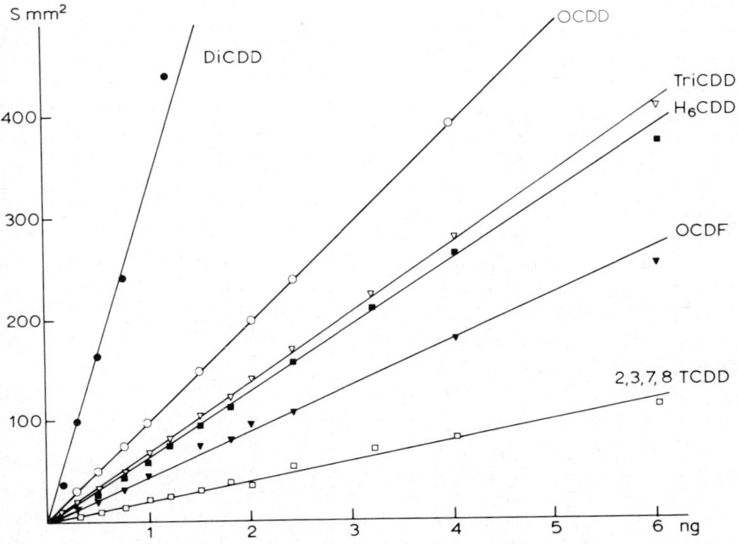

Fig. 11.3. Calibration graphs for several PCDDs using an ECD. (Courtesy of A. Cecinato.)

Fig. 11.3 shows calibration graphs for most PCDDs, with linear behaviour over an extended range. The response of the ECD towards PCDDs differs greatly according to the chlorine atoms in the molecule, as could be expected. However, Fig. 11.4 shows that the increase in the ECD response with number of chlorine atoms is linear, except for the di- and trichloro compounds.

The detection limits for PCDDs depend on the sample purity, and can be as low as 20 pg for TCDD or about 2 pg for OCDD in a pure sample. However, in the routine analysis of environmental samples containing pesticides and PCBs in the original sample, significant quantitative figures can be obtained only if the total amount of single PCDDs actually injected is 200 pg or more [24]. The level of purification is also very important if a low detection limit is to be obtained.

Fig. 11.5 shows the effect of purification of an artificial sample of TCDD+PCBs. In chromatogram A the TCDD peak is almost completely hidden and could be identified only by comparing it with a blank. After purification (chromatogram B) TCDD yields the major peak in the mixture and can be measured quantitatively.

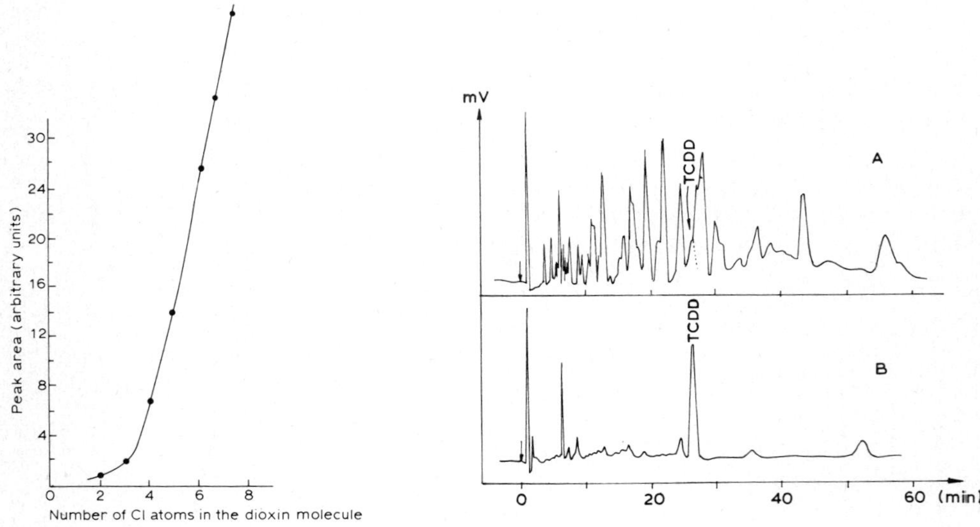

Fig. 11.4. ECD response toward the same weights of several PCDDs. (Courtesy of A. Cecinato.)

Fig. 11.5. Purification of TCDD from PCBs. (A) Aroclor 1254 (5.00 μg) + Aroclor 1260 (5.00 μg) + TCDD (0.32 μg) in 1 ml of *n*-hexane. Sample injected, 2 μl. (B) Same mixture after purification, concentrated to 0.1 ml. Sample injected: 1 μl. (From ref. 24.)

11.5. GAS CHROMATOGRAPHIC COLUMNS FOR DIOXIN ANALYSIS

The non-polar character of dioxins and related compounds (PCBs, PCBFs, etc.) makes the choice of the stationary phase fairly easy. However, owing to the small chemical differences between the various isomers and to the similar volatilities of the interfering compounds, a careful choice of the stationary phase is important.

For packed columns, the most selective stationary phase is obtained by mixing OV-17 and the fluorinated silicone QF-1 [25]. The best selectivity is obtained with 1.5% OV-17 + 1.95% QF-1 on Chromosorb G AW. This liquid stationary phase ensures the separation of the 13 most common chlorinated pesticides and TCDD, as shown in Fig. 11.6 [24]. Columns made with this stationary phase also have a long life with constant retention times. Recently, Langhorst and Shadoff [16] and Lamparski and Nestrick [17] used a 210 cm × 2 mm I.D. column packed with 0.60% OV-17 and 0.40% Poly S 170 on 80–100 mesh Permabond methylsilicone.

Nowadays glass or fused silica capillary columns and gas chromatographs dedicated to their use are commercially available. However, because of the high boiling range of dioxins and the need for reliable quantitative analysis, "cold on-column injection" is very useful. This consists in inserting the syringe needle directly into the capillary column kept at a low temperature as in the "on-column injector" shown in Fig. 11.7 [26], where

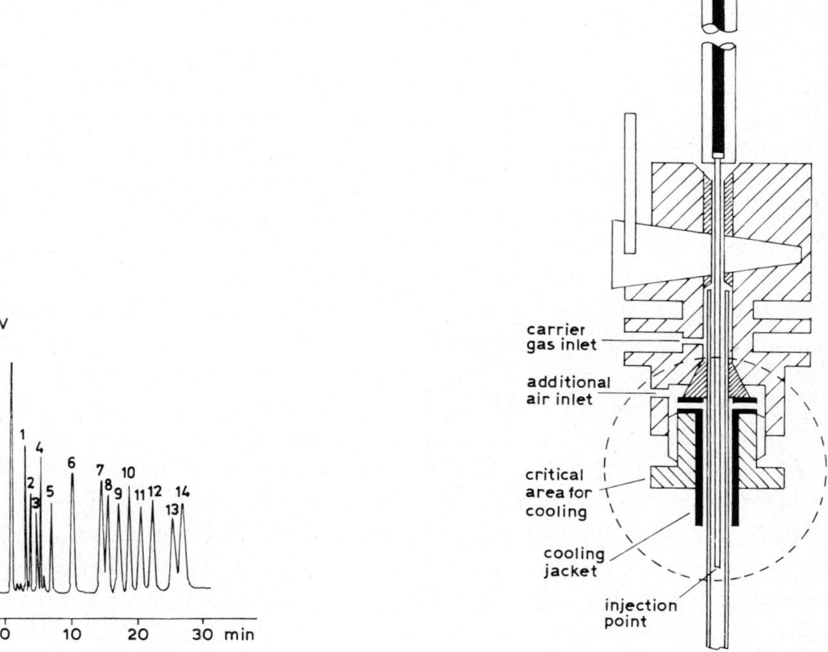

Fig. 11.6. GC separation of TCDD from chlorinated pesticides in an artificial mixture. Column: 2 m × 2 mm I.D. glass. Supelcoport (110–120 mesh) + 1.5% SP 2250 + 1.95% SP 2401. Temperature: 205°C; pressure: 3.0 kg/cm^2. Peaks: 1 = α-BHC; 2 = lindane; 3 = β-BHC; 4 = heptachlor; 5 = aldrin; 6 = heptachlor epoxide; 7 = *p,p'*-DDE; 8 = dieldrin; 9 = TDE; 10 = endrin; 11 = *o,p'*-DDT; 12 = *p,p'*-DDD; 13 = TCDD; 14 = *p,p'*-DDT. (From ref. 24.)

Fig. 11.7. Schematic diagram of the "on-column" injector. (Courtesy of Carlo Erba Strumentazione, Milan.)

the cooled zone of the "on-column injector" is filled in black. With this device, the sample is transferred inside the capillary without any solvent evaporation. Moreover, any spreading of high-boiling compounds on the walls of the injector is prevented, so that tailing due to the presence of dead volumes is eliminated and the column efficiency remains high.

The best application of capillary columns for the separation and identification of all 22 TCDD isomers has recently been reported by Buser and Rappe [22], who used three different capillary columns. Some pairs that were not separated on one column were separated on one of the other two. A mass spectrometer was used as a specific detector, but in the absence of interfering compounds the ECD could also be used. The most important feature of this work is that on the first two columns 2,3,7,8-TCDD is well separated from any other isomer, the separation being almost complete on the Silar 10C column used. On this column the determination of 2,3,7,8-TCDD at very low concentration in a mixture of all TCDDs and other chlorinated compounds in a fly ash sample from a municipal incinerator was carried out.

11.6. APPLICATIONS OF THE ELECTRON-CAPTURE DETECTOR TO THE ANALYSIS OF POLYCHLORODIBENZODIOXINS

The wide use of the ECD in the analysis of dioxins in environmental and biological samples is severely limited owing to interferences from other chlorinated compounds and their low concentrations in comparison with the interferents. No paper published after 1974 has described such analyses with the use of electron-capture detection only.

There are some instances in which a large number of routine analyses have to be carried out on samples with well known compositions and in which the dioxin concentration is higher than those of the interfering compounds. In this event an ECD can be used as an auxiliary detector, as has been pointed out before. The ECD can also be used for "screening" purposes, as a negative detector. If an efficient and reproducible column is available, the absence of any peak with a retention time corresponding to that of the PCDDs in the sample may be sufficient to exclude the presence of the particular compounds. This implies either the absence of interfering compounds in the original sample or a very good clean-up prior to the analysis. The former case should be excluded because, as discussed before, dioxins are always accompanied by at least trace amounts of interfering compounds. The latter situation has apparently been reached recently, as shown by Langhorst and Shadoff [16] and Lamparski and Nestrick [17]. In this case, after the clean-up procedure, the sample undergoes further purification by preparative high-performance liquid chromatography and the final analysis for 2,3,7,8-TCDD is carried out by GC–mass spectrometry (MS).

It is the author's opinion that in routine analysis, when many samples of the same kind have to be checked, using the procedure described in the papers cited the ECD could be used for most samples. In the conclusion of their paper, Lamparski and Nestrick state: "This procedure, utilizing packed column GC–low resolution MS, has provided reliable results for several heavily contaminated matrices where the combination of a less sophisticated clean-up followed by both packed and capillary column GC–high resolution MS has failed". We agree with this statement, but owing to the very low resolution of both the packed column they used and the mass spectrometer, it makes sense after the sophisticated clean-up procedure. If the authors had used very efficient capillary columns, the mass spectrometer could have been replaced with an ECD without much loss in the overall specificity. An example of the possibility of using an ECD in the analysis of dioxin-containing samples is given in ref. 24.

The Air Pollution Institute of the Italian National Research Council was deeply involved in the experimental decontamination of the Seveso soil and houses from 2,3,7,8-TCDD using UV radiation in the presence of appropriate sensitizers, according to the suggestions of Crosby and Wood [27] and Liberti [28], and after preliminary experiments good results were obtained [29].

In this work, the analytical problem was relatively simple because of the following circumstances:

(1) It was clear that 2,3,7,8-TCDD was the only PCDD present in the matrices to be analysed.

(2) The only interfering compounds present were chlorinated pesticides and small amounts of PCBs.

(3) The initial concentration of 2,3,7,8-TCDD was relatively high.

(4) All samples belonged to the same matrix, so that the chromatographic pattern was constant and well known.

(5) The analytical problem was reduced to the need to follow the changes in the TCDD concentration after different times of exposure to solar radiation.

In spite of this, some samples were analysed by GC–MS. Fig. 11.8a shows a chromatogram of an extract of Seveso soil after purification but before irradiation. In spite of the presence of many chlorinated pesticides in the sample, TCDD is well separated. Fig. 11.8b shows the analysis of the same sample to which was added a mixture of ethyl oleate and xylene (2:1), after solar irradiation for 5 days. The TCDD peak is greatly reduced.

Fig. 11.9 shows the change in the TCDD concentration in the Seveso soil and in an artificial sample placed in a room irradiated with UV light. The results reported in the

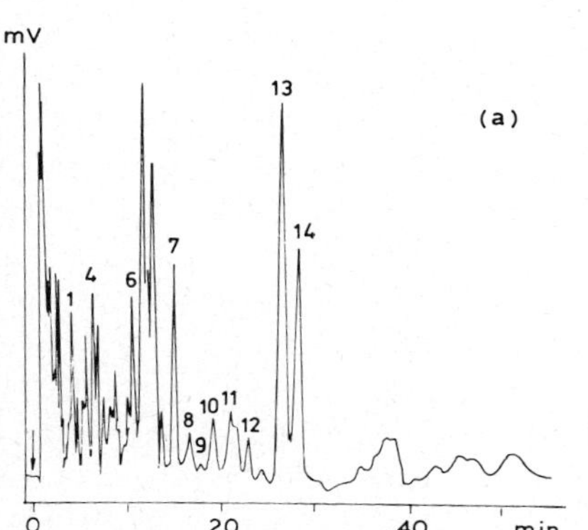

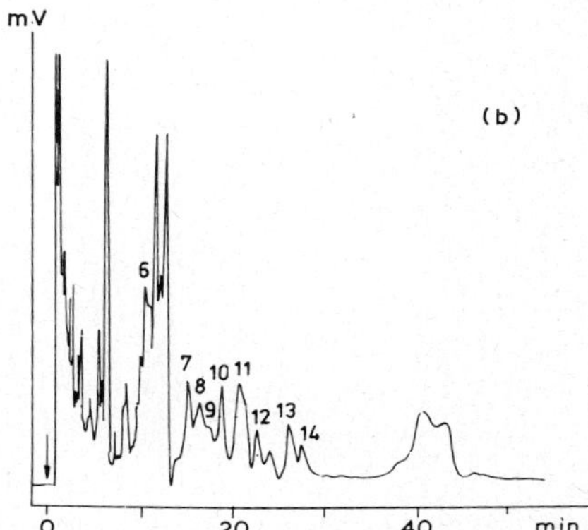

Fig. 11.8. GC analysis of an extract from a sample of Seveso soil polluted by TCDD to which was added a solution of ethyl oleate and xylene: (a) purified but before solar irradiation; (b) after solar irradiation for 5 days. (From ref. 24.)

work cited were checked using a mass spectrometer and the agreement was within a few percent of the quantitative data.

As an example of more difficult problems that cannot be solved with the use of the ECD, Fig. 11.10 shows the chromatogram obtained for an extract of fly ashes from an urban incinerator using an ECD. The separation is poor and positive identification of the compounds is impossible. In such instances more specific detectors should be used. GC–MS in the mass fragmentographic mode ensures the identification of PCDDs from PCDFs because of the difference in the *m/e* values at 16 a.m.u.

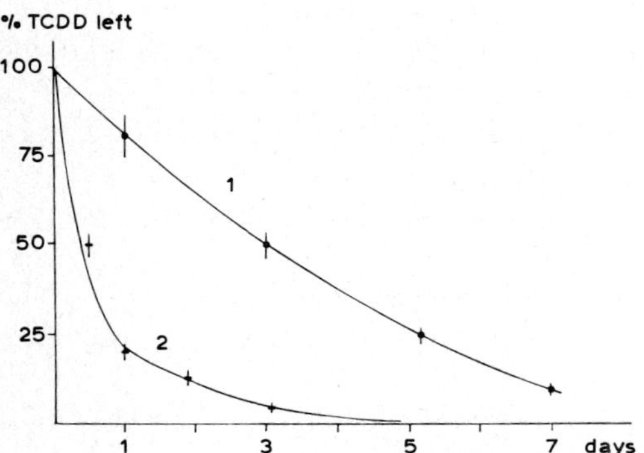

Fig. 11.9. Disruption curves of TCDD. (1) Seveso soil spread with a mixture of ethyl oleate and xylene and exposed to natural solar radiation. (2) Artificial TCDD sample placed in a room spread with ethyl oleate–xylene mixture and irradiated with a Philips MLU 300-W lamp. Irradiation power on the sample: 20 $\mu W/cm^2$. (From ref. 24.)

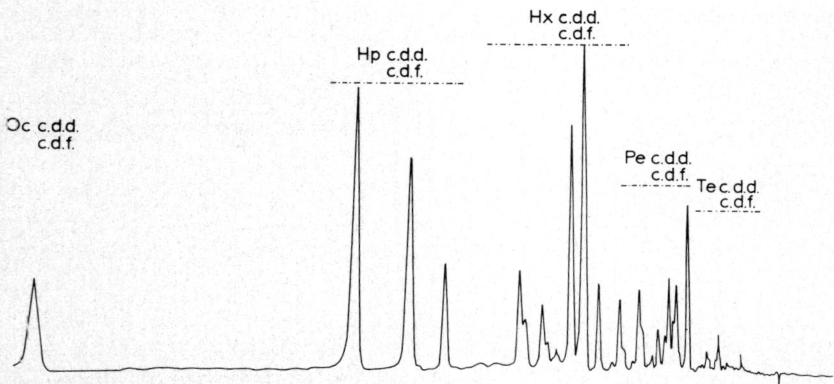

Fig. 11.10. ECD chromatogram of a fly ash extract. Column: 25-m glass capillary coated with OV-17 Temperature: 220°C.

11.7. CONCLUSIONS

The analytical problems involved in searching for, identifying and determining PCDDs in environmental, biological and industrial matrices cannot be solved using an ECD alone because of the presence of interfering compounds. However, if highly selective and reliable purification and clean-up of the sample is possible, an ECD can be of use as an auxiliary detector in routine work.

Once the sample has been purified, the use of a very good capillary column with an ECD can be competitive with and is perhaps better than an inefficient packed column followed by a low-resolution mass spectrometer. This is a particular case of the general rule that should always be followed: the more efficient the separation step, the less specific can be the detector, and vice versa. Recent achievements in mass spectrometry, such as MS–MS [30], make it possible to state that when an extremely specific detection technique is available, the separation step can even be avoided. It is our opinion that either a capillary column with 10^6 theoretical plates or the extremely expensive and sophisticated techniques of MS–MS should be avoided, and that a capillary column (with 10^4–10^5 theoretical plates coupled with a medium-resolution (3000–10,000) mass fragmentographic system is the best solution to the dioxin problem at present.

REFERENCES

1 B.A. Schwetz, J.M. Norris, G.L. Sparschu, V.K. Rowe, P.J. Gehring, J.L. Emerson and G.C. Gerbig, *Environ. Health Perspect.*, 5 (1973) 87.
2 H. Buser, *J. Chromatogr.*, 107 (1975) 275, and references cited therein.
3 IARC, *Monographs on the Evaluation of the Carcinogenic Risk to Man*, Vol. 15, Lyon, 1977, p. 41, and references cited therein.
4 A. Poland, E. Glover and A.S. Karde, *J. Biol. Chem.*, 251 (1976) 4936.
5 K. Olie, P.L. Vermeulen and O. Hutzinger, *Chemosphere*, 8 (1977) 455.
6 H.R. Buser and H.P. Bosshardt, *Mitt. Geb. Lebensmittelunters. Hyg.*, 69 (1978) 2343.
7 A. Liberti, D. Brocco, A. Cecinato and M. Possanzini, *Microchim. Acta*, I (1981) 271.
8 S. Facchetti, CCR Ispra, Italy, personal communication, 1980, to be published.
9 D.S. Millington, V. Parr and K. Hall, *VG Organic Tech. Literature, Insight*, No. 9, January 1980.
10 T.J. Nestrick and L. Lamparski, *Anal. Chem.*, 53 (1981) 122.
11 R.L. Harless, E.O. Oswald, M.K. Wilkinson, A.E. Dupuy, Jr., D.D. McDaniel and H. Tal, *Anal. Chem.*, 52 (1980) 1239.
12 H.R. Buser, *Anal. Chem.*, 49 (1978) 918, and references cited therein.
13 P.A. Mills, *J. Ass. Offic. Anal. Chem.*, 44 (1961) 171.
14 D. Firestone, J. Ress, N.L. Brown, R.P. Barron and J.N. Damico, *J. Ass. Offic. Anal. Chem.*, 55 (1972) 85.
15 H.R. Buser and H.P. Bosshardt, *J. Chromatogr.*, 90 (1974) 71.
16 M.L. Langhorst and L.A. Shadoff, *Anal. Chem.*, 52 (1980) 2037.
17 L.L. Lamparski and T.J. Nestrick, *Anal. Chem.*, 52 (1980) 2045.
18 R.M.M. Kooke, J.W.A. Lustenhouwer, K. Olie and O. Hutzinger, *Anal. Chem.*, 53 (1981) 461.
19 D.W. Phillipson and B.J. Puma, *Anal. Chem.*, 52 (1980) 2328.
20 A. Di Domenico, F. Merli, L. Boniforti, I. Camoni, A. Di Muccio, F. Taggi, L. Vergori, G. Colli, G. Elli, A. Gorni, P. Grassi, G. Invernizzi, A. Jemma, L. Luciani, F. Cattabeni, L. De Angelis, G. Galli, C. Chiabrando and R. Fanelli, *Anal. Chem.*, 51 (1979) 735.

21 H.R. Buser, *J. Chromatogr.*, 114 (1975) 95.
22 H.R. Buser and C. Rappe, *Anal. Chem.*, 52 (1980) 2257.
23 A. Cecinato, Istituto sull'Inquinamento Atmosferico del CNR, Area della Ricerca di Roma, personal communication, February 1981.
24 G. Bertoni, D. Brocco, V. Di Palo, A. Liberti, M. Possanzini and F. Bruner, *Anal. Chem.*, 50 (1978) 732.
25 J.R. Tomson (Editor), *Analysis of Pesticides of Human and Environmental Samples*, U.S. EPA, Research Triangle Park, NC, 1974, Section 9E, p. 4.
26 M. Galli, S. Trestianu and K. Grob, Jr., *J. High Resolut. Chromatogr. Chromatogr. Commun.*, 1 (1978) 263.
27 D.G. Crosby and A.S. Wood, *Science*, 173 (1971) 748.
28 A. Liberti, *Decontamination of a Polluted Area by Photodegradation of Chlorodioxin*, paper presented at the International Symposium on Photodegradation of Chlorodioxin, Rome, August 1976.
29 I. Allegrini, G. Bertoni, D. Brocco and M. Possanzini, *Chim. Ind. (Milan)*, 59 (1977) 541.
30 F.W. McLafferty, P.J. Todd, D.C. McGilveray and M.A. Baldwin, *J. Amer. Chem. Soc.*, 102 (1980) 3360.

Chapter 12

The electron-capture detector as a monitor of halocarbons in the atmosphere

P.G. SIMMONDS

CONTENTS

12.1. INTRODUCTION

Halocarbons of both natural and anthropogenic origin are distributed throughout the global environment. They are comparatively stable compounds with potentially long life-times, especially when fully substituted with halogen, and therefore of special interest because of their possible accumulation in the environment. The great majority enter the environment through their widespread use as industrial solvents, aerosol propellants, and as degreasing and dry cleaning fluids. World-wide production of many of these industrial halocarbons is approximately 0.5 megatons per annum. Gaseous halogenated methanes and ethanes are some of the most abundant and stable halocarbons. The two chloro-fluoromethanes, CF_2Cl_2 and $CFCl_3$ in particular, have attracted considerable attention due to their postulated role in the depletion of stratospheric ozone [1–4], and their possible effects on climate [5, 6].

The electron-capture detector (ECD) has acquired an important role as a monitor of halocarbons because many members of this class of compound can be detected with very high sensitivity, due to the strong electron affinity of individual halogen atoms. Complete substitution with halogen enhances the lower limit of detection dramatically, so that the direct determination of carbon tetrachloride, for example, in a 5-ml sample of ambient air is possible in the low parts per 10^{12} by volume (ppt). It is perhaps fortuitous that those compounds which are predicted to have the greatest environmental impact are often most sensitively measured by the ECD. However, many other important atmospheric halocarbons, such as methyl chloride, are only weak electron adsorbers and ordinarily would not be detected without substantial preconcentration of the sample. Fortunately, the deliberate addition of dopents, such as oxygen and nitrous oxide, to the normal carrier gas amplifies the response of the ECD to weak adsorbers and permits their direct determination in ambient air [7–9]. This promising technique is described in Chapters 4 and 5.

Although this chapter is primarily concerned with the measurement of atmospheric halocarbons by ECD it is important to recognise that gas chromatography–mass spectrometry (GC–MS) techniques, using cryotrapping and single ion monitoring are also making a major contribution to the measurement of trace halocarbons [10–16]. This is especially true for compounds such as carbon tetrafluoride (CF_4) which do not attach thermal electrons appreciably, and hence give little or no response in the ECD.

One of the important bonuses of the measurement of atmospheric halocarbons is their potential use as almost ideal tracers in the study of mass transport process in the atmosphere [17], and as area source urban tracers [18]. However, the use of electron capturing gases as tracers in plume dispersion and other atmospheric transport studies goes back to 1965 [19], when both CF_2Cl_2 and sulphur hexafluoride (SF_6) were found to be excellent tracers. In the past 15 years the deployment of halogenated compounds as gaseous meteorological tracers has developed extensively, and this specialised field of research will be covered in later sections of this chapter.

12.2. MEASUREMENTS OF ATMOSPHERIC HALOCARBONS

The first application of the ECD and indication that the atmosphere contains small but measurable quantities of halocarbons was in 1971, when Lovelock [17] reported the presence of trichlorofluoromethane ($CFCl_3$) in the air over Southern Ireland. Subsequent measurements by Lovelock et al. [20,21], and other workers [22–26] soon revealed the presence of other halogenated species in background air, including dichlorodifluoromethane (CF_2Cl_2), methyl chloroform (CH_3CCl_3), and carbon tetrachloride (CCl_4) as well as methyl iodide (CH_3I), a product of marine biological synthesis. Not surprisingly, those compounds which are products of chemical industry are also found in much greater concentrations in the air over large cities [27–30]. These urban concentrations often exceed their global background concentrations by at least an order of magnitude. However, it was not until 1974 that there was a sudden and obvious concern for the concentrations and global distribution of atmospheric halocarbons, when Molina and Rowland [3], and Cicerone et al. [4] suggested that the photolysis of chlorofluorocarbons in the stratosphere could initiate the chlorine atom catalysed destruction of ozone.

In the last five years there have been a very large number of halocarbon measurements reported by different investigators [31–40]. With improvements in the methods of analysis, and by freeze-out concentration of large volumes of air many more halocarbons have been added to the list of halocarbons observed in the atmosphere [41,42]. Estimates of halocarbon production, their concentrations, and known sources are listed in Table 12.1.

Methyl chloride is the most abundant and dominant natural halocarbon present in the atmosphere. Known sources include the smouldering combustion of vegetable matter [43, 44], probable conversion of marine methyl iodide by chloride ion [45], and emission from rotting wood by fungal decay [46]. With multiple sources its atmospheric abundance varies from 500 ppt to about 2000 ppt. Recently, Rasmussen et al. [47] have reported an average tropospheric concentration of 600 ± 60 ppt, and also made the

TABLE 12.1

CONCENTRATIONS, SOURCES, AND GLOBAL PRODUCTION ESTIMATES OF ATMOSPHERIC HALOCARBONS*

Halocarbon	Source	Average mixing ratio (v/v) (10^{-12})	Reference	Global production** (10^6 metric tons/year)
CH_3Cl	Marine biology, biomass combustion, fungal decay	500–2000	47, 85, 97	5–20
CH_3Br	Marine biology	≈ 5	36, 48, 54, 97	0.077
CH_3I	Marine biology and chemistry	< 5	36, 54, 97	0.74
CH_2Cl_2	Chemical industry	30***	1, 32, 33	0.42
$CHCl_3$	Chemical industry, water chlorination, and paper bleaching	20***	1, 32, 33, 54	0.99§
CCl_4	Chemical industry, atmospheric chemistry?	120–142	2, 32, 33, 97	0.77
CF_2Cl_2	Chemical industry	280	95, 97	0.4
$CFCl_3$	Chemical industry	170	95, 97	0.31
CH_3CCl_3	Chemical industry	115	97	0.55
$CHCl:CCl_2$	Chemical industry	20***	1, 32, 33	1.0
$CCl_2:CCl_2$	Chemical industry	30***	1, 32, 33	1.0
$CFCl_2CF_2Cl$	Chemical industry	18***	1, 33	0.055
CHF_2Cl	Chemical industry	45	1, 2, 103	0.128§§
$CHFCl_2$	Chemical industry	1–2	66, 116	unknown
CF_2ClCF_2Cl	Chemical industry	11***	33	0.02
CF_4	Chemical industry, aluminium industry	65	15	0.00001

*Values are for mid-1979, unless otherwise noted.
**Precise release estimates for many halocarbons are not available, however for CF_2Cl_2 and $CFCl_3$, releases are about 85% of production. For other industrial halocarbons releases are generally less than 50% of production.
***Values are for mid-1978.
§Yung et al. (1975).
§§Chemical Manufacturers Association (1979 estimate).

interesting observation that emission from volcanic fumeroles can be 1 to 2 orders of magnitude greater than the background concentration. Methyl bromide and methyl iodide are only found in trace concentrations in the atmosphere (ca. 5 ppt); both are believed to be products of marine biological activity [36,48]. Although other natural halocarbons such as bromoform ($CHBr_3$) are known to be synthesised by marine red algae of the genus Asparagopsis taxiformis [49], these compounds remain undetected in the atmosphere, probably as a result of their rapid photolysis or destruction by hydroxy radicals.

All of the other halocarbons in Table 12.1 are primarily of industrial origin. Methyl chloroform, trichloroethylene, and perchloroethylene are used extensively as solvents and dry cleaning fluids. The two chlorofluorocarbons, CF_2Cl_2 and $CFCl_3$, are both used as refrigerants and aerosol propellants, although $CFCl_3$ is also used as a blowing agent in the production of foams. The source of carbon tetrachloride is mainly as an intermediate in the manufacture of the two aerosol propellants [50–53], although Lovelock et al. have argued that CCl_4 may also be a by-product of atmospheric chemistry [21]. Similarly, Yung et al. [54] have suggested that chloroform may have other sources besides its use as a solvent and intermediate in the chemical industry. The chlorination of water, bleaching of paper pulp, and atmospheric decomposition of perchloroethylene [55], have all been cited as significant additional sources of atmospheric chloroform. It is significant that very little is known about the ultimate fate of these various halocarbons, and the changes and interactions they may undergo as a consequence of tropospheric and stratospheric chemistry. Ionization, photochemical dissociation, and heterogeneous catalytic processes may all play important roles in the eventual destruction of atmospheric halocarbons [56–60].

There is little doubt that the list of halocarbons observed in the atmosphere will continue to expand as the sensitivity of the analytical methods improves. For example, dichlorobromomethane ($CHBrCl_2$), dibromochloromethane ($CHClBr_2$), and bromoform are all known to be formed in domestic water supplies during the chlorination process [61, 62], although they have as yet to be measured in the free atmosphere even though they must eventually escape into the air. Furthermore sufficient evidence has now accumulated to believe that the oceans are an important and major source of natural halocarbons [63, 64], particularly those containing bromine and iodine, whose source strengths have yet to be adequately defined. A much larger survey of natural halocarbon sources is still required to obtain a more comprehensive understanding of the atmospheric halogen budget. This is especially true when it is remembered that methyl chloride alone accounts for about 25% of the total chlorine budget of the atmosphere.

12.3. ANALYTICAL METHODS

12.3.1. Chromatography

The gas chromatographic separation of individual halocarbons is relatively straightforward, and a variety of different columns and stationary phases have been used for this purpose. Conventional gas–liquid chromatography (GLC) using silicone fluids as stationary phases provides the most useful general purpose columns for halocarbon separations. Separation is essentially in order of their boiling points as shown in Table 12.2. However there are some exceptions: for example, CH_3Cl (b.p. −24.2°C) elutes after CF_2ClCF_2Cl (b.p. 3.8°C). Some groups of halocarbons with similar boiling points, such as CH_3I, CH_2Cl_2, and $CFCl_2CClF_2$ are difficult to resolve isothermally and subambient temperature programming is required [41, 42]. Mention should also be made of the improved selectivity achieved by Bruner and co-workers [65,66] with a column of Carbopack B coated with 0.5% SP-1000 as a surface modifier. This unusual material combines the benefits of gas–liquid–solid chromatography, where molecular polarisability is the driving force of the chromatographic process. A summary of some of the different types of chromatographic column which have been used for halocarbon separations is given in Table 12.3, together with the relevant literature references.

TABLE 12.2

RETENTION TIMES AND BOILING POINTS FOR SELECTED HALOCARBONS ON SILICONE COLUMN

Column 3 m × 6 mm packed with 10% SF-96 silicone oil on 100–120 mesh Chromosorb W. Programmed from −10°C to 70°C at 8°C/min, with 1 min delay at −10°C, and held at 70°C until last component of interest eluted. Argon–methane (95:5) at 50 ml/min. (From D.E. Harsch, D.R. Cronn and W.R. Slater, *J. Air Pollut. Contr. Assoc.*, 29 (1979) 975.)

Halocarbon	Retention time (min)	Boiling point (°C) (760 Torr)
$CHClF_2$	<2.5	−40.8
CF_2Cl_2	3.0	−29.8
$CClF_2CClF_2$	4.0	3.8
CH_2:CHCl	4.5	−14.0
CH_3Cl	5.0	−24.2
CH_3Br	6.0	3.6
$CHCl_2F$	6.4	8.9
CH_3CH_2Cl	6.5	12.7
CCl_3F	8.0	23.8
CH_3I	9.5	42.4
CCl_2FCF_2Cl	10.0	47.7
CH_2Cl_2	10.8	40.0
cis-CHCl:CHCl	12.0	60.1
$CHCl_3$	12.5	61.7
CH_2ClCH_2Cl	13.0	83.5
CH_3CCl_3	14.0	74.1
CCl_4	15.0	76.5
CHCl:CCl_2	16.5	87.0
CCl_2:CCl_2	27.0	121.0
$CHBr_3$	29.5	150.0

Retention time for oxygen is 1.5 min.

TABLE 12.3

GAS CHROMATOGRAPHY OF HALOCARBONS

Column type	Halocarbons separated	Ref.
3% SE-52 on 80–100 mesh Chromosorb W	CF_2Cl_2, CH_3CCl_3, CCl_4, $CHCl_3$, C_2HCl_3, C_2Cl_4	26
20% DC-200 on 30–60 mesh Chromosorb W	CF_2Cl_2, $CFCl_3$, $CFCl_2CF_2Cl$, CH_3CCl_3, CCl_4, $CHCl_3$, CH_3I, C_2HCl_3, C_2Cl_4, CH_3Br, CH_3Cl	31 34 81
10% SF-96 on 100–120 mesh Chromosorb W	See Table 12.2	41
0.4% Carbowax 1500 on 80–100 mesh Chromosorb W	CF_2Cl_2, $CFCl_3$, CCl_4, $CHCl_3$, CH_3CCl_3, C_2Cl_4, C_2HCl_3	31
Alumina 80–100 mesh	CF_2Cl_2, $CFCl_3$	31
Chromosil 310, 80–100 mesh	CF_2Cl_2, $CFCl_3$, $CFCl_2CF_2Cl$, CH_3Cl, CH_3I, CCl_4, CH_3CCl_3, CH_3Br, C_2HCl_3, C_2Cl_4	31
Durapak *n*-octane/Porasil C 100–120 mesh	CF_2Cl_2, $CFCl_3$, CH_3Cl, $CHFCl_2$, $CFCl_2CF_2Cl$, CH_2Cl_2, $CHCl_3$, CH_3CCl_3, CCl_4, C_2HCl_3	10 12
Durapak Carbowax 400/Porasil F (low k')	CH_3Cl, CH_3CCl_3, C_2Cl_4	10
6% Poly-*m*-phenyl ether on Tenax GC, 60–80 mesh	CH_2BrCH_2Br, CH_2ClCH_2Cl	107
0.5% SP-1000 on Carbopack B, 80–100 mesh	CH_3Cl, CF_2Cl_2, $CFCl_3$, $CHCl_3$, $CHFCl_2$, $CFCl_2CF_2Cl$, CH_3CCl_3, CCl_4, C_2HCl_3, CH_2Br_2, CH_2Cl_2	65 66 79
10% Na_2SO_4 on Porasil A, 80–100 mesh	CF_2Cl_2, $CFCl_3$, C_2HCl_3 CCl_4, C_2Cl_4	28 29
Porapak S + T (80:20)	CH_2:CHCl	104
30% Didecyl phthalate on 100–120 mesh Chromosorb P	CF_2Cl_2, $CFCl_3$, $COCl_2$	105
Porapak Q, 80–100 mesh	CF_2Cl_2, $CFCl_3$	23 106
Porasil D, 80–100 mesh	CF_2Cl, $CFCl_3$	See Fig. 12.4

12.3.2. Electron-capture detector

Most recent measurements of halocarbons employ electron-capture detectors with constant current electronics, which have now been adopted by almost all commercial manufacturers of gas chromatographs. However, many of the earlier measurements, particularly by Lovelock and co-workers, preferred the constant frequency mode of operation where at least for strong electron adsorbers the ECD can be treated as a gas phase coulometer, and hence its own internal standard [67–69]. For detailed discussions of both the constant current and constant frequency modes of operation of the ECD, reference should be made to the theoretical papers by Lovelock [70], Lovelock and Watson [71] and Gobby et al. [72].

The relative sensitivities of different halocarbons as measured by the ECD constitute an important factor which determines the lower limit of detection for individual halocarbons. Relative electron attachment coefficients for various halocarbons are discussed by Clemons and Altshuller [73], and have been reviewed by Pellizzari [74]. In addition, the molar ECD responses of many halocarbons of current interest have been reported by Miller and Grimsrud [75], and some of these in increasing order of response are listed in Table 12.4. Moreover, it is important to recognise that many halocarbons undergo dissociative electron attachment and their response will therefore increase with an increase in detector temperature [76–78]. Consequently, most halocarbon measurements can be optimised with an ^{63}Ni ECD operating between 250 and 300°C.

In a more practical sense these relative detector sensitivities also identify those halo-

TABLE 12.4

RELATIVE RESPONSES OF THE ELECTRON-CAPTURE DETECTOR TO ATMOSPHERIC AND OTHER HALOCARBONS

(From D.A. Miller and E.P. Grimsrud, *Anal. Chem.*, 51 (1979) 851.)

Halocarbon	Relative molar response (normalised to CH_3Cl at 300°C)	Halocarbon	Relative molar response (normalised to CH_3Cl at 300°C)
$CHCl:CHCH_3$	0.0031	*trans*-$CHCl:CHCl$	3.7
$CH_2:CHCl$	0.0068	$CH_2:CHCH_2Cl$	4.9
$CH_2:CClCH_3$	0.19	$ClCH_2CH_2Cl$	5.1
CH_3Cl	1.0	CF_3Cl	9.3
$(CH_3)_3CCl$	1.3	CH_2Cl_2	10.0
$CH_3CH_2CH_2Cl$	1.3	CH_3Br	24
CHF_2Cl	1.35	$CH_2:CCl_2$	37.3
$CH_3CHClCH_3$	1.4	CF_2Cl_2	253
CH_3CH_2Cl	1.5	*trans*-$ClCH_2CH:CHCH_2Cl$	500
$CH_3CH_2CH_2CH_2Cl$	1.6	$CHCl_3$	662
$CH_3CHClCH_2CH_3$	2.2	$CHCl:CCl_2$	732
cis-$CHCl:CHCl$	2.3	$CCl_2:CCl_2$	4880
$ClCH_2CH_2CH_2CH_2Cl$	3.2	$CFCl_3$	5210
$ClCH_2CH_2CH_2Cl$	3.6	CCl_4	10,500

carbons which can be determined directly in background air without the need for concentration of the sample. At present all of the following halocarbons can be detected in ambient air by EC–GC without enrichment: CF_2Cl_2, $CFCl_3$, $CFCl_2CCl_2F$, CH_3CCl_3, CCl_4, CH_3I, $CHCl{:}CCl_2$, and $CCl_2{:}CCl_2$. A variety of concentration methods have evolved to detect other trace halocarbons including cryogenic freeze-out [26, 41, 42], adsorption on Tenax [79, 80], porous polymers [79–82], and graphitised carbon black [66, 79].

Finally, mention should be made of the important discovery that carrier gas dopents, such as oxygen and nitrous oxide, substantially enhance the ECD response to weak electron adsorbers. This point is illustrated in Fig. 12.1, which shows the direct determination

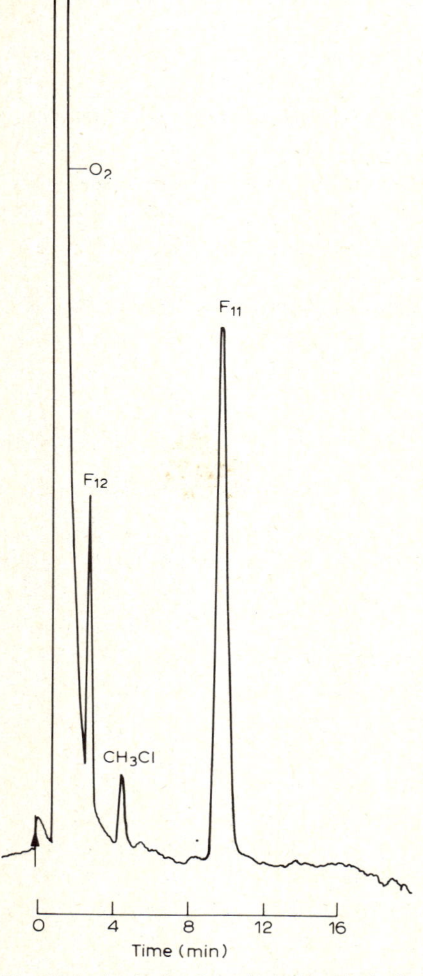

Fig. 12.1. Analysis of methyl chloride in 5 ml of ambient air, column 5 ft. × 1/4 in., 1.5% Poly-MPE on 80–100 mesh Porasil E. Isothermal 22°C. Carrier gas argon–methane (95:5) + 2000 ppm oxygen, flow-rate 27 ml/min.

of CH_3Cl in a 5-ml sample of ambient air on a special column of Porasil E, modified with polymetaphenoxylene (see also Rasmussen et al. [47]).

12.3.3. Calibration

The absolute calibration of halocarbon measurements in the range one part in 10^9–10^{12} parts of air is probably the most singly difficult problem associated with halocarbon ECD analysis. Even amongst those laboratories with considerable experience of making routine halocarbon measurements, several recent interlaboratory calibration studies have revealed uncomfortably large discrepancies in the analysis of the same chlorofluorocarbon standard [83, 84]. However, as pointed out by Lovelock and Simmonds [85], the constant current ECD method used by most of the participating laboratories will give an analytical precision of better than 0.5% for replicate analyses, but can have serious systematic inaccuracies.

Serial dilution of gravimetrically prepared parts per 10^9 (ppm) standards is one of the most commonly used methods for preparing halocarbon standards, down to about the 1 ppt level [28, 86, 87]. Permeation tubes have also been applied with some success in preparing halocarbon standards, although a long equilibration time is often necessary to achieve high accuracy [88]. Bruner et al. [89] have used a clever combination of permeation tube and exponentional dilution to prepare trace standards. In order to try and offset some of the difficulties and accumulative errors inherent in serial dilution techniques, Lovelock and Watson [71] have recently proposed the use of a large scale exponential chamber in which accurate standards can be prepared down to 1 ppt in one single error-free step.

12.4. LONG TERM MONITORING OF ATMOSPHERIC HALOCARBONS

The continuing debate on the possible harmful effects of chlorofluorocarbons on the ozone layer implies that these compounds will continue to accumulate in the atmosphere. So far stratospheric dissociation has been identified as the only major sink for their destruction, in which case the calculated lifetimes of CF_2Cl_2 and $CFCl_3$ would be 90 years and 50 years, respectively [1, 2]. It has, however, been pointed out that earlier chlorofluorocarbon observations were not inconsistent with lifetimes as short as 15 years, when the overall variability and accuracy of the data were taken into proper account [90, 91]. Moreover, even small undetected tropospheric sinks would significantly reduce the lifetimes of the chlorofluorocarbons and hence their long term impact on the ozone layer [92]. To determine more accurately halocarbon lifetimes requires careful calibrated measurements of their atmospheric concentrations on a global basis over a period of several years. Five years of halocarbon measurements in the Northern Hemisphere have already been reported at Adrigole, Ireland from 1970–1975 [93]. A similar two year time series of measurements have also been made at a site in Southern England [87], while a much shorter series of continuous measurements have been reported for the Southern Hemisphere by Frazer and Pearman [39]. However, only a long time series of measurements on a coordinated global basis would minimise the overall uncertainties associated with atmospheric variability, systematic instrumental errors, and poorly defined halocarbon release statistics, which collectively makes the determination of atmospheric lifetimes such a formidable task. Cunnold et al. [94] have therefore developed a rationale and statistical methodology for establishing halocarbon lifetimes based on trends in atmospheric concentrations obtained from a small network of global monitoring stations. This experiment known as the Atmospheric Lifetime Experiment (ALE), was started in November 1977, and now involves five stations making daily calibrated measurements of CF_2Cl_2, $CFCl_3$, CH_3CCl_3, CCl_4 and nitrous oxide (N_2O). A preliminary report on the present status of this program has been given to the Chemical Manufacturers Association (CMA) who have been responsible for funding this program [95]. These five stations are located at Adrigole, Ireland (52°N 10°W); Cape Mears, Oregon, U.S.A. (45°N 124°W); Ragged Point, Barbados, W.I. (13°N 59°W); American Samoa (14°S 171°W); and Cape Grim, Tasmania (41°S 145°E). At each coastal site there is a small temperature regulated building, which houses all of the instrumentation. The role of the ECD as a reliable analytical instrument is crucial to the operation of these stations as each station is equipped with an automated electron-capture–gas chromatograph, where the required standard deviation (precision) for determining the chlorofluorocarbons is better than ±2.5%.

12.5. ATMOSPHERIC LIFETIME EXPERIMENT

12.5.1. Station instrumentation and methodology

Each station samples relatively clean oceanic air and is equipped with a microprocessor-controlled Hewlett-Packard 5840A dual-channel electron-capture–gas chromatograph. The two channels are operated independently with separate chromatographic columns. Channel A contains a 5 ft. × 1/4 in. column packed with Porasil D (80–100 mesh), which separates N_2O, CF_2Cl_2, and $CFCl_3$ using a 2-ml air sample. Channel B analyses a 5-ml air sample for $CFCl_3$, CH_3CCl_3, and CCl_4 on a 6 ft. × 1/4 in. column packed with 10% OV-101 silicone on 100–120 mesh Chromosorb W. Both columns are maintained isothermally at an oven temperature of 50°C. The carrier gas is a specially purified grade of argon–methane (95:5), controlled by a two-stage diffusion-resistant pressure regulator. The respective flow-rates are 30 and 40 ml/min. The electron-capture detectors are operated at 320°C for channel A and 250°C for channel B. The injector temperature is 270°C, and each heated injection tube is filled with palladised asbestos, which serves to catalytically destroy trace halocarbons in the carrier gas. The carrier gas is passed through an 18 in. × 1 in. stainless steel pipe filled with a mixture of 5Å and 13X molecular sieves. Further 5Å molecular sieve traps (6 in. × 1/4 in.) are in line after the flow controllers. These precautions have proven to be necessary to ensure that the quality of the carrier gases is maintained consistently at each station and from tank to tank. A block diagram of the experimental arrangement is shown in Fig. 12.2.

Outside air is drawn into each of the stations by means of a non-contaminating metal bellows pump. Filters in the intake lines are essential to keep out insects and dust. The incoming air is first reduced to a nominal flow-rate of 50 ml/min by means of a flow divider

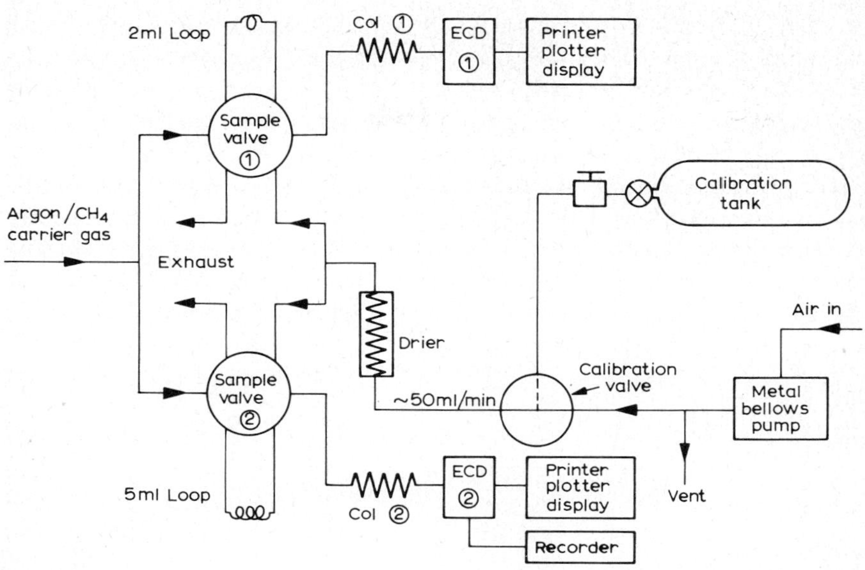

Fig. 12.2. Experimental arrangement of the HP 5840A dual EC–GC instrument for the atmospheric lifetime experiment.

and vent valve, it then passes through a three-way valve followed by a special drier fabricated from type 815 Nafion® tubing (DuPont Plastics Division), which reduces the water content of the atmosphere to a constant low relative humidity [96]. Dry ambient air enters each of the gas sampling valve loops which are thermostatted at 50°C within the chromatographic oven. The three-way valve also connects the calibration tank with the gas sample loops so that they can be filled with a preprogrammed instruction from the microprocessor. The high pressure in the calibration tank (300–500 p.s.i.) is reduced to 10 p.s.i. without contamination via an ultra clean Veriflow stainless steel regulator (UHP 660-580-2-SS-MI). The gas flows are stopped before injection providing a precise and reproducible injection. Flow meters are incorporated on the exits of each channel for visual daily checks by the station caretaker to be sure that flows are within the specified limits.

It is important to note that the ECD signal outputs from channels A and B are each processed separately. The ECD of channel A (Porasil column) is operated with the constant current mode electronics supplied by Hewlett-Packard. This is a popular mode of operation for commercial gas chromatographs because of its wide dynamic range. However, some caution is necessary, since compounds which strongly attach electrons can produce a response which is non-linear with concentration other than at very high carrier gas flow-rates. This problem can be largely overcome by the separate analysis of a calibrated standard as close to the analytical measurement as possible and under the same conditions of operation. This procedure is routinely followed in the ALE program so that calibration and ambient air analyses are never more than one hour apart. On the other hand the constant frequency (4 kHz) mode of operation which is used on the silicone column (channel B) produces a slightly noisier signal but is optimised for the coulometric method of analysis. This is potentially a more accurate and absolute method, which allows the direct calculation of halocarbon concentrations from a fundamental knowledge of the actual reaction conditions within the detector [71]. Furthermore, halocarbon measurements have been made at Adrigole since 1970 using the coulometric technique. Accordingly, it was desirable to continue with this method to maintain continuity in this long time series of measurements. We have also included in the design of the constant frequency electronics additional relay circuits for programmed operation of ancillary equipment, such as the on-off cycles of the metal bellows pump, and the back-up

TIME	RUN	CHANNEL
0001 - 0100	6	CB
0100 - 0200	7	CA
0200 - 0300	8	A
0300 - 0500		Wait 2 hours
0500 - 0600	9	B
0600 - 0700	10	CB
0700 - 0800	11	CA
0800 - 0900	12	A
0900 - 1100		Wait 2 hours
1100 - 1200	13	B
1200 - 1300	14	CB
1300 - 1400	15	CA
1400 - 1500	16	A
1500 - 1600		Wait 1 hour
1600 - 1700	17	B (PROGRAM RESET)
1700 - 1800	1	B
1800 - 1900	2	CB
1900 - 2000	3	CA
2000 - 2100	4	A
2100 - 2300		Wait 2 hours
2300 - 2400	5	B

C = Calibration run

A = Porasil column

B = Silicone column

Fig. 12.3. Atmospheric lifetime experiment, 24-h program sequence at Adrigole and Barbados stations.

potentiometric recorder, under control of the microprocessor. A frequency meter is also used to monitor the frequency of the constant current channel A. This provides a sensitive measure of the overall cleanliness and stability of the detector. In addition to the chromatographic apparatus each station is equipped with a barothermohydrograph, which provides a continuous record of the station pressure, humidity, and temperature. The actual 24-hour program routine for the Adrigole and Barbados stations is shown in Fig. 12.3.

12.5.2. Data calibration

The Hewlett-Packard 5840A instruments are programmed for an automatic calibration run just before or after each ambient air analysis. Calibration gas is supplied from a secondary calibration tank containing about 10^3 liters. These calibration tanks are specially prepared at the Oregon Graduate Center by Dr. Rasmussen and are stainless steel tanks cleaned by the Summa® process and filled on the Oregon coast with clean maritime air at about 500 p.s.i. The mixing ratios of all gases in these secondary tanks are measured relative to two primary calibration tanks [97]. Enough measurements are carried out (usually about 6) to ensure that the 2σ error in the mean values assigned to the secondary tank is $<1\%$. This level of accuracy is necessary in order to attain sufficient precision for the reliable determination of atmospheric lifetimes from measured trends.

The two primary calibration tanks have mixing ratios assigned to them based on a 1977–1978 inter-calibration program involving the Oregon Graduate Center (R.A. Rasmussen) and National Oceanic and Atmospheric Agency (E. Dutton).

As an independent check on these primary standards one of the scientists in the ALE program (J.E. Lovelock) has constructed a $7.14 \cdot 10^7$ cm^3 dilution chamber at his laboratory facilities in Cornwall. Within this chamber, accurate halocarbon concentrations in the range of interest in the atmosphere can be prepared in a single, error-free step. The halocarbons are measured as they undergo exponentional decay, with time from injection into the chamber. Extrapolation to zero time then provides the absolute calibration.

12.5.3. Data transmission

Data are printed out automatically by the Hewlett-Packard 5840A on an hourly-daily basis. In addition, a back-up potentiometric recorder is used on channel B to preserve data in the event of a failure of the thermal printer. Two typical examples of the automatic chromatogram print-outs from the Barbados station are illustrated in Figs. 12.4 and 12.5. Fig. 12.4 shows the resolution of N_2O, CF_2Cl_2, and $CFCl_3$ in ambient air on the Porasil D column, while Fig. 12.5 shows the separation of $CFCl_3$, CH_3CCl_3, and CCl_4 on the OV-101 column. Data received from each of the stations on a weekly basis are transcribed onto daily data sheets, which are then further processed by the data analysts to obtain daily and monthly means with standard deviations. In general integrated peak areas printed out by the chromatograph are converted into mixing ratios by means of the calibrated standard. However the integration of certain small peaks, in particular CH_3CCl_3 and CCl_4, often lacks the instrumental precision required, and for this reason manual measurements of the relevant peak heights are used instead.

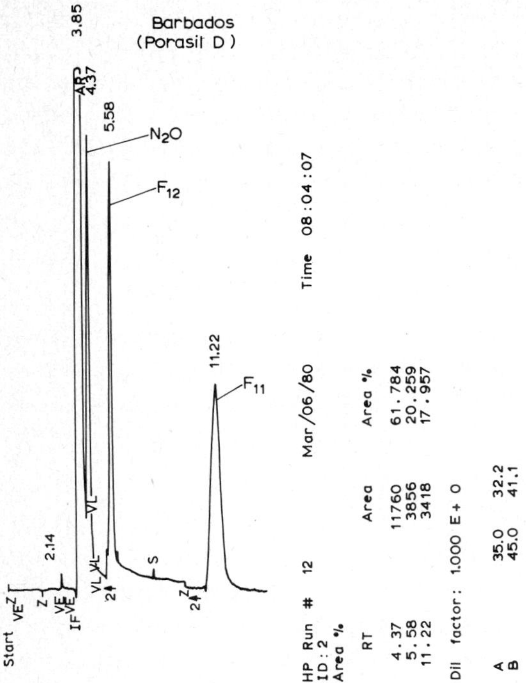

Fig. 12.4. Chromatogram of 2 ml of ambient air on 5 ft. × 1/4 in. column of 80–100 mesh Porasil D. Isothermal 50°C. Carrier gas argon–methane (95:5), flow-rate 32.2 ml/min. F12 = CF_2Cl_2, F11 = $CFCl_3$.

12.5.4. Station site visits and special procedures

Each station is visited on a quarterly basis by one of the principal scientists for detailed servicing and exchange of the secondary calibration tanks. During these visits all molecular sieve filters are replaced with new units containing freshly activated sieve. The Nafion drier is also replaced by a second unit filled with fresh sieve. Measurements of both the pulse and d.c. standing currents of the two ECDs are also taken for reference and for computing coulometric values. The most important aspect of the site visit is the exchange of the secondary calibration gas tank with a new tank of calibration gas. Cross calibrations of both tanks are performed on site and the used tank which still contains 200–300 p.s.i. of gas is returned to Oregon Graduate Center for remeasurement against the primary standard.

Each site has a local caretaker responsible for making a daily visit to the station, and also transmitting all of the data to the principal scientist. The caretaker also reinitiates the 24-hour program cycle and checks for any obvious abnormalities in the data record. He has also been trained to correct minor faults in the program which is occasionally disrupted by extended power failures.

Adrigole amongst all the ALE stations has the most variable data. This was anticipated since Adrigole is situated midway between the two major source areas of halocarbons,

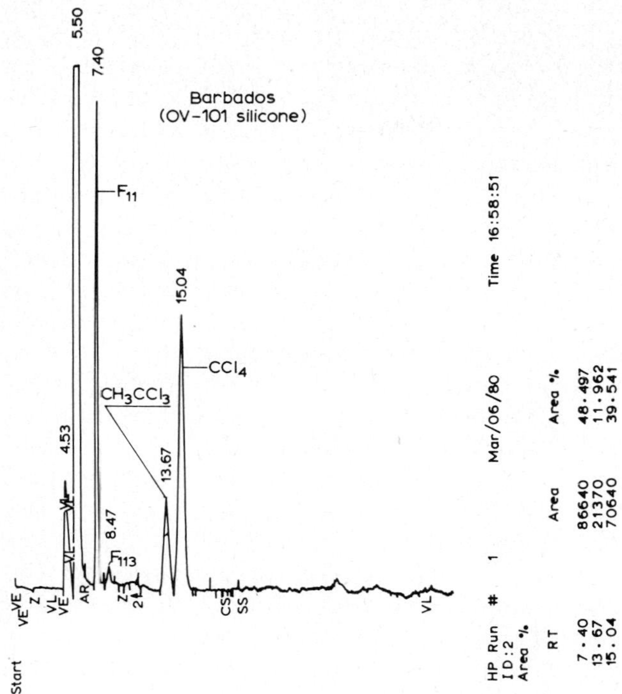

Fig. 12.5. Chromatogram of 5 ml of ambient air on 6 ft. × 1/4 in. column of 10% OV-101 on 100–120 mesh Chromosorb W. Isothermal 50°C. Carrier gas as in Fig. 12.4, flow-rate 41.1 ml/min. F113 = $CF_2ClCFCl_2$.

namely Western Europe and the United States. In order to identify the component of the variability measurements of both trichloroethylene and perchloroethylene are recorded at Adrigole, as both of these compounds are sensitive indicators of regional pollution. The presence of increased levels of these compounds is used to discard observations which have considerable variability due to pollution episodes, and therefore do not truly represent average Northern Hemispheric mixing ratios.

At the present time two complete years of data have been obtained from the ALE network of stations, although at least one more full year of data is required to accurately determine halocarbon lifetimes. Nevertheless the instrumental precision of better than 2.5% has been maintained for most of the last two years, and a self-consistent continuous data record has been acquired. Additional details of the experiment can be found in ref. 95.

12.6. HALOCARBONS AS ATMOSPHERIC TRACERS

The special advantage of the ECD for atmospheric dispersion studies lies in its unrivalled sensitivity towards certain polyhalogenated compounds, which fortuitously also possess properties that make them excellent tracers. Although not strictly a halocarbon, no discussion of atmospheric tracers would be complete without reference to sulphur hexafluoride (SF_6). Historically, this was the first tracer to be used with the ECD [19]. As an odourless, non-toxic, and chemically and physiologically inert gas, it is almost the ideal tracer and can be routinely detected at concentrations of 10^{-11}—10^{-12} by volume in 5 ml of ambient air [98,99]. Furthermore for most practical purposes its ECD response is coulometric, and it can be chromatographically separated with considerable advantage on columns of 5Å molecular sieve, since it is eluted before oxygen [100]. Not only has SF_6 found general use as a meteorological tracer, it has also been used successfully to characterize the ventilation systems of buildings [101]. While SF_6 can be determined at concentrations as low as 10^{-14} by volume by concentration on activated charcoal [102], its usefulness at these levels is limited by the background concentration of SF_6, which has been steadily increasing through its industrial use as a gaseous dielectric. The global average concentration of SF_6 is reported to be $3 \cdot 10^{-13}$ by volume [33]. Nevertheless, where sufficient SF_6 can be dispersed to overpower the background, it is still a very attractive and economic tracer for short-range atmospheric experiments.

Fortunately, there are a substantial number of alternative halogenated compounds which are potentially useful as atmospheric tracers. Successful tracer experiments have been conducted with octafluorocyclobutane (C_4F_8) and bromotrifluoromethane (CF_3Br) [98]. Also a series of long-range atmospheric diffusion experiments employed dibromodifluoromethane (CF_2Br_2) and dibromotetrafluoroethane ($C_2F_4Br_2$) as tracers [108, 109]. Obvious requirements for a candidate tracer, in addition to a very low background concentration, non-toxicity, and inertness, are ease of separation from other atmospheric halocarbons, and a large cross-section for electron attachment.

Perfluorocarbons meet all of these criteria and are some of the most useful tracers, since nature is not known to participate in fluorine chemistry to any great extent. Compounds such as perfluoromethylcyclohexane, perfluorodimethylcyclohexane, and perfluorodecalin are all attractive tracers which can be conveniently separated from most other atmospheric halocarbons on conventional silicone columns. A typical separation is shown in Fig. 12.6. However in order to avoid any possible interference due to co-chromatography with other halocarbons, a heated palladium catalyst can be used to selectively transmit only the perfluorocarbons, while destroying all unwanted halocarbons such as $CFCl_3$ and CCl_4 [110, 111]. Palladised molecular sieve or asbestos is particularly effective as catalyst and will also remove trace oxygen when a small quantity of hydrogen is deliberately added to the carrier gas. Using this technique an apparatus has been described for the continuous and ultrasensitive measurement of perfluorinated air-borne tracer substances, based on direct analysis with an ECD which uses ambient air as the carrier gas following the chemical conversion of indigenous oxygen to water [112, 113]. This modified ECD provides a continuous signal in real-time over the concentration range 10^{-9} to $4 \cdot 10^{-14}$ parts by volume without the need for chromatographic separation of SF_6 and oxygen. Recent improvements in the design of this apparatus have included replacement of the original condenser with a specially designed Nafion water permeation device, and microprocessor-controlled fluidic switching of the detector flow between the sample and a reference gas stream. This produces a very stable baseline with the small penalty of a slightly reduced overall sensitivity [114].

Finally special mention should be made of two unusual applications of the ECD and halogenated tracers. The first involves the micro-encapsulation of halogenated compounds for the tagging and subsequent unambiguous identification of oil spills [115]. The second involves the use of perfluorinated taggants in blasting caps and explosives for later detection by electron-capture monitors [111]. It is clear that other highly specialised uses of the ECD will be reported in the future, as the applications of this ultrasensitive and practical detector are only limited by the imagination of the investigator.

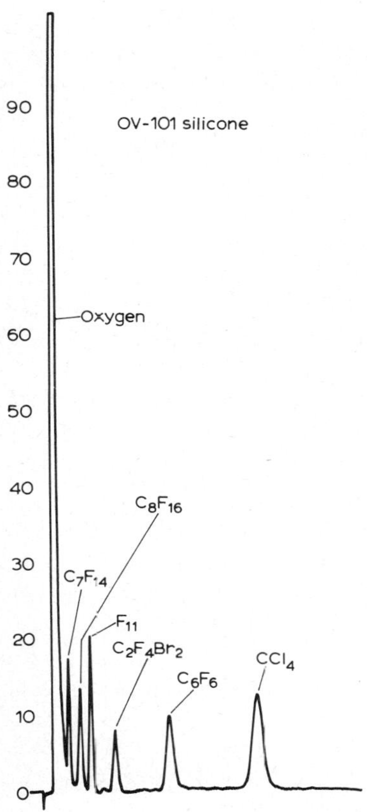

Fig. 12.6. Chromatogram of some candidate meteorological tracers on 5 ft. × 1/4 in. column of 10% OV-101 on 100–120 mesh Chromosorb P. Carrier gas as in Fig. 12.4, flow-rate 30 ml/min. Isothermal 23°C. Chart speed 1 in. in 6 min. C_7F_{14} = perfluoromethylcyclohexane, C_8F_{16} = perfluorodimethylcyclohexane, $C_2F_4Br_2$ = dibromotetrafluoroethane, C_6F_6 = hexafluorobenzene.

12.7. ACKNOWLEDGEMENTS

I am indebted to my two colleagues, Dr. J.E. Lovelock and Dr. R.A. Rasmussen for many helpful discussions during the preparation of this manuscript, as well as their major collaborative effort in establishing the ALE network of monitoring stations. The cooperation of my other co-investigators in the Atmospheric Lifetime Experiment, Dr. R. Prinn, Dr. F. Alyea, Dr. D. Cunnold, Dr. R. Rosen, Dr. P. Frazer and Mr. S. Crawford, is gratefully acknowledged.

REFERENCES

1 *Chlorofluorocarbons and their Effects on Stratospheric Ozone*, Pollut. Pap. No. 15, HMSO, london, 1979.
2 *Stratospheric Ozone, Depletion by Halocarbons, Chemistry and Transport*, National Academy of Sciences, Washington, DC, 1979.
3 M.J. Molina and F.S. Rowland, *Nature (London)*, 249 (1974) 810.
4 R.J. Cicerone, R.S. Stolarski and S. Walters, *Science*, 185 (1974) 1165.
5 V. Ramanathan, *Science*, 190 (1975) 50.
6 W.C. Wang, Y.L. Yung, A.A. Lacis, T. Mo and J.E. Hansen, *Science*, 194 (1976) 685.
7 E.P. Grimsrud and D.A. Miller, *Anal. Chem.*, 50 (1978) 1141.
8 E.P. Grimsrud and R.G. Stebbins, *J. Chromatogr.*, 155 (1978) 19.
9 R.E. Sievers, M.P. Phillips, R.M. Barkley, M.A. Wizner, M.J. Bollinger, R.S. Hutte and F.C. Fehsenfeld, in A. Zlatkis (Editor), Advances in Chromatography, 1979, Elsevier, 1979, p. 3.
10 E.P. Grimsrud and R.A. Rasmussen, *Atmos. Environ.*, 9 (1975) 1014.
11 E.P. Grimsrud and R.A. Rasmussen, *Atmos. Environ.*, 9 (1975) 1010.
12 D.R. Cronn and D.E. Harsch, *Anal. Lett.*, 12 (1979) 1489.
13 B.J. Tyson *Anal. Lett.*, 8 (1975) 807.
14 D.R. Cronn and D.E. Harsch, *Anal. Lett.*, 9 (1976) 1015.
15 S.A. Penkett, N.J.D. Prosser, R.A. Rasmussen and M.A.K. Khalil, *Nature (London)*, 286 (1980) 793.
16 R.A. Rasmussen, S.A. Penkett and N.J.D. Prosser, *Nature (London)*, 277 (1979) 549.
17 J.E. Lovelock, *Nature (London)*, 230 (1971) 379.
18 P.J. Frazer, B.L. Sawford and P.C. Manins, *Atmos. Environ.*, 11 (1977) 1025.
19 G.F. Collins, F.E. Bartlett, A. Turk, S.H. Edmunds and H.L. Mark, *Air Pollut. Contr. Assoc. J.*, 15 (1965) 109.
20 J.E. Lovelock, *Atmos. Environ.*, 6 (1972) 917.
21 J.E. Lovelock, R.J. Maggs and R.J. Wade, *Nature (London)*, 241 (1973) 194.
22 A.J. Murray and J.P. Riley, *Nature (London)*, 242 (1973) 810.
23 C.W. Su and E.D. Goldberg, *Nature (London)*, 245 (1973) 27.
24 P.E. Wilkness, R.A. Lamontagne, R.E. Larson, J.W. Swinnerton, C.R. Dickson and T. Thompson, *Nature (London)*, 245 (1973) 45.
25 P.E. Wilkness, J.W. Swinnerton, R.A. Lamontagne and D.J. Bresson, *Science*, 187 (1975) 834.
26 A.J. Murray and J.P. Riley, *Anal. Chim. Acta*, 65 (1973) 261.
27 P.G. Simmonds, S.L. Kerrin, J.E. Lovelock and F.H. Shair, *Atmos. Environ.*, 8 (1974) 209.
28 N.E. Hester, E.R. Stephens and O.C. Taylor, *Air Pollut. Contr. Assoc. J.*, 24 (1974) 592.
29 N.E. Hester, E.R. Stephens and O.C. Taylor, *Atmos. Environ.*, 9 (1975) 603.
30 T. Ohta, M. Morita and I. Mizoguchi, *Atmos. Environ.*, 10 (1976) 557.
31 H.B. Singh, L.J. Salas and L.A. Cavanagh, *Air Pollut. Contr. Assoc. J.*, 27 (1977) 332.
32 H.B. Singh, L.J. Salas, H. Shigeishi and E. Scribner, *Global Distribution of Selected Halocarbons, SF_6, and N_2O*, Publ. EPA 600/3-78-100, Environmental Protection Agency, Research Triangle Park, NC, 1978.
33 H.B. Singh, L.J. Salas, H. Shigeishi and E. Schribner, *Science*, 203 (1979) 899.
34 B.J. Tyson, J.C. Arvesen and D. O'Hara, *Geophys. Res. Lett.*, 5 (1978) 535.
35 R.A. Rasmussen, D. Pietrotti, J. Krasnec and B. Halter, *Cruise Report of the Alpha Helix*, Washington State University, 1977.
36 J.E. Lovelock, *Nature (London)*, 256 (1975) 193.
37 E.P. Grimsrud and R.A. Rasmussen, *Atmos. Environ.*, 10 (1976) 305.
38 P.W. Krey, R.J. Lagomarsino and L.E. Toonkel, *J. Geophys. Res.*, 82 (1977) 1753.
39 P.J. Frazer and G.I. Pearman, *Atmos. Environ.*, 12 (1978) 839.
40 J.E. Lovelock, *Nature (London)*, 252 (1974) 292.
41 D.E. Harsch, D.R. Cronn and W.R. Slater, *Air Pollut. Contr. Assoc. J.*, 29 (1979) 975.
42 R.A. Rasmussen, D.E. Harsch, P.H. Sweany, J.P. Krasnec and D.R. Croon, *Air Pollut. Contr. Assoc. J.*, 27 (1977) 579.

43 J.R. Newsome, V. Norman and C. Keith, *Tob. Sci.*, 9 (1965) 602.
44 P.J. Crutzen, L.E. Heidt, J.P. Krasnec, W.H. Pollock and W. Seiler, *Nature (London)*, 282 (1979) 253.
45 O.C. Zafiriou, *J. Mar. Res.*, 33 (1975) 73.
46 M.I. Cowan, A.T. Geln, S.A. Hutchinson and M.E. MacCartney, *Trans. Brit. Mycol. Soc.*, 60 (1973) 247.
47 R.A. Rasmussen, L.E. Rasmussen, M.A.K. Khalil and R.W. Dalluge, *J. Geophys. Res.*, in press.
48 S.C. Wofsy, M.B. McElroy and Y.L. Yung, *Geophys. Res. Lett.*, 2 (1975) 215.
49 B.J. Burreson, R.E. Moore and P. Roller, *Tetrahedron Lett.*, 7 (1975) 473.
50 H.B. Singh, D.P. Fowler and T.O. Peyton, *Science*, 192 (1976) 1231.
51 M.J. Molina and F.S. Rowland, *Geophys. Res. Lett.*, 1 (1974) 309.
52 W. Brock Neely, *Sci. Total Environ.*, 8 (1977) 267.
53 A.P. Altshuller, *Environ. Sci. Technol.*, 10 (1976) 596.
54 Y.L. Yung, M.B. McElroy and S.C. Wofsy, *Geophys. Res. Lett.*, 2 (1975) 397.
55 A. Appleby, J. Kazazis, D. Lillian and H.B. Singh, *J. Environ. Sci. Health*, A11 (1976) 711.
56 Y. Correia, G.J. Martens, F.H. VanMensch and B.P. Whim, *Atmos. Environ.*, 11 (1977) 1113.
57 S. Gäb, J. Schmitzer, H.W. Thamm, H. Palar and F. Korte, *Nature (London)*, 220 (1977) 331.
58 R.J. Donovan, K. Kaufmann and J. Wolfrum, *Nature (London)*, 262 (1976) 204.
59 R.E. Rebbert and P.J. Ausloos, *J. Photochem.*, 6 (1976) 265.
60 R. Atkinson, G.M. Breuer, J.N. Pitts, Jr. and H.L. Sandoval, *J. Geophys. Res.*, 81 (1976) 5765.
61 J.J. Rook, *Water Treat. Exam.*, 23 (1974) 234.
62 A.A. Nicholson, O. Meresz and B. Lemyk, *Anal. Chem.*, 49 (1977) 814.
63 R.E. Moore, *Accounts Chem. Res.*, 10 (1977) 40.
64 G.A. Helz and R.Y. Hsu, *Limnol. Oceanogr.*, 23 (1978) 858.
65 F. Bruner, G. Bertoni and G. Crescentini, *J. Chromatogr.*, 167 (1978) 399.
66 G. Crescentini and F. Bruner, *Nature (London)*, 279 (1979) 311.
67 J.E. Lovelock, R.J. Maggs and E.R. Adlard, *Anal. Chem.*, 43 (1971) 1962.
68 E.P. Grimsrud and S.H. Kim, *Anal. Chem.*, 51 (1979) 537.
69 D. Lillian and H.B. Singh, *Anal. Chem.*, 46 (1974) 1060.
70 J.E. Lovelock, *J. Chromatogr.*, 99 (1974) 3.
71 J.E. Lovelock and A.J. Watson, *J. Chromatogr.*, 158 (1978) 123.
72 P.L. Gobby, E.P. Grimsrud and S.W. Warden, *Anal. Chem.*, 52 (1980) 473.
73 C.A. Clemons and A.P. Altshuller, *Anal. Chem.*, 38 (1966) 133.
74 E. Pellizzari, *J. Chromatogr.*, 98 (1974) 323.
75 D.A. Miller and E.P. Grimsrud, *Anal. Chem.*, 51 (1979) 851.
76 W.E. Wentworth, E. Chen and J.E. Lovelock, *J. Phys. Chem.*, 70 (1966) 445.
77 W.E. Wentworth, R.S. Becker and R. Tung, *J. Phys. Chem.*, 71 (1967) 1652.
78 P.G. Simmonds, D.C. Fenimore, B.C. Pettitt, J.E. Lovelock and A. Zlatkis, *Anal. Chem.*, 39 (1967) 1428.
79 C. Vidal-Madjar, M.F. Gonnord, F. Benchah and G. Guiochon, *J. Chromatogr. Sci.*, 16 (1978) 190.
80 T. Tanaka, *J. Chromatogr.*, 153 (1978) 7.
81 J.W. Russell and L.A. Shadoff, *J. Chromatogr.*, 134 (1977) 375.
82 B. Versino, M. deGroot and F. Geiss, *Chromatographia*, 7 (1974) 302.
83 R.A. Rasmussen, *Atmos. Environ.*, 12 (1978) 2505.
84 E.E. Hughes, W.D. Dorko and J.K. Taylor, *NBSIR 78-1480*, National Bureau of Standards, Washington, DC, 1978.
85 J.E. Lovelock and P.G. Simmonds, in T.M. Sugden and T.F. West (Editors), *Chlorofluorocarbons in the Environment*, Ellis Horwood, Chichester, 1980.
86 D. Lillian, H.B. Singh, A.A. Appleby, L. Lobban, R. Arnts, R. Grumper, R. Hague, J. Toomey, J. Kazazsis, M. Antell, D. Hansen and B. Scott, *Environ. Sci. Technol.*, 9 (1975) 1042.
87 S.A. Penkett, B.M.R. Jones and A.E.J. Eggleton, *Atmos. Environ.*, 13 (1979) 123.
88 H.B. Singh, L. Salas, D. Lillian, R. Arnts and A.A. Appleby, *Environ. Sci. Technol.*, 11 (1977) 511.

89 F. Bruner, M.C. Canulli and M. Possanzini, *Anal. Chem.*, 45 (1973) 1790.
90 N.D. Sze and M.F. Wu, *Atmos. Environ.*, 10 (1976) 1117.
91 J.P. Jesson, P. Meakin and L. Glasgow, *Atmos. Environ.*, 11 (1977) 499.
92 F.N. Alyea, D.M. Cunnold and R.G. Prinn, *Atmos. Environ.*, 12 (1978) 1009.
93 D.H. Pack, J.E. Lovelock, G. Cotton and C. Curthoys, *Atmos. Environ.*, 11 (1977) 329.
94 D.M. Cunnold, F.N. Alyea and R.G. Prinn, *J. Geophys. Res.*, 83 (1978) 5493.
95 R.G. Prinn, F.N. Alyea, D.M. Cunnold, R.A. Rasmussen, P.G. Simmonds, J.E. Lovelock and R. Rosen, *Atmospheric Lifetime Experiment*, Preliminary Report to the Chemical Manufacturers Association, Washington, DC, 1980.
96 B.F. Foulger and P.G. Simmonds, *Anal. Chem.*, 51 (1979) 1089.
97 R.A. Rasmussen and M.A.K. Kahlil, *Proc. of the NATO Advanced Study Institute on Atmospheric Ozone, Portugal, 1979.*
98 B.E. Saltzman, A.I. Coleman and C.A. Clemons, *Anal. Chem.*, 38 (1966) 753.
99 R.N. Dietz and E.A. Cote, *Environ. Sci. Technol.*, 7 (1973) 338.
100 P.G. Simmonds, G.R. Shoemake, J.E. Lovelock and H.C. Lord, *Anal. Chem.*, 44 (1972) 860.
101 P.J. Drivas, P.G. Simmonds and F.H. Shair, *Environ. Sci. Technol.*, 6 (1972) 609.
102 C.A. Clemons, A.I. Coleman and B.E. Saltzman, *Environ. Sci. Technol.*, 2 (1968) 551.
103 R.A. Rasmussen, M.A.K. Khalil, S. Penkett and N. Prosser, *Geophys. Res. Lett.*, in press.
104 D. Krockenberger, H. Lorkowski and L. Rohrschneider, *Chromatographia*, 12 (1979) 787.
105 H.B. Singh, *Nature (London)*, 264 (1976) 428.
106 A.E. O'Keeffe and G.C. Ortman, *Anal. Chem.*, 38 (1966) 760.
107 A. Jonsson and S. Berg, *J. Chromatogr.*, 190 (1980) 97.
108 M. McCooeye, G. Gardner and L. Elias, *Lab. Tech. Report, LTR-UA-31*, Nat. Res. Council Canada, 1975.
109 G.J. Ferber and R.J. List, *Long Range Transport and Diffusion Experiments, Final Report to Air Resources Lab.*, NOAA, MD, 1973.
110 F.W. Williams and H.G. Eaton, *Anal. Chem.*, 46 (1974) 179.
111 R.N. Dietz, R.W. Goodrich and E.A. Cote, *Proc. New Concepts Symposium and Workshop on Detection and Identification of Explosives, Reston, VA*, U.S. Depts. of Treasury, Energy, Justice and Transportation, 1978.
112 P.G. Simmonds, A.J. Lovelock and J.E. Lovelock, *J. Chromatogr.*, 126 (1976) 917.
113 E.L. Thomsen and J.E. Lovelock, *Atmos. Environ.*, 10 (1976) 3.
114 J.E. Lovelock, P.G. Simmonds and A.J. Lovelock, *Prelim. Report to Air Resources Lab.*, NOAA, MD, 1980.
115 B.L. Mitchell, P.G. Simmonds and F.H. Shair, *Environ. Sci. Technol.*, 7 (1931) 121.
116 S. Penkett, N. Prosser, R. Rasmussen and M.A.K. Khalil, *Nature (London)*, 286 (1980) 793.

Chapter 13

Biomedical applications of the electron-capture detector

JÖRGEN VESSMAN

CONTENTS

13.1. INTRODUCTION

The applications of the electron-capture detector (ECD) in the biomedical field reflect to a very high degree the important developments that have taken place during the last 15 years in, for example, clinical pharmacology, biopharmaceutics, pharmacokinetics and clinical chemistry, all of which are important in the development of new therapeutic principles. The number of papers published describing the use of gas chromatography (GC) with electron-capture detection is impressive and the possibilities for introducing new approaches have not yet been exhausted. Although the first applications go back to the early 1960s, it was not until 1967 or 1968 that the number of papers reported in the chromatographic, analytical chemistry and pharmacological literature really began to grow. In parallel with the development of better electron-capture detectors, other detection systems have appeared that are competitive, e.g., gas chromatography with mass spectrometric (MS) or alkali flame ionization detection and liquid chromatography. Of these techniques, GC–ECD is probably second only to GC–MS, which, owing to a higher selectivity, can be used more reliably for analyses at the low trace level. For the analysis of one or two compounds, such as in biopharmaceutical studies and in drug monitoring, the ECD is the reliable workhorse as the concentrations requiring detection are well within its operating limits. It is also probable that ECD will be used more efficiently when capillary columns and column switching devices are used more regularly. The latter ancillary technique is of importance in simplifying the removal of excess of reagent and to reduce the need for lengthy clean-up procedures. Applications are often seen where a special approach has been made to fit the ECD, such as in studies of enzyme activity, where an electrophore substrate has been chosen to facilitate the determination of its disappearance or the appearance of the product at low concentrations.

Applications of the ECD to biomedical problems have been reviewed in part previously. Thus, Cummins [1] presented an early review on the use of ECD for the determination of halogenated derivatives of organic compounds. There one can find some of the early experiences regarding detector response and derivative structures. Hattori [2] presented a review (in Japanese) on the application of ECD to biological specimens.

This chapter reviews applications of the ECD in the biomedical field. The major part deals with the quantitation of drugs and related compounds in trace amounts in biological fluids, an area where there has been and still is a great need for sensitive and selective methods. The applications have been grouped either according to a functional group or to a special structure. The list of papers could without difficulty have been longer, but the papers chosen reflect to a large extent the author's interest in this field.

13.2. PRIMARY AMINES

Many drugs and several endogenous compounds are primary amines. It was therefore natural that very early a great deal of attention was focused on the possibility of determining minute amounts of such compounds in various biological fluids. Amphetamine, for instance, was the subject of several studies in which the effects of various derivatizing reagents were compared with respect to the highest response in the ECD. Änggård and

Hankey [3] found that pentafluorobenzoylated amphetamine gave the largest response but, owing to the bad chromatographic performance of this derivative, the trichloroacetyl compound was preferred. All sensitivity values were expressed relative to lindane and no temperature effects were given. The monochloroacetyl derivative gave a response that was not much better than that for the trifluoroacetamide. On the other hand, a good response was given by both the monomaleamide and the *p*-trifluoromethylbenzamide derivatives.

Matin and Rowland [4] reported a study of clinically important primary and secondary amines. The derivatives from amphetamine that were most promising were, in order of electron-capture sensitivity, pentafluorobezamide > a pentafluorobenzylidene compound > *p*-nitrobenzamide > heptafluorobenzamide. From their study it was also clear that derivatives with a carbonyl group or a C=N bond in conjunction with the perfluoroalkyl chain were superior to those with one of these bonds saturated.

Noonan et al. [5] favoured the use of the trichloroacetamide of amphetamine and were able to follow the fate of the stimulant in horses. Bruce and Maynard [6] claimed that the heptafluorobutyramides gave the best results. Some related amines, such as chlorphentermine and fenfluramine, were also studied. It is of interest that the responses obtained for these three compounds differed considerably, indicating that not only the electrophore introduced is of importance for the final electron-capture response. This was also pointed out by Matin and Rowland [4]. Bruce and Maynard found it necessary to pre-condition the column with heptafluorobutyric anhydride (HFBA) in order to obtain a satisfactory response. This treatment for this kind of derivative is not mentioned elsewhere. A similar comparison was reported by Wilkinson for amphetamine and the ephedrines [7]. The derivative of choice in this case was the pentafluorobenzamide, which was about 40 times better than the heptafluorobutyryl (HFB) derivative. However, among the various ephedrines the sensitivity values varied considerably, e.g., the smallest detectable amount on the column was 25 pg for ephedrine and 60 pg for norephedrine. This relationship between primary and secondary amine derivatives was also observed by Matin and Rowland [4].

The relative merits of some methods for the assay of amphetamine in biological fluids were recently evaluated. The gas chromatographic method for trace amounts made use of the trichloroacetyl derivative [8]. This method was considered to be the most sensitive and reliable. For pharmacokinetic studies this method was the best, together with nitrogen-selective detection of the non-derivatized compound.

Walle [9] reported on the use of 2,4-dinitrophenyl derivatives of some amines of low molecular weight, including amphetamine. The minimum detectable amount of this derivative was around 20 pg, but in biological fluids such as urine the limit was 5 ng/ml, largely owing to interference from other constituents in the sample. For the secondary amine diethylamine, the lower limit of determination was as high as 50 ng/ml.

Walle and Ehrsson [10,11] made an important contribution to the understanding of the process of derivatization of amines and alcohols when they published their study on the preparation of heptafluorobutyryl derivatives. Twenty-one compounds were investigated, most of them of biological interest. The studies confirmed earlier observations by Clarke et al. [12] and Änggård and Hankey [3] that the trifluoroacetamides were less sensitive than the heptafluorobutyramides. It was also clear that the alcohol derivatives of HFB were about five times more sensitive to the ECD than the amine derivatives. If an aromatic nucleus was in the vicinity the ECD response was increased for both kinds of derivative. The derivatives of ephedrine were clearly identified as bis-perfluoroacylated, in contrast to reports by Wilkinson [7] and Cummins and Fourier [13]. The interesting observation of the high sensitivity for the bis-trifluoroacetyl (TFA) derivatives will be discussed in a later section.

The important question of the temperature dependence of the detector response to these derivatives was first studied by Walle and Ehrsson [10], who reported a limited temperature effect for the range from 120 to 180°C in an ECD with a tritium source. One of their most important findings was that the poor quality of the anhydride and the presence of water could to some extent be compensated for by the use of a tertiary aliphatic amine as catalyst (acid scavenger). Pyridine was not useful as it introduced severe disturbances. The removal of excess of reagent was best achieved with an aqueous rinse, preferably at pH 10. This treatment also had the benefit that the HFB derivative of a tertiary amino alcohol such as tropine could be chromatographed in high yields. Without this wash there seemed to be some complex formation between the tertiary amine and heptafluorobutyric acid that could not easily be broken. The second paper in this series dealt with the application of these procedures to plasma and urine samples [11].

Of the primary amine drugs, mexiletine, an antiarrythmically active compound, has been quantified using ECD [14,15]. The first report made use of an HFB derivative. The preparation was made without a catalyst at 90°C for 30 min, in contrast to the conditions used by Walle and Ehrsson [10]. Recently another study, also using HFB acylation for the quantitation of mexiletine, was published. The monoheptafluorobutyryl derivative provided maximum response at a detector temperature of 295°C. The detection limit was 5 pg.

Amantadine is a primary amine with antiviral properties, and has been quantified as the trichloroacetyl derivative [16]. Concentrations down to 10 ng/ml were determined in a procedure in which excess of reagent was removed by an alkali wash. In an older method with the reagent left in the mixture the limit was 25 ng/ml [17].

Tocainide is another antiarrythmic agent derived from lidocaine. The primary amine function was derivatized with HFBA giving rise to a mono-derivative [18]. However, as this compound also has an amide function, it would be possible to introduce a second acyl group. This may be the reason for the rather narrow time interval, 45–55 min, for the optimum yield to be obtained. The compound used for quantitation was identified by GC–MS as the mono-HFB derivative. Excess of reagent was removed by evaporation, as treatment with aqueous media resulted in degradation of the derivative.

Primary amines can be diperfluoroacylated, as shown by Ehrsson and Brötell [19]. These derivatives, prepared with amine catalysis, exhibit excellent gas chromatographic properties and have a very high response in the ECD. The bis-TFA derivatives are of greater interest than the HFB derivatives even if the latter give a better response. This preference for TFA is based on the fact that they can give a more selective reaction owing to the considerably lower response shown by the mono-TFA derivatives compared with the bis-TFA derivatives. Excess of reagent had in this instance to be removed by extraction of the reaction mixture with buffer of pH 6. The application of this procedure to low level analyses was demonstrated for amphetamine and phenylethylamine.

Amongst other applications the pentafluoropropionyl (PFP) amidation of 1-(2,6-

dimethylphenoxy)-2-aminopropane can be mentioned. Surprisingly, this method was reported not to have any advantage over the flame ionization method [20]. Chrysenamine, a carcinogenic compound, was determined in biological samples after reaction with trifluoroacetic anhydride (TFAA) [21]. For this derivative the minimum detectable amount was 1 ng, compared with the limit for the non-derivatized polyaromatic amine which was about 25 times less. This indicates that the polyaromatic skeleton makes a contribution to the detector response.

Primary aromatic amines are also found among some drugs. Procainamide has been quantified by ECD using PFP derivatization of the amine group [22]. Using a dipropylamino analogue as internal standard it was possible to assay 0.1-ml samples with concentrations down to 1 μg/ml. N-Acetylprocainamide, a metabolite, did not interfere and the method compared favourably with an older colorimetric technique.

Another compound with an aromatic primary amino group is metoclopramide, which has antiemetic properties. Like procainamide it is also a tertiary amine, giving some problems upon gas chromatography at low levels. A recent publication, however, showed that nanogram amounts could be quantified as the mono-HFB derivative [23]. Derivatization was performed with HFBA at 55°C for 20 min, although a shorter time would suffice. Amounts down to 1 pg could be detected on-column and practical limits in biological samples were around 75 ng/ml in the rat for 50 μl of plasma. A similar procedure was recently proposed in which heptafluorobutyrylimidazole (HFBI) was used as reagent [24]. Pharmacokinetic studies in humans gave a lower limit of detection of 5 ng/ml.

The de-ethylated metabolite of metoclopramide was recently quantified according to the same principle [25]. In this instance the bis-HFB derivative was formed. The derivative seems to be rather volatile as severe losses were reported when excess of reagent was removed by evaporation.

The very high ECD response of the derivatives from primary aromatic amines is in line with the studies by Walle and Ehrsson on HFB derivatives of amines [10]. Compared with bensylamine aniline gave a derivative with at least a 10-fold higher response.

Phenylethylamine is an endogenous amine whose quantitative analysis has been of interest for the past 20 years. Clarke et al. [26] published one of the first reports of its determination using a GC–ECD technique. They found that the TFA derivatives were relatively insensitive to the ECD, as were the monochloroacetylated derivatives. Responses several hundred times greater were exhibited by the PFP and the HFB derivatives. If the substrate also contained an esterified hydroxyl group the response was among the highest attainable, even if only TFA derivatization was performed.

Moffat and Horning [27] published an extensive study on the derivatization of phenylethylamine with 13 reagents. The most sensitive derivative was the Schiff base with perfluorooctanaldehyde, closely followed by that of pentafluorobenzaldehyde. The HFB derivative was the third and the 2,4-dinitrophenyl compound the fourth most responsive to the detector. The derivative of choice was the pentafluorobenzylidene compound, which enabled 10 pg of phenylethylamine to be detected.

Martin and Baker [28] described the analysis of phenylethylamine in rat brain. The procedure, which consisted in homogenization, ion-exchange clean-up and acetylation in the aqueous phase, ended with a pentafluoropropionylation step and quantitation with

ECD. The N-acetyl derivative was made to react with PFPA in ethyl acetate and resulted in a derivative that had excellent gas chromatographic properties and a high ECD response, as less than 5 pg could be detected. Later studies revealed this derivative to be N-acetyl-N-pentafluoropropionyl-2-phenethylamine [29]. This reaction was performed at 60°C without a catalyst, in contrast to studies by Ehrsson and Mellström [30].

Blau et al. [31] have reported the determination of phenylethylamine in urine using ECD of the pentafluorobenzoyl derivative. Analysis of 1-ml urine samples with tolylethylamine as an internal standard were performed with a relative standard deviation of 10% at the 1 μg/l level. All-glass derivatization vessels were used, as screw-caps lined with polytetrafluoroethylene (PTFE) resulted in losses and a large number of interfering peaks.

Edwards and Blau [32] also used the 2,4-dinitrophenyl (DNP) derivative of phenylethylamines for quantitation in tissues such as brain and liver. After homogenization, extraction and back-extraction, the amines were made to react in the aqueous phase with 2,4-dinitrobenzenesulphonic acid (DNBS), which has the advantage of not reacting with hydroxyl groups. About 20 ng/g of tissue could be quantified. Clean-up steps using ion-exchange columns were not recommended for the submicrogram range as the recovery of the amines decreased.

The corresponding 2,6-dinitro-4-trifluoromethylbenzene (DNT) derivatives were studied and exhibited about double the ECD response. However, as they could not be prepared via the sulphonic acid, this approach was not useful. Some years later, however, Doshi and Edwards [33] reported on the use of 2,6-dinitro-4-trifluoromethylbenzene-sulphonic acid for the determination of primary amines, including phenylethylamines. The DNT derivatives were more volatile and showed less tailing than the DNP derivatives. The reactivity of this reagent was also much greater than that for 2,4-dinitrobenzene-sulphonic acid. In addition, there seemed to be at least the same specificity towards primary amines as that shown with 2,4-DNBS. Hydroxyl groups were not reactive either and, owing to the rapid reaction, catecholamines did not degrade, which makes this reagent very interesting. The method was applied successfully to histamine and 1-methylhistamine in urine.

An application of this principle to the quantitation of 1-methylhistamine in rat and human urine was recently published [34]. Various analogues were tested as internal standards, the best choice being 2-pyridylethylamine. The amines were extracted from sodium chloride-saturated urine, back-extracted and lyophilized. The residue was finally treated with the DNT–sulphonic acid reagent.

There has been a great interest in di- and polyamines in biological fluids because elevated levels of, e.g., putrescine, spermidine and spermine have been found in the urine of patients with cancer. Makita et al. [35] reported a method for the determination of such amines using derivatization in the aqueous phase with pentafluorobenzoyl (PFB) chloride. The derivatives were then extracted, dried and chromatographed. Picogram amounts of these high-boiling derivatives were detected. However, spermine could not be quantified in trace amounts. This derivatization procedure for these polar compounds is most attractive.

13.3. SECONDARY AMINES

The reagents used for acylation of primary amines were also evaluated for secondary amines. Heptafluorobutyrylation was found by both Walle and Ehrsson [10] and Matin and Rowland [4] to give derivatives with higher response for secondary amines (tertiary amides) than those for the corresponding primary amines (secondary amides). On the other hand, pentafluorobenzoylation gave more sensitive derivatives with primary amines, as reported by Matin and Rowland [4]. They explained these observations as depending on differences in resonance possibilities between the carbonyl group and the pentafluorophenyl ring. It was found for both HFB and PFB derivatives that electron-donating substituents in the aromatic ring resulted in a lower response.

One of the first applications of heptafluorobutyrylation of a therapeutically effective secondary amine was reported for protriptyline, a tricylic antidepressant [36]. HFBI was used to give a derivative with 50 pg as the detection limit on-column. Applications to plasma, erythrocytes and urine had a detection limit of 10 ng/ml. The work-up procedure included one re-extraction of the amine before reaction with HFBI at 60°C for 15 min. The use of the acylimidazole was preferable to the anhydride.

Walle and Ehrsson [11] applied their HFB acylation procedure to the determination of phenmetrazine. Levels down to 0.5 ng/ml could be determined with a relative standard deviation of 6%. When re-extraction was used as a purification step 0.05 ng/ml could be detected. If the latter step was omitted the lower limit for quantitative determination was 5 ng/ml. This principle was also applied to methamphetamine, methoxyphenamine, chlorphentermine and nortriptyline.

Desipramine has been shown by Ervik et al. [37] to undergo an unusual reaction with perfluoroacyl anhydrides, in that the monoacyl derivative reacts further. In the diacyl derivative formed the second acyl group was situated in the side-chain after a dehydrogenation had taken place. This strongly conjugated β-perfluoroacylenamine structure gave a very high ECD response for the derivative. With trifluoroacylation the bis-derivative was about 600 times more sensitive than the mono-derivative. Even if the di-HFB compound was more sensitive (detection limit 0.2 ng), the use of TFAA is a more selective approach. Samples from patients could be analysed at concentrations down to 1 ng/ml even if only 1% of the biological sample was used in the chromatographic step.

The introduction of the second TFA group requires considerably more time than that for the first group. The corresponding HFB acylation was also found to be slower. This very selective reaction can, however, be a drawback, as pointed out by Lundgren et al. [38]. They had to determine desipramine as a metabolite of lofepramine, an antidepressant drug, and chose the mono-HFB derivative as it was not possible to find an internal standard that would form the corresponding bis-TFA derivative of desipramine.

Nortriptyline and some of its hydroxylated metabolites were made to react with HFBA and quantified in plasma and urine at levels down to 10 ng/ml [39]. The 10-hydroxylated derivatives lost water upon reaction with HFBA, giving rise to 10,11-dehydro derivatives. These were shown by other techniques not to be metabolites, thus avoiding interferences. This dehydration reaction is not uncommon for hydroxyl groups in a benzylic position.

The detection limit for the pure nortriptyline HFB derivative was 10 pg, with the other compounds having responses within ±25%. The 10-hydroxylated metabolites reacted more slowly than nortriptyline in a non-quantitative manner. Maprotiline, another tricyclic antidepressant compound, was used as internal standard in this study. The determination of this drug in biological fluids was reported by Geiger et al. [40]. The response in both studies was reported to be 70% of that of nortriptyline.

A related compound, but without the side-chain, is metapramine. Comparative studies using nitrogen-selective detection of the non-derivatized drug and ECD of the HFB derivative showed that the latter was more selective with respect to interference from metabolites [41]. Nitrogen detection was more sensitive, more reproducible and more rapid. The methods were considered to be complementary. In the derivatization procedure reaction with HFBA was performed for 1 min at 122°C.

Methylphenidate has been used as an antidepressant and also abused as a "stimulant" in horse racing. A method was reported for its determination in plasma using pentafluoropropionylation [42]. As a means to increase the sensitivity, so-called "reductive fragmentation" was reported. Treatment with $LiAlH_4$ converted the carboxylate moiety to an alcohol, which upon derivatization gave a di-PFP derivative, that was five times more sensitive. This approach has also been used for some tertiary amines (see below). The increased sensitivity of bisperfluoroacyl derivatives such as those from 1,2- or 1,3-diols will be discussed under amino alcohols.

Formation of a mono-PFP derivative was used in a method for tracing methylphenidate and methylamphetamine in the control of doping [43]. Phacetoperane was hydrolysed and the amino alcohol formed then reacted to give a bis-PFP derivative.

An early report described gas chromatography and ECD of some chlorpromazine metabolites, viz., the desmethyl- (*sec*.-amine) and didesmethyl (primary amine) [44]. Of interest was the observation that the TFA derivatives increased the sensitivity about 100-fold over the non-derivatized metabolites, whereas PFP derivatives were similar to the TFA compound. This is related to the inherent electrophore properties of the chlorophenothiazine moiety. The HFB derivatives were a further 5–10 times more sensitive.

Another compound with an inherent ECD response which was determined after reaction with TFA is 6-chloro-2-(1-piperazinyl)pyrazine [45]. This compound has serotonin-like activity and therefore it was of interest to quantify it in biological fluids. TFA derivatization was stated to give the required sensitivity (i.e., around 100 ng/ml with acceptable precision). The internal standard was the 6-trifluoromethyl analogue, which implies that even this structural element contributes to the electrophore properties of aromatic structures.

Fenfluramine has been used as an appetite suppressant and contains a trifluoromethyl substituent in the aromatic ring. A method for its quantitation in biological fluids was based on the formation of a trichloroacetyl derivative [46]. The lower level determined was 20 ng/ml. A simple extraction and a short derivatization step without special purification of the reaction mixture was reported not to be subject to interferences. Fenfluramine and its metabolite norfenfluramine are present as racemates. As the enantiomers have different pharmacological properties it is of importance to be able to measure the isomers when the racemate has been administered. A resolving agent for amines is N-trifluoroacetyl-1-prolyl chloride. For fenfluramine, it was reported that the resulting amides did not have the required ECD response [47]. If, however, a PFP group was substituted for the TFA group, sufficient sensitivity was obtained (detection limit 2 pg). Even the resolution was reported to be better for this derivative. Applications were reported for brain homogenates, plasma and red blood cells. The distribution of the optical isomers was not found to be identical. An alternative reagent with an excellent response, N-pentafluorobenzoyl-S-(−)-prolyl 1-imidazolide was evaluated by Matin et al. [48].

Another molecule with effect on hunger is tiflorex, which is twice as potent as fenfluramine. Instead of a trifluoromethyl group, tiflorex has a trifluoromethylthioether substituent. The compound has electrophore properties but for several reasons a derivatization procedure with trichloroacetyl chloride was preferred [49]. The sensitivity was doubled (1 ng/ml) and separation from the demethylated metabolite was improved.

Two isomeric secondary amines (structures derived from indane) were derivatized with PFB chloride and HFBA. The exo isomer showed the same high ECD response for both derivatives, whereas the endo isomer, Kabi 1847, gave a more sensitive HFB amide [50]. Moreover, the PFB amide in this instance was temperature dependent, again showing how the entire structure influences the response and its behaviour. The HFB derivative (trimethylamine catalysis) was used in a method in which the sample size from rats was at most 200 μl in order to allow for a pharmacokinetic study. By performing the reactions on the microscale and injecting a larger portion of the reaction mixture, as little as 35 pg could be determined. Moreover, it was also shown that considerable interferences could be eliminated if the acylation was optimized with respect to the amount of anhydride used in the presence of the plasma sample.

Another acylating reagent that has been used for secondary amines in the presence of their tertiary homologues is trichloroethyl chloroformate. An application has been reported for norpethidine [51].

Most acylations with HFB were found to be very rapid when base catalysis was used [10]. An exception was reported for terodiline, which is sterically hindered around the secondary amino group. Using high concentrations of trimethylamine and the HFBA it was possible to acylate the drug quantitatively. In order to perform reliable and reproducible quantitative determinations it was found that an internal standard of the same steric arrangement around the nitrogen atom was necessary [52]. This approach compensated for small changes in the reaction conditions, e.g., content of water or HFB acid. For this reason heptane was the best solvent for extraction and the subsequent acylation. This method was used in a comparison with the following degradative method. The HFB derivative had an ECD response with a fairly small temperature dependence.

Terodiline was originally quantified by Vessman and Strömberg [53] in biological fluids using the so-called "benzophenone" method. This is based on the formation of benzophenone from diphenylmethane-substituted compounds. Benzophenone has a fairly good response to the ECD and this permits quantitation of about 5 ng/ml. The oxidation to benzophenone was not quantitative, but the use of a *p*-chlorophenyl-substituted analogue could compensate for this and made the method practically useful when it was introduced at the start of the reaction. Metabolites might interfere if they differ in the aliphatic part of the molecule but not in the ring structure. Comparisons with the HFB method mentioned above indicated that the interfering metabolites contributed only slightly [52]. The selectivity was achieved through a combination of ion-pair and base extraction as well as the oxidation step.

Pipradrol, a mild psychostimulant, was oxidized with chromic acid to give 6-benzoyl-1,2,3,4-tetrahydropyridine. This unusual oxidation product (a benzophenone was expected) could be used in a bioanalytical method as the selectivity would be unusually high for this degradation procedure. Hartvig [54] reported a high ECD response [comparable to benzophenone or about 20 pg as minimum detectable amount (MDA)] due to an internal hydrogen bond between the carbonyl and the N–H bond.

Amines derived from some pesticides via hydrolysis were quantified by Crosby and Bowers [55] as trifluoromethyl-substituted nitroanilines. These derivatives were more volatile and also showed a better ECD response than the corresponding DNP compounds. It is surprising that so few applications of this approach have been reported. Piperidine was, however, quantified as the derivative with 3,5-dinitro-4-chlorobenzotrifluoride in brain samples. Hexamethyleneimine was added to the brain homogenate and after centrifugation the derivatization was performed in the aqueous brain extract. Picomoles per gram of tissue were quantified with acceptable precision [56].

Acylation with trichloroacetic acid (TCA) was reported recently [57]. 2-Ethylpiperidine was used as internal standard and the amines were extracted, re-extracted and derivatized with the acyl chloride. Down to 3 ng in a brain aliquot (about 35 pmole) could be analysed and the regional distribution of the amine could be studied. The results were validated with a GC–MS procedure.

Nitsche et al. [58] studied the reaction of primary and secondary amines with isocyanates. The ureas formed with 3-trifluoromethylphenyl isocyanate had sufficient response to permit the detection of 0.1 ng.

Methindione, a secondary amine of indandione structure, could be analysed without derivatization due to the conjugated dione system, as described by Vessman et al. [59]. This made the bioanalysis of this drug very simple and clean chromatograms were obtained even with only a few ng/ml. The advantage of using natural electrophores is considerable as no new interfering compounds are produced.

13.4. INDOLE AMINES

This group of compounds occur at trace levels in the central nervous system and it has long been of interest to quantify them. Melatonin and some related compounds were determined by Degen et al. using HFB derivatives [60]. For melatonin (N-acetylmethoxytryptamine) 35 pg could be measured. The derivative was formed from HFBI and resulted in the exchange of one HFB for the acetyl group. The acyl exchange was apparently catalysed by the triethylamine present.

Christian et al. [61] reported the determination of tryptamine and methoxytryptamine in cerebrospinal fluid with detection limits around 1 pg for their bis-HFB derivatives. PTFE closures were avoided as they introduced a number of interfering peaks. The very low levels of these compounds in complex mixtures might have restricted the application of ECD. A recent study of metalonin with negative ion chemical ionization mass spectrometry used PFP acylation, which resulted in another derivative structure [62]. Levels in the femtogram range were reported, which is below the limit for ECD.

Bosin et al. [63] used PFP derivatives of some indoles and their benzo[*b*]thiophene and 1-methylindole analogues.

13.5. TERTIARY AMINES

Halogen-substituted drugs of the tertiary amine type were investigated early for the possibility of determining trace amounts in biological fluids. A typical example is chlorpromazine, which was thoroughly studied by Curry [64–66]. The possibility of following the treatment of patients by monitoring plasma levels of the drug revolutionized therapeutic procedures as a standard dose regimen was no longer followed but instead recommended concentrations in this biological fluid. Initial problems in the determination of this base at the trace level were reported [65], together with some still not fully understood phenomena, such as variation in sensitivity to chlorpromazine among certain ECDs if the sample is contaminated with biological material. Christoph et al. [67] modified Curry's method by using a high-volume injection port and introducing an internal standard, which was claimed to improve the reproducibility significantly. The large volumes injected (100 μl) created problems regarding solvent purity. To reduce adsorption losses Curry recommended the use of polycarbonate tubes [64], which Christoph et al. found to give unacceptable blanks.

Some of the numerous metabolites of chlorpromazine could be determined simultaneously, whereas others such as chlorpromazine N-oxide and its sulphoxide had to be reduced prior to extraction [68].

Although chlorpromazine was studied very early using the ECD it was not until recently that the temperature dependence of the response was studied [69]. It was found to give higher response with an increase in detector temperature.

Some other phenothiazine-derived drugs were studied by Curry [70]. Only a few were found to give a sufficient response to the ECD, surprisingly including promazine, which is chlorpromazine without the halogen. The trifluoromethyl analogues showed very low sensitivity.

Noonan et al. [71] proposed a bromination reaction to convert some phenothiazine tranquillizers into sensitive derivatives. Of the several approaches studied the use of 1% of bromine in an iodine-saturated hydrocarbon was the most suitable. The derivatives required high column temperatures. Promazine formed a dibromo compound whereas acepromazine and triflupromazine gave monobromo derivatives.

Bromhexine is a mucolytic agent and a very sensitive assay of this 2-amino-3,5-dibromobenzylammonium derivative involving TFA derivatization was reported recently [72,73]. Levels as low as 1 ng/ml could be detected. Triethanolamine was added to the alkalinized sample and was claimed to improve the recovery significantly. Acylation with pentafluoropropionic anhydride (PFPA) or HFBA did not result in derivatives with better sensitivity than with TFAA, mainly owing to the two bromine atoms originally present.

Lorcainide is an antiarrhythmic agent. For the determination of the N-dealkylated and hydroxylated metabolites silylation was preferred [74]. Owing to the presence of a *p*-chloro substituent in the benzene acetamido moiety a sufficient response was obtained to allow 5 ng/ml of lorcainide to be determined.

Nicotine was one of the first alkaloids to be gas chromatographed but attempts to use the ECD were not reported until the mid-1970s [75]. Neelakantan and Kostenbauder [75] used catalytic hydrogenation of nicotine to form octahydronicotine, which contains two secondary amino functions which were derivatized with PFPA. With the ECD, 100 pg/ml could be detected.

A similar approach was used by Lundgren et al. [38] to determine lofepramine via the reduction product, *p*-chlorobenzaldehyde. A bromo analogue was used as an internal standard. Only the desmethyl metabolite interfered. Cocaine has also been determined after reductive cleavage of the ester functions to 2-hydroxymethyltropine, which was acylated to a bis-HFB derivative [76]. This diol ester structure gave reasonable signals for about 20 ng/ml of sample. A reductive reaction [77] followed by PFP acylation was recently reported, with about the same sensitivity as reported above. Problems were encountered with the stability of this diester, especially in biological samples.

Other degradative reactions have been described for some tertiary amines. Formation of benzophenone derivatives by oxidation was applied to methadone [78] and the proposed method gave values for serum levels that were identical with those from a GC–MS determination. Very clean chromatograms were obtained and differences in oxidation rates with the peroxide system used eliminated interferences from the main metabolite. Brompheniramine was determined as the permanganate oxidation product 2-benzoylpyridine after removal of the metabolites [79]. Less than 10 ng/ml could be determined.

Amitryptiline has been oxidized to an anthraquinone in a procedure that co-determined the metabolite nortriptyline [80]. Hartvig et al. [81] separated the tertiary amine from metabolites in a chromatographic step and determined 2 ng/ml in plasma. The results compared fairly well with those from nitrogen-selective detection. The oxidation yield was not quantitative but by using 2-chloroamitryptiline as an internal standard 50 ng/ml was determined with a relative standard deviation of 4%.

The demethylated metabolites of chlorpheniramine were removed via an acetylation reaction [82]. This resulted in a more rapid procedure, with permanganate oxidation as the final pre-chromatographic step.

Clemastine has a marked antihistaminic activity in low doses. A pharmacokinetic study was based on analytical results from a chromic acid oxidation procedure that gave rise to *p*-chlorobenzophenone [83]. The lowest level that could be determined was 0.5 ng/ml in plasma. The sensitivity of the *p*-chlorobenzophenone method was better than that of the direct determination of the drug, which was only 30 ng/ml. This sensitivity difference has been discussed from the point of view of ECD response and temperature dependence [50].

Dealkylation reactions of tertiary amines with chloroformates have been studied and applied during the last 8–9 years. In the first attempt, reported by Degen and Riess [84], the hypnotic N,N-dimethyldibenzo[*b,f*]thiepin-10-methylamine did not give the expected secondary amine by demethylation, but rather N,N-dimethylcarbamate and the corresponding chloromethylene compound were formed. This reaction could be utilized as this compound had a sufficient ECD response to allow the determination of a few ng/ml.

Imipramine, a frequently used antidepressant, was dealkylated to desipramine in a reaction in which the intermediate carbamate was hydrolysed to the secondary amine [85]. This and the degradation product from the internal standard, 2-chloroimipramine, were determined after TFA acylation. Desipramine gave the bis-TFA derivative whereas 2-chlorodesipramine was monoacylated.

The finding that carbamates had excellent gas chromatographic properties led to the application of electrophore carbamates such as pentafluorobenzyl (PFBz) and trichloroethyl (TCE) carbamate [86,87]. The former were generally more sensitive but TCE chloroformate reacted faster and was easier to obtain in an analytically acceptable quality. Using this approach diphenhydramine could be quantitatively converted into a PFBz carbamate [88]. The problem of removing excess of reagent hampered the utility of this approach, but treatment with strong alcoholic alkali overcame this difficulty [89]. The procedure was applied to the determination of trimipramine.

A comparison of the two chloroformates was reported by Hartvig et al. [87]. It was clear that the structure near the tertiary amine determined if deamination or the more useful carbamate formation occurred. Compounds of therapeutic interest that gave carbamates in quantitatively useful yields and not described above were chlorpromazine, promazine, alimemazine, methdilazine, cyproheptadine and chlorprothixene. The responses were such that 3–5 pg of the pure derivatives could be detected.

Pethidine has been in use for many years for the relief of acute pain. The use of TCE carbamates for their quantitative determination in biological fluids allowed the study of the pharmacokinetics of this drug [90].

In all applications of the dealkylation of tertiary amines in biological samples one must remember that the corresponding secondary amines form exactly the same carbamate. A pre-separation step is therefore essential and in most instances chromatography has been used, as for instance in the determination of pethidine and norpethidine [51] and for amitriptyline and nortriptyline [91]. The latter could be determined at levels down to 1 ng/ml. A comparison with GC–MS was reported [92].

Separately from the series of studies reported above, a method using TCE carbamate for the determination of doxepin was published [93]; no separation of the antidepressant drug and its demethylated metabolite was used.

Studies of compounds such as the pheniramines, with an additional pyridine ring, were not reported in the above-mentioned papers. However, it could be expected that the pyridine would also react. This was also noted in a report on the limitation of the chloroformate reaction for the analysis of tertiary amines [94].

An example of the susceptibility of two different tertiary amino groups is shown by the reaction of nicotine with TCE chloroformate [95]. The pyrrolidine ring was opened and the carbamate formed was upon heat treatment partly dehydrohalogenated to the corresponding olefin. Pyridine was required in the reaction, otherwise very low yields were obtained. This derivatization reaction could be used for determinations at levels down to 10 ng/ml in plasma but, owing to nicotine blanks and the heat lability of the derivative, this method would not be competitive with, for instance, the direct detection of nicotine with a nitrogen-selective detector.

Yamaguchi et al. [96] reported recently on the quantitation of an antianginal compound, ANP-4364, that was de-ethylated with TCE chloroformate. Although the compound has a 2,3-dichloro-4-methoxyphenyl substituent, the halogenated chloroformate was chosen. The PFBz carbamate was 2.5 times more sensitive but reacted considerably slower. Of interest was their approach to remove the de-ethylated compound. As secondary amines react rapidly with chloroformates at room temperature but the tertiary analogues require treatment at elevated temperatures (110°C for 30 min), it was possible to convert the dealkylated metabolite selectively into an ethyl carbamate and then to separate this from the tertiary amine by simple solvent extraction. No interference was observed even if concentrations 10 times higher than peak levels were analysed.

A similar approach was reported by Barnhart and Massad [97] for the determination of dextromethorphan in serum. PFBz chloroformate was preferred owing to the higher sensitivity of the derivative. In this instance the dealkylated metabolite was removed by reaction with acetic anhydride. Detection limits were below 1 ng/ml. Applications to the study of the urinary excretion of dextromethorphan were recently reported by Barnhart [98].

An unusual derivatization reaction of tertiary aromatic amines was described by Walle [99]. Trifluoroacetylation of the aromatic ring in N,N-dimethylaniline occurred, giving trifluoroacetophenone derivatives with extremely high electron-capture responses, which also were very sensitive to temperature changes. Trimethylamine was used as a catalyst as TFA acid inhibited the reaction. Again, the advantage of fewer interferences when using TFA instead of HFB was pointed out.

13.6. TERTIARY AMINES WITH HYDROXYL AND CARBONYL GROUPS

Morphine has been the subject of several studies. Wallace et al. [100] reported the use of TFAA to form a bis-TFA derivative with a detection limit of less than 0.1 ng or about 25 ng/ml. Codeine, with only one TFA group, was considerably less sensitive, again showing the importance of selecting the right perfluoroacylating reagent. Prior to use the column had to be treated with microgram amounts of the derivative, which also was found to be sensitive to high humidity. Nalorphine was used as an internal standard.

PFP derivatization of morphine was performed by Dahlström and Paalzow [101]. These derivatives were more stable and had a higher sensitivity than the TFA derivatives. About 5 pg could be detected and 500 pg/ml could be determined in plasma. In a 30-mg brain sample 100 pg could be determined.

Dahlström et al. [102] later reported the simultaneous determination of codeine and morphine in biological samples using pentafluoropropionylation. A 0.75-ng amount of morphine and 7.5 ng of codeine per sample could be determined. The ECD sensitivity was sufficient for the elucidation of the pharmacokinetics of these drugs when codeine was administered. One internal standard of relevant structure was used for each of the compounds and it was claimed that this was necessary in order to obtain a low coefficient of variation, especially for morphine.

The detection of trace-levels of morphine in heroin was studied by Moore [103]. To determine morphine, codeine and [^{16}O]acetylmorphine at levels below 0.1%, several halogenated acyl reagents were studied, with HFBA being preferred. It was claimed that HFB chloride reacted more slowly than its anhydride. The morphine derivative containing two HFB groups was approximately 7 times more sensitive (20 pg) than the other two with only one HFB group. At least 0.001% of morphine and 0.01% of [O^{16}]acetylmorphine could be detected. Codeine was found at levels dependent on the amount of morphine present. This procedure was used to differentiate the origins of various samples.

A micro-method for the determination of morphine in urine was proposed by Nicolau et al. [104] as a confirmatory procedure in large-scale screening. HFB acylation was chosen and the results agreed well with those of thin-layer chromatography and radioimmunoassay.

Christophersen and Rasmussen [105] performed on-column acylation of some opiates and used a heart-cutting technique to remove excess of reagent. Down to 20 pg of morphine could be detected and the technique was proposed for application to bioanalysis.

Felby [106] described a method for the determination of morphine in blood samples taken post mortem. The samples were often of poor quality and in addition to extraction a purification step with adsorption on to a column of acid aluminium oxide was used. PFP derivatives were prepared and made possible the determination of 1 ng/g in 0.1 g of post mortem blood.

Some narcotic antagonists have a structural resemblance to the morphine alkaloids and have been analysed according to the same principles. Naltrexol and naltrexone were thus determined as the PFP derivatives with sufficient sensitivity to follow an oral dose over 24 h [107]. Sams and Malspeis [108] investigated three perfluorinated anhydrides in the reaction with naloxone and naltrexone. Owing to their poor stability the TFA derivatives were of less interest. The tris-HFB ester formed under catalysis with pyridine was chosen. Without pyridine the less sensitive diester was obtained. From 40 to 70 pg of the compounds could be detected. The tris-TFA derivative was reported to have a response of the same magnitude. Naltrexone prodrugs were evaluated after hydrolysis to naltrexone and followed by reaction with PFPA and dimethylaminopyridine as catalyst [109].

Hydrocodone has antitussive properties. In order to shed some light on its pharmacokinetics a method was developed by Barnhart and Caldwell [110] involving formation of pentafluorophenylhydrazone derivative which permitted as little as 1 ng/ml to be determined. The demethylated metabolite could be chromatographed but the response was substantially less than that of the original drug derivative. The relative standard deviations in the analysis of some control sera were 3–4% for concentrations from 7 to 25 ng/ml.

Apomorphine has been used as a stimulant in horse racing and has been shown to be a dopamine receptor antagonist. Several methods have been proposed for its determination. A gas chromatographic approach used heptafluorobutyrylation, giving a bis-HFB derivative with excellent response to the ECD [111]. If the concentration of HFBA was optimized in the reaction using triethylamine instead of pyridine as catalyst, the derivatization of extraneous material could be reduced. This effect is seldom investigated in bioanalytical studies (cf., p. 283).

Pentazocine is a highly active analgesic compound. Brötell et al. [112] reported a method involving pentafluorobenzylation in an extractive-alkylation reaction to convert the compound into an electrophore derivative. Concentrations down to 1 ng/ml could be determined. Solvent extraction was used to remove the excess of reagent. The major problem was caused by adsorption loss of the derivative, a tertiary amine. Considerable improvements in detection limit and reproducibility were achieved by using a so-called "solvent flush" technique, with desipramine present in a large excess. The relative standard deviation was reduced from 21 to about 2% with this approach. The PFBz ether derivative was shown to have excellent stability.

A modification of this method was reported in which levallorphan was used as an internal standard, and some simplifications to the tedious work-up procedure reduced the total preparation time [113].

A procedure using HFB esterification of pentazocine was reported to reduce the levels that can be determined (10–20 ng/ml) [114].

Naloxone has also been measured using extractive alkylation with pentafluorobenzyl bromide. Meffin and Smith [115] claimed a detection limit of 5 ng/ml but reported that the calibration graphs were non-linear. The use of desipramine to reduce adsorption losses as described by Brötell et al. [112] was not successful.

Measurements of amino alcohol metabolites of 1-α-acetylmethadol are of interest in the maintenance treatment of heroin addicts with acetylmethadol. Lau and Henderson [116] evaluated some derivatization reagents and found that trichloroacetyl chloride was the best choice for quantitation in complex biological fluids. However, there was no reagent that gave maximum responses for all the metabolite derivatives studied, indicating again the importance of the overall structure in determining the electron affinity properties.

Indicine and its N-oxide are pyrrolizidine alkaloids that are of interest because of antitumor activity. Both can be analysed by GC–ECD after conversion to PFP derivatives [117]. The N-oxide, however, must be reduced to indicine as otherwise polymerization will occur during the acylation reaction. Although more than one PFP group was introduced the overall sensitivity of the method was not especially high, but from a selectivity point of view the method appeared to have advantages.

Some phenothiazine derivatives have a hydroxyl group in the side-chain, which make these tertiary diamines amendable to reaction with the usual acylating agents. Larsen and Naestoft [118] described such a procedure for perphenazine and fluphenazine. The HFB derivative was prepared and permitted the determination of 10 ng/ml. To determine the sulphoxide silylation was preferred to acylation [119]. Cooper et al. [120] separated perphenazine, its sulphoxide and some phenolic metabolites after silylation. Detection was by ECD but the sensitivity on-column was considerably lower (ca. 20 ng).

Vitamin B_6 derivatives such as pyridoxine, pyridoxal and pyridoxamine have been analysed as HFB derivatives with ECD, permitting 10 pg to be detected [121]. The stability was acceptable except for the pyridoxal derivative, which had to be converted into its methyl acetal.

Haloperidol and trifluperidol are butyrophenones with potent neuroleptic properties. They have been analysed without derivatization by ECD in the nanogram range, despite the amino alcohol structure [122,123]. The response is probably not due only to the halogen content as the *p*-fluoroacetophenone moiety might be suspected to contribute. Forsman et al. [124] reported a method in which a chloro analogue was used as an internal standard. The levels in patients' sera varied between 0.5 and 10 ng/ml. Only thioridazine and dibenzepim were found to interfere among a number of drugs often administered simultaneously.

13.7. QUATERNARY AMMONIUM COMPOUNDS

Gas chromatography is not the first method of choice for drugs of this kind as they have to be degraded in some way prior to analysis. However, some applications were described during the 1970s. The first, described by Vessman et al. [125] for the *gem*-diphenyl derivative emepronium, involved oxidative conversion to benzophenone. The extraction was based on ion-pair partition with perchlorate as counter ion and the degradation was effected by chromic acid oxidation. The detection limit was ca. 1 ng/ml. With another oxidizing agent, barium peroxide, in combination with small extraction and reaction volumes, concentrations in the range 0.2–0.8 ng/ml could be detected [126]. This approach was also reported for the chromic acid oxidation to give acceptable results in samples containing 0.5 ng/ml of drug [50]. Oxyphenonium is an anticholinergic drug which, after extraction as a perchlorate ion pair, was hydrolysed to cyclohexylphenylglycolic acid. This was esterified with PFBz bromide in an extractive alkylation reaction that permitted the determination of 2 ng/ml [127]. The alkylation of the tetrapentylammonium ion pair was performed in dichloromethane and had the advantage of requiring only a small excess of the reagent, which could be removed by evaporation.

Another approach for the determination of quaternary compounds was proposed by Pohlmann and Cohan [128], who utilized the thermal degradation of quaternary ammonium iodides to methyl iodide, the latter being detected at levels down to as low as 10^{-14} mole. Neostigmine and pyridostigmine were used as model compounds to study the application of the method to serum samples. Hofmann degradation is often a useful approach for converting quaternary ammonium compounds to volatile fragments. Haloperidol, droperidol and penfluoridol (all tertiary amines) have been analysed in this way after a methylation reaction to produce the quaternary ammonium ion. The fragments gave characteristically more symmetrical peaks and shorter retention times than the parent butyrophenones [129]. Although no ECD applications were reported, these would be possible as both fragments contain electrophore elements (cf., the parent compounds).

Bretylium tosylate, *o*-bromobenzylethyldimethylammonium tosylate, is an antiarrhythmic agent that recently was analysed by GC–ECD. The procedure was based on Jendens' procedure for degrading acetylcholine with sodium benzenethiolate [130]. The *o*-bromobenzyl phenyl thioether gave a response to as little as 1 ng/ml of bretylium. An iodo analogue was used as an internal standard.

13.8. AMINO ALCOHOLS

This group includes many compounds of great therapeutic interest, which also is reflected in the number of publications on, e.g., the β-blocking drugs.

Cummins and Fourier [13] reported the formation of a mono-HFB derivative for the determination of pseudoephedrine and related compounds. There were some differences in the responses between these derivatives. Pseudoephedrine and norephedrine were not separated by this procedure. Lin et al. [131] modified the procedure and were able to detect 0.7 ng.

Midha et al. [132] presented a method for the determination of ephedrine and norephedrine in plasma and urine. The derivatization reagent was PFB chloride, which gave an N-PFB derivative after an alkaline wash that hydrolysed the O-acyl function. This derivative was nearly 15 times more sensitive than the N-HFB compound. Pentane extraction gave fewer interfering peaks than the more polar solvents. The method was capable of determining 2 ng/ml.

Clarke et al. [12] made interesting observations on the ECD response of perfluoroacylated diols and amino alcohols. It was found that bis-TFA derivatives of 1,2-diols exhibited about the same sensitivity as the corresponding bis-HFB compounds. For the TFA derivatives the response decreased in a regular fashion with increasing number of methylene groups between the trifluoroacylated groups. The strong response of bis-TFA-ephedrine was reported by Walle and Ehrsson [10]. As little as 2 pg could be detected, compared with about 1 pg for the bis-HFB derivative. This observation was confirmed in the first application of the gas chromatographic determination of a β-receptor blocking agent, alprenolol, as reported by Ervik [133]. Practolol, the first selective β-blocking agent, which contains a *p*-acetamido group, was determined as a tris-TFA derivative after the isolated drug had been deacylated [134]. The detection limit in samples was 5 ng/ml, whereas 25 pg could be detected on-column.

Propranolol was first determined as a bis-HFB derivative by Di Salle et al. [135], but the advantage of using bis-TFA derivatives was later pointed out by Walle [136] in a study covering most of the early β-blocking drugs. Concentrations down to 0.1 ng/ml were detectable. Using amine catalysis the reaction was complete in less than 5 min. Applications were later reported for the determination of propranolol and some of its metabolites in brain tissue of mice [137].

Kates and Jones [138] claimed that PFP derivatives were to be preferred with respect to speed of analysis and removal of contaminant peaks. Although HFB derivatization has been claimed to be necessary for some structures, e.g., S-2-(3-*tert*.-butylamino-2-hydroxypropoxy)-3-cyanopyridine [139], several applications have now been published in which TFAA has been used, for instance with oxprenolol [140,141], metoprolol [142,143] and befunolol [144]. Atenolol, a hydrophilic compound, has been the subject of many studies using HFBA [145–147]. The advantage of a TFA reaction in this instance was clearly demonstrated by Ervik et al. [148], who obtained much cleaner chromatograms from the same extract after TFA treatment. Timolol, a very potent drug, was quantified as a bis-HFB derivative using HFBI as reagent [149]. This drug has also found an interesting application in the treatment of glaucoma. To monitor systemic absorption after ocular application a modified method able to measure 1 ng/ml was proposed [150].

Bufuralol, a benzofuran derivative with a *tert*.-butylamino side-chain, was found to decompose upon HFBA treatment (dehydration) [151]. In this case the N-PFP-O-trimethylsilyl derivative was prepared, with a detection limit of 20 pg or 10 ng/ml. A mass fragmentographic assay was needed for studying the later samples in pharmacokinetic studies. Acebutolol was also reported to undergo decomposition when derivatized with TFAA or other TFA donors [152], probably owing to thermal instability. A stable trimethylsilyl-TFA (TMS-TFA) derivative with a moderate response to the ECD was prepared with the TFA exchanging for a TMS at the amino nitrogen.

Nofedone has the characteristic amino hydroxylated "β"-chain and has antiarrhythmic properties. Heusse et al. [153] used HFBA as the derivatizing reagent and reported levels down to 5 ng/ml. The sensitivity was found to increase with the number of biological extracts injected, indicating that the column performance varied owing to some saturation effect.

Diols of the 1,2- and 1,3-type have been studied as boronates using boronic acids containing electron-capturing substituents. Poole et al. [154] investigated the use of this approach with 2,4-dichlorobenzeneboronic acid as a reagent for the β-blocking drugs and found that the derivatives were formed in the injection zone by an on-column transboronation reaction. Reagent impurities interfered in the chromatograms. This method would be suitable for screening purposes above trace levels owing to the very simple work-up procedure.

Another possibility, which to date has been reported only for the determination of phenylpropanolamine, is reaction with pentafluorobenzaldehyde (or its bisulphite addition compound) to form a pentafluorophenyloxazolidine derivative [155]. This structure was tentatively claimed for the reaction product. The advantages over other reactions are that the derivative can be formed directly in the plasma sample and that the stability is excellent. Being a pentafluorophenyl-substituted compound, a high ECD response was found. Calibration graphs were constructed for concentrations down to 1 ng per 3 ml.

Ethambutol is an important drug in the treatment of tuberculosis. It has the structure of a bis vicinal amino alcohol, and it was therefore not surprising that an ECD procedure using TFAA as reagent was proposed [156]. Using pyridine catalysis the reaction was complete after 2 h. Excess of reagent was removed by washing with an acidic aqueous solution, like the procedure used for mono-HFB derivatization of ephedrine. The structure of the ethambutol derivative was not defined but in plasma as little as 100 ng per 0.2 ml could be detected.

A recent modification omitted the treatment with acid and instead removed TFAA by evaporation [157]. As a result the sensitivity increased at least 10-fold. This was partly attributed to less decomposition of the derivative.

13.9. HETEROCYCLIC COMPOUNDS

Theophylline (1,3-dimethylxanthine) is used in the treatment of asthma, and gas chromatography after an alkylation reaction has often been used for its pharmacokinetic evaluation. Arbin and Edlund [158] used PFBz bromide in an extractive alkylation procedure, with theobromine as an internal standard. The hydrophilic compounds were isolated by column extraction and then back-extracted into an alkaline aqueous solution. After addition of tetrapentylammonium hydroxide the alkylation of the ion pair in methylene chloride was complete after 10 min. Rat brains and plasma samples were analysed with recoveries of 100 ± 2.5% at levels of 200 ng per sample of brain. At the 20-ng level the recovery was 100 ± 8%. Owing to the protolytic properties of the derivative it could be purified by an extraction step. The PFB derivatives chromatographed well on XE-60 but after prolonged use the columns exhibited adsorption properties, that were approximately inversely related to the polarity of the solvent used for injection. Schwertner et al. [159] analysed plasma for theophylline using solvent extraction with salting-out before derivatization with PFB chloride. The derivative had an MDA of 50 pg, giving a practical detection limit of 10 ng/ml.

Two other xanthine derivatives, proxyphylline and dyphylline, contain hydroxypropyl side-chain groups. The electrophore used in this instance for determination in capillary blood samples from children was the HFB group [160]. These derivatives were detectable in amounts of 0.9 and 0.3 pg, respectively.

5-Fluorouracil is used in the treatment of certain forms of cancer, and a sensitive method was recently reported for its determination in pharmacokinetic studies using a chloromethyldimethylsilyl derivative [161]. It was possible to detect 10 pg of 5-fluorouracil (20 ng/ml in plasma), which is a surprisingly good response for this type of derivative. The structure of the derivative was not elucidated. The excess of reagent was treated with ethyl acetate to prevent saturation of the detector by chlorinated by-products.

A similar approach was used by Stadler [162] to measure thymine and uracil. The content of DNA could be related to the amount of thymine present in the sample. The gas chromatographic procedure involved switching techniques with two capillary columns, which had the advantage that the excess of chloromethyl silylating reagent did not have to be removed. The short capillary pre-column allowed nearly splitless injection of large sample volumes. Five picograms of a thymine standard were detected, equivalent to 0.4–2 ng of DNA.

Goitrin (1,5-vinyloxazolidine-2-thione) is similar to thiouracil in antithyroid activity and a method for its determination in food was reported by McLeod et al. [163]. Reaction with HFBA in the presence of trimethylamine resulted in a mono-HFB compound that chromatographed well and exhibited satisfactory sensitivity in several detectors, the ECD being the most sensitive.

Pemoline (2-amino-5-phenyl-2-oxazolin-4-one), a mild stimulant, could be determined in biological fluids using an unusual conversion to 5-phenyl-2,4-oxazolidindione, which had a sufficient ECD response [164]. The selectivity was the important aspect in this instance, the detection limit being 0.1 μg/ml in urine or plasma and 0.5 ng for the pure derivative.

Sulfinpyrazone, originally introduced as a uricosuric agent, has recently been shown to be of value in certain thromboembolic conditions. A selective and sensitive method based on ECD was reported recently [165]. This compound had to be methylated in order to obtain a single chromatographic peak. A flash-alkylation procedure was used with injector temperatures above 270°C. Under these conditions the sulphoxide moiety undergoes an elimination reaction and forms a compound structurally related to phenylbutazone. This compound (an anti-inflammatory agent), which has electrophore properties, has been determined in biological material using ECD. Sioufi et al. [166] described an assay in human plasma with a sensitivity of 10 ng/ml. The simple work-up procedure with the non-derivatized sample resulted in very clean chromatograms.

Clonidine, an imidazolidine compound, is a very potent drug for the treatment of hypertension and Chu et al. [167] described a method for its determination in biological fluids. Reaction with HFBA resulted in a bis-HFB derivative, the structure of which was not fully elucidated. Owing to interferences produced by the HFBA treatment, a purification step using a short Pasteur pipette filled with silica gel was introduced. This resulted in extracts that were clean enough to detect 25 pg/ml in a 4-ml plasma sample. At 60 pg/ml the relative standard deviation was 8%. The derivative was sensitive to adsorption on the support, but this could be decreased by vapour-phase silanization of the column.

Edlund [168] used pentafluorobenzylation to improve the detector response of clonidine. The derivative formed was probably a cyclized form of the principal reaction product. The proposed structure was in keeping with the observed high ECD response, which increased with higher temperatures. The first step in extraction from plasma used silica columns, capillary columns being used for the final separation. The method was applied to samples down to 100 pg/ml and a 200-μg dose of clonidine could be monitored for 24 h.

An antiallergic compound, 3-(5-tetrazolyl)thioxanthone 10,10-dioxide, was determined after reaction with HFBA [169]. The tetrazolyl ring was converted into a 1,3,4-oxadiazolyl analogue with a heptafluoropropyl side-chain. The response of this derivative was not competitive with that of the HFB derivatives as only 1 ng could be detected and concentrations in the μg/ml range were reported. The chromatograms from plasma and urine showed few disturbances, but faecal extracts did.

Pilocarpine is used to lower intraocular pressure. As it is an imidazole derivative, acylation with HFBA was used in an attempt to determine the drug in 50–100-μl samples from the aqueous humor of rabbits [170]. A mono-HFB derivative was formed, the structure of which was not fully elucidated as the exact position of the acyl residue in the imidazole ring was not assigned. The response was excellent, the detection limit being about 10 pg.

The antifungal agents clotrimazole and miconazole are applied topically to treat certain infections. To monitor the penetration of the drugs from formulations ECD was used as the drugs contain one and four chlorine atoms, respectively [17]. Clotrimazole was determined at levels down to about 10 ng per 2 mg of skin. Miconazole was injected together with tetramethylammonium hydroxide in methanol, which improved the chromatographic properties, but probably a breakdown product was eluted.

Iodochlorhydroxyquin or clioquinol, used for treatment of amoebiasis, was determined in biological material using an ECD procedure. Jack and Riess [172] acetylated the phenol and were able to determine 50 ng/ml. A later procedure used extractive methylation with tetrahexylammonium at pH 13 [173]. The methods gave results that agreed fairly well. Clioquinol and chloroquinaldol were determined by a modified procedure by Hartvig and Fagerlund [174]. It was claimed that the methyl derivative of clioquinol was unstable and that the drug itself showed poor stability in plasma. The reaction conditions were modified to include the less lipophilic tetrabutylammonium ion and a lower pH of 11, which was a compromise between stability and reactivity. The derivatives had an excellent sensitivity (0.5 pg detectable) and permitted the determination of 5 ng/ml in plasma.

Chloroquine, used in treatment of malaria, has a quinoline ring system with a tertiary amine side-chain attached to the aromatic amine. Analysis of biological materials such as urine with ECD using a single extraction was reported with a detection limit of a few nanograms [175].

Nitromethaqualone, a non-barbituric hypnotic with a quinazolinone structure, is an excellent electrophore due to the nitro group. Van Boven et al. [176] described a procedure involving a simple extraction and determination by ECD, levels down to 1 ng/ml in blood being measured without difficulty. The non-nitro compound methaqualone was apparently less sensitive. De Silva et al. [177] determined a monochloro-substituted benzothiazepinone in biological fluids. The compound, which also had a tertiary amino group, had a sufficient ECD response to permit the determination of 15 ng/ml in a 4-ml plasma sample (10 ng of pure compound). Adsorption problems were reduced by adding diethylamine to the solution to be chromatographed.

Griseofulvin has been the subject of many bioanalytical papers. Shah et al. [178] measured the antibiotic compound in skin using diethyl ether extracts. The method was applied in biopharmaceutical studies by Jamali and Axelson [179]. Subnanogram amounts of this monochloro-substituted gris-2-en-3,4′-dione were easily detected. Schwarz et al. [180] stated the detection limit to be 30 pg and determined 25 ng/ml in human plasma. A recent publication described in addition the determination of the 6-desmethylated metabolite [181]. To be able to perform the analysis simultaneously the metabolite was butylated with diazobutane. The sum of the metabolite and griseofulvin was also measured in a procedure using diazomethane.

Furylfuramide is a fungicide and its measurement in food is of interest. As it is a nitrofuryl derivative, ECD was a natural approach. The acrylamide function was converted into a nitrile in a reaction with HFBA and used to improve the chromatographic performance [182]. The detection limit was stated to be about 1 ng.

Warfarin, a hydroxycoumarin derivative with important use as an anticoagulant, has been analysed by GC after various derivatization procedures. Deckert [183] proposed that ECD of the non-derivatized molecule would be possible but that reaction with TFA would be a better choice. Odam and Townsend [184] presented a method in which the hydroxyl group was methylated with diazomethane. In animal tissues the detection limit was 5 ng/g. Treatment with diazomethane gave a constant proportion of the 2- and 4-methyl derivatives, with the 4-isomer having the better response. Reaction with HFBA and its smaller homologous anhydrides resulted in an intramolecular dehydrated product, which also had a good ECD response.

Kaiser and Martin [185] used thin-layer chromatography (TLC) to purify the extract before forming the PFBz derivative in acetone with potassium carbonate as a catalyst. Down to 20 ng/ml in human plasma could be determined.

Related to warfarin is acenocoumarin, which contains a nitrobenzyl moiety. A procedure similar to that of Kaiser and Martin using a PFBz derivative was applied to human plasma samples, although the structure would suggest that the use of such a derivative is unnecessary. However, no peak was found for the underivatized compound. The detection limit was stated to be 0.5 ng and the lowest level detected in plasma was about 10 ng/ml [186].

8-Methoxypsoralen has been used for the treatment of psoriasis after oral administration and exposure to UV light. Ehrsson et al. [187] made a thorough study of the conditions necessary for determining this electrophore in human plasma. A purification step made use of opening and closure of the lactone. The conjugated structure of the 8-alkoxypsoralens gave the compound a high ECD response with an MDA of 6 pg at 200°C. The detection limit was 10 ng/ml. Higher temperatures decreased the response. Schmid and Koss [188] omitted the purification step and claimed that extraction from an acidified sample had the same effect. Blanks were observed, however, corresponding to 5–10 ng/ml.

A recent paper independently described a similar procedure with a single extraction and claimed that 5 ng/ml could be detected [189]. Dihydro-8-methoxypsoralen was used as an internal standard, whereas the other workers used the 5-methoxy analogue. Diphenadione, a prothrombopenic agent, has an indandione structure which would indicate some electron-capturing ability. However, owing to chromatographic difficulties, a method was developed in which a chromium oxide oxidation converted the compound into benzophenone [190]. For plasma samples the limit of detection was 10 ng/g.

Nifedipine is a dihydropyridine derivative used to control hypertension and angina pectoris. Its determination is simplified by the presence of the nitro substituent, which gives the molecule good electrophore properties. On the other hand, it is easily oxidized to the corresponding pyridine derivative. Moreover, the compound is sensitive to light, forming a nitroso derivative. Jakobsen et al. [191] found the latter to be insensitive to the ECD, whereas nifedipine and the pyridine analogue could be determined at the 1 ng/ml level. Kondo et al. [192] determined nifedipine and two of its metabolites (carboxylic acids or lactones) and stated that the pyridine analogue was not formed metabolically, which was contrary to the report of Jakobsen et al. Kondo et al. deliberately oxidized the dihydropyridine to facilitate the handling of the compound in the assay [192].

Another dihydropyridine derivative, nicardipine (YC-93), was also determined as a pyridine compound [193]. A dihydropyridine derivative with a 2,3-dichlorophenyl substituent (felodipine, H 154/82) was determined by Ahnoff without conversion to a pyridine compound [194]. Although less light-sensitive, oxidation to the pyridine compound could occur if the work-up procedure and the chromatographic system were not optimized. The pyridine was found to be a normal constituent in samples from some species, however.

13.10. CATECHOLAMINES AND RELATED COMPOUNDS

The catecholamines are important transmittor substances of biomedical interest. Owing to the high sensitivity that can be obtained with an ECD after appropriate derivatization much attention has been paid to this technique. However, the methods of choice include radioenzymatic methods, liquid chromatography with electrochemical detection and GC–MS. However, some papers of great interest involving the ECD have been published and these will be briefly considered.

The first application of ECD to the determination of dopamine was reported by Clarke et al. in 1967 [195]. The extremely high sensitivity of the tris-TFA derivative allowed the detection of nanogram amounts after isolation from the sample by adsorption to alumina.

Kawai and Tamura [196] used the trifluoroacetate derivatives of norepinephrine and dopamine for their detection in urine. Isodrin was introduced early in the method as an internal standard and reportedly gave more reliable results. The urinary output of these endogenous compounds in patients differed from those in normal persons.

Sakauchi et al. [197] proposed the use of HFB derivatives of catecholamines for their determination in biological materials as they were more sensitive to ECD than the TFA and TMS-ether HFB derivatives. There seem to have been no applications of this procedure.

Martin and Ansell [198] reported a procedure for the determination of dopamine and norepinephrine in rat brain. The catecholamines were isolated by adsorption to alumina and further purified by ion-pair extraction of the hydrophilic compounds with di(ethylhexyl) phosphate. The TFA derivatives were prepared in acetonitrile, which required only 5 min at room temperature, whereas previous reactions were performed for 30–45 min in ethyl acetate. Down to 3 pg of the derivatives could be detected, which made the practical sensitivity about 5 ng per sample.

Arnold and Ford [199] preferred the PFP derivatives for the determination of catecholamines in brain tissue. The alumina was lyophilized prior to elution, which reduced the losses and provided anhydrous conditions for the derivatization. PFP derivatives were chosen owing to their to stability, ECD sensitivity and chromatographic properties. Interestingly, the aliphatic methyl ether derivative of norepinephrine was more sensitive than the fully acylated derivative; 50 ng/g could be determined.

Dopa decarboxylase activity in tissue was monitored by analysing in vitro preparations for dopamine [200]. Dopamine was determined in picogram amounts. This approach was claimed to give more specific information than previously available methods.

Dopamine and norepinephrine were determined in tissues and organs by an improved ECD method by Kawano et al. [201]. The procedure used PFP derivatives of these compounds and the α-methylated analogues as internal standards. The eluates from alumina were lyophilized before acylation. The concentrations in rat tissues varied from 13 to 1400 ng/g.

Bigdeli and Collins [202] determined, in addition to catecholamines, tetrahydroisoquinolines such as salsolinol and tetrahydropapaveroline. The latter are postulated as in vivo cyclization products of catecholamines with aldehydes. Salsolinol was found to be extremely sensitive to the ECD as an HFB derivative with an MDA of 0.2–0.5 pg. For dopamine the MDA was 1–3 pg and for norepinephrine about 20 pg. Tetrahydropapaveroline, with a dihydroxybenzyl group instead of a methyl group, was surprisingly less sensitive (about 200 pg). For this compound mass fragmentography was recommended.

A study of the selective determination of salsolinol using the above-mentioned method was presented by O'Neill and Rahwan [203]. They claimed 10 pg as the on-column detection limit with a lower level of 5–10 ng/g in brain tissue.

Schiff base formation with pentafluorobenzaldehyde was proposed by Moffat and Horning [204] for primary amines. For epinephrine, a secondary amine, two tetrahydroisoquinolines were formed. Hydroxyl groups were silylated after the Schiff base reaction. Down to 10 pg of each derivative could easily be determined.

Using this approach, Lhuguenot and Maume [205] determined six primary amines in samples such as rat adrenal extracts. The pentafluorobenzylimine was formed first. Care was taken to optimize the reaction with respect to both the aldehyde and silylating reagents. The stabilities of the derivatives were excellent. The method was compared with a GC–MS procedure, which was concluded to be more specific.

Important catecholamine metabolites are the O-methylated derivatives such as metanephrine. Several studies on these compounds have been reported. Änggård and Sedvall [206] studied the formation and stability of the perfluoroacyl derivatives. The stabilities of PFP and HFB derivatives were superior to those of TFA derivatives. The ECD response was found to be very dependent on the presence of acylated hydroxyl groups. When only a primary amine was available for reaction the derivative sensitivity was 10 times lower. Down to 20 pg of the metabolites could be measured. The requirements for a mass fragmentographic assay were discussed.

An application of these observations to human urine was reported by Bertani et al. [207]. Nelson et al. [208] determined normetanephrine, metanephrine and 3-methoxytyramine in urine using the bis-N,O-PFP-β-O-ethyl derivatives of the first two catecholamines and the bis-N,O-PFP derivative of 3-methoxytyramine. Purification was achieved by ion-exchange chromatography and ion-pair extraction with di(ethylhexyl) phosphate. The β-hydroxyl group was ethylated with ethanolic hydrogen chloride. An interfering compound in this procedure was reported later to be *p*-sympatol [209]. The method could be modified to exclude this interference by performing the alkylation with *n*-propanol instead of ethanol.

Phenylephrine has been used for many years as a vasoconstrictor. A method for its determination in biological fluids used TFA derivatization after isolation by ion-exchange chromatography [210]. The formation of a fully trifluoroacylated derivative was presumed, but the conditions for the reaction were reported to be sensitive to slight changes. About 25 ng per 2 ml of plasma could be easily determined.

Metabolites of amines such as amphetamine often include phenols. The determination of aminophenols can often be performed using approaches like those mentioned above. The TFA derivatives have the required sensitivity but are unstable. Belvedere et al. [211] used TFAA for *p*-hydroxyamphetamine and *p*-hydroxynorephedrine. Levels around 70 ng/g in brain tissue were determined with acceptable precision.

13.11. AMIDES

A few examples can be found where an amide function has been used to prepare a compound for ECD. Ehrsson and Mellström [30] studied the preparation and properties of some amide TFA and PFB derivatives. Primary amides were found to undergo dehydration to nitriles, a common reaction in drug derivatization. Both reagents were used together with high concentrations of trimethylamine in order to provide a reaction of practical speed. PFB derivatization of secondary amides required about 60 min. The PFB derivatives were preferred over TFA derivatives owing to their greater stability. The ECD response to compounds such as methyprylone, phenacetin and acetanilide was such that a few picograms gave symmetrical peaks well above the noise level.

Anthranilamide, a primary amide, has been determined in urine. Upon treatment with TFAA a cyclic derivative, 2-trifluoromethyl-3,4-dihydro-4-quinazolone, was formed [212]. A moderate ECD response was found and linear calibration graphs were obtained from 0.1 to 0.4 μg of derivative. The selectivity was excellent.

Carbamazepine and the structurally related dihydrooxazepine carboxamide (Squibb 10996) are used in the treatment of epilepsy. Attempts were made to determine these compounds using ECD after derivatization with HFBA. Both compounds were converted into nitriles without electrophore properties [213]. The results for carbamazepine reported by Schwertner et al. [214] were based on the reaction with PFB chloride at 120°C. No catalyst was used and GC–MS data indicated that the derivative was a bis-N-PFB compound. Compound 10996 was included as an internal standard. In comparison with the data of Ehrsson and Mellström the response seemed to be lower as the detection limit was stated to be 0.5 mg/l [214].

The dehydration of an amide to a nitrile was used to improve the chromatographic properties of an experimental compound, Kabi 1702, which had a saccharin-like moiety as an excellent electrophore [50]. These reactions are fairly straightforward and the removal of excess of reagent is easily accomplished by evaporation.

Rafoxanide is an anthelmintic agent with a salicylanilide structure. To follow the fate of the compound in plasma Talley et al. [215] formed the TMS derivative with bis-silylacetamide in large excess. This was especially important for the derivatization of the amide function. This derivative, in contrast to the mother compound, was stable to light. The presence of two iodine and two chlorine atoms rendered the derivative electron capturing with a detection limit of about 1 ng on a regular column. The use of a shorter column (10 cm) resulted in a better response (0.2 ng or 10 ng/ml), indicating that the bis-silylated amide acted as a silyl donor on-column.

A similar structural element is found in niclosamide, used in the treatment of schistosomiasis and also as an anthelmintic. An assay using extractive methylation was recently described by Churchill and Ku [216]. One secondary aromatic amide group and a phenol were methylated, with N-methylation being the slower process. Microgram amounts in urine were determined with a relative standard deviation of 2%. The sensitivity was not stated but from the clean chromatograms it seemed that lower levels could be determined.

13.12. IMIDES INCLUDING BARBITURATES

Although free barbituric acids were shown to possess some electron affinity [217], only recently have some practical applications to human plasma been reported [218]. Thiamylal and thiopental were methylated using flash alkylation with trimethylanilinium hydroxide. ECD was preferred to flame photometric detection because of its broader linear range and greater sensitivity. Thiopental was 10 times more sensitive than thiamylal, the minimum detectable amount being 10 ng. No degradation of the derivatives was found. Down to 100 ng/ml of thiopental could be determined.

Walle [219] studied the formation of PFBz derivatives of seven barbituric acids. The chromatographic performance was excellent and the ECD response was among the highest found, the bis derivative of barbital having an MDA of 0.05 pg. The mono derivatives were only slightly less sensitive. The detector response was only slightly temperature dependent (decrease of a factor of 2 over the range 220–290°C). The reaction was performed at 60°C for 50 min in methanol and it was observed that the concentration of triethylamine had to be more than 0.3 mole/l in order for the reaction to proceed to completion. Other conditions were less crucial. Diphenylhydantoin, an anticonvulsive drug with only one position for derivative formation, had an MDA of 0.25 pg.

The application of this procedure to serum samples was reported by Sun and Chun [220]. Pentobarbital gave a bis derivative upon reaction with PFBz bromide in ethanol–water at 85°C for 4 h. Excess of reagent was removed by evaporation to dryness. The MDA for this derivative was 0.2 ng and 100 ng/ml could be determined in serum.

Gyllenhaal et al. [221] used the same derivative for the determination of phenobarbital in saliva and confirmed the excellent response found by Walle. The derivative was formed in an extractive alkylation reaction at pH 9 with tetrabutylammonium as counter ion. To remove the excess of the strongly electron-capturing reagent a pre-column venting system was used by means of which the reaction mixture could be introduced directly into the chromatograph.

Dilli and Pillai [222] studied the response of 2-chloroethyl derivatives of ten barbiturates and five anticonvulsant drugs. For the barbiturates the MDAs ranged from 70 to 160 pg, whereas only diphenyl hydantoin among the anticonvulsant drugs gave a useful response. The reasons for the electrophore properties of this kind of derivative were discussed. An application of this principle to the determination of amobarbital, pentobarbital and phenobarbital in saliva was later reported [223]. The reaction was performed in ethyl acetate under reflux at 97°C for 5 min with bischloroethyl sulphate and triethylamine catalysis. Removal of excess of reagent by evaporation enabled 0.1–1 μg/ml of derivative to be determined.

Primidone can be regarded as a prodrug to phenobarbital and both are used as antiepileptics. Wallace et al. [224] used PFB chloride to form the stable bispentafluorobenzamide derivatives. The reaction was performed at 120°C for 20 min and excess of reagent removed under a stream of dry air. Phenobarbital and phenylethylmalonamide, both metabolites, did not react under these conditions. The detection limit was stated to be less than 10 ng/ml.

Phenylethylmalonamide was determined after treatment with TFAA at 120°C for 30 min, giving a derivative which left one amide group intact and the other dehydrated to a nitrile [225]. This structure unexpectedly showed a medium ECD response, making the selective determination of 1–30 μg/ml possible.

Ethosuximide is an anticonvulsant which Wallace et al. [226] determined as a derivative with PFB chloride. A high reaction temperature (160°C) was used and after 30 min and cooling excess of reagent was removed in a stream of dry air. The ECD response was not stated, but using 2 μl from 5 ml of final solution 50 μg per 0.1 ml of serum could be determined.

The last three methods all made use of internal standards, which were structural analogues and thus facilitated the performance of the integrated method.

13.13. CARBAMATES AND UREAS

There are relatively few drugs that have these structural elements. Owing to the thermal instability of these compounds the primary reason for derivatization is to make them stable. Introduction of suitable electrophore groups by such reactions has been used for trace analysis purposes.

Mellström and Ehrsson [227] described a procedure for trifluoroacylation of an experimental carbamate drug candidate (see Table 13.1, p. 340). Using trimethylamine as a catalyst a TFA derivative of the carbamate N–H bond with a reasonably good ECD response was obtained (MDA ca. 20 pg). The choice of stationary phase was critical, a propionitrile silicone being preferred. The electrophore properties of an aromatic carbamate were demonstrated by Lewis and Paris [228], who were able to measure 0.2–0.5 ng of carbaryl.

VandenHeuvel et al. [229] gave some examples in which carbamates were converted into unexpected products upon reaction with HFBA. The need to establish the structure of a given derivative was stressed.

Evans [230] studied the possibility of reacting some simple aliphatic ureas with TFAA. He concluded that a TFA group could only be introduced on a nitrogen that had both a free hydrogen and a substituent. Saunders and Vanatta [231] studied the reaction of TFAA with a trisubstituted urea with herbicidal activity. The ECD response was such that picogram amounts could be determined. Diflubenzuron was detected as a TFA derivative (MDA ca. 10 pg) by Worobey and Webster [232]. Later studies revealed that the correct structure of the derivative (a degradation product) was a perfluoroamide which was formed during derivatization [233]. Alternative derivatives for ureas that are more stable have been proposed, e.g., ethylation of the N–H bonds as described by Lawrence et al. [234].

13.14. HYDRAZINE-DERIVED COMPOUNDS

The analysis of hydrazine itself in various samples has been of interest owing to its toxicity. Liu et al. [235] trapped hydrazine from tobacco smoke by reaction with pentafluorobenzaldehyde to form the corresponding azine and claimed a detection limit for hydrazine of 0.1 ng per cigarette.

Drugs which are derivatives of hydrazine include hydralazine, which is used to treat hypertension. Jack et al. [236] presented a procedure in which the hydrophilic compound was reacted with nitrous acid in the plasma sample. The product, a tetrazolophthalazine (a heteroaromatic system), could be determined with ECD at concentrations of 10 ng/ml. This procedure does not differentiate between hydralazine and its hydrazone derivative, which led Degen [237] to use another approach, viz., reaction with 2,4-pentanedione directly in plasma. The determination made use of a nitrogen-selective detector, which was as sensitive as the above method [237].

Propildazine, which has vasodilating properties, wss isolated by ion exchange and then subjected to reaction with HFBA in ethyl acetate using trimethylamine as a catalyst [238]. The derivative, a 1,2,4-triazolpyridazine, had an HFB ester group and a heptafluoropropyl group on the triazole ring. The MDA for the derivative was 30 pg, but owing to interfering endogenous compounds plasma levels lower than 0.4 μg/ml could not be determined. This was insufficient for monitoring plasma levels of propildazine in humans.

Isoniazid, an antitubercular drug, was reacted in the aqueous phase with *p*-chlorobenzaldehyde. Owing to the number of nitrogen atoms in the compound an alkali flame ionization detector was preferred, but the structure would be electron capturing [239].

13.15. GUANIDINES, BIGUANIDES AND RELATED COMPOUNDS

Guanido-containing drugs such as guanethidine, guanoxan, debrisoquin and bethanidine are used in the treatment of hypertension. Bioanalytical methods are of importance in controlling variability in dose–response relationships. Hengstmann et al. [240] hydrolysed the guanidines to the corresponding primary amines and reacted them with PFPA and HFBA. The ECD response was too low to determine therapeutic levels.

Erdtmansky and Goehl [241], working with a guanido-substituted metabolite of an amidoxime, found a simple approach to the determination of this group in biological fluids. The procedure involved a two-phase reaction with the aqueous biological sample and hexafluoroacetylacetone in benzene. The reaction took place at 100°C under reflux and resulted in stable cyclic derivatives. This simple, sensitive (25 ng/ml) and specific assay could be improved as suggested by the authors (the MDA was 5 pg). For guabenzodioxan this was found to be necessary, but the choice in this instance was GC–MS [242].

Pellizzari and Seltzman [243] used a hydrolysis procedure for guanethidine in plasma and determined the HFB derivative by two-dimensional GC. In this way 0.7 ng/ml could be determined but below this level the peak symmetry was poor. Without the second column the detection limit was 10 ng/ml.

Methylguanidine and guanidine were studied as the hexafluoroacetone condensation products [244]. The ECD response (as MDA) was reported to be 50 pg.

Diminazene is a bis-amidine-substituted compound with anti-infective properties. Condensation of its reduction product with hexafluoroacetylacetone gave derivatives with an ECD response but not sufficient for the low levels found after therapeutic doses [245]. GC–MS was used instead for levels down to 100 ng/ml.

Guanfacine, an amidino compound with antihypertensive activity, was also reacted with hexafluoroacetone to give a derivative with an MDA of 4 pg, corresponding to 125 pg/ml; 500 pg/ml could be determined with a relative standard deviation of 5% [246].

The experimental compound PF-257 is a psychotropic agent containing an acetamidoxime moiety. Reaction with HFBA resulted in a cyclic derivative, a 1,2,4-oxadiazole substituted with a heptafluoropropyl group [247]. The reaction was not quantitative but sufficiently reproducible to permit measurements in plasma at levels down to 0.1 ng/ml. The chromatographic performance was excellent, as was the ECD response (MDA = 1 pg).

Biguanides possess hypoglycaemic activity. Cyclization reactions have been presented for these compounds. Matin et al. [248] prepared the corresponding 1,3,5-*s*-triazines after reaction with monochlorodifluoroacetic anhydride. The monochlorodifluoro moiety made them sufficiently sensitive for the measurement of therapeutic levels (less than 1 ng/ml) of, e.g., phenformin, buformin and metformin. Another biguanide, moroxydine, which has a morpholine moiety, suffered from extraction problems [249]. The use of a large counter ion, bromothymol blue, partially solved this problem in that serum samples of less than 0.5 ml could be extracted after protein precipitation. The derivatization was then carried out as for phenformin. Alkalay et al. [250] investigated the use of other anhydrides for the reaction and separation.

Chlorhexidine, a widely used germicide, could be determined after hydrolysis to *p*-chloroaniline and conversion of the latter to *p*-chloroiodobenzene after diazotization

[251]. The recovery was low (ca. 25% at 50–200 ppm), but the number of steps was large. The ECD response was high (MDA ca. 5 pg).

13.16. SULPHONAMIDES AND SULPHONYLUREAS

The gas chromatographic determination of sulphonamides (important chemotherapeutic agents) without degradation was described recently. Gyllenhaal and Ehrsson [255] studied acylation reactions (TFA and HFB) and also alkylation with PFBz bromide using N-ethylbenzenesulphonamide as a model substance. All derivatives had a high ECD response, with a low response ratio between the TFA- and HFB-substituted compounds. This indicated that the sulphonamide group itself contributed to the electron affinity of the derivative when properly conjugated.

Bye and Land [253] made use of this in a method for the determination of sulphadiazine and its N^4-acetyl metabolite in plasma and urine. The sulphonamide group was methylated with diazomethane to give a derivative that chromatographed well and had an ECD response giving an MDA of less than 1 ng. Interferences, which were encountered at low levels in urine with the well known Bratton-Marshall procedure, were not observed.

Gyllenhaal et al. [254] continued their studies with sulphonamides using an extractive alkylation with methyl iodide and tetrabutylammonium as counter ion at pH 10. Twenty-four structurally related sulphonamides of therapeutic interest were tested and could be quantitatively alkylated in less than 100 min for most of the compounds. N^4-Acylated derivatives reacted more slowly than the parent compounds. Dichloromethane was usually used as the solvent but methyl isobutyl ketone gave a more rapid reaction. An application was described for the determination of sulphapyridine in serum [255]. With 10 pg as the minimum detectable amount and a 0.1-ml sample volume, serum levels of 1 μg/ml were determined with a relative standard deviation of 2.3% for sulphapyridine and 5.4% for the acetylated metabolite.

Nose et al. [256] tested dimethylformamide dimethyl acetal as a derivatizing reagent on 14 compounds. From the mass spectra it was clear that a dimethylacetal was introduced into the N^4-position and a methyl group into the N^1-position [256]. The response was sufficiently high to permit the determination of 10–20 ng/g, depending on the structure. The chromatographic performance was excellent.

An ECD and a liquid chromatographic (LC) method were compared for the determination of sulphapyridine and its acetyl analogue [257]. Flash heater methylation for preparing the derivative for ECD was tried but gave several peaks, so diazomethane was used instead. The LC method was less sensitive than the GC method (50 ng on-column compared with 10 pg), but as the therapeutic concentrations are high enough (20–50 μg/ml) the LC method was preferred owing to its speed of analysis.

Saccharin has a sulphonamide structure which after N-alkylation chromatographs well and has a high ECD response. Daun [258] reported the determination of saccharin in biological material using TLC purification, N-methylation with methyl iodide in dimethyl sulphoxide and ECD. The use of diazomethane resulted in a mixture of N-methyl and O-methyl derivatives, the latter of which had a much greater response. The proposed procedure was applied to samples such as blood, urine, faeces and tissues with reliable quantitation at levels below 0.1 μg/g.

Hartvig et al. [259] applied extractive methylation to the determination of saccharin in urine. As the reaction rate was considerably slower (1 h was required) than for most other sulphonamides a high concentration of methyl iodide was used. Excess of reagent was removed by evaporation and residues of tetrabutylammonium iodide (formed during the reaction) were removed by extraction with silver sulphate solution. Without the latter step degradation products from the quaternary compound occur as a tailing front on the chromatogram. With this method 10 ng/ml could be determined. Saccharin had a high ECD response, the MDA being 2 pg.

Chlorthalidone belongs to a group of sulphonamide derivatives that are used in the treatment of hypertension because of their saluretic activity. Gas chromatographic methods were not used until these polar molecules could be converted into hydrophobic derivatives. A study by Ervik and Gustavii [260] clearly showed the great potential of extractive alkylation for the determination of trace amounts of compounds such as chlorthalidone in plasma. The partition ratio of the drug as an ion pair with tetrahexylammonium as counter ion into dichloromethane is low, but owing to the continuous consumption of the extracted fraction in the alkylation process, the overall reaction did not require more than 50 min at 20°C. A few picograms gave symmetrical peaks which allowed the analysis of plasma samples containing less than 1 ng/ml. The compound had to be extracted from plasma before the alkylation process, as direct reaction resulted in interferences from endogenous substances and consumption of the participating reagents.

Degen and Schweizer [261] applied this technique to (neo)sulfalepsine and chlorthalidone. The use of N-butylchlorthalidone as an internal standard prior to any manipulations improved the method considerably. The major part of chlorthalidone in blood is found in the erythrocytes, which can create problems if the sampling procedure is not under control, as pointed out by Fleuren and Van Rossum [262] using the same procedure but with nitrogen-selective detection.

Lindström and co-workers [263,264] applied the extractive alkylatioen principle to the determination of furosemide and hydrochlorothiazide in various biological fluids, whereas VandenHeuvel et al. [265] used on-column methylation for the latter drug. In both instances tetramethyl derivatives were formed and could be detected in plasma at levels of about 10 and 50 ng/ml, respectively, but the extractive alkylation technique involved fewer steps. VandenHeuvel et al. reported that this electrophore drug did not give useful derivatives with either diazomethane or silylating reagents. However, with proper control of the trimethylanilinium hydroxide concentration (0.005–0.01 *M*) a single peak was observed. They used a bromo analogue as an internal standard, as did also Redalieu et al. [266] in a modification of the method of Lindström et al. This allowed levels as low as 5 ng/ml to be determined.

Acetazolamide has been used mainly in the treatment of glaucoma. A method for its determination in blood, plasma and saliva made use of flash methylation with trimethylanilinium hydroxide after solvent extraction from the sample [267]. Levels down to 100 ng/ml in plasma and 5 ng/ml in saliva could be determined. Fagerlund et al. [268] applied the extractive methylation approach directly to serum samples containing acetazolamide. Their study included ten other sulphonamides with related structures, all of which formed trimethyl or tetramethyl derivatives. Furosemide and bumetanide were reported to give two derivatives. It was also observed that if the quaternary ammonium iodide was not removed entirely the methyl ester of furosemide was converted into either the butyl or the hexyl ester. The ECD response was as high as reported for the previous compounds. The response for the trimethylacetazolamide derivative was shown to be dependent on temperature as it increased about 4-fold from 180 to 350°C.

VandenHeuvel et al. [269] determined a bissulphonamide with anthelmintic activity using extractive methylation. The compound gave multiple products with trimethylanilinium hydroxide, but with methyl iodide and tetrabutylammonium as counter ion conversion to a hydrophobic derivative with excellent ECD response was achieved (MDA 20 pg).

Sulphonylureas such as tolbutamide and chlorpropamide are used as hypoglycaemics. First attempts to determine them by ECD used the hydrolysis products, propyl- and butylamine, followed by reaction with 2,4-dinitrofluorobenzene. Matin and Rowland [270] reported successful determinations below 1 μg/ml.

Braselton et al. [271] studied the formation of thermally stable derivatives of some sulphonylureas. A two-step procedure using diazomethane to produce an N-methyl derivative (of the sulphonamide group) followed by acylation of the remaining N–H bond (TFA, PFP or HFB) was found to be suitable. In addition to their stability (even with tolazamide), the resulting derivatives showed a high sensitivity to electron-capture detection (MDA ca. 1 pg) and excellent chromatographic properties. The ECD response was suggested to be a result of both the electron delocalized system of the aromatic sulphone and the perfluoroacyl group. An application of this interesting principle was reported for tolbutamide and chlorpropamide [272]. The results with the ECD method were confirmed by mass fragmentography, although the latter was about 10 times less sensitive. Close agreement was found, and validated the use of the simpler ECD method for routine work. In addition, derivatives of two tolbutamide metabolites could be determined.

Schlicht et al. [273] simultaneously analysed five antidiabetic sulphonylureas, e.g., tolazamide and glycodiazin, in blood using reaction with diazomethane followed by TFAA. The drugs were isolated in an adsorption step on XAD-2 and desorbed with ethyl acetate. This work-up procedure could be used for haemolized or putrid samples. Concentrations in the μg/ml range could be determined.

Hartvig et al. [274] extended their studies on extractive alkylation to sulphonylureas. Methylation of the sulphonamide nitrogen resulted in a derivative that was stable for the eight compounds investigated except tolazamide and glibornuride. The alkylation process had to be performed at low pH values in order to avoid hydrolysis. Application of the method to plasma samples was reported. Glipicide levels as low as 20 ng/ml could be determined owing to the high detection selectivity.

13.17. BENZODIAZEPINES

The general structure of the 1,4-benzodiazepines is sufficiently electron capturing to permit their determination in the low nanogram range in biological fluids. Drugs belonging to this class have been used because of their anti-anxiolytic, sedative, muscle relaxant and anticonvulsant activity. This wide range of therapeutic applications has resulted in a large number of papers on their bioanalysis. Gas chromatography with ECD has been the dominant technique from the beginning, the first paper being on diazepam analysis by De Silva et al. in 1964 [275]. Curiously, the title of this interesting contribution did not indicate that the ECD was used, nor did the abstract. The use of ECD in the determination of 1,4-benzodiazepines used as antiepileptics was reviewed by De Silva [276].

At the time of the first application of the ECD to the determination of diazepam only the ^{3}H detector was available as well as very few stationary phases with high temperature stability [275]. This favoured an approach using acid hydrolysis of the benzodiazepine structure to a substituted benzophenone, which was stated to increase the sensitivity 10-fold over that of the original compound. In this way both diazepam and the demethylated metabolite nordiazepam could be determined as 2-methylamino-5-chlorobenzophenone and 2-amino-5-chlorobenzophenone, respectively, at levels down to 20 ng/ml in blood. Problems with interferences from endogenous compounds co-extracted with diethyl ether required extensive washing of some extracts. In addition, different metabolites could give rise to the same benzophenone. With the advent of the ^{63}Ni ECD and the phenylsilicones the parent benzodiazepines and their 2-ones extracted into diethyl ether could be chromatographed and determined in nanogram amounts, as demonstrated by De Silva and Puglisi [277] for medazepam, its metabolite diazepam and other biotransformation products. Owing to the number of metabolites the direct or absolute calibration technique had to be used. A simplified procedure for the determination of diazepam and nordiazepam was presented by Berlin et al. [278], who used a single extraction into benzene from 100 μl of plasma mixed with the same volume of a saturated KCl solution. The organic phase was evaporated nearly to dryness and an external standard added. This method was used for bioavailability studies after repeated doses with levels from 200 to 400 ng/ml.

The need for simple extraction procedures and direct quantitation resulted in alternative methods, such as ethyl benzoate extraction [279] and a modified benzene extraction method [280].

For the determination of medazepam, Baird et al. [281] proposed the use of prazepam as an internal standard and a back-extraction for clean-up purposes. Concentrations down to 5 ng/ml in plasma could be determined.

Later Weinfeld et al. [282] modified their procedure for diazepam and nordiazepam to include benzene extraction and no clean-up procedure. With methylnitrazepam as an internal standard 10 ng/ml of nordiazepam in plasma could be determined with OV-17 as the stationary phase. This method was automated. An interesting observation was made that the nordiazepam response and its peak shape were greatly improved in extracts from biological samples. The lower level of detection was approximately 5 ng/ml. Wallace et al. [283] reported on the direct analysis of diazepam and nordiazepam using a stationary phase, SP 2250-DB, which was stated to be superior to OV-17, especially for nordiazepam. The method compared favourably with a liquid chromatographic procedure, which also had a limit of detection of about 50 ng/ml. This was also the case with a recent study by Raisys et al. [284]. The agreement was better for diazepam than for nordiazepam. They used a simple toluene extraction.

Clorazepate is a precursor drug to nordiazepam, and owing to its extreme hydrolytic susceptibility it has been analysed only as the latter compound. Hoffman and Chun [285] demonstrated the presence of unchanged clorazepate in dog blood by making an immediate extraction of nordiazepam and then hydrolysing the remaining clorazepate in the plasma sample to nordiazepam. The lower detection limit was 10 ng/ml. It was reported that clorazepate could not be demonstrated in man. Brooks et al. [286] presented a similar procedure to determine both compounds in human plasma. They were able to show that clorazepate had peak levels of about 20 ng/ml by immediately adding alkali to the drawn and oxalated blood sample. Nordiazepam was determined intact at levels as low as 4 ng/ml. Wretlind et al. [287] followed clorazepate levels as nordiazepam using an extractive methylation reaction giving diazepam.

Analysis for nitrazepam followed similar lines to those for the determination of nordiazepam. Thus the early procedures used the hydrolysis approach, making the determination of 0.1 ng possible [288]. Ehrsson and Tilly [289] applied the extractive methylation procedure. Using benzene as extracting solvent, which was necessary in order to obtain clean extracts as the acid-labile compound could not be back-extracted, 5 ng/ml could be determined with acceptable precision [289].

Kangas [290] compared the direct method with the hydrolysis approach. The methods were clinically comparable. However, the direct method gave satisfactory results only if the column had been treated by repeated injections of plasma extracts, in line with the observations made for nordiazepam. The direct method could determine 1 ng/ml and the acid hydrolysis method was more sensitive (0.2–0.5 ng/ml). In another study Kangas [291] compared nitrogen-selective detection with an ECD. The former responded similarly to nitrazepam, the 7-amino metabolite and its 7-acetamide derivative. With ECD only nitrazepam gave a useful response, about 60 times higher than with the nitrogen detector, whereas 7-acetamidonitrazepam gave the same response to both detectors. The low response for 7-aminonitrazepam seems to be the only one reported for a benzodiazepine structure [289] (see Table 13.1, p. 341).

De Boer et al. [292] used a capillary support-coated open-tubular (SCOT) column and were able to determine 1 ng/ml of the parent compound. A solid injection system was used and the stationary phase was Carbowax 20M.

In several of the applications for the determination of nitrazepam, clonazepam was used as an internal standard. It is therefore evident that this drug can also be determined according to this method [293–297]. De Silva and Bekersky [298] applied the extractive methylation procedure to the determination of clonazepam at levels down to 0.5 ng/ml. Benzene–methylene chloride (9:1) was used to extract the drug and the reference standard nitrazepam. As mentioned before, priming the column with biological extracts seemed to deactivate "reactive" sites and improve the peak shape and response. Flunitrazepam has a similar structure and could also be determined by a similar procedure. Faber et al. [299] reported a single extraction procedure with benzene and the determination of the parent compound using 8% SE-30 as the stationary phase. The detection limit was 1 ng/ml in serum [299].

Bromazepam was one of the first benzodiazepines to be determined by ECD. A hydrolysis procedure was used for concentrations down to ca. 85 ng/ml in blood [300]. Methods employing the parent compound were later developed [301] and also hydrolysis procedures with improved sensitivity (5 ng/ml), as reported by Cano et al. [302].

Oxazepam and lorazepam have structures that make them prone to a thermal rearrangement [303]. If nordiazepam with an amide N–H bond is hydroxylated in position 3 the resulting metabolite, oxazepam, upon heat treatment undergoes a ring contraction to a quinazoline carboxaldehyde. This compound at first created some chromatographic problems upon analysis at low levels. Early analyses for oxazepam and lorazepam were therefore performed via the acid hydrolysis route as described by Knowles and co-workers [304,305].

Both compounds are excreted mainly as glucuronides and the interference from metabolites is negligible. The oxazepam method had a detection limit of 20 ng/ml in biological fluids, compared with 50 ng/ml for the direct injection approach [304]. The lorazepam method was claimed to have a detection limit of 10 ng/ml [305]. An alternative method for the determination of oxazepam [306] optimized the extraction conditions and the hydrolysis to aminochlorobenzophenone, and used lorazepam as an internal standard. In a micro-scale procedure 500 pg could be detected and 0.8 ng/ml determined. From this it is clear that the ECD response is high, viz., 40 pg for the aminochlorobenzophenone on-column. Sánchez et al. [307] determined pivalyl esters of oxazepam and used the hydrolysis method, comparing ECD with GC–MS. Of interest was their observation that the TFA derivative of the hydrolysis product had an ECD response about 20 times higher than that of the free compound. The response of the HFB derivative was a further factor of 2 better. GC–MS was able to detect about 0.6 ng/ml compared with 7 ng/ml for ECD.

Even if the hydrolysis procedures resulted in clean chromatograms from oxazepam samples, it was attractive to determine the parent drug via its thermal degradation product. Giles et al. [308] were able to detect 10 ng/ml of oxazepam after a simple extraction into benzene–diethyl ether (5:1). The quinazoline carboxaldehyde was chromatographed on OV-17. Similar procedures were used for lorazepam by De Silva et al. [309] and Howard et al. [310], with a detection limit of about 5 ng/ml in blood. Greenblatt et al. [311] used oxazepam as an internal standard in this procedure, with benzene–1.5% isoamyl alcohol as extracting solvent. The simple extraction and the minimal clean-up required were accompanied by low coefficients of variation, e.g., 5.1% at 2.5 ng/ml. The addition of isoamyl alcohol and the use of oxazepam as an internal standard were claimed to make an important contribution to this low figure.

De Silva and Puglisi [277] used silylation to make possible the determination of intact oxazepam among diazepam metabolites. The same approach was reported not to yield reproducible derivatization for lorazepam [309]. On the other hand, Higuchi et al. [312] reported the use of the bis-TMS derivatives of both oxazepam and lorazepam in a mass fragmentographic assay capable of determining 2 ng/ml. Vessman et al. [313] used extractive alkylation to produce the N^1,O^3-dimethyl derivatives of oxazepam and the internal standard, lorazepam. The derivatives showed good chromatographic properties although high column temperatures were required. Down to 60 pg could be detected on-column, permitting at least 1 ng/ml to be determined. No metabolites interfered. It was also demonstrated that oxazepam was present in serum after single doses of both diazepam and clorazepate, which could not be demonstrated by the hydrolysis method and probably not without problems by the thermal rearrangement approach. The importance of control of the extractive alkylation conditions was emphasized as it was shown that a rearrangement to unidentified isomers occurred at too high a concentration of the quaternary ammonium hydroxide in the organic phase.

De Silva et al. [314] described a procedure in which most of the above-mentioned benzodiazepine derivatives could be analysed after grouping them according to their extractability and need for derivatization. Extractive methylation of the amide N–H bond was utilized, and also silylation of the hydroxyl groups.

Additional drugs with the benzodiazepine structure have been introduced and usually the above-mentioned principles for bioanalysis were used.

Pinazepam [315] and camazepam [316] were analysed as diazepam. An iodo analogue to flunitrazepam was analysed in its intact form. It was again observed that the extracts from biological samples improved both the peak shape and the sensitivity [317]. Imidazo-1,4-benzodiazepines were studied by Puglisi et al. [318].

A prodrug (so-called peptido-aminobenzophenone) giving chlorodiazepam as a metabolite was determined after conversion to an aminoquinolone in alkaline media [319]. This derivatization was performed directly in the plasma sample. The limit of determination was 9 ng per 0.5 ml of plasma.

A screening procedure for benzodiazepines in blood made use of an oxidation reaction with dilute chromic acid [320]. The products were not identified but the product profiles were characteristic of the commoner benzodiazepines present in nanogram amounts.

A 1,5-benzodiazepine, clobazepam, was determined both intact and as the N-desmethyl metabolite [321]. A simple benzene extraction was used, which with ECD permitted the determination of down to 10 ng/ml.

Chlordiazepoxide was early found to create chromatographic problems due to a thermal rearrangement, as outlined by Sadée and Van der Kleijn [303]. Recently an acetylation and rearrangement step of the N^4-oxide group was proposed and shown to improve the chromatographic performance greatly. An MDA of 50 pg was reported [322]. Only one paper has reported the temperature dependence of the ECD response for some benzodiazepines. Lindley [69] found that the detector response to diazepam decreased up to 360°C but was almost constant at higher temperatures. For flurazepam the ECD response was constant up to 300°C but decreased from 300 to 420°C [69]. This observation was stated to be consistent with non-dissociative electron capture. The nitro-substituted drugs clonazepam and desmethylflunitrazepam, when determined as the N-ethyl derivatives, showed little temperature dependence of the ECD response over the same temperature range.

13.18. NITRATE ESTERS

Nitro compounds were recognized early as being extremely sensitive to the ECD. It was therefore natural that the nitrate esters were among the first physiologically active compounds to be determined with ECD. Thus ethylene glycol dinitrate and nitroglycerin were determined in blood and urine by Williams et al. in 1966 [323]. Using a home-made detector they determined less than 40 pg of the dinitrate and less than 1 ng of nitroglycerin in hexane extracts. Conditioning the column by priming it with microgram amounts at regular intervals was found to be necessary. Litchfield [324] also determined the mononitrates of ethylene glycol and found their response to be about half that of dinitrates. Nanogram amounts were determined in blood after extraction with diethyl ether. Rosseel and Bogaert [325] determined nitroglycerin and isosorbide dinitrate in plasma. With 5 ml of sample and extraction into ethyl acetate they were able to detect 0.5 ng/ml of nitroglycerin with isosorbide dinitrate as an internal standard. Isosorbide ester could be determined at the 0.2 ng/ml level 80 min after sublingual administration using isoidide nitrate as the internal standard.

Neurath and Dünger [236], in addition to nitroglycerin, also determined pentaerythritol tetranitrate and their denitration products. The latter compounds were silylated to permit low-level determinations. The detection limit was claimed to be 0.1–2 ng/ml in blood, depending on the compound. The original drugs were not detected in blood after oral administration. To prolong the action of nitroglycerin, sustained-release formulations have been of great interest. To follow this biopharmaceutical development sensitive assays were needed. Givant and Sulman [327] used a method which, after a simple hexane extraction, allowed 50 pg to be determined, corresponding to about 0.2 ng/ml.

Isosorbide dinitrate has gained increasing interest in the treatment of angina pectoris, as demonstrated by the large number of publications on this topic. As the dinitrate disappears very rapidly from the blood many methods have been proposed to follow this decline. Malbica et al. [328] modified the above-mentioned method of Rosseel and Bogaert and claimed that 0.5 ng/ml in 1-ml samples of plasma could be determined, although the coefficients of variation were around 20%. The detector exhibited a slight response maximum at 175°C. Isoidide dinitrate was used as an internal standard.

An improved method was proposed by Laufen et al. [329]. A one-step extraction with pentane made possible the detection of 50 pg/ml. A similar procedure was described by Doyle et al. [330].

As the mononitrates of isosorbide have pharmacological activity and contribute to the total effect of the drug, it is desirable to measure their concentrations. Richard et al. [331] and Chin et al. [332] determined the intact mononitrates after extraction into diethyl ether. Chin et al. claimed that highly loaded SE-30 columns were needed in order to reduce adsorption losses.

Another approach was presented by Smith and Besic [333]. Extraction from plasma with diethyl ether and a clean-up step involving heptane–methanol partitioning was followed by conversion to the *tert.*-butyldimethylsilyl ethers to reduce the pronounced adsorptive properties. The choice of silyl derivative was made from both a separation and a stability point of view.

Using wide-bore glass capillary columns Rosseel and Bogaert [334] were able to determine both isosorbide dinitrate and its two mononitrates. The introduction of a capillary column increased the specificity considerably. With isoidide mononitrate and isomannide dinitrate as internal standards and ethyl acetate extraction the detection limits were 0.5 ng/ml for the dinitrate and 2 ng/ml for the 2-mononitrate. The 5-isomer was about 10 times less sensitive.

13.19. AMINO ACIDS

This important group of compounds were the subject of a very early study to evaluate the ECD for their trace level analysis. Landowne and Lipsky [335] used the N-2,4-dinitrophenyl (DNP) derivatives of the natural amino acids and found that, after methylation with diazomethane, they all gave the same molar response, even those which form bis-derivatives. They expressed the minimum detectable amount on-column in units of mole/s, which for the DNP derivatives were of the order of $3 \cdot 10^{-16}$ mole/s, corresponding to a few picograms of a particular amino acid. As derivatives were formed mainly with neutral and acidic amino acids this seems to have limited the interest in applying them to biological samples [336,337].

Some other derivatives for the study of amino acids and peptides with ECD were reported by Landowne [339], the most interesting being the cyclic imides with phthalic acid, which exhibited the same ECD sensitivity as DNP derivatives. In addition, they were more volatile than the heavy DNP derivatives.

The problem with the low volatility of the amino acid derivatives was solved by using N-TFA *n*-butyl esters, which were studied in detail by Gehrke and co-workers. Most of the studies dealt with samples for which high sensitivity was not the major problem. However, for the analysis of picogram amounts Zumwalt et al. used ECD [339]. For amino acid derivatives containing more than one TFA group the minimum detectable amounts were of the order of 2–5 pg (e.g., cysteine, serine and hydroxyproline). This is in line with the observations mentioned for bifunctional compounds such as amino alcohols. For derivatives of compounds such as alanine and phenylalanine the corresponding figure was about 50 pg. If the HFB derivatives were used the detection limits for a few amino acids were reported to be 1–2 pg. These figures were given for derivatives that were determined directly after derivatization, indicating problems with the stability of the derivatives in diethyl ether. One means of increasing the overall sensitivity of this technique was to use a solvent venting chromatographic system to allow the injection of 100-μl volumes and to by-pass excess of the acylating reagent from the detector.

Although a very sensitive response can be obtained with this approach there have been surprisingly few reports on the use of ECD for the analysis of the protein amino acid group. One reason might be the wide temperature range required to elute the derivatives from the column and the effect that this change has on the detector baseline stability. This was touched upon by Bengtsson and Odham [340], who introduced a micro method for the determination of free amino acids in biological systems using HFB acylation and isobutylation, capillary columns and ECD. They claimed that less than 10 pg of the individual amino acids could be detected. The amino acids were isolated on a cation-exchange column and then derivatized in glass capillary tubes to provide considerable pre-concentration of the sample. Care was taken to prevent oxidation of some amino acids with HFBA by adding di-*tert*.-butylhydroxytoluene. The method was also used with mass fragmentographic detection, with about 10 times better sensitivity. In a recent publication the same authors applied these methods to samples from biological microenvironments [341]. ECD was preferred to mass fragmentography when a large number of amino acids had to be determined in trace amounts.

An ECD approach has also been used in the amino acid sequencing of peptides.

Robertson et al. [342] evaluated the pentafluorophenylthiohydantoins formed from the peptides in a reaction with pentafluorophenyl isocyanate. The response was three orders of magnitude greater than the corresponding flame ionization detector signal, enabling subnanogram amounts to be detected. The less volatile cyclic amino acid derivatives were difficult to determine owing to disturbances of the baseline generated by temperature programming.

The thyroid hormones thyroxine (T_4) and triiodothyronine (T_3) have been investigated by ECD. Stouffer and co-workers [343,344] introduced the thermally stable N,O-dipivalyl methyl ester derivatives and were able to detect these two iodinated compounds in human serum. The amino acids were extracted into acidified butanol and purified by back-extraction followed by esterification with thionyl chloride in methanol and acylation with pivalic anhydride in the presence of triethylamine. Later a work-up procedure using Sephadex LH-20 was proposed, making 20 pg readily detectable [345].

Silylation procedures have been employed by some workers. Thus, Hansen [346] used N,O-bis(trimethylsilyl)acetamide to derivatize T_3 and T_4 as well as the physiologically inactive reverse T_3 (isomeric to T_3) and diiodothyronine in hydrolysates from thyroid preparations. Backer and Pileggi [347] used a similar procedure but were faced with difficulties in applying it to serum extracts.

Nihei et al. [348] reported on the use of the Stouffer procedure with the bispivalyl derivatives. After extraction and derivatization the derivatives were purified in an anion-exchange step, which made possible the determination of 50 pg of T_3 and 100 pg of T_4 in 1 ml of serum from both normal subjects and hyperthyroid patients. Slow temperature programming was reported to be essential. A minor conversion of T_4 to T_3 during the derivatization reaction was observed.

The heavy derivatives of T_3 and T_4 have good ECD responses as expected, but require high temperatures for their elution. Petersen et al. [349] introduced the N,O-(bis-HFB)-methyl esters of T_3, T_4 and reverse T_3. This not only improved the gas chromatographic properties but also lowered the detection limit to about 0.2 pg. This permitted the determination of the free fractions of these amino acids, in contrast to earlier methods which measured the total concentration. The free or unbound compounds were isolated by dialysis and further purified by ion exchange. An important finding was the fact that the cation exchanger could be deactivated by the addition of 15 pg of diiodothyronine, which then followed the other analytes in the work-up and derivatization procedure. A 5′-chloro-T_3 analogue was tried as an internal standard and gave excellent reproducibility but, owing to interference with reverse T_3 which was found in pathological sera, it could not be used.

Hušek [350] introduced a very interesting derivatization reaction for α-amino acids, which were cyclized in a one-step reaction to oxazolidinones with 1,3-dichloro-1,1,3,3-tetrafluoroacetone in acetonitrile. These derivatives exhibited a good ECD response comparable to that of lindane. Amino acids with an additional functional group, such as tyrosine, monoiodo- and diiodotyrosine, had to be acylated. It was found that TFA, HFB and acetyl derivatives of the tyrosine oxazolidine had about the same response and that they were stable, particularly the HFB derivatives [351]. The latter derivative had an ECD response of the same magnitude as that of the N,O-bis-HFB derivatives. This derivatization approach was used by Davis et al. [352] to determine mono- and diiodotyrosine in serum. The isolation involved protein precipitation, extraction into butanol and evaporation. The detection limit was said to be 20 ng/ml but owing to an elevated baseline this varied from sample to sample. The compounds of interest could only be detected in hypothyroid patients.

L-3,4-Dihydroxyphenylalanine (L-DOPA) is a precursor for catecholamines. Imai et al. [353] reported its determination in urine using a TFA derivative which was reported to be more sensitive and selective than a TFA-oxazolone. The urine was defatted by shaking with diethyl ether–benzene, then DOPA was isolated on alumina and after desorption esterified with *n*-butanol at 100°C for 20 min. Acylation was performed in ethyl acetate for 5 min. The method had a detection limit of about 5 ng/ml.

Kynurenine is a major metabolite of tryptophan. Joseph [354] reported on the determination of this amino acid via the alkaline degradation product, *o*-aminoacetophenone. This moderately strong electrophore was subjected to a two-phase reaction with TFAA in butyl acetate and a borate buffer. The reaction seemed to be instantaneous and the excess of reagent was easily removed. Very clean chromatograms were reported with blanks less than 5 ng. Kynurenine was determined in various samples such as brain tissue, cerebrospinal fluid, plasma and urine. The possibility for interferences from related compounds giving the same acetophenone was considered and found to be of minor importance.

γ-Aminobutyric acid (GABA) is a transmittor substance present in brain tissue. Pearson and Sharman [355] described an ECD method based on the formation of hexafluoroisopropyl N-TFA-4-aminobutyrate. The procedure was comparatively simple and showed high specificity and sensitivity. The supernatant from the brain homogenate was purified on an ion-exchange column and, after evaporation to dryness, subjected to reaction with TFAA and hexafluoroisopropanol for 1 h at ambient temperature. It was shown that the GABA derivative was extremely sensitive to the ECD. Of 26 other amino acids studied, β-alanine and methionine showed the same response whereas leucine, proline, serine and threonine gave lower signals but did not interfere chromatographically. The other amino acids were either eluted earlier or did not show any response. An improvement in this method was proposed by Schmid and Karobath [356], who used a homologue as an internal standard and isolated this and GABA on a cation-exchange paper strip prior to derivatization. The method was validated by a mass fragmentographic procedure and was found to be specific and as sensitive (MDA ca. 10 fmole or 3.5 pg). The derivatives had excellent stability but, owing to the high fluorine content, problems were encountered with evaporation losses in the final work-up procedure. Glutamic acid gave three derivatives, one of which could interfere with low GABA levels.

Baclofen is a β-*p*-chlorophenyl-substituted γ-aminobutyric acid having muscle-relaxing properties. This compound did not react like GABA [355]. Degen and Riess [357] used a three-step procedure with esterification in methanolic hydrochloric acid followed by transesterification in acidified butanol. The butyl ester was made to react with HFBI, giving a stable derivative [357]. The isolation procedure was of interest as adsorption on charcoal was used in a batch procedure. Urine and plasma samples containing down to ca. 50 ng/ml were analysed with a precision of 10%. The recovery was about 50% for both types of sample, as shown by experiments using a ^{14}C-labelled compound.

The single-step reaction for the determination of GABA was also applied to a series of γ-glutamyl amino acids which are involved in the regulation of glutathione. Wilk and

Orlowski [358] determined a series of these peptides in subnanogram amounts in tissues and body fluids with acceptable precision.

Mefenamic acid, an aromatic amino acid, is an anti-inflammatory and antipyretic drug and its use in horses prompted Bland et al. [359] to develop a method for screening it in urine. The compound was extracted into methylene chloride and subjected to reaction with PFPA in the presence of zinc chloride in ethyl acetate [359]. The structure of the derivative was not elucidated but both N-acylation and reaction with the carboxyl group were proposed. The response was moderate in that 50 ng/ml could be determined.

Tranexamic acid is a synthetic amino acid with antifibrinolytic properties. An ECD procedure was presented by Vessman and Strömberg [360] which allowed both therapeutically and biopharmaceutically interesting concentrations to be determined. The ω-amino group was made to react in a mixture of dimethyl sulphoxide (DMSO)–buffer (pH 9.4) with fluoronitrobenzotrifluoride, followed by ethylation in an extractive alkylation step. The use of not more than 0.2 ml of plasma or serum allowed this first step to be performed directly in the biological sample after pH adjustment and addition of DMSO. The derivative is sufficiently volatile and can be used to detect at least 40 pg of tranexamic acid using an ECD. The derivatization procedure did not give rise to isomerization, as HFBA was reported to do. The simple α-amino acids gave much more volatile derivatives, whereas polyfunctional acids gave non-volatile products. The retention time of the ϵ-aminocaproic acid derivative was four times longer than that of the corresponding α-amino acid, norleucine.

Amino acid structures are common among drug metabolites, e.g., pethidinic and norpethidinic acid. Lindberg et al. [361] studied the conditions for the alkylation of these two major metabolites of pethidine with PFBz bromide. Norpethidinic acid formed a bis-N,O-PFBz derivative, whereas pethidinic acid gave the corresponding ester. It was important, however, to optimize the conditions in order to avoid quaternization of the tertiary amine function. The corresponding amines had to be removed before the extractive alkylation procedure as they otherwise could undergo re-esterification and thus interfere. Although the response of these derivatives in the ECD was good, it was not possible to clean up the reaction mixture in a routine manner and mass fragmentography was preferred for analytical use.

13.20. CARBOXYLIC ACIDS INCLUDING HYDROXY AND KETO ACIDS

A large number of pharmacologically active compounds are carboxylic acids. Usually they have to be derivatized to permit low-level determinations. The preparation of electron-capturing derivatives became possible with the introduction of PFBz bromide. Ehrsson [362] introduced an extractive alkylation technique, which usually resulted in rapid reactions. Hydrophilic acids can be alkylated by the use of more lipophilic counter ions.

Gyllenhaal et al. [363] demonstrated this principle for the determination of free fatty acids in minute serum samples. The excess of reagent, which seriously disturbs the detector, was removed by reaction with an aminophenol in a subsequent extraction step. The presence of fatty acids in the reagents resulted in blanks, thus making the use of the high response of the derivative (MDA ca. 5 pg) impossible. The response was only slightly temperature dependent.

Greving et al. [364] made use of this principle in the determination of some drugs which are esters of carboxylic acids, e.g., anticholinergics such as atropine, oxyphenonium, analgesics such as pethidine and tranquillizers such as benactyzine. As direct analysis presented problems they investigated a hydrolysis approach, the acids being converted into PFBz esters. Extraction of the acids from the hydrolysed biological sample was performed by ion-pair extraction. At a pH between 7 and 8 the fatty acids did not interfere. The ion pair with tetrapentylammonium was made to react with a low concentration of the alkylating agent in dichloromethane. The MDA was reported to be 0.7 pg. Special care had to be devoted to interference from palmitic acid, which was found to be present in large amounts in reagents such as tetrapentylammonium hydroxide and dichloromethane as well as on glassware. Davis [365] used the crown ether 18-crown-6 to promote the reaction with PFBz for some hydrophilic acids. By choosing a suitably basic anion such as a potassium salt, the derivatization could be made selective for carboxylic acids or for acids and phenols. Amounts less than 500 pg were difficult to determine owing to interferences from artefacts formed during the reaction in benzene.

Sioufi and Pommier [366] described a method for the determination of benzoic acid in plasma and urine based on reaction at 50°C for 2 h with PFBz in the presence of potassium carbonate in acetone; 10 ng/ml could be determined.

Trichloroethyl esters have also been proposed as useful derivatives for "transparent" acids. Reaction with 2,2,2-trichloroethanol was performed in TFAA [367]. The alcohol had to be removed on a silica column. Detection limits for acids such as benzoic, phenylacetic and *trans*-cinnamic acid were in the range 80–400 pg. Alley et al. [368] prepared the trichloroethyl esters of short-chain acids using HFBA as a catalyst, which was better than TFAA plus TFA acid. Excess of reagent was removed by addition of palmitic acid. The method was applied to samples of synovial fluid which contained acetic and butyric acid.

Trifluoroacetic acid is a metabolite of halothane, an anaesthetic. Witte et al. [369] used a headspace procedure to prepare the trichloroethyl ester, as the acid itself was not sufficiently sensitive to the ECD. Some fluorinated alcohols did not give derivatives with as high a response. Using sample volumes of 200 μl, lyophilization and reaction in the presence of concentrated sulphuric acid, 1 μg/ml of TFA acid could be detected.

Valproic acid is an antiepileptic drug resembling a fatty acid. It was recently reported that the phenacyl ester had a fairly high ECD response, as 100 pg could be measured [370]. This interesting observation formed the basis of a simple method consisting in precipitation of protein with acetonitrile, derivatization in the supernatant with crown ether catalysis and extraction. Octanoic acid was used as an internal standard. The high therapeutic levels (ca. 100 μg/ml) gave very clean chromatograms.

Monoethylhexyl phthalate is found as a degradation product in blood plasma contaminated with the diester and in infusion solutions stored in poly(vinyl chloride) bags. Simple esterification with diazomethane was used to determine the acid in blood products [371], whereas a rapid reaction with hexyl iodide in a phase-transfer catalysed process was used for the infusion solutions [372]. The latter approach permitted the simultaneous determination of diethylhexyl phthalate (DEHP). This widely occurring contaminant has been determined in blood products and in placenta with ECD-based methods [373,374]. DEHP and other phthalate esters have an inherent ECD response, which was not considered sufficient to determine phthalate ester concentrations in biological samples. Takeshita et al. [375] introduced a procedure in which a bistrifluoroethyl ester was formed with the phthalic acid prepared in an initial hydrolysis step followed by column purification. This resulted in a considerably higher (ca. 100 times) sensitivity, as 0.1 pg could be detected. BF_3 catalysis was used. Examples of the analysis of a muscle from a bird were given.

Perfluorooctanoic acid, after treatment with diazomethane, was determined in blood (ca. 10 ng/ml) [376]. Extraction of the homologous internal standard from blood and plasma presented some problems, whereas aqueous samples did not.

Dichloroacetic acid, a metabolite of the anaesthetic compound methoxyflurane, was determined in plasma and urine after lyophilization and methylation [377]; 50 pg could be detected and 40 ng/ml quantified. Trichloroacetic acid gave a higher response, as expected. Lukas et al. [378] prepared the methyl ester in a mixture of the aqueous sample and BF_3 in methanol; 10 min at 100°C resulted in a quantitative yield. The detection limit was 50 ng/ml.

Trichloroacetic acid, a metabolite of chloral hydrate and trichloroethylene, was determined in 0.1 ml of serum as its methyl ester, prepared with BF_3 in methanol, with a detection limit of 10 μg/ml [379]. β-Nitropropionic acid was reported to be determined as its ester with PFBz in mould filtrates [380].

Indolylcarboxylic acids are of biological interest, especially 5-hydroxyindoleacetic acid. The ECD response of the methyl esters as TFA derivatives was reported to be good, with a detection limit of 50 pg [381].. Another study presented data for the mono-, di- and trichloroethyl esters of indol-3-ylacetic acid [382]. The response increased with the number of chlorine atoms, as expected, permitting the detection of 10 pg of the trichloroethyl ester. The dichloro ester was about 25 times less sensitive.

The simultaneous derivatization of carboxyl and hydroxyl groups is attractive. Brooks et al. [383] used a mixture of HFBA and pyridine to achieve this, in a very rapid reaction. Anhydrous conditions were necessary. Alcohols such as ethanol, butanol and trichloroethanol could be used. Salicylic acid and lactic acid were studied in synovial fluid.

Naphthoxylactic acid is a metabolite of the widely used β-blocking agent propanolol. Its determination in plasma was carried out in a two-step procedure consisting in treatment with diazomethane and acylation with HFBA [384]. The latter reaction was performed with amine catalysis. This very stable derivative resulted in an MDA of 2 pg and could be used to determine 2 ng/ml in plasma.

Chlorophenoxyacetic acids, used mainly as herbicides, have a low ECD response. Mierzwa and Witek [385] prepared 2-chloroethyl-, trichloroethyl- and trifluoroethyl esters of the 4-chloro-2-methyl and 2,4-dichloro acids. They found the trichloroethyl esters to be 100 times more sensitive than the 2-chloroethyl derivatives, with the trifluoro ester in between. Further, the trichloro ester was more sensitive for the 4-chloro-2-methylphenoxyacetic acid (150 pg vs. 240 pg). The low response for the monochlorophenoxyalkanoic acids is the main reason why chlorophenoxyisobutyric acid was by preference determined with flame ionization detection [386,387]. This once widely used drug (a metabolite of clofibrate) for the treatment of hyperlipaemia has only been determined with ECD after derivatization with PFBz and separation by two-dimensional gas chromatography [388]. With *p*-chlorophenoxyacetic acid as an internal standard at least 10 ng per 0.25 ml could be determined in a simple procedure.

2,4,5-Trichlorophenoxyacetic acid, on the other hand, could be determined as a methyl ester [389]. For pharmacokinetic studies in the mouse with sample volumes in the microlite range the method could be used to detect 10 ng of derivative. A silica column clean-up was required for the lower concentrations.

Krebs cycle α-keto acids were studied as methyl esters by Simmonds et al. [390]. Multiple ester products were sometimes formed but structures with conjugated carbonyl groups had a high ECD response [390]. However, this approach does not seem to have been utilized bioanalytically. Harrison and Collins [391] demonstrated the presence of α-keto acids in beer using ECD. Most studies on α-keto acids are now based on the formation of quinoxalinols. Frigerio et al. [392] demonstrated the preparation and structures of some halogenated derivatives intended for ECD evaluation. Another approach was reported by Kallio and Linko [393], who prepared the 2,4-dinitrophenyl hydrazones of keto acid methyl esters. The ECD response was 100 times higher than that for the flame ionization detector and 0.1 ng of ketomonocarboxylates could be determined. The dicarboxylates were about 5 times less sensitive.

A special approach was demonstrated for the determination of citrate, which was converted into bromoform [394]. This selective derivative made possible the analysis of nanogram amounts. Tocco et al. [395] presented a method for the determination of oxalic acid in urine based on precipitation of the calcium salt and esterification with 2-chloroethanol in the presence of BF_3. Oxalic acid in concentrations down to 5 μg/ml were determined and the results agreed well with those obtained by GC–MS. Application of the technique to plasma required special precautions owing to the generation of oxalic acid.

Prostaglandins are fatty acid derivatives, many of which have strong biological effects. As they are active in minute concentrations, sensitive and selective assays are needed. The E prostaglandins, after dehydration to the B type, esterification, purification by TLC and silylation, could be determined with the ECD as the dehydrated product had inherent electron-capturing properties [396]. With nor- and homo-derivatives as internal standards 1 ng/g could be detected in various rat tissues. The bromomethyldimethylsilyl ether methyl esters were reported to give a similar detector response. An application of the conversion of the E to B prostaglandins followed by ECD of the silylated methyl esters after separation on capillary columns was reported by Korteweg et al. [397].

Levitt et al. [398] prepared the tri-HFB methyl esters of prostaglandins $F_{1\alpha}$ and $F_{2\alpha}$. These derivatives exhibited sensitivity in the picogram range and were reasonably stable during storage. However, peak broadening and multiple peak formation made this approach less attractive. The PFBz ester was introduced as an electrophore group into prostaglandin $F_{2\alpha}$ [399]. After silylation a thermally stable derivative with a high volatility and excellent ECD properties was obtained. The reaction with PFBz was performed in acetonitrile in the presence of diisopropylamine. The on-column detection limit was 12.5 pg, which was high enough (ca. 0.2 ng/ml) to measure $F_{2\alpha}$ in monkey plasma after subcutaneous administration. Skrinska and Butkus [400] analysed for prostaglandin E_1 and E_2 (after conversion to B forms), $F_{1\alpha}$ and $F_{2\alpha}$ and A_1 and A_2 in rabbit kidney medulla. Witn the TMS ether PFBz esters the ECD response permitted the detection of 1 pg of F and 10 pg of B prostaglandins. The interference from biological components could be reduced in a liquid chromatographic step on silica. For the B_1 and B_2 compounds the TMS ether PFBz esters gave better resolution and sensitivity than the TMS ether methyl esters. Tromboxane B_2, biosynthetically related to the prostaglandins, was analysed as a methyl ester, PFBz oxime silyl ether, by Fitzpatrick et al. [401]. Reaction of the free aldehyde form of the acetal ring with PFBz hydroxylamine resulted in a stabilization of the compound that was not possible with the PFBz ester. In studies of platelet aggregation, nanogram amounts were determined although sometimes with poor reproducibility. The derivative existed in *syn-* and *anti-*forms.

Fitzpatrick et al. also showed that the PFBz oxime PFBz ester derivative was twice as sensitive as the PFBz oxime methyl ester for a series of prostaglandins. The latter derivative was preferred for tromboxane B_2 as the PFBz ester was unstable [402]. In most instances 15–60 pg could be detected. To follow the release of prostacyclin, 6-oxo-prostaglandin $F_{1\alpha}$ was determined as the methoxime, PFBz ester bis-silyl ether, in perfused rabbit and rat hearts [403]. The synthesis of prostacyclin in cell cultures was followed by determining 6-oxo-$F_{1\alpha}$ as either the PFBz oxime methyl ester or methoxime PFBz ester on glass capillary columns; 100 pg could be determined on-column [404].

Recently an antisecretory trimethylprostaglandin E_2 analogue was determined by GC–MS using negative chemical ionization [405]. A blank level of 5 pg was obtained in a procedure that consisted in extraction, PFBz ester formation and silylation. A monofluorinated analogue was added as a carrier. The ECD response was claimed to be five times lower than that in the negative ion mode but five times higher than that in the positive ion mode.

Anti-inflammatory agents such as indomethacin interact with prostaglandin synthesis. These widely used agents are carboxylic acids. Indomethacin has inherent electron-capturing properties, so that only esterification is necessary. The acid is easily hydrolysed in weakly alkaline media. Ferry et al. [406] used diazoethane, as methylation could result in interference from a possible metabolite. Determination of levels down to 0.1 ng on-column was possible. Helleberg [407] also used diazoethane after having rejected the use of BF_3 in methanol and alkylation in alkaline media; 50 ng/ml could be determined in 0.5-ml samples of serum. Arbin [408] reported on three propylation methods using either diazopropane, extractive alkylation or solid–liquid phase transfer-catalysed alkylation. The last method was the mildest, as reflected in the low relative standard deviation for the catalysed alkylation reaction (3.5%). The extraction procedure was optimized, resulting in a method able to determine 5 ng/ml of indomethacin. The mild conditions with a dissolved quaternary ammonium bicarbonate catalyst were later shown to be of value in the derivatization of acetylsalicylic acid [409]. Møller Jensen [410] used extractive ethylation directly for serum samples, 50 ng/ml of indomethacin was detectable. Although indomethacin has good electron-capturing properties, the introduction of electrophoretic groups has been reported. Sibeon et al. [411] used PFBz bromide and claimed a detection limit of 10 ng/ml. The reaction was performed with potassium carbonate in acetone. The use of a 5-fluoro analogue as an internal standard facilitated the analytical procedure. Evans [412] proposed the use of pentafluoropropanol in PFPA to give a pentafluoropropyl ester. This enhanced the sensitivity, as 1 ng/ml was reported to be determined.

After indomethacin a number of compounds with anti-inflammatory activity were produced. De Silva et al. [413] developed a spectrofluorimetric and an ECD method for 6-chloro-α-methylcarbazole-2-acetic acid (Carprofen). The latter procedure involved TLC purification and reaction with diazomethane but had a similar sensitivity to the fluorimetric assay.

Sennello et al. [414] determined alclofenac with ECD and flame ionization detection. The 3-chlorophenylacetic acid derivative did not render the compound sufficiently sensitive to the ECD and flame ionization detection was preferred. Flurbiprofen was determined in plasma [415] and ibuprofen in serum [416], both after conversion to PFBz esters. Fenclorac has one aromatic and one aliphatic chlorine atom, which is probably the reason why its methyl ester could be detected in plasma at levels down to at least 200 ng/ml [417]. Benoxaprofen has one aromatic chlorine atom, but also has a conjugated heterocyclic ring system, which made its determination with ECD very simple and also very sensitive (10 ng/ml), as reported by Chatfield and Woodage [418]. Tolmetin is a derivative of benzoylpyrrolacetic acid and as such it should be easy to detect with ECD; however, a procedure involving formation of a PFBz ester was described [419]. The analysis of micro-samples of plasma (25 μl) could be carried out with excellent sensitivity using this derivative (MDA 10 pg, corresponding to 100 ng/ml). The derivatization could be performed rapidly with extractive alkylation but in the absence of the tetrabutylammonium ion as a counter ion the reaction required 4 h. The latter unusual procedure was preferred as the major metabolite, a dicarboxylic acid, was excluded. Extractive methylation was used in the determination of ketoprofen. This compound has a benzophenone structure which was used for a selective and simple assay, as reported by Stenberg et al. [420]. The response at 240°C permitted the detection of 0.25 μg/ml.

Two diuretic compounds, both of which are *o*-dichlorobenzene derivatives and have a carbonyl group in conjugation, were determined after methylation. Ticrynafen was determined at levels down to 10 ng/ml [421], whereas for the indanyloxyacetic acid derivative the lower level was 5 ng/ml [422].

13.21. PHENOLIC ACIDS

These acids include some which are products of catecholamine metabolism, e.g., homovanillic acid (HVA), vanillylmandelic acid (VMA) and 3,4-dihydroxyphenylacetic acid (DOPAC). Numerous papers have dealt with the determination of these acids in various types of biological samples. ECD has been used to some extent. The first report by Dziedzic et al. [423] used a hexafluoroisopropyl (HFIP) ester of trifluoroacetylated HVA, which was prepared in a two-step reaction. HVA was extracted with ethyl acetate from sodium chloride saturated and acidified urine and subjected to reaction with TFAA followed by BF_3-catalysed esterification at 100°C. The last step was considered to be a reaction of HFIP with the mixed anhydride formed in the first step. The derivative had excellent gas chromatographic properties and the sensitivity permitted the detection of less than 0.1 ng on-column. The change in excretion of HVA in urine from normal subjects and patients with different diseases was demonstrated. Sjöquist and Änggård [424] used another approach to determine HVA in cerebrospinal fluid using an HFB-methyl ester. After reaction with HFBA the derivative was purified by chromatography on lipophilic Sephadex, a step which was reported to be important as many background disturbances were removed. The method showed good agreement with a GC–MS assay. The capacity of the latter method was higher owing to the reduced influence of interferences from solvents and endogenous compounds.

HVA is accompanied by minor amounts of an isomeric metabolite, isohomovanillic acid (iso-HVA), the presence of which has been of much concern. Dziedzic et al. [425] described a modification of the earlier HVA method. The preparation of the HFIP-TFA derivative of HVA was made in a single rapid reaction using a 1:3 mixture of HFIP and TFAA [425]. They demonstrated the presence of iso-HVA in urine in amounts of about one tenth of that of HVA. In addition, some other phenolic acids such as DOPAC and *p*- and *m*-hydroxyphenylacetic acid were found. The same procedure was applied to cerebrospinal fluid [426].

Watson and Wilk [427] assessed DOPAC and HVA in cerebrospinal fluid using the PFP pentafluoro-1-propanol derivatives, which were prepared in a one-step reaction. The response was high and it was possible to detect 5 pg. The acids were extracted with diethyl ether, which was claimed to be better than ethyl acetate as few uninteresting compounds appeared in the chromatogram. Appropriate internal standards also contributed to the good reproducibility of the method.

Weg et al. [428] utilized the catechol structure of DOPAC in the isolation step with alumina. This selective adsorption step eliminated considerable amounts of unwanted material. The derivatization procedure consisted in esterification with hydrochloric acid in dry ethanol and reaction of the phenolic groups with PFPA. It was also found necessary to prime the column with 3 ng of the DOPAC derivative (a bis-PFP ester) in order to obtain a reproducible and linear response. The working range of the method was 5–70 pg per injection. To permit the determination of DOPAC and HVA in rat striatum in samples at various times after killing the animals, Watson et al. [429] investigated alternatives to the alcohol they used before, viz., pentafluoro-1-propanol. HFIP was not suitable but a monochloro analogue, 1-chloro-1,1,3,3-pentafluoro-2-propanol, gave esters which as PFP derivatives separated from endogenous peaks and permitted the simultaneous determination of both acids [429]. The presence of VMA could not be demonstrated, but it was reported that the response of the pentafluoro-1-propanol derivative of VMA was twice that of the HFIP monochloro analogue.

Pearson and Sharman [430] presented data for the determination of DOPAC, HVA and iso-HVA in nerve tissue using TFA-HFIP derivatives. Small samples could be analysed but care had to be taken as sometimes unknown substances could interfere, as for example with DOPAC in aqueous humour. Markianos and Rüther [431] determined HVA and VMA in the serum of normal subjects and psychotic patients and found VMA in only a small number of serum samples. They reported that the response of the VMA derivative with PFP-HFIP was five times lower than that of HVA. Walter et al. [432] described the routine measurement of HVA in rat brain using ethyl acetate extraction and derivatization with trifluoroethanol and TFAA. With amounts of HVA below 2.5 μg a single derivative was obtained with a reportedly low recovery.

Chauhan and Darbre [433] analysed urine for HVA, iso-HVA and VMA using the TFA-HFIP derivatives. A combination of glass capillary columns with different liquid phases revealed that interferences might be more common than at first believed. The derivatization procedure was optimized and sealed glass ampoules were recommended. Of the nine liquid phases tested Dexil 300 GC showed no interfering peaks, although it was sometimes necessary to check this for iso-HVA, which was present in amounts about one fortieth of that of HVA.

One of the very first papers on the determination of these catecholamine metabolites in urine was an assay of VMA by Wilk et al. [434], who converted VMA to vanillin in an oxidative reaction with periodate. Vanillin was then made to react with TFAA under anhydrous conditions at or below room temperature. The derivative was inert and permitted less than 1 ng to be detected. The response of the TFA derivative was about 300 times that of the unreacted vanillin. Yoshiue et al. [435] analysed urine for VMA and HVA. They used adsorption by XAD-4 for isolation and then prepared the TFA-HFIP derivatives, which were determined at levels down to 90 pg on-column.

Makita et al. [436] reported on the use of the O-pentafluorobenzoyl methyl esters for the determination of phenolic acids in urine. The PFB reaction was performed in the aqueous phase (bicarbonate solution) and after extraction the carboxyl group was treated with diazomethane. Urine samples were first extracted with diethyl ether. The derivatives were very stable and exhibited extremely high ECD responses, the detection limits being between 1 and 10 pg. The simultaneous determination of HVA, VMA and 5-hydroxyindoleacetic acid was possible.

The presence of *m*- and *p*-hydroxyphenylacetic acid in rat brain was demonstrated by Davis et al. [437]. Homogenization, protein precipitation, extraction into ethyl acetate and TLC purification was performed before the derivatization with PFP and HFIP took place. This mixture was washed with an aqueous buffer (pH 6), which was claimed to be better than evaporation of the excess of reagent. The sensitivity was excellent, with MDA values lower than 50 pg. The response of the PFP methyl ester was reported to be lower. The use of capillary columns clearly demonstrated the presence of *m*-hydroxyphenylacetic acid, hitherto not distinguished from the *p*-isomer.

The use of chloromethyldimethylsilyl ethers and esters was demonstrated by Morita and Montgomery [438]. The derivatives had only moderate responses (MDA ca. 4 ng) and were also reported to be unstable.

13.22. PHENOLS AND PHENOLIC ALCOHOLS

Few phenols have native electron-capturing properties. Morimoto et al. [439], in a study of the determination of ubiquinones in human blood, reported that α-tocopherol could be detected by ECD. The *trans*-stilbene structure is an electrophore of moderate sensitivity. The hydroxylated metabolites (especially 4- and 4,4′-) were studied by Sinsheimer and Smith [440], who reported that the methylated phenols were less sensitive than *trans*-stilbene, whereas the acetylated and silylated compounds gave a higher response. For the bis derivatives the increase was 7- to 8-fold. However, for the determination of these metabolites the use of chloromethyldimethylsilyl derivatives was recommended, with detection limits around 10 ng.

Numerous derivatization reactions for phenols have been studied. Kawahara [441] introduced PFBz bromide as a suitable reagent for determining subnanogram amounts of simple phenols and mercaptans. The alkylation took place in acetone with potassium carbonate. Ehrsson et al. [442] thoroughly studied the conditions for the preparation of HFB derivatives. They used trimethylamine as a catalyst and removed the excess of reagent by solvent extraction (buffer, pH 6.0). The derivatives were stable to this treatment whereas higher pH values of the aqueous phase, ammonia or solvent with a higher water uptake than benzene caused degradation of the ester. The response of the HFB esters was very high, with an MDA of about 1 pg. The response was reported to be only slightly influenced by structural changes, in contrast to the HFB derivatives of alcohols and amines.

McCallum and Armstrong [443] reported on seven derivatives of thymol and found the PFB esters to be the most sensitive, with an MDA of about 1 pg. The HFB derivative was about seven times less sensitive and comparable to the PFBz ether, which was also less volatile than the other two.

Diethylstilboestrol is a synthetic oestrogen also used in agriculture. To determine trace amounts the bisphenol has been derivatized. King et al. [444] used the PFP derivative, which was formed after methanol extraction of the animal feed and clean-up on a column of Sephadex LH-20. This derivative was only slightly better than HFB and TFA derivatives, but about 100 times better than the dichloroacetyl derivative prepared in a two-phase reaction. Isomerization occurred during the reaction. The MDAs were 0.2 and 0.3 pg for the *cis*- and *trans*-PFP derivatives, respectively. Ryan and Lawrence [445] included diethylstilboestrol in a comparison of some perfluoro derivatives and found the HFB derivative to be preferable. Another study with chloroacetyl derivatives showed that the monochloro- was slightly better than the dichloroacetate derivative [446]. The trichloro derivative was not only less sensitive but also appeared to undergo degradation on the column.

Laitem et al. [447] reported the determination of diethylstilboestrol and the related compounds dienoestrol and hexoestrol in meat and tissue of cattle. The derivatives of choice were the PFB esters, which were very stable and exhibited a response comparable to the HFB esters but, owing to their lower volatility, they separated well from co-extractants. Excess of reagent was removed on a Sephadex LH-20 column. In a 5-g sample 2.5 ng was detected.

Paracetamol is a widely used analgesic. An ECD assay using a PFBz derivative was presented by Chan and McCann [448]. They used Kawahara's conditions with acetone and anhydrous potassium carbonate under reflux for 30 min at 60°C. The internal standard, *n*-butyryl-*p*-aminophenol, was added prior to the extraction with diethyl ether–dichloromethane. Evaporation was used to remove excess of reagent. The sensitivity of the method, less than 50 ng/ml in plasma, was enough to permit calculation of various pharmacokinetic parameters.

Δ^9-Tetrahydrocannabinol is the major active component of marihuana and has been the subject of several analytical studies. Fenimore et al. [449] made use of ECD and an HFB ester in a procedure with dual column separation. A packed column was used for initial clean-up and the fraction of interest was trapped in a cold trap and then introduced into a capillary column. With a 1-ml serum sample tetrahydrocannabinol was determined at levels below 100 pg/ml. They also compared various electron-capturing derivatives. The TFA and HFB esters had MDAs of 5 and 2 pg, respectively, whereas both the PFB and chlorodifluoroacetate derivatives had an MDA of 1 pg. The latter were not used, however, owing to their too long retention times, which was also the case with the PFBz ether. Garrett and Hunt [450] reported a study in which the PFB esters were used after clean-up by liquid chromatography [450]. The method allowed the determination of 1 ng/ml, and compared favourably with a liquid scintillation assay.

Triclosan, a bacteriostatic agent, was determined in human plasma and urine by Sioufi et al. [451]. The trichloro-substituted phenol was acetylated after an initial extraction and back-extraction. Conjugates, not only glucuronides, were split enzymatically and were also determined in plasma. The method had a detection limit of 2 ng/ml.

A compound of related structure and activity, hexachlorophene, has been determined in biological fluids. Bachmann and Shetlar [452] used benzene extraction of the lyophilized tissue or serum sample and injected the phenol extract on to the column. The chromatographic properties were such that 0.2 ng could be detected. Clean-up procedures were omitted even if detector washing and column reconditioning had to be carried out more often. Porcaro and Shubiak [453] used silylation, which improved both the recovery and the sensitivity (MDA ca. 100 pg). They reported studies of the adsorption of hexachlorophene on the skin. A study of the determination of this bisphenol in blood and urine was reported by Browning et al. [454]. Acetylation was found to be superior to silylation with respect to preparation and chromatographic properties. It was possible to determine 50 ng/ml after ethyl acetate extraction. Porcaro et al. [455] later used diethyl ether for extraction from whole blood. Washing of the phenolate of hexachlorophene with *n*-hexane removed lipophilic components which interfered with the silylation reaction. Ferry and McQueen [456] claimed that the methyl ethers were superior and were able to determine 1 pg on-column. The extraction into diethyl ether from plasma was enhanced if a citrate buffer was used. The methylation was performed with diazomethane.

A major metabolite of epinephrine and norepinephrine is 3-methoxy-4-hydroxyphenylethylene glycol (MHPG) and there have been several reports of the use of ECD for its determination. Wilk et al. [457] used TFAA to form a tris derivative with excellent sensitivity and chromatographic properties. When applied to urine samples extraction was made with ethyl acetate after clean-up on an ion-exchange column. Sharman [458] used acetylation of phenolic hydroxyl groups in the aqueous phase, thereby facilitating the extraction into dichloromethane. The remaining alcohol groups were treated with HFBA. The resulting derivative had a response which was extremely high, but no figures for comparison were given. However, in mouse hypothalamic tissue MHPG 40 ng/g could easily be determined. To confirm the identity of the peak, hydrolysis of the acetyl HFB derivative with concentrated ammonia was carried out, followed by peracylation with HFBA. The procedure could also be applied to the dihydroxy analogue (DHPG) but failed with norepinephrine and epinephrine.

Karoum et al. [459] analysed cerebrospinal fluid, urine and brain tissue for phenolic alcohols such as MHPG, 4-hydroxy-3-methoxyphenylethanol and 3,4-dihydroxyphenylethanol. Of the perfluoroacyl derivatives PFP was preferred, owing to the suitable volatility of the derivative and reagent [459]. It was possible to determine 5 ng/ml, the limit being set by interferences from solvents and enzymatic agents used to hydrolyse conjugates. Bond [460] determined MHPG in urine and cerebrospinal fluid with a combination of ion-exchange clean-up and acetylation in aqueous solution. The phenolic acetate was extracted into dichloromethane and, after evaporation was treated with TFAA at 70°C for 15 min. This combined clean-up procedure made possible the determination of both free and conjugated MHPG in the low ng/ml range. O'Keefe and Brooksbank [461] used extraction into ethyl acetate, acetylation in aqueous solution and purification of the phenol acetate by TLC. The electrophore reagent was HFBA, giving an excellent response to a few picograms on-column. Fellows et al. [462] presented a rapid assay of MHPG in urine using an ion-exchange column in the acetate form to isolate the analyte, followed by lyophilization and treatment with PFPA. Owing to the presence of several interfering peaks the retention time was as long as 20 min.

The first method for the determination of MHPG in plasma was reported by Dekirmenjian and Maas [463]. The tris-TFA derivative was formed after extraction and washing steps.

Halaris et al. [464] determined MHPG in plasma with some unusual isolation steps. The plasma sample was heat treated (boiling water-bath) to coagulate the proteins. The supernatant was ultrafiltered and the filtrate extracted with ethyl acetate. The PFP derivatives were more stable than the TFA esters and had an MDA less than 1 pg. The derivative was shown to have a higher response as the detector temperature was increased, indicating a dissociative electron-capture process. Several advantages of the pulse mode detector were reported to contribute to the good results obtained. Several anticoagulants were compared in the sampling step. Citrate–phosphate–dextrose buffer gave the most reliable results whereas heparin interfered in the assay.

Lombrozo et al. [465] presented a modified procedure, which consisted in the use of an internal standard, 3-methoxy-4-hydroxyphenethanol. This compound was present in small amounts in each urine sample. However, this disturbance was circumvented by supplementing the urine with various amounts of MHPG to make each urine sample serve as its own set of standard solutions for a calibration graph. This approach also compensated for possible degradation losses. Extraction into ethyl acetate with 10% isoamyl alcohol was stated to reduce losses upon evaporation before the treatment with TFAA.

Davis and Fenimore [466] recently described a semi-automated assay of MHPG in urine. Using a venting system and back-flushing, the chromatograms of the TFA derivative were very clean, permitting reliable determinations at levels in the low ng/ml range. Extraction was made into ethyl acetate using salting-out conditions, which also precipitated the proteins present. The very volatile TFA derivative was preferred to the corresponding derivative with chlorodifluoroacetic anhydride. The ECD response was shown to increase markedly with increasing temperature and the detector was kept at 310°C. It was necessary to maintain the temperature of the detector within close limits.

13.23. ALCOHOLS

There have been many applications of ECD to analyses for alcohols, but the alcohol functional group is usually present together with other groups. Usually there is a need to introduce an electrophore using the same principles as for, e.g., amines and phenols. Ethchlorvynol, a hypnotic-sedative drug, is an example of a chlorovinyl-substituted tertiary alcohol that gave a sufficient response to permit the determination of down to 250 ng/ml, as reported by Cummins et al. [467]. This was enough for the fate of the drug to be followed for 48 h in serum. Chlorobutanol, a tertiary alcohol with sedative and hypnotic properties, is also used in greyhound racing. The determination of the alcohol in dog urine was described by Smith and Thorpe [468]. Amyl alcohol containing benzaldehyde as an internal standard was used for extraction. Trichloroethanol and its glucuronide are metabolites of chloral hydrate, an established hypnotic drug, and can easily be determined by ECD, as shown by Garrett and Lambert [469]. They used chlorobutanol as an internal standard for the determination of trichloroethanol in urine in a pharmacokinetic study. Extraction with diethyl ether and injection of 0.25% of the extract was sufficient for the urinary levels that obtained. Breimer et al. [470] described a head-space determination of trichloroethanol and chloral hydrate. The alcohol could be detected at levels down to 0.5 μg/ml in blood or urine with this simple approach. Whole blood had to be treated with lead acetate to prevent the conversion of chloral hydrate into the alcohol. The isolation procedure yielded very clean chromatograms. Chloral hydrate was measured as trichloroacetyldehyde, the more volatile form of this compound. Berry [471] presented a rapid method for the measurement of trichloroethanol in blood and urine. The sample (0.1 ml) was diluted with ethanol, followed by direct injection of an aliquot on to the column. This considerably diluted sample conserved the column and introduced only very small amounts of contaminants. The therapeutic levels ranged from 2 to 120 μg/ml in plasma. Chloral hydrate was shown by Cole et al. [472] to be a metabolite of the anaesthetic agent trichloroethylene [472] by identifying the presence of trichloroethanol and trichloroacetaldehyde (chloral) by GC–MS.

Chloramphenicol was one of the early compounds to be studied with ECD as its structure contains a dichloroacetylamide group and a *p*-nitro substituent. Resnick et al. [473] extracted the antibiotic together with a monochloro analogue as an internal standard into amyl acetate and then formed a bis-silyl ether. For verification purposes the TFA derivatives were sometimes used. The response of chloramphenicol was eight times higher than that of the internal standard, which is within the expected range. Using dry conditions it was possible to detect less than 1 ng of the drug. Yamamoto et al. [474] chromatographed the compound without derivatization and claimed a detection limit for the overall procedure of 10 ng. A recent publication described the use of a bis-HFB derivative of chloramphenicol. Wal et al. [475] used this approach for the analysis of milk after an initial acetonitrile extraction followed by clean-up with liquid–liquid partitioning. The ECD response of the diester was excellent, with an MDA of less than 5 pg. The monoesters were formed if the reaction with HFBA was performed at room temperature, and 30 min at 75°C was therefore recommended. It was of interest that the columns had to be treated with HFBA in order to increase the range of the linear response at lower concentrations.

Thiamphenicol is related to chloramphenicol, having a methylsulphonyl group in place of the nitro group, which limits the side-effects. The electrophore properties of the molecules are, however, about the same, which can be seen in a paper by Aoyama and Iguchi [476], who determined concentrations down to 10 μg/ml without derivatization and lower levels using the silyl derivative. Plomp et al. [477] used a ^{63}Ni detector, making possible the use of higher column and detector temperatures for the bis-silyl derivatives. The small samples used (0.1–0.5 ml of plasma or amniotic fluid) gave a detection limit of 100 ng/ml. Both of these procedures involved the use of ethyl acetate extraction and drying with anhydrous sodium sulphate prior to silylation.

In order to determine the monosaccharide composition of submicrogram amounts of glycoproteins, Pritchard and Niedermeier [478] developed a gas chromatographic procedure. Methanolysis of the glycoproteins in a single step gave methyl glycosides, which are more stable than the reducing sugars. Acylation was performed with N-methylbis(trifluoroacetamide), which gave derivatives with a high ECD response. This permitted the determination of carbohydrates in as little as 0.1 μg of glycoprotein. Methyl glycosides gave rise to up to four anomers in different proportions. One important point in the procedure was the use of small capillary tubes for the derivatization reaction, one tube for all steps with one sample. An anti-inflammatory carbohydrate-derived drug tribenoside is a mixture of the α- and β-anomers in the proportions 1:2, and Sioufi and Pommier [479] described a procedure for its determination in human plasma. The tribenzyl-substituted compounds were extracted into benzene and the monohydroxylated drugs made to react for 10 min with HFBI after evaporation of the solvent. The use of these derivatives permitted the determination of 5 and 10 ng/ml of the α- and β-isomers, respectively.

Guaiacol glyceryl ether (or guaifenesin) has been used as an expectorant. Maynard and Bruce [480] determined this diol in blood using extraction into dichloromethane and conversion into an HFB ester. Mephenesine, a structurally related drug, was used as an internal standard. The derivative was claimed to be a monoester of the primary hydroxyl group. With this method about 50 ng/ml in blood could be determined. A metabolite of propranolol with a glycol structure was determined in biological samples by Saelens et al. [137]. They used a bis-TFA derivative, which exhibited an excellent response (MDA = 0.2 pg). The use of the diol structure for the preparation of boronate derivatives with electrophore properties was proposed by Singhawangcha et al. [481]. The boronic acids do not form stable derivatives with monofunctional compounds, thus making the reaction with diols specific. For guaifenesin the 2,4-dichlorobenzeneboronic acid was chosen, as the derivative eluted in a relatively empty region of the chromatogram. The reaction in ethyl acetate required 15 min at room temperature or 5 min at 50°C. This is in contrast to the reaction with the amino alcohol alprenolol, which had to be injected together with the reagent. The column temperature was increased at the end of each chromatographic run in order to remove less volatile components. The derivative was detected at the 25 pg level and showed a dissociative electron-capture mechanism. The method could determine guaifenesin in plasma at levels as low as 15 ng/ml.

Analysis of a diol structure was described in a study of the epoxide of 7,12-dimethylbenz[*a*]anthracene [482]. The epoxide was hydrolysed in acetone and water to the 5,6-dihydrodiol, which upon treatment with HFBA either dehydrated to a monophenol or first reacted with HFBA and then dehydrated to a mono-HFB derivative. The epoxide itself gave the same mono derivative. The individual components were therefore first separated on an alumina column. The method could easily determine nanogram amounts of the epoxide.

Polyethylene glycols are themselves not readily amendable to gas chromatography. Smith and Dawson [483] used the Zeisel reaction with hydrogen bromide to produce dibromoethane and dibromopropane, which were subsequently determined with ECD. In this way 1–10 μg of the glycol could be determined if calibration standards were analysed in parallel with the unknowns.

13.24. STEROIDS

Related to the alcohols are the very important and well studied steroids. These physiologically active compounds were among the first to be studied in therapeutic investigations. A great deal of effort was devoted to improving the sensitivity of gas chromatographic methods. Lovelock et al. [484] reported that some steroids with suitable conjugated structural elements could capture electrons, e.g., 4-androstene-3,17-dione. This approach was used by De Jong and Van der Molen [485] in the determination of dehydroepiandrosterone, which after isolation was oxidized to 4-androstene-3,6,17-trione, a reaction specific for steroidal 5-en-3-ols. The detection limit was 100 pg on-column, corresponding to 4 ng/ml in extracts. After extraction into diethyl ether, both paper and thin-layer chromatographic purification were used before oxidation with chromium trioxide in acetic acid. The oxidation mixture was purified by TLC. The trione was reported to be five times more sensitive than the 3β-HFB derivative.

Most studies on the determination of steroids with ECD have utilized halogenated derivatives, most of which have had impact on the development of analytical procedures in other fields. In the following paragraphs only a few examples will be mentioned. A good example is the first report on the use of HFB derivatives by Clark and Wotiz [486]. They were able to detect down to 100 pg of several steroids such as testosterone, androsterone and dehydroepiandrosterone. The volatility was superior to that of the chloroacetates studied by Landowne and Lipsky [487]. An early review of the use of ECD of steroids isolated from biological samples was given by Van der Molen [488]. Poole et al. [489] reviewed the techniques for determining steroids with and without natural electrophores by GC–ECD. A study of the structure–ECD sensitivity relationship for some 17α-acetoxyprogesterones by Koshy [490] should also be mentioned. That derivatization of steroids is not without complications was pointed out by Poole and Morgan [491], who showed that attempts to chromatograph a cholesterol HFB ester were unsuccessful as the intended derivative eliminated HFB-acid, giving cholesta-3,5-diene. This reaction was common to other steroids such as ergosterol, oestradiol and pregnanetriol. The use of pentafluorophenyldimethylsilyl ethers was proposed instead.

Chlormadinone acetate, a contraceptive drug, was determined in plasma by Bopp et al. [492]. Extraction from plasma into hexane was followed by an alkali wash and then concentration into a small volume of benzene. One eighth of this extract was injected on to the column. By use of the corresponding caproate as an internal standard reliable quantitative data were obtained at levels down to about 1 ng/ml. The combination of the conjugated carbonyl group with a halogen atom contributes to the high response of this compound with an MDA around 200 pg. To determine low levels of steroids, including chlormadinone acetate, Leitch et al. [493] recommended lecithin as a priming agent even if plasma extracts were analysed, as this treatment gave a reproducible and higher response at the beginning of a series of runs. Down to 30 pg of chlormadinone acetate could be detected.

Medroxyprogesterone acetate, used in the control of fertility and in the treatment of certain types of carcinoma, was determined in human plasma by Rossi et al. [494]. A 3-enol HFB ester was prepared in a reaction with HFBA in acetone–acetonitrile at room temperature for 60 min. The plasma was extracted with cyclohexane and no further purification was carried out. Owing to interference from an endogenous compound (suggested to be a cholesterol HFB derivative) the detection limit was 5 ng/ml. A similar procedure capable of detecting 1 ng/ml in plasma was first presented by Kaiser et al. [495]. Kirschner and Taylor [496], in addition to HFBA, also evaluated the derivatives of hexadecafluorononanoic (HFN) acid and eicosafluoroundecanoic acid with testosterone. Both had a higher detector response and longer retention times than that with HFBA. Using the HFN derivative a simplified procedure for the determination of testosterone in plasma was possible with fewer purification steps than with HFB derivatives. In addition the ECD response was more than twice that of HFB. Devaux and Horning [497] showed that treatment of testosterone with HFBA and HFBI led to two isomeric enol esters. The presence of a basic catalyst such as pyridine influenced the ratio between the isomeric esters. Dehennin et al. [498] evaluated some halogenated diesters of testosterone for the determination of testosterone in plasma. The derivatives of choice were either the 3-HFB-17β-pentadecafluorooctanoate or the 3-HFB-17β-PFB esters being 115 and 127%, respectively, as sensitive as the bis-HFB compound.

11-Deoxycorticosterone was determined in human plasma as an HFB derivative by Wilson and Fraser [499]. The work-up procedure included dichloromethane extraction, washing with three aqueous solutions of different pH, acetylation, TLC, extraction from the adsorbent, hydrolysis of the acetate, another TLC separation and finally reaction with HFBA. As HFB-progesterone was used as an internal standard and introduced in the HFB reaction it was necessary to remove endogenous plasma progesterone. The use of a detector by-pass valve improved the determination of trace amounts of the analyte. In this way the influence of impurities in the solvents and reagents could be reduced. The bis-HFB derivative gave the steroid a high ECD response and permitted the determination of concentrations down to about 25 pg/ml using plasma samples of 5–25 ml.

18-Hydroxycorticosterone was determined in human peripheral plasma by Wilson et al. [500] using a similar approach. As the analyte might exist in two different forms an oxidation step to an 11-oxo derivative was included after extraction and aqueous washes. In addition to this conversion to a single molecular species many contaminants were destroyed. After a paper chromatographic separation, derivatization was performed with HFBA. The detection limit was 0.7 pg on-column, corresponding to 20 pg/ml in extracts.

Musial and Williams [501] reported a method for the determination of conjugated and free bile acids as HFB derivatives. After an initial extraction with chloroform–methanol (2:1), both the free and conjugated bile acids were subjected to reaction with HFBA and pyridine at 105°C for 2 h to give the same derivative for both types, i.e., both glycine and taurine conjugates were deconjugated. In addition to the HFB esters of the alcohols, it was proposed that the carboxylic acid was converted into an anhydride with

HFB (high-resolution MS). Excess of reagent was removed by shaking the benzene solution with ammonia solution. Although the two conjugates gave various yields the method correlated fairly well with an enzymatic analysis with bile acids at concentrations from 4 to 60 ng/ml.

Pulsinelli et al. [502] reported on a method for the determination of pregnenolone in biological samples. Extraction into diethyl ether–chloroform together with a tritiated marker was followed by paper and thin-layer chromatographic (Al_2O_3) purification. The 20-keto group was reduced enzymatically, then the 20β-hydroxypregnenolone was subjected to reaction with chloroacetic anhydride for 12 h. The reaction mixture containing the derivatives was washed and finally purified by TLC (silica). For gas chromatographic purposes the 20β-hydroxymonochloroacetate of progesterone was added as an internal standard. This compound seemed to be more sensitive in the ECD than the 3,20-bis derivative formed above, which was mainly prepared for purification purposes. Levels between 50 and 200 ng were accurately determined, with a detection limit of 10 ng (20-ml sample).

Fabre et al. [503] determined aldosterone and its tetrahydro analogue in blood by oxidation with periodate to their respective γ-lactones, which chromatographed without degradation. The ^{14}C-labelled compound was added to the sample in order to correct for losses during extraction, TLC, oxidation and another TLC step. The conjugated carbonyl group in the androsterone γ-lactone gave the compound a sufficient response to the ECD to be determined in adrenal venous blood. For peripheral plasma samples this sensitivity was inadequate. Instead, the concentration of tetrahydroandrosterone was followed using the same approach plus the conversion of the lactone thereof into a 3α-monochlorodifluoroacetyl derivative, which was 100 times more sensitive, as described by Nicolis et al. [504]. The periodate oxidation introduced a useful purification as only compounds able to form the lactone were left in the chromatographic region of interest. Concentrations below 100 pg/ml could be detected.

25-Hydroxycholecalciferol was determined in plasma by Sklan et al. [505]. They separated the analyte from accompanying compounds by TLC and prepared the TFA esters, which rendered the compound ECD sensitive with an MDA of about 50 pg. A single derivative was formed in contrast to the cholecalciferol derivatives.

Keto steroids have been analysed as oximes. Koshy et al. [506] introduced the pentafluorobenzylhydroxylamine reagent for the determination of both mono- and diketo steroids. The reaction in pyridine had to be optimized with respect to the amount of reagent as an excess of reagent gave broad peaks. Geometrical isomers were formed but their retention times coincided on polar columns. The derivatives had good stabilities and the testosterone PFBz oxime derivative had an MDA of about 5 pg. This derivative was used by Nambara et al. [507] for the gas chromatography of some ketosteroids (silylation was also required), such as androsterone, dehydroepiandrosterone and 11-ketoandrosterone. Plasma proteins were precipitated with ethanol and the steroids subsequently extracted into ethyl acetate from an acidic aqueous phase followed by derivatization. 11-Keto groups did not react. Dehydroepiandrosterone was determined at levels down to 0.6 ng per 0.5 ml of plasma.

Wehner and Handke [508] presented a simple and rapid method for the determination of progesterone in rat plasma using a column extraction procedure with Extrelut as the only purification step. The derivatization procedure of Koshy et al. was adopted with a minor modification. The 3,20-di-PFBz oxime derivative was stable and permitted the analysis of down to 20 ng/ml in rat samples from 1.5 to 3 ml.

Estradiol and estrone were simultaneously determined in peripheral plasma samples as pentafluorophenylhydrazone derivatives by Mead et al. [509]. The compounds were extracted into dichloromethane, back-extracted with alkali and acidified and re-extracted with dichloromethane before methylation with dimethyl sulphate. After a TLC step to separate the two methyl esters, oxidation and pentafluorophenylhydrazone formation was followed by two TLC separations and finally by gas chromatography. On-column 0.15 ng could be detected, permitting the determination of about 1 ng in 20 ml of plasma.

Another approach to the analysis of plasma for testosterone was described by Muni et al. [510], who prepared the O-methyldichlorotetrafluoroketal after extraction into diethyl ether–chloroform (3:1) and TLC purification. Derivatization of the 17-hydroxyl group was carried out in benzene with 1,3-dichloro-1,1,3,3-tetrafluoroacetone and the hemiketal formed was methylated with diazomethane. As little as 0.5 ng/ml of testosterone in plasma could be detected with this stable derivative (10-ml sample).

From these examples it is clear that methods requiring less clean-up must be preferable if the same sensitivity can be obtained as with ECD. This now seems to be the case for many steroids, where mass fragmentographic methods are used for the most specific approaches and immunological methods for large series of samples.

13.25. CARBONYL COMPOUNDS

A very early application of ECD was reported by Dugan and Lundgren [511] for the detection of menadione, coenzyme Q_6 (or ubiquinone) and some related quinoid compounds in biological materials such as cell cultures. It was not possible to determine vitamin K_1 by GC at that time, but many other unidentified compounds appeared. α-Tocopherylquinone gave a single well defined peak at the 10-μg level with an early commercial detection device.

Morimoto et al. [439] reported that even α-tocopherol could be detected by ECD, indicating some electrophore configuration in common, e.g., the phenolate anion as suggested by Crow [512]. Their main interest was ubiquinone-10, which they extracted from human serum with acetone–ethanol (1:1) at 70°C. The residue after evaporation was purified twice by TLC. Blood was extracted with large volumes of solvent and purified by TLC. Both extracts contained large amounts of cholesterol, which interfered with the ECD and necessitated the TLC purification procedure. In normal persons the mean value of 0.5 μg/ml was easily measured, as the method could detect 0.1 μg/ml. By analogy with tocopherol the response of ubichromenol-9 was as good as that of the quinones.

A recent publication by Bechtold and Jähnchen [513] described the analysis of vitamin K_1 and its 2,3-epoxide in plasma. The naphthoquinone structure had a sufficient response to permit the detection of 3 ng/ml in plasma. The use of vitamin K_2 (menaquinone-4) as an internal standard compensated for the variation in recoveries. The 2,3-epoxide was only detected if the anticoagulant phenprocoumon had been given to the patient.

Other applications to be mentioned are more peripheral but still of interest. The bird repellant anthraquinone was determined in, e.g., peas [514]. The excellent sensitivity reported by Hartvig et al. [81] was confirmed. Acetone in butter was determined as a 2,4-dinitrophenylhydrazone derivative [515]. Diacetyl and 2,3-pentanedione were determined easily in wine after trapping the vapours in cold methanol [516]. The presence of acetoin was demonstrated after oxidation to diacetyl. It was confirmed that α-hydroxy acids did not interfere in the white and red wines studied. Diacetyl was determined in milk and the simplicity of the ECD assay was emphasized [517]. The ECD response was 1200 times higher than that with flame ionization detector and permitted the determination of 2 ng/ml.

Phanquone, an antiamebic, was determined in biological material using an interesting formation of an oxime, as reported by Degen et al. [518]. The phenanthroline dione was not easily extracted, and an aqueous reaction with methoxyamine was performed, yielding a derivative that was stable, extractable and had electrophore properties. For the derivative the detection limit was 15 ng/ml. The hydroquinone form of the drug was co-determined. A 10-methyl analogue was used as an internal standard. The optimal conditions for the formation of the oxime and its extraction were evaluated.

A gas chromatographic method was used to determine vinyl epoxide synthetase activity using trichloroethylene as a model substrate [519]. As mentioned under alcohols, chloral hydrate is a metabolite of trichloroethylene. In this case trichloroacetaldehyde was the subject of the analysis. The stability of this carbonyl compound in comparison with normal aldehydes explains the choice of this substrate. One approach made use of selective conversion to chloroform (the haloform reaction), which gave the highest sensitivity. With optimized conditions 3 ng/ml in an incubation mixture could be determined. This method was simpler and quicker than the other method, which was more generally applicable and used the reaction with pentafluorophenylhydrazine. The hydrazone formed apparently lost 1 mole of hydrogen chloride to give a conjugated system involving the perfluorinated aromatic ring. This product was an excellent electrophore as it was easily possible to detect 30 ng/ml.

The pentafluorophenylhydrazones have also been used to detect the low-molecular-weight aliphatic carbonyl compounds [520,521] but the PFBz oximes, as described for ketosteroids were found to be more suitable [522]. This derivatization reaction could be carried out in aqueous solution.

13.26. METABOLICALLY MODIFIED NITROGEN COMPOUNDS: N-OXIDES, N-HYDROXY AND N-NITROSO DERIVATIVES

In bioanalytical procedures indicine N-oxide was reduced with zinc–acetic acid in plasma [117] and chloropromazine N-oxide was reduced with iron(II) sulphate [68], both to their corresponding tertiary amines (see Section 13.5).

N-Hydroxyamines are found as labile intermediates in the metabolism of primary and secondary amines. They cannot be determined by direct gas chromatography because of their instability. Beckett and Achari [523] proposed derivatization procedures to stabilize some aliphatic primary hydroxylamines, with a preference for the silyl derivatives. The TFA derivatives, in which the acyl group reacted at the hydroxyl centre, also gave sharp and symmetrical peaks as the active hydrogen atom was no longer present [523]. The ECD response was not tested but a similar derivatization approach with HFBA was found to give a stable and very sensitive derivative with N-hydroxydesmethylpromethazine. This metabolite was found to interfere with desmethylpromethazine chromatographically in an ECD analysis of plasma [524].

An ECD application was reported for the determination of N-hydroxy-2-fluorenylacetamide in tests for N-hydroxylase activity in a rat-liver microsomal system. The product is known to be involved in chemical carcinogenesis. Razzouck et al. [525] developed a procedure with N-hydroxy-4-biphenylacetamide as an internal standard, which was added after the incubation. The first step in the derivatization was treatment with concentrated hydrochloric acid and extraction with cyclohexane, followed by reaction with either TFAA or monochloroacetic anhydride, giving N-chloro-2-fluorenyl-TFA-amide or the corresponding chloroacetyl derivative [525]. Using capillary columns the MDA was 20 pg, compared with 50 pg with a packed column. The difference was mainly due to better discrimination of the chemical background from the extracts with the capillary column.

N-Nitrosamines are of considerable concern as they can be formed metabolically and are suspected carcinogens. The use of perfluoroacylation of some nitrosamines was first described by Brooks et al. [526], who used HFBA mixed with pyridine to derivatize N-nitrosodiethylamine in extracts from urine. The structure of the derivative was not fully elucidated, but the response was shown to be very high. N-Nitrosopyrrolidine, N-nitrosopiperidine and N-nitrosodimethylamine had MDAs of about 17 ng. The flame ionization detector response for the last compound was about 6000 times lower. An application to the determination of nitrosamines in urine was reported by the same group [527]. Another approach to N-nitrosamine analysis was given by Alliston et al. [528], who determined such steam-volatile compounds in foodstuffs. After isolation by steam distillation the N-nitrosamines were reduced electrochemically to the corresponding amines, which were converted into HFB derivatives, with sensitivities as expected in the low picogram range. To differentiate from the amines in the sample, distillation had to be performed from alkaline, acidic and neutral solution. A third possibility was presented by Telling, who oxidized the nitroso group with peroxytrifluoroacetic acid to give a nitramine [529]. This electrophore had a sufficient response to permit the detection of 25 pg of dimethylnitramine and 110 pg of nitropyrrolidine. Nitrosamines with a phenyl substituent and N-nitrosopiperidine did not give the expected product upon oxidation. The critical step in this procedure was the distillation, whereas the oxidation step and the two clean-up steps on alumina and magnesium oxide–alumina were reproducible. The nitramine approach was used by Walker et al. for food [530].

13.27. MISCELLANEOUS SAMPLE TYPES

Parathion, an organophosphate compound with a *p*-nitrophenyl substituent, is used as an insecticide. It acts as an irreversible cholinesterase inhibitor, as also does its metabolite paraoxon. The latter has been used in the treatment of glaucoma. The determination of parathion in mouse plasma by ECD was reported by Vukovich et al. [531]. The acidified plasma sample was extracted with hexane containing methyl parathion as an internal standard. Plasma levels in the mouse ranged from 100 to 300 ng/ml. Paraoxon was not found in the samples although it had sufficient sensitivity to be detected if present.

4,4′-Diaminodiphenylsulphone is used in the treatment of leprosy. One metabolite of importance is 4-acetamidophenyl-4′-aminophenylsulphone, as this can also be hydrolysed back to the original sulphone. Burchfield et al. [532] reported a method for its determination in plasma. The *meta*-derivatives were added as internal standards prior to the extraction with ethyl acetate. After aqueous washes the compounds were back-extracted into sulphuric acid, treated with sodium nitrite and finally with iodine to yield the iodo derivatives. The acetyl group was then hydrolysed. These derivatives were chosen as both compounds gave the same derivative upon perfluoroacylation. The iodo derivatives required a high column temperature of 285°C but exhibited high sensitivity, with the diiodo derivative being superior. An MDA of 30 pg for the latter permitted the determination of a few ng/ml.

Cyclophosphamide, an anticancer drug, was determined in biological specimens from the mouse by Pantarotto et al. [533]. Isophosphamide was added as an internal standard and extraction with diethyl ether was followed by treatment with TFAA in ethyl acetate at 70°C for 20 min. One TFA group was introduced into the substrate, as shown by GC–MS. The presence of two chloroethyl groups and the TFA substituent provided a high ECD response for the derivatives (MDA ca. 25 pg) and permitted the detection of 10 ng/ml in a 0.1-ml sample.

Chloroform, a long established anaesthetic and also of concern as a contaminant in drinking water, has been analysed in rat blood and tissues by Vogt et al. [534]. The homogenized tissue was extracted with methylcyclohexane and an aliquot injected without any further treatment. The solvent peak and the effluent after the chloroform peak were vented, which kept the detector free of contamination for a longer time. The response was such that 2 pg could be detected, corresponding to 2.5 ng/g or 2.5 ng/ml of sample. Chloroform was found in several tissues, especially fat-containing tissues, after exposure to chlorinated water.

The interest in N-nitroso compounds has focused attention on the presence of nitrites in various biological samples, as nitrites might react with amines and amides to form N-nitroso compounds. Funazo et al. [535] recently presented a method based on the reaction of nitrite with *p*-bromoaniline and coupling the diazotized intermediate with copper(II) chloride to form *p*-bromochlorobenzene. The reaction was performed in a two-phase system with the sample, e.g., saliva acidified with hydrochloric acid and a toluene layer containing *p*-dichlorobenzene as an internal standard. The second step in the reaction required 2 h. A number of ions normally found in biological samples did not interfere, including the nitrate anion. The calibration graph was linear in the range 0.01–0.5 ppm (10–500 ng/ml). Lower concentrations could be detected if another aniline compound was used, e.g., nitroaniline. Another approach was used by Tesch et al. [536], who analysed for nitrate and nitrite after reaction with benzene to nitrobenzene in a two-phase system. Blood had to be deproteinized and the supernatant was then treated with concentrated sulphuric acid in the presence of benzene. Nitrate was thus determined directly whereas nitrite had first to be oxidized with hydrogen peoxide to nitrate. Nitrite is not converted to nitrobenzene without the oxidation step. Owing to reductive activity in saliva, this type of sample had to be treated with a preservative immediately after collection. Using small samples such as finger-tip blood the detection limit was 0.1 ppm. The analysis of saliva and blood was complicated by the presence of interfering ions such as chloride and thiocyanate. The addition of silver(I) sulphate removed these ions. In the analysis of nitrite, oxidation with peroxide removed these interferences. In a recent study nitrate was made to react with either benzotrifluoride or tetrafluoroanisole to give nitro compounds with a higher response [537]. The halide interference was counteracted with silver acetate, permitting the analysis of as little as 10 ppb.

Other ions that can be derivatized include cyanide and thiocyanate. Cyanide was converted to cyanogen chloride in a reaction with chloramine-T (sodium *p*-toluenesulphonchloramide) [538]. Cyanide was isolated from the biological sample using microdiffusion into sodium hydroxide. This method permitted the detection of 25 ng/ml with an MDA of 12.5 pg of cyanide.

A method for the determination of thiocyanate in tissues and body fluids was presented by De Brabander and Verbeke [539]. This ion is normally present in body fluids and has a thyreostatic effect. The proposed method made use of the reaction of thiocyanate with bromine to form cyanogen bromide. Interference from cyanide could be eliminated if the sample was treated with alkaline formaldehyde. The product was extracted with toluene containing 2-bromopropane as an internal standard. The detection limit was 0.2 ppm. Protein-containing samples had to be deproteinized as otherwise false-positive values were obtained. The most efficient reagent was methanol. The method was used to detect the anion in plasma, urine, liver and kidney.

Ever since the discovery that selenium(IV) could complex with phenylenediamines to form the corresponding piazselenols, this process has been used both for fluorescence and chromatographic analysis. Young and Christian [540] used the reaction with 2,3-diaminonaphthalene, the on-column detection limit being 0.5 ng, corresponding to a concentration of 10 ng/ml in extracts.

Shimoishi [541] investigated 13 piazselenols and found that the 4,6-dibromopiazselenol gave the highest response with an ECD temperature of 280°C. The 5-nitro compound was one third as sensitive [541]. Selenium(VI) was determined in liver samples using a reduction step with concentrated hydrochloric acid. It was possible to determine 1 ng of selenium extracted as the piazselenol into 1 ml of toluene. Poole et al. [542] used the 5-nitropiazselenol to determine selenium in placenta, hair, blood and urine. The optimized digestion procedure used nitric acid–magnesium nitrate, which resulted in very clean final toluene extracts. They studied the temperature dependence of the ECD response and found that all piazselenols exhibited higher responses with increasing detector temperature. Using a detector temperature of 350°C the 5,6-dichloropiazselenol and the 5-nitro compound could be used to detect 1 pg of selenium.

To determine selenium in biological materials Bycroft and Clegg [543] used 5-chloro- and 5-bromopiazselenols, the MDAs being 27 and 13 ng, respectively. Kurahashi et al. [544] recently described a procedure for the determination of selenium in human blood. The practical detection limit was 2 ng.

Some elements can form volatile metal β-diketonates. An example is beryllium, which was determined in biological samples as its trifluoroacetylacetonate, the ECD response of which permitted the determination of 20 ng/ml in blood [545]. Blood and aqueous solutions gave similar results except at lower concentrations, where the recovery was higher for samples of blood. EDTA was preferred to heparin as an anticoagulant fewer interferences were introduced.

13.28. SOME SPECIAL APPLICATIONS

13.28.1. Purity control

The determination with ECD of trace amounts of organic compounds as impurities in synthesized substances has been described in a few instances. When a single impurity is of interest, derivatization procedures can be used but an impurity pattern can be more difficult to produce in this way. Cyclohexylamine, an impurity in the artificial sweetener cyclamate, could be determined at the ppm level by ECD. Gunner and O'Brien [546] used picryl chloride to form a N-trinitrophenyl derivative. The amine together with trimethylcyclohexylamine was separated from the main component by base extraction into chloroform. The derivative was prepared under reflux and the residue taken up in benzene. The MDA was 0.1 ng of the derivative, corresponding to 30 pg of cyclohexylamine. Practical levels of 1 ppm could be determined. Weston and Wheals [547] used the DNP derivative and were able to detect 1 ppm of cyclohexylamine. Excess of reagent could interfere at low levels but treatment of the reaction mixture with 6 *N* sodium hydroxide solution at 60°C for a further 20 min hydrolysed the excess of 2,4-dinitrofluorobenzene. Solomon et al. [548] used the same approach to analyse for cyclohexylamine in sodium cyclamate but used sodium bicarbonate to remove the excess of reagent; 10 pg could be detected permitting the determination of 10 ppb.

Chlorodibenzo-*p*-dioxins are of concern as impurities in chlorophenoxyalkanoic acids and in pentachlorophenol. As examples of the importance of knowing impurity levels down to ppb levels they are outstanding. Edmunds et al. [549] used an alumina column clean-up procedure to separate the dioxins from 2,4,5-trichlorophenoxyacetic acid and found the levels of 2,3,7,8-tetrachlorodibenzo-1,4-dioxin to be from 5 to 500 ppb. Mieure et al. [550] analysed for the chlorinated benzodioxins in pentachlorophenol. They used a two-column clean-up procedure to eliminate phenols and diphenyl ethers. In the final fraction taken for analysis both chlorodibenzodioxins and chlorodibenzofurans could be determined. It was claimed that ppb to ppm concentrations could be monitored.

Tranexamic acid is the *trans*-isomer of 4-aminomethylcyclohexanecarboxylic acid, and the *cis*-isomer is present as an impurity. To analyse for this isomer in concentrations below 0.1% the same procedure as described for the determination of tranexamic acid in biological fluids was used [360], which involves coupling with 4-fluoro-3-nitrobenzotrifluoride. To detect the low levels of the *cis*-isomer (*trans*:*cis* ratio = 10,000:1) it was difficult to achieve a quantitative reaction as the large sample required consumed the reagent. However, with the use of ECD it was possible to work with more reasonable amounts (50 mg). This is an example showing the advantage of a selective and sensitive detector in purity control [551]. See also Section 13.6 on morphine in heroin [103].

13.28.2. Test for biological activity

A microassay for the determination of monoamine oxidase activity in serum and platelets using ECD was presented by Farris et al. [552]. They selected *m*-iodobenzylamine as a substrate and after incubation obtained *m*-iodobenzaldehyde. The *m*-iodobenzaldehyde was found to be resistant to further oxidation, was easily extracted and could be sensitively detected (MDA < 1 pg). Serum samples from 5 to 200 μl could be used and would allow routine monitoring of this important enzyme in patients. An in vitro assay for microsomal thioether oxidation was presented by Nigg et al. [553]. They used *p*-chlorothioanisole as substrate and after incubation analysed for the corresponding sulphoxide and sulphone. Only the sulphoxide was present in microsomal preparations from the mouse and housefly. It was possible to follow a few nanomoles per milligram of protein per hour. Sulphur hexafluoride is an excellent electrophore and an application was evaluated in which this compound was used to determine pulmonary clearance [554]. With calibration graphs the dynamic range was between 10^{-10} and 10^{-8} ml/ml in samples. Some examples have also been given in previous sections, e.g., a test for vinyl epoxide synthetase activity (see carbonyl compounds, Section 13.25) and a test for N-hydroxylase activity (see metabolically modified nitrogen compounds, Section 13.26). Dopa decarboxylase activity was monitored by analysing in vitro preparations for dopamine (see catecholamines, Section 13.10).

13.28.3. Confirmation of identity

It is often important to be able to confirm the identity of a certain component by an alternative technique or another derivative. The presence of *p*-nitrophenol, a metabolite of parathion, in urine was demonstrated in both exposed and normal persons by Kirby et al. [555]. For confirmation they used a peak shift technique that involved reduction of the nitro group with chromium(II) chloride and reaction of the amino group with HFBA. This derivative had a shorter retention time and gave a 15-fold higher response. A few ng/ml could be verified and quantified. In order to identify unknown components of ECD profiles using GC–MS, King et al. [556] developed a technique that utilized a homologous series of *n*-alkylamines as perfluoroacyl derivatives. These compounds functioned as a reference grid of retention parameters and were applicable to both detection systems. It was also possible to monitor a specific fluorine-containing fragment ion and to correlate this with the profile obtained with ECD. With this reference grid it would be easier to identify an unknown component by GC–MS, as sometimes the mass spectrometric behaviour of a derivative might differ considerably from that in the ECD.

13.29. APPLICATIONS OF ELECTRON-CAPTURE NEGATIVE ION CHEMICAL IONIZATION MASS SPECTROMETRY

In 1978 Hunt and Crow [557] described a study in which the above technique was used. They proposed that the determination of trace amounts of many compounds would be possible by the use of GC–MS with ionization by resonance electron capture. Even if very few molecules capture thermal electrons and form stable molecular ions or high

TABLE 13.1

MINIMUM DETECTABLE CONCENTRATION VALUES FOR SOME COMPOUNDS OR DERIVATIVES

Functional group	Compound or derivative	MDC (mole/s × 10^{16})	Detector type ^{3}H	Detector type ^{63}Ni	Detector temperature (°C)	Temperature dependence studied	Reference
Primary amine (2)	DNP isopropylamine	64	×		225		Walle 1968 [9]
	DNP aniline	17					"
	DNP benzylamine	210					"
	DNP amphetamine	200					"
	Mono-TFA dodecylamine	670	×		150 (foil)		Ehrsson and Brötell 1971 [9]
	Di-TFA dodecylamine	38					"
	Mono-HFB dodecylamine	100					"
	Di-HFB dodecylamine	6					"
	Di-TFA phenethylamine	23					"
	Di-TFA benzylamine	20					"
	Di-TFA amphetamine	13					"
	HFB amphetamine	0.92	×		165		Walle and Ehrsson 1970 [10]
	HFB phentermine	1.4					
	HFB phenethylamine	3.3*		×	300		Moffat and Horning 1970 [27]
	DNP phenethylamine	6.4					
	PFBz ald. SB phenethylamine	1.4*					
	HFB ald. SB phenethylamine	4.6*					
Secondary amines (3)	DNP diethylamine	57	×		225		Walle 1968 [9]
	DNP N-methylaniline	27					"
	HFB methamphetamine	0.55	×		165		Walle and Ehrsson 1970 [10]
	HFB methoxyphenamine	2.4					
	HFB phenmetrazine	0.13					
	TFA nortriptyline	1900		×	260		Ervik et al. 1970 [37]
	HFB nortriptyline	2.6					"
	Mono-TFA desipramine	1000					"
	Di-TFA desipramine	1.6					"
	Mono-HFB desipramine	7.6					"
	Di-HFB desipramine	0.6					"

	HFB terodiline	12		×	190 (foil)	+	Hartvig et al. 1974 [52]
	TFA terodiline	770					
	Methindione	3.6	×		180	+	Vessman et al. 1974 [59]
Secondary and tertiary amines (5)	PFBz carbamate of diphenhydramine	1.4		×	190	+	Hartvig and Vessman 1974 [86]
	PFBz carbamate of imipramine	1.6		×	320	+	Hartvig et al. 1976 [89]
	PFBz carbamate of trimipramine	1.8					
	PFBz carbamate of 2-chloroimipramine	1.6					
	TCE carbamate of promazine	3.9		×	320	+	Hartvig et al. 1976 [87]
	PFBz carbamate of promazine	1.4					
	TCE carbamate of chlorpromazine	4.1					
	PFBz carbamate of chlorpromazine	1.7					
	TCE carbamate of chlorprothixene	3.9					
	PFBz carbamate of chlorprothixene	1.3					
	TCE carbamate of pethidine	3.0		×	300		Hartvig et al. 1976 [51]
	TCE carbamate of ANP-4364	3.3		×	300		Yamaguchi 1979 [96]
Tertiary amines (5)	Ring TFA derivative of N,N-dimethylaniline	3.0		×	140	+	Walle 1975 [99]
(6)	PFBz pentazocine	0.8		×	275		Brötell et al. 1973 [112]
Amino alcohols (8)	Di-TFA ephedrine	10	×		165		Walle and Ehrsson 1970 [10]
	Di-HFB ephedrine	5					
	Di-TFA alprenolol	2.8		×	270		Walle 1974 [136]
	Di-TFA propranolol	2.0					”
	Di-TFA oxprenolol	4.3					”
	Di-TFA dichloroproterenol	0.5					”
	Di-TFA desisopropylpropranolol	1.5					”
Sulphonamides and sulphonyl-ureas (16)	TFA N-ethylbenzenesulphonamide	9.0		×	270		Gyllenhaal and Ehrsson 1975 [252]
	HFB N-ethylbenzenesulphonamide	5.0					”
	PFBz N-ethylbenzenesulphonamide	2.0					
	N^1-Methylsulphapyridine	8.0		×	350		Gyllenhaal et al. 1978 [254]
	N^1-Methylsulphamerazine	3.9					”
	N^1-Methylsulphalene	2.2					”
	N^1-Methylsulphadimidine	3.5					”

(Continued on p. 340)

TABLE 13.1 *(continued)*

Functional group	Compound or derivative	MDC (mole/s × 10^{16})	Detector type		Detector temperature (°C)	Temperature dependence studied	Reference
			^{3}H	^{63}Ni			
	N^1-Methylsulphaisodimidine	3.6					Gyllenhaal et al. 1978 [254]
	N^1-Methylsulphadiazine	4.2					,,
	N^1-Methyl-N^4-acetylsulphadiazine	6.7					,,
	N-Methylsaccharin	2.8		×	350		Hartvig et al. 1978 [259]
	Trimethylfurosemide	70		×	275		Lindström et al. 1974 [263]
	Trimethylacetazolamide	2.3		×	270	+	Fagerlund et al. 1979 [268]
	N-Methyl-TFA tolbutamide	2.8		×	275		Braselton 1977 [271]
	N-Methyl tolbutamide	1.8		×	330	+	Hartvig et al. 1980 [274]
Biguanides and related compounds (15)	Chlorodifluoroacetyl condensation product of moroxydine	8.0		×	280		Gröningsson et al. 1981 [249]
	Benzisoxazoleacetamidoxim	1.4		×	200		Yamaguchi et al. 1978 [247]
Carbamates (13)	2-Methoxyethylcarbamic acid propyl ester TFA derivative	24	×		150 (foil)		Mellström and Ehrsson 1974 [227]
Imides (12)	Di-PFBz barbital	0.22		×	250	+	Walle 1975 [219]
	Di-PFBz phenobarbital	0.26					,,
	Mono-PFBz hexobarbital	0.34					,,
	Mono-PFBz diphenylhydantoin	0.82					,,
	Di-PFBz phenobarbital	0.50		×	300		Gyllenhaal et al. 1976 [221]
Amides (11)	PFB methyprylon	0.7	×		150 (foil)		Ehrsson and Mellström 1972 [30]
	PFB phenacetin	1.7					
	PFB acetanilide	1.5					,,
Catecholamines (10)	TMS PFBz ald. SB tyramine	3.3*		×	300		Moffat and Horning 1970 [27]
	TMS PFBz ald. SB dopamine	2.5*					
	TMS PFBz ald. SB normetanephrine	2.7*					,,
	TMS PFBz ald. SB norepinephrine	2.1*					,,

Heterocyclics (9)	PFBz theophylline	4.0	×		225		Arbin and Edlund 1975 [158]
	HFB proxyphylline	0.3	×		225		Arbin et al. 1976 [160]
	HFB dyphylline	0.2					,,
	O-Methylclioquinol	1.6		×	300		Hartvig and Fagerlund 1977 [174]
	8-Methoxypsoralen	20		×	200	+	Ehrsson et al. 1977 [187]
	Griseofulvin	5.9		×	290		Ehrsson and Tilly 1973 [289]
Amino acids (19)	DNP glycine methyl ester	4.5*	×		200		Landowne and Lipsky 1963 [335]
	N-Chloroacetylamino acid methyl ester	200*	×				Landowne 1965 [338]
	N-(3-Nitrophthalyl)amino acid methyl ester	7.5*					,,
	N-Phthalylamino acid methyl ester	4.2*					,,
Benzodiazepines (17)	Methylnitrazepam	5.2		×	290		Ehrsson and Tilly 1973 [289]
	Amino metabolite of nitrazepam	2000					
	Acetamido metabolite of nitrazepam	60					,,
	Amino-3-hydroxy metabolite of nitrazepam	10					,,
	N^1,O^3-Dimethyloxazepam	70		×	300		Vessman et al. 1977 [313]
Phenols (22)	HFB methyl-*p*-hydroxybenzoate	4.7		×	150 (foil)		Ehrsson et al. 1971 [442]
	HFB nitrophenol	2.2					,,
	HFB hexestrol	1.6					,,
	HFB *p*-hydroxystilbene	5.8					,,
	TFA *p-tert.*-butylphenol	3700					,,
	PFB *p-tert.*-butylphenol	5.9					,,
	HFB *p-tert.*-butylphenol	5.3					,,
	TFA *p*-bromophenol	2.1					,,
	PFB *p*-bromophenol	1.7					,,
	HFB *p*-bromophenol	1.6					,,

(Continued on p. 342)

TABLE 13.1 *(continued)*

Functional group	Compound or derivative	MDC (mole/s × 10^{16})	Detector type		Detector temperature (°C)	Temperature dependence studied	Reference
			^3H	^{63}Ni			
Carboxylic acids (20)	PFBz palmitate	1.7		×	300	+	Gyllenhaal et al. 1976 [363]
	PFBz tropicate	0.86		×	300		Greving et al. 1978 [364]
	PFBz benzilate	1.2					"
	PFBz diphenylacetate	0.39					"
	PFBz cyclohexylphenylacetate	0.30					"
	Dinonylphthalate	520	×		220 (foil)		Vessman and Rietz 1974 [373]
	Indomethacin, methyl ester	1.1		×	300		Arbin 1977 [408]
	Ketoprofen, methyl ester	6.0		×	240		Stenberg et al. 1979 [420]
Carbonyl compounds (25)	Benzophenone	31	×		104 (foil)	+	Vessman and Hartvig 1971 [565]
	p-Chlorobenzophenone	38				+	"
	2-Benzoylpyridine	48				+	"
	p-Chloroacetophenone	13			100 (foil)	+	"
	Acetophenone	1300				+	"
	2-Amino-5-chlorobenzophenone	70	×		220		Vessman et al. 1972 [306]
	(hydrolysis product of nordiazepam and oxazepam)	50	×		170		Vessman and Hartvig 1971 [565]
	Anthraquinone	47	×		200 (foil)		Hartvig et al. 1976 [81]
				×		+	
Alcohols (23)	di-TFA propranololglycol	0.7		×	280		Saelens et al. 1976 [137]
	HFB benzylalcohol	9.0	×		165		Walle and Ehrsson 1970 [10]
	HFB 2-phenethylalcohol	23					"
	HFB 3-phenylpropanol	47					"
	HFB octanol	54					"
	HFB benzhydrol	4					"

*Values with have been corrected from the original figure due to different definition of the signal-to-noise level (from two to three times).

mass fragment ions, use of derivatization reactions as in ECD methods are possible. This was exemplified in their paper, where femtogram amounts (10–25 fg) were detected of dopamine as a pentafluorobenzaldehyde condensation product, tetrahydrocannabinol and amphetamine as PFB derivatives and amphetamine as a tetrafluorophthalimide.

The first application of this powerful technique was demonstrated by Lewy and Markey [62], who determined melatonin at concentrations as low as 1 pg/ml in human plasma. The derivative obtained after treatment with PFPA and trimethylamine was a stable spirocyclic compound. The sensitivity for this electron-capturing compound was increased 150-fold compared with ECD using this detection device. Plasma polyamines were determined as TFA derivatives by Shipe et al. [558]. Deuterated analogues were used as internal standards and "carriers". The more TFA groups were present, the higher was the response in the negative ion mode.

Garland and Min [559] described a method for the determination of clonazepam in human plasma using gas chromatography–negative ion chemical ionization mass spectrometry (GC–NICIMS). As clonazepam exhibited a good response when analysed by ECD it was of interest to explore the negative ion mode. The molecular ion and the $(M-H)^-$ ion were the principal ions. As the internal standard, a $^{15}N^{18}O_2$ analogue, contributed to the ion from clonazepam, the practical detection limit was 0.1 ng/ml. The same approach was applied to the determination of a trimethylprostaglandin E_2 analogue by Min et al. [405]; this method was commented on under carboxylic acids. Using capillary columns the sensitivity limit of this assay was 100 pg/ml. Garland et al. [560] determined ipronidazole, a nitro-substituted compound, for confirmatory purposes by GC–NICIMS. In skin and turkey muscle it was possible to determine 2 ppb. Likewise it has been possible to determine polychlorinated dibenzo-*p*-dioxins in biological samples [561] and to screen for residues of chlorophenol and chlorobiphenyls at ppb levels in human adipose tissue [562]. The use of GC–NICIMS will probably be of great importance for those problems where GC–ECD will be inadequate. Much of what has been gained in experience with ECD derivatization can probably be used. However, the fragmentation can sometimes be so extensive that the selectivity and sensitivity are lost, as for the pentafluorophenyldimethylsilyl derivatives of alcohols as discussed by Crow [512].

13.30. COMPARATIVE SENSITIVITY VALUES FOR THE ELECTRON-CAPTURE DETECTOR RESPONSE OF SOME COMPOUNDS IN THIS CHAPTER

The response of the ECD towards different compounds has been expressed in various ways. In the previous part of this chapter the practical figures reported in the literature have been used, e.g., the minimum detectable amount in picograms or nanograms. For purposes of comparison this is not the best approach. Very early Landowne and Lipsky gave an expression for the minimum detectable amount in moles per second in papers on the response of the ECD towards isomeric butyl halides [563], steroid haloacetates [487] and the DNP derivatives of amino acid methyl esters]335]. This approach was followed by other workers [9,27]. Thus Moffat and Horning [27] computed the sensitivity values from the quotient Q/w_b, where Q is the amount of derivative (in moles) to give a peak twice as high as the noise and w_b is the peak width in seconds (from the intersection of the tangents with the baseline). Walle [9] used the same approach but defined Q as the amount of derivative (in moles) that gave a peak three times the noise level.

The sensitivity value in moles per second has in many papers been taken to be the minimum detectable amount (MDA), but in recent years the same figures have been expressed as the minimum detectable concentration (MDC) [254]. This is preferred to the former abbreviation as in fact the sensitivity value, MDC, expresses a value related to the maximum concentration in the eluted peak (see for instance a discussion by Schill [564]).

As most of the reported sensitivity values following those of Walle and co-workers have used his definition with a peak height three times the average noise level, it was considered appropriate to use this in Table 13.1. Other values have been corrected to this. In Table 13.1 MDC values are compiled for compounds mentioned in papers, which are referred to or have some connection with papers discussed in this chapter.

Table 13.1 gives an idea of the response to be expected from a given derivative. It must be emphasized, however, that the overall structure is of importance to the final response, e.g., there are clear differences between aliphatic and aromatic amino derivatives and halogen-containing phenols add to the response of their TFA derivatives. Moreover, some of the older values from ^{3}H detectors could probably be improved with the high-temperature detectors available today. It is of interest to see how high the response is for bis-TFA derivatives in comparison with both mono- and di-HFB derivatives, e.g., for primary amines, amino alcohols and the very special case of desipramine. This has been pointed out earlier as a very useful means for selective derivatization compared with the use of HFB derivatives [19,37,148].

The low sensitivity of the metabolite of nitrazepam with the nitro group reduced to an amino group was commented on before. It is also evident from Table 13.1 that the amino-3-hydroxy metabolite has a sensitivity comparable to that of nitrazepam. This is related to the rearrangement to a quinazolinone structure, which thus would be a better electrophore than the aminobenzodiazepine.

13.31. ABBREVIATIONS USED

DNP	dinitrophenyl
ECD	electron-capture detector; electron-capture detection
GC	gas chromatography
GC–MS	gas chromatography–mass spectrometry
HFB	heptafluorobutyryl
HFBA	heptafluorobutyric anhydride
HFBI	heptafluorobutyrylimidazole
MDA	minimum detectable amount, usually given as the amount injected on to column that gives a response two or three times the average noise. The retention time should preferably be in the range 3–7 min
MDC	minimum detectable concentration (in mole/s). This is a sensitivity value that is independent of retention time and column performance. The response should be given for the amount injected that gives a signal three times the average noise
PFB	pentafluorobenzoyl

PFBz pentafluorobenzyl
PFP pentafluoropropionyl
PFPA pentafluoropropionic anhydride
SB Schiff base
TCE trichloroethyl
TFA trifluoroacetyl
TFAA trifluoroacetic anhydride
TLC thin-layer chromatography
TMS trimethylsilyl

REFERENCES

1 L.M. Cummins, in I.I. Domsky and J.A. Perry (Editors), *Recent Advances in Gas Chromatography*, Marcel Dekker, New York, 1971, pp. 313–340.
2 M. Hattori, *Takeda Kenkyusho Nempo (Annu. Rep. Takeda Res. Lab.)*, 32 (1973) 231–250.
3 E. Änggård and A. Hankey, *Acta Chem. Scand.*, 23 (1969) 3110–3119.
4 S.B. Matin and M. Rowland, *J. Pharm. Sci.*, 61 (1972) 1235–1240.
5 J.S. Noonan, P.W. Murdick and R.S. Ray, *J. Pharmacol. Exp. Ther.*, 168 (1969) 205–209.
6 R.B. Bruce and W.R. Maynard Jr., *Anal. Chem.*, 41 (1969) 977–979.
7 G.R. Wilkinson, *Anal. Lett.*, 3 (1970) 289–298.
8 R.O. Bost, C.A. Sutheimer and I. Sunshine, *Clin. Chem.*, 22 (1976) 789–801.
9 T. Walle, *Acta Pharm. Suecica*, 5 (1968) 367–384.
10 T. Walle and H. Ehrsson, *Acta Pharm. Suecica*, 7 (1970) 389–406.
11 T. Walle and H. Ehrsson, *Acta Pharm. Suecica*, 8 (1971) 27–38.
12 D.D. Clarke, S. Wilk and S.E. Gitlow, in H.A. Szymanski (Editor), *Biochemical Applications of Gas Chromatography*, Vol. 2, Plenum Press, New York, 1968, pp. 137–163.
13 L.M. Cummins and M.J. Fourier, *Anal. Lett.*, 2 (1969) 403–409.
14 S. Willox and B.N. Singh, *J. Chromatogr.*, 128 (1976) 196–198.
15 A. Frydman, J.-P. Lafarge, F. Vial, R. Rulliere and J.-M. Alexandre, *J. Chromatogr.*, 145 (1978) 401–411.
16 A. Sioufi and F. Pommier, *J. Chromatogr.*, 183 (1980) 33–39.
17 P. Biandrate, G. Tognoni, G. Belvedere, A. Frigerio, M. Rizzo and P.M. Morselli, *J. Chromatogr.*, 74 (1972) 31–34.
18 R. Venkataramanan and J.E. Axelson, *J. Pharm. Sci.*, 67 (1978) 201–205.
19 H. Ehrsson and H. Brötell, *Acta Pharm. Suecica*, 8 (1971) 591–598.
20 R.J. Perchalski, B.J. Wilder and R.H. Hammer, *J. Pharm. Sci.*, 63 (1974) 1489–1491.
21 G. Belvedere, G. Franchi and A. Frigerio, *J. Chromatogr.*, 84 (1973) 171–174.
22 T.M. Ludden, D. Lalka, M.G. Wyman, B.N. Goldreyer, K.D. Haegele, D.T. Brooks, I. Davila and J.E. Wallace, *J. Pharm. Sci.*, 67 (1978) 371–373.
23 Y.K. Tam and J.E. Axerlson, *J. Pharm. Sci.*, 67 (1978) 1073–1077.
24 L.M. Ross-Lee, M.J. Eadie, F. Bochner, W.D. Hooper and J.H. Tyrer, *J. Chromatogr.*, 183 (1980) 175–184.
25 Y.K. Tam and J.E. Axelson, *J. Chromatogr.*, 170 (1979) 157–163.
26 D.D. Clarke, S. Wilk and S.E. Gitlow, *J. Gas Chromatogr.*, 4 (1966) 310–313.
27 A.C. Moffat and E.C. Horning, *Anal. Lett.*, 3 (1970) 205–216.
28 I.L. Martin and G.B. Baker, *J. Chromatogr.*, 123 (1976) 45–50.
29 I.L. Martin and G.B. Baker, *Biochem. Pharmacol.*, 26 (1977) 1513–1516.
30 H. Ehrsson and B. Mellström, *Acta Pharm. Suecica*, 9 (1972) 107–114.
31 K. Blau, I.M. Claxton, G. Ismahan and M. Sandler, *J. Chromatogr.*, 163 (1979) 135–142.
32 D.J. Edwards and K. Blau, *Anal. Biochem.*, 45 (1972) 387–402.

33 P.S. Doshi and D.J. Edwards, *J. Chromatogr.*, 176 (1979) 359–366.
34 P.S. Doshi and D.J. Edwards, *Life Sci.*, 26 (1980) 1947–1953.
35 M. Makita, S. Yamamoto and M. Kono, *Clin. Chim. Acta*, 61 (1975) 403–405.
36 S.F. Sisenwine, J.A. Knowles and H.W. Ruelius, *Anal. Lett.*, 2 (1969) 315–324.
37 M. Ervik, T. Walle and H. Ehrsson, *Acta Pharm. Suecica*, 7 (1970) 625–634.
38 R. Lundgren, A. Olsson and G. Plym Forshell, *Acta Pharm. Suecica*, 14 (1977) 81–94.
39 O. Borgå and M. Garle, *J. Chromatogr.*, 68 (1972) 77–88.
40 U.P. Geiger, T.G. Rajagopalan and W. Riess, *J. Chromatogr.*, 114 (1975) 167–173.
41 A.R. Viala, J.-P. Cano, A.G. Durand, T. Erlenmaier and R.M. Garreau, *Anal. Chem.*, 49 (1977) 2354–2357.
42 R. Huffman, J.W. Blake, R. Ray, J. Noonan and P.W. Murdick, *Anal. Chem.*, 12 (1974) 382–384.
43 F.T. Delbeke and M. Debackere, *J. Chromatogr.*, 106 (1975) 412–417.
44 C.-G. Hammar and B. Holmstedt, *Experientia*, 24 (1968) 98–100.
45 A.G. Zacchei, L.L. Weidner and T.I. Wishousky, *J. Chromatogr.*, 162 (1979) 39–45.
46 G. Belvedere, G. Tognoni and P.L. Morselli, *Eur. J. Clin. Pharmacol.*, 5 (1972) 62–64.
47 S. Caccia and A. Jori, *J. Chromatogr.*, 144 (1977) 127–131.
48 S.B. Matin, M. Rowland and N. Castagnoli, Jr., *J. Pharm. Sci.*, 62 (1973) 821–823.
49 G. Bianchetti, M. Mitchard and P.L. Morselli, *J. Chromatogr.*, 152 (1978) 87–95.
50 J. Vessman, in E. Reid (Editor), *Methodological Surveys in Biochemistry, Vol. 7, Blood Drugs and Other Analytical Challenges*, Ellis Horwood, Chichester, 1978, pp. 119–127.
51 P. Hartvig, K.-E. Karlsson, C. Lindberg and L.O. Boreus, *Eur. J. Clin. Pharmacol.*, 11 (1977) 65–69.
52 P. Hartvig, G. Freij and J. Vessman, *Acta Pharm. Suecica*, 11 (1974) 97–108.
53 J. Vessman and S. Strömberg, *Acta Pharm. Suecica*, 6 (1969) 505–518.
54 P. Hartvig, *Acta Pharm. Suecica*, 11 (1974) 109–114.
55 D.G. Crosby and J.B. Bowers, *J. Agr. Food Chem.*, 16 (1968) 839–843.
56 C. Barsuhn, *Life Sci.*, 18 (1976) 419–422.
57 A.G. Zacchei and L.L. Weidner, *Anal. Biochem.*, 87 (1978) 586–593.
58 I. Nitsche, F. Selenka and K. Ballschmiter, *J. Chromatogr.*, 94 (1974) 65–73.
59 J. Vessman, S. Strömberg and G. Freij, *J. Chromatogr.*, 94 (1974) 239–244.
60 P.H. Degen, J.R. Do Maral and J.D. Barchas, *Anal. Biochem.*, 45 (1972) 634–644.
61 S.T. Christian, F. Benington, R.D. Morin and L. Corbett, *Biochem. Med.*, 14 (1975) 191–200.
62 A.J. Lewy and S.P. Markey, *Science*, 201 (1978) 741–743.
63 T.R. Bosin, A.R. Buckpitt and R.P. Maickel, *J. Chromatogr.*, 94 (1974) 316–320.
64 S.H. Curry, *Anal. Chem.*, 40 (1968) 1251–1255.
65 S.H. Curry, in I.S. Forrest, C.J. Carr and E. Usdin (Editors), *The Phenothiazines and Structurally Related Drugs*, Raven Press, New York, 1974, pp. 335–345.
66 S.H. Curry, *Psychopharmacol. Commun.*, 2 (1976) 1–15.
67 G.W. Christoph, D.E. Schmidt, J.M. Davis and D.W. Janowsky, *Clin. Chim. Acta*, 38 (1972) 265–270.
68 S.H. Curry and S. Evans, *J. Pharm. Pharmacol.*, 28 (1976) 467–468.
69 T.N. Lindley, *J. Anal. Toxicol.*, 3 (1979) 18–20.
70 S.H. Curry, in E. Reid (Editor), *Methodological Surveys in Biochemistry, Vol. 7, Blood Drugs and Other Analytical Challenges*, Ellis Horwood, Chichester, 1978, pp. 109–117.
71 J.S. Noonan, J.W. Blake, P.W. Murdick and R.S. Ray, *Life Sci.*, 11 (1972) 363–373.
72 A.P. De Leenheer and L.M.R. Vandecasteele-Thienpont, *J. Chromatogr.*, 175 (1979) 301–306.
73 A.P. De Leenheer and L.M.R. Vandecasteele-Thienpont, *J. Pharm. Sci.*, 69 (1980) 99–100.
74 R. Woestenborghs, M. Michiels and J. Heykants, *J. Chromatogr.*, 164 (1979) 169–176.
75 L. Neelakantan and H.B. Kostenbauder, *Anal. Chem.*, 46 (1974) 452–454.
76 J.W. Blake, R.S. Ray, J.S. Noonan and P.W. Murdick, *Anal. Chem.*, 46 (1974) 288–289.
77 J.J. Javaid, H. Dekirmenjian, J.M. Davis and C.R. Schuster, *J. Chromatogr.*, 152 (1978) 105–113.
78 P. Hartvig and B. Näslund, *J. Chromatogr.*, 111 (1975) 347–354.
79 R.B. Bruce, J.E. Pitts and F.M. Pinchbeck, *Anal. Chem.*, 40 (1968) 1246–1250.
80 J.E. Wallace, H.E. Hamilton, L.K. Goggin and K. Blum, *Anal. Chem.*, 47 (1975) 1516–1519.

81 P. Hartvig, S. Strandberg and B. Näslund, *J. Chromatogr.*, 118 (1976) 65–74.
82 J.W. Barnhart and J.D. Johnson, *Anal. Chem.*, 49 (1977) 1085–1086.
83 R. Tham, B. Norlander, Ö. Hägermark and L. Fransson, *Arzneim.-Forsch.*, 28 (1978) 1017–1020.
84 P. Degen and W. Riess, *J. Chromatogr.*, 85 (1973) 53–63.
85 P. Hartvig and L. Maukonen, *Farm. Aikak.*, 83 (1974) 141–147.
86 P. Hartvig and J. Vessman, *Anal. Lett.*, 7 (1974) 223–231.
87 P. Hartvig, N.-O. Ahnfelt and K.E. Karlsson, *Acta Pharm. Suecica*, 13 (1976) 181–189.
88 P. Hartvig and J. Vessman, *J. Chromatogr. Sci.*, 12 (1974) 722–724.
89 P. Hartvig, W. Handl, J. Vessman and C.M. Svahn, *Anal. Chem*, 48 (1976) 390–393.
90 P. Hartvig, K.E. Karlsson, L. Johansson and C. Lindberg, *J. Chromatogr.*, 121 (1976) 235–242.
91 P. Hartvig and B. Näslund, *J. Chromatogr.*, 133 (1977) 367–371.
92 C. Lindberg, M. Berg, L.O. Boreus, P. Hartvig, K.E. Karlsson, L. Palmér and A.-M. Thörnblad, *Biomed. Mass Spectrom.*, 5 (1978) 540–543.
93 J. Wallace, H.E. Hamilton, R. Olivares and S.C. Harris, *J. Anal. Toxicol.*, 2 (1978) 44–49.
94 L.A. Sternson and A.D. Cooper, *J. Chromatogr.*, 150 (1978) 257–258.
95 P. Hartvig, N.-O. Ahnfelt, M. Hammarlund and J. Vessman, *J. Chromatogr.*, 173 (1979) 127–138.
96 T. Yamaguchi, S. Yoshikawa, Y. Sekine and M. Hashimoto, *J. Chromatogr.*, 173 (1979) 147–154.
97 J.W. Barnhart and E.N. Massad, *J. Chromatogr.*, 163 (1979) 390–395.
98 J.W. Barnhart, *Toxicol. Appl. Pharmacol.*, 55 (1980) 43–48.
99 T. Walle, *J. Chromatogr.*, 111 (1975) 133–139.
100 J.E. Wallace, H.E. Hamilton, K. Blum and C. Petty, *Anal. Chem.*, 46 (1974) 2107–2111.
101 B. Dahlström and L. Paalzow, *J. Pharm. Pharmacol.*, 27 (1975) 172–176.
102 B. Dahlström, L. Paalzow and P.O. Edlund, *Acta Pharmacol. Toxicol.*, 41 (1977) 273–279.
103 J.M. Moore, *J. Chromatogr.*, 147 (1978) 327–336.
104 G. Nicolau, G. Van Lear, B. Kaul and B. Davidow, *Clin. Chem.*, 23 (1977) 1640–1643.
105 A.S. Christophersen and K.E. Rasmussen, *J. Chromatogr.*, 168 (1979) 216–220.
106 S. Felby, *Forensic Sci. Int.*, 13 (1979) 145–150.
107 K. Verebey, M.J. Kogan, A. De Pace and S.J. Mulé, *J. Chromatogr.*, 118 (1976) 331–335.
108 R.A. Sams and L. Malspeis, *J. Chromatogr.*, 125 (1976) 409–420.
109 G.L. Burce, H.B. Bhat and T. Sokoloski, *J. Chromatogr.*, 137 (1977) 323–332.
110 J.W. Barnhart and W.J. Caldwell, *J. Chromatogr.*, 130 (1977) 243–249.
111 J.R. Miller, J.W. Blake and T. Tobin, *Res. Commun. Chem. Pathol. Pharmacol.*, 15 (1976) 447–455.
112 H. Brötell, H. Ehrsson and O. Gyllenhaal, *J. Chromatogr.*, 78 (1973) 293–301.
113 S.E. Swezey, T.F. Blascke and P.J. Meffin, *J. Chromatogr.*, 154 (1978) 256–260.
114 K.A. Pittman and C. Davison, *J. Pharm. Sci.*, 62 (1973) 765–769.
115 P.J. Meffin and K.J. Smith, *J. Chromatogr.*, 183 (1980) 352–356.
116 D.H.M. Lau and G.L. Henderson, *J. Chromatogr.*, 129 (1976) 329–338.
117 M.M. Ames and G. Powis, *J. Chromatogr.*, 166 (1978) 519–526.
118 N.-E. Larsen and J. Naestoft, *Med. Lab. Technol.*, 30 (1973) 129–132.
119 N.-E. Larsen and J. Naestoft, *J. Chromatogr.*, 109 (1975) 259–264.
120 S. Cooper, J.-M. Albert, R. Dugal and M. Bertrand, *J. Chromatogr.*, 150 (1978) 263–265.
121 A.K. Williams, *J. Agr. Food Chem.*, 22 (1974) 107–109.
122 F. Marcucci, L. Airoldi, E. Mussini and S. Garattini, *J. Chromatogr.*, 59 (1971) 174–177.
123 I.A. Zingales, *J. Chromatogr.*, 54 (1971) 15–24.
124 A. Forsman, E. Martensson, G. Nyberg and R. Öhman, *Naunyn-Schmiedebergs Arch. Pharmakol. Exp. Pathol.*, 286 (1974) 113–124.
125 J. Vessman, S. Strömberg and G. Rietz, *Acta Pharm. Suecica*, 7 (1970) 363–372.
126 P. Hartvig, B. Näslund and J. Vessman, *J. Pharm. Sci.*, 65 (1976) 1707–1709.
127 J.E. Greving, J.H.G. Jonkman, F. Fiks, R.A. de Zeeuw, L.E. van Bork and N.G.M. Orie, *J. Chromatogr.*, 142 (1977) 611–619.
128 J.L.W. Pohlmann and S.L. Cohan, *J. Chromatogr.*, 131 (1977) 297–301.
129 J. Rosenfeld, M. Kawai, J.R.A. Rigg and J.K. Khandelwal, *J. Chromatogr.*, 129 (1976) 387–392.
130 E. Patterson, P. Stetson and B.R. Lucchesi, *J. Chromatogr.*, 181 (1980) 33–39.

131 E.T. Lin, D.C. Brater and L.Z. Benet, *J. Chromatogr.*, 140 (1977) 275–279.
132 K.K. Midha, J.K. Cooper and I.J. McGilveray, *J. Pharm. Sci.*, 68 (1979) 557–560.
133 M. Ervik, *Acta Pharm. Suecica*, 6 (1969) 393–400.
134 B. Scales and M.B. Cosgrove, *J. Pharmacol. Exp. Ther.*, 175 (1970) 338–347.
135 E. di Salle, K.M. Baker, S.R. Bareggi, W.D. Watkins, C.A. Chidsey, A. Frigerio and P.L. Morselli, *J. Chromatogr.*, 84 (1973) 347–353.
136 T. Walle, *J. Pharm. Sci.*, 63 (1974) 1885–1891.
137 D.A. Saelens, T. Walle and P.J. Privitera, *J. Chromatogr.*, 123 (1976) 185–192.
138 R.E. Kates and C.L. Jones, *J. Pharm. Sci.*, 66 (1977) 1490–1492.
139 S. Vickers and C.A.H. Duncan, *J. Pharm. Sci.*, 68 (1979) 1559–1560.
140. D.B. Jack and W. Riess, *J. Chromatogr.*, 88 (1974) 173–176.
141 P.H. Degen and W. Riess, *J. Chromatogr.*, 121 (1976) 72–75.
142 M. Ervik, *Acta Pharmacol. Toxicol. Suppl. V*, 36 (1975) 136–144.
143 C.P. Quarterman, M.J. Kendall and D.B. Jack, *J. Chromatogr.*, 183 (1980) 92–98.
144 K. Kawahara and T. Ofuji, *J. Chromatogr.*, 168 (1979) 266–272.
145 B. Scales and P.B. Copsey, *J. Pharm. Pharmacol.*, 27 (1975) 430–433.
146 S. Han Wan, R.F. Maronde and S.B. Matin, *J. Pharm. Sci.*, 67 (1978) 1340–1342.
147 J.O. Malbica and K.R. Monson, *J. Pharm. Sci.*, 64 (1975) 1992–1994.
148 M. Ervik, K. Kylberg-Hanssen and P.-O. Lagerström, *J. Chromatogr.*, 182 (1980) 341–347.
149 D.T. Tocco, F.A. de Luna and A.E.W. Duncan, *J. Pharm. Sci.*, 64 (1975) 1879–1881.
150 G. Alvan, B. Calissendorff, P. Seideman, K. Widmark and G. Widmark, *Clin. Pharmacokinetics*, 5 (1980) 95–100.
151 R.J. Francis, P.B. East, S.J. McLaren and J. Larman, *Biomed. Mass Spectrom.*, 3 (1976) 281–285.
152 P.J. Meffin, S.R. Harapat and D.C. Harrison, *Res. Commun. Chem. Pathol. Pharmacol.*, 15 (1976) 31–51.
153 D. Heusse, P. Populaire, A. Renard, P. Pasquier and J. Gregoire, *J. Pharm. Sci.*, 69 (1980) 273–275.
154 C.F. Poole, L. Johansson and J. Vessman, *J. Chromatogr.*, 194 (1980) 365–377.
155 L. Neelakantan and H.B. Kostenbauder, *J. Pharm. Sci.*, 65 (1976) 740–742.
156 C.S. Lee and L.Z. Benet, *J. Chromatogr.*, 128 (1976) 188–192.
157 C.S. Lee and L.H. Wang, *J. Pharm. Sci.*, 69 (1980) 362–363.
158 A. Arbin and P.-O. Edlund, *Acta Pharm. Suecica*, 12 (1975) 119–126.
159 H.A. Schwertner, T.M. Ludden and J.E. Wallace, *Anal. Chem.*, 48 (1976) 1875–1878.
160 A. Arbin, M. Berg and L.O. Boréus, *Acta Pharm. Suecica*, 13 (1976) 235–240.
161 H.W. van den Berg, R.F. Murphy, R. Hunter and D.T. Elmore, *J. Chromatogr.*, 145 (1978) 311–314.
162 J. Stadler, *Anal. Biochem.*, 86 (1978) 477–489.
163 H.A. McLeod, G. Benns, D. Lewis and J.F. Lawrence, *J. Chromatogr.*, 157 (1978) 285–290.
164 S.-Y. Chu and L.T. Senello, *J. Chromatogr.*, 137 (1977) 343–350.
165 P. Jakobsen and A. Kirstein Pedersen, *J. Chromatogr.*, 163 (1979) 259–269.
166 A. Sioufi, F. Caudal and F. Marfil, *J. Pharm. Sci.*, 67 (1978) 243–245.
167 L.-C. Chu, W.F. Bayne, F.T. Tao, L.G. Schmitt and J.E. Shaw, *J. Pharm. Sci.*, 68 (1979) 72–74.
168 P.O. Edlund, *J. Chromatogr.*, 187 (1980) 161–169.
169 A. Bye and G. Land, *J. Chromatogr.*, 115 (1975) 93–99.
170 W.F. Bayne, L.-C. Chu and F.T. Tao, *J. Pharm. Sci.*, 65 (1976) 1724–1728.
171 S.M. Wallace, V.P. Shah, S. Riegelman and W.L. Epstein, *Anal. Lett.*, 11 (1978) 461–468.
172 D.B. Jack and W. Riess, *J. Pharm. Sci.*, 62 (1973) 1929–1932.
173 P.H. Degen, W. Schneider, P. Vuillard, U.P. Geiger and W. Riess, *J. Chromatogr.*, 117 (1976) 407–413.
174 P. Hartvig and C. Fagerlund, *J. Chromatogr.*, 140 (1977) 170–173.
175 K. Murayama and A. Nakajima, *Yakugaku Zasshi (J. Pharm. Soc. Jap.)*, 97 (1977) 445–449.
176 M. van Boven, P. Daenens and G. Vandereycken, *J. Chromatogr.*, 182 (1980) 435–438.
177 J.A.F. de Silva, N. Munno and R.E. Weinfeld, *J. Pharm. Sci.*, 62 (1973) 449–452.

178 V.P. Shah, S. Riegelman and W.L. Epstein, *J. Pharm. Sci.*, 61 (1972) 634–636.
179 F. Jamali and J.E. Axelson, *J. Pharm. Sci.*, 66 (1977) 1540–1543.
180 H.J. Schwarz, B.A. Waldman and V. Madrid, *J. Pharm. Sci.*, 65 (1976) 370–372.
181 H. Kamimura, Y. Omi, Y. Shiobara, N. Tamaki and Y. Katogi, *J. Chromatogr.*, 163 (1979) 271–279.
182 N. Nose, A. Tanaka, S. Kobayashi, T. Suzuki, Y. Hoshino and A. Watanabe, *J. Hygienic Chem.*, 20 (1974) 194–199.
183 F.W. Deckert, *J. Chromatogr.*, 69 (1972) 201–203.
184 E.M. Odam and M.G. Townsend, *Analyst (London)*, 101 (1976) 478–484.
185 D.G. Kaiser and R.S. Martin, *J. Pharm. Sci.*, 63 (1974) 1579–1581.
186 G. Bianchetti, R. Latini and P.L. Morselli, *J. Chromatogr.*, 124 (1976) 331–335.
187 H. Ehrsson, S. Eksborg, I. Wallin, N. Kållberg and G. Swanbeck, *J. Chromatogr.*, 140 (1970) 157–164.
188 J. Schmid and F.W. Koss, *J. Chromatogr.*, 146 (1978) 498–502.
189 M. Van Boven and P. Daenens, *J. Pharm. Belg.*, 35 (1980) 103–106.
190 R.W. Bullard, G. Holguin and J.E. Peterson, *J. Agr. Food Chem.*, 23 (1975) 72–74.
191 P. Jakobsen, O. Lederballe Pedersen and E. Mikkelsen, *J. Chromatogr.*, 162 (1979) 81–87.
192 S. Kondo, A. Kuchiki, K. Yamamoto, K. Akimoto, K. Takahashi, N. Awata and I. Sugimoto, *Chem. Pharm. Bull.*, 28 (1980) 1–7.
193 S. Higuchi, H. Sasaki and T. Sado, *J. Chromatogr.*, 110 (1975) 301–307.
194 M. Ahnoff, *J. Chromatogr.*, in preparation.
195 D.D. Clarke, S. Wilk, S.E. Gitlow and M.J. Franklin, *J. Gas Chromatogr.*, 5 (1967) 307–310.
196 S. Kawai and Z. Tamura, *Chem. Pharm. Bull.*, 16 (1968) 1091–1094.
197 N. Sakauchi, S. Kumaoka and Y. Hanawa, *Endocrinol. Jap.*, 19 (1972) 589–592.
198 I.L. Martin and G.B. Ansell, *Biochem. Pharmacol.*, 22 (1973) 521–533.
199 E.L. Arnold and R. Ford, *Anal. Chem.*, 45 (1973) 85–89.
200 K.P. Wong and M. Sandler, *Clin. Chim. Acta*, 50 (1974) 119–128.
201 T. Kawano, M. Niwa, Y. Fujita, M. Ozaki and K. Mori, *Jap. J. Pharmacol.*, 28 (1978) 168–171.
202 M.G. Bigdeli and M.A. Collins, *Biochem. Med.*, 12 (1975) 55–65.
203 P.J. O'Neill and R.G. Rahwan, *J. Pharm. Sci.*, 66 (1977) 893–895.
204 A.C. Moffat and E.C. Horning, *Biochim. Biophys. Acta*, 222 (1970) 248–250.
205 J.-C. Lhuguenot and B.F. Maume, *J. Chromatogr. Sci.*, 12 (1974) 411–418.
206 E. Änggård and G. Sedvall, *Anal. Chem.*, 41 (1969) 1250–1256.
207 L.M. Bertani, S.W. Dziedzic, D.D. Clarke and S.E. Gitlow, *Clin. Chim. Acta*, 30 (1970) 227–233.
208 L.M. Nelson, F.A. Bubb, P.M. Lax, M.W. Weg and M. Sandler, *Clin. Chim. Acta*, 92 (1979) 235–240.
209 P.M. Lax, G.S. King, B.R. Pettit and M. Sandler, *Clin. Chim. Acta*, 96 (1979) 269–272.
210 L.J. Dombrowski, P.M. Comi and E.L. Pratt, *J. Pharm. Sci.*, 62 (1973) 1761–1763.
211 G. Belvedere, S. Caccia, A. Frigerio and A. Jori, *J. Chromatogr.*, 84 (1973) 355–360.
212 H. Kato, K. Yamada, S. Kawai, T. Ohno, Y. Masada and K. Hashimato, *J. Chromatogr.*, 82 (1973) 323–329.
213 A.O. Niedermayer and R.G. Koski, *J. Pharm. Sci.*, 64 (1975) 1531–1534.
214 H.A. Schwertner, H.E. Hamilton and J.E. Wallace, *Clin. Chem.*, 24 (1978) 895–899.
215 C.P. Talley, N.R. Trenner, G.V. Downing, Jr. and W.J.A. VandenHeuvel, *Anal. Chem.*, 43 (1971) 1379–1382.
216 F.C. Churchill and D.N. Ku, *J. Chromatogr.*, 189 (1980) 375–388.
217 B.J. Gudzinowicz and S.J. Clark, *J. Gas Chromatogr.*, 3 (1965) 147–151.
218 R.H. Smith, J.A. MacDonald, D.S. Thompson and W.E. Flacke, *Clin. Chem.*, 23 (1977) 1306–1309.
219 T. Walle, *J. Chromatogr.*, 114 (1975) 345–350.
220 S.-R. Sun and A.H.C. Chun, *J. Pharm. Sci.*, 66 (1977) 477–480.
221 O. Gyllenhaal, H. Brötell and B. Sandgren, *J. Chromatogr.*, 122 (1976) 471–477.
222 S. Dilli and D.N. Pillai, *J. Chromatogr.*, 137 (1977) 111–117.
223 S. Dilli and D. Pillai, *J. Chromatogr.*, 190 (1980) 113–118.

224 J.E. Wallace, H.E. Hamilton, E.L. Shimek, Jr., H. Schwertner and K. Blum, *Anal. Chem.*, 49 (1977) 903–906.
225 J.E. Wallace, H.E. Hamilton, E.L. Shimek, Jr., H.A. Schwertner and K.D. Haegele, *Anal. Chem.*, 49 (1977) 1969–1973.
226 J.E. Wallace, H.A. Schwertner, H.E. Hamilton and E.L. Shimek, Jr., *Clin. Chem.*, 25 (1979) 252–255.
227 B. Mellström and H. Ehrsson, *Acta Pharm. Suecica*, 11 (1974) 91–96.
228 D.L. Lewis and D.F. Paris, *J. Agr. Food Chem.*, 22 (1974) 148–149.
229 W.J.A. VandenHeuvel, V.B. Gruber and R.W. Walker, *J. Chromatogr.*, 87 (1973) 341–350.
230 R.T. Evans, *J. Chromatogr.*, 88 (1974) 398–402.
231 D.G. Saunders and L.E. Vanatta, *Anal. Chem.*, 46 (1974) 1319–1321.
232 B.L. Worobey and G.R.B. Webster, *J. Ass. Offic. Anal. Chem.*, 60 (1977) 213–217.
233 B.L. Worobey and G.R.B. Webster, *J. Chromatogr.*, 153 (1978) 423–431.
234 J.F. Lawrence, C. van Buuren, U.A.T. Brinkman and R.W. Frei, *J. Agr. Food Chem.*, 28 (1980) 630–632.
235 Y.-Y. Liu, I. Schmeltz and D. Hoffmann, *Anal. Chem.*, 46 (1974) 885–889.
236 D.B. Jack, S. Brechbühler, P.H. Degen, P. Zbinden and W. Riess, *J. Chromatogr.*, 115 (1975) 87–92.
237 P.H. Degen, *J. Chromatogr.*, 176 (1979) 375–380.
238 P. Ventura, M. Zanol, M. Visconti and G. Pifferi, *J. Chromatogr.*, 161 (1978) 237–242.
239 J.A. Timbrell, J.M. Wright and C.M. Smith, *J. Chromatogr.*, 138 (1977) 165–172.
240 J.H. Hengstmann, F.C. Falkner, J.T. Watson and J. Oates, *Anal. Chem.*, 46 (1974) 34–39.
241 P. Erdtmansky and T.J. Goehl, *Anal. Chem.*, 47 (1975) 750–752.
242 T. Brecht, J.H. Hengstmann and G.E. von Unruh, in M. Eggstein and H.M. Liebich (Editors), *Proc. 1st Int. Symp. Mass Spectrometry and Combined Techniques in Medicine, Clinical Chemistry and Clinical Biochemistry, Tübingen, November 14–15, 1977*, University of Tübingen, Tübingen, 1977, pp. 391–400.
243 E.D. Pellizzari and T.P. Seltzman, *Anal. Biochem.*, 96 (1979) 118–125.
244 T. Kawabata, H. Oshshima, T. Ishibashi, M. Matsui and T. Kitsuwa, *J. Chromatogr.*, 140 (1977) 47–56.
245 H.G. Fouda, *Biomed. Mass Spectrom.*, 5 (1978) 72–75.
246 M. Guerret, D. Lavene, J. Longchampt and J.L. Kiger, *J. Pharm. Sci.*, 68 (1979) 219–221.
247 T. Yamaguchi, Y. Utsui and M. Hashimoto, *J. Chromatogr.*, 150 (1979) 147–154.
248 S.B. Matin, J.H. Karam and P.H. Forsham, *Anal. Chem.*, 47 (1975) 545–549.
249 K. Gröningsson, M. Johansson and J. Vessman, *J. Chromatogr.*, in preparation.
250 D. Alkalay, J. Volk and M.F. Bartlett, *J. Pharm. Sci.*, 65 (1976) 525–529.
251 K. Siefert, D. Casagrande and H. Silberman, *J. Chromatogr.*, 109 (1975) 193–198.
252 O. Gyllenhaal and H. Ehrsson, *J. Chromatogr.*, 107 (1975) 327–333.
253 A. Bye and G. Land, *J. Chromatogr.*, 139 (1977) 181–185.
254 O. Gyllenhaal, U. Tjärnlund, H. Ehrsson and P. Hartvig, *J. Chromatogr.*, 156 (1978) 275–283.
255 O. Gyllenhaal, B. Näslund and P. Hartvig, *J. Chromatogr.*, 156 (1978) 330–334.
256 N. Nose, S. Kobayashi, A. Hirose and A. Watanabe, *J. Chromatogr.*, 123 (1976) 167–173.
257 M.E. Sharp, S.M. Wallace, K.W. Hindmarsh and M.A. Brown, *Can. J. Pharm. Sci.*, 15 (1980) 35–38.
258 R.J. Daun, *J. Ass. Offic. Anal. Chem.*, 54 (1971) 1140–1145.
259 P. Hartvig, O. Gyllenhaal and M. Hammarlund, *J. Chromatogr.*, 151 (1978) 232–236.
260 M. Ervik and K. Gustavii, *Anal. Chem.*, 46 (1974) 39–42.
261 P.H. Degen and A. Schweizer, *J. Chromatogr.*, 142 (1977) 549–557.
262 H.L.J.M. Fleuren and J.M. van Rossum, *J. Chromatogr.*, 152 (1978) 41–54.
263 B. Lindström and M. Molander, *J. Chromatogr.*, 101 (1974) 219–221.
264 B. Lindström, M. Molander and M. Groschinsky, *J. Chromatogr.*, 114 (1975) 459–462.
265 W.J.A. VandenHeuvel, V.F. Gruber, R.W. Walker and F.J. Wolf, *J. Pharm. Sci.*, 64 (1975) 1309–1312.
266 E. Redalieu, V.V. Tipnis and W.E. Wagner, Jr., *J. Pharm. Sci.*, 67 (1978) 726–728.

267 S.M. Wallace, V.P. Shah and S. Riegelman, *J. Pharm. Sci.*, 66 (1977) 527–530.
268 C. Fagerlund, P. Hartvig and B. Lindström, *J. Chromatogr.*, 168 (1979) 107–116.
269 W.J.A. VandenHeuvel, V.F. Gruber, R.W. Walker and F.J. Wolf, *J. Agr. Food Chem.*, 25 (1977) 389–392.
270 S.B. Matin and M. Rowland, *J. Pharm. Pharmacol.*, 25 (1973) 186–187.
271 W.E. Braselton, Jr., E.D. Bransome, Jr., H.C. Ashline, J.T. Stewart and I.L. Honigberg, *Anal. Chem.*, 48 (1976) 1386–1394.
272 W.E. Braselton, Jr., E.D. Bransome, Jr. and T.A. Huff, *Diabetes*, 26 (1977) 50–57.
273 H.-J. Schlicht, H.-P. Gelbke and G. Schmidt, *J. Chromatogr.*, 155 (1978) 178–181.
274 P. Hartvig, C. Fagerlund and O. Gyllenhaal, *J. Chromatogr.*, 181 (1980) 17–24.
275 J.A.F. de Silva, M.A. Schwartz, V. Stefanovic, J. Kaplan and L. D'Arconte, *Anal. Chem.*, 36 (1964) 2099–2105.
276 J.A.F. de Silva, in C.E. Pippenger, J.K. Penry and H. Kutt (Editors), *Antiepileptic Drugs: Quantitative Analysis and Interpretation*, Raven Press, New York, 1978, pp. 111–138.
277 J.A.F. de Silva and C.V. Puglisi, *Anal. Chem.*, 42 (1970) 1725–1736.
278 A. Berlin, B. Siwers, S. Agurell, Å. Hiort, F. Sjöqvist and S. Ström, *Clin. Pharmacol. Ther.*, 13 (1972) 733–744.
279 A.G. Howard, G. Nickless and D.M. Hailey, *J. Chromatogr.*, 90 (1974) 325–329.
280 J.A.S. Gamble, R.A.E. Assaf, J.S. Mackay, M.S. Kennedy and P.J. Howard, *Anaesthesia*, 30 (1975) 159–163.
281 E.S. Baird, D.M. Hailey,and S. Malcolm, *Clin. Chim. Acta*, 48 (1973) 105–108.
282 R.E. Weinfeld, H.N. Posmanter, K.-C. Khoo and C.V. Puglisi, *J. Chromatogr.*, 143 (1977) 581–595.
283 J.E. Wallace, H.A. Schwertner and E.L. Shimek, Jr., *Clin. Chem.*, 25 (1979) 1296–1300.
284 V.A. Raisys, P.N. Friel, P.R. Graaf, K.E. Opheim and A.J. Wilensky, *J. Chromatogr.*, 183 (1980) 441–448.
285 D.J. Hoffman and A.H.C. Chun, *J. Pharm. Sci.*, 64 (1975) 1668–1671.
286 M.A. Brooks, M.R. Hackman, R.E. Weinfeld and T. Macasieb, *J. Chromatogr.*, 135 (1977) 123–131.
287 M. Wretlind, Å. Pilbrant, A. Sundwall and J. Vessman, *Acta Pharmacol. Toxicol. Suppl. I*, 40 (1977) 28–39.
288 G.P. Beharrell, D.M. Hailey and M.K. McLaurin, *J. Chromatogr.*, 70 (1972) 45–52.
289 H. Ehrsson and A. Tilly, *Anal. Lett.*, 6 (1973) 197–210.
290 L. Kangas, *J. Chromatogr.*, 136 (1977) 259–270.
291 L. Kangas, *J. Chromatogr.*, 172 (1979) 273–278.
292 A.G. de Boer, J. Röst-Kaiser, H. Bracht and D.D. Breimer, *J. Chromatogr.*, 145 (1978) 105–114.
293 J. Naestoft and N.-E. Larsen, *J. Chromatogr.*, 93 (1974) 113–122.
294 J. Naestoft, M. Lund and N.-E. Larsen, *Acta Neurol. Scand.*, 49 (1973) Suppl. 53, 103–108.
295 G.J.G. Parry and D.G. Ferry, *J. Chromatogr.*, 128 (1976) 166–168.
296 J.P. Cano, J. Guintrand, C. Aubert and A. Viala, *Arzneim.-Forsch.*, 27 (1977) 338–342.
297 M. Gerna and P.L. Morselli, *J. Chromatogr.*, 116 (1976) 445–450.
298 J.A.F. de Silva and I. Bekersky, *J. Chromatogr.*, 99 (1974) 447–460.
299 D.B. Faber, R.M. Kok and E.M. Rempt-Van Dijk, *J. Chromatogr.*, 133 (1977) 319–326.
300 J.A.F. de Silva and J. Kaplan, *J. Pharm. Sci.*, 55 (1966) 1278–1283.
301 J.A.F. de Silva, I. Bekersky, M.A. Brooks, R.E. Weinfeld, W. Glover and C.V. Puglisi, *J. Pharm. Sci.*, 63 (1974) 1440–1445.
302 J.P. Cano, A.M. Baille and A. Viala, *Arzneim.-Forsch.*, 25 (1975) 1012–1016.
303 W. Sadée and E. van der Kleijn, *J. Pharm. Sci.*, 60 (1971) 135–137.
304 J.A. Knowles and H.W. Ruelius, *Arzneim.-Forsch.*, 22 (1972) 687–692.
305 J.A. Knowles, W.H. Comer and H.W. Ruelius, *Arzneim.-Forsch.*, 21 (1971) 1055–1065.
306 J. Vessman, G. Freij and S. Strömberg, *Acta Pharm. Suecica*, 9 (1972) 447–456.
307 M.C. Sánchez, J. Colomé and E. Gelpi, *J. Chromatogr.*, 126 (1976) 601–613.
308 H.G. Giles, T. Fan, C.A. Naranjo and E.M. Sellers, *Can. J. Pharm. Sci.*, 13 (1978) 64–65.

309 J.A.F. de Silva, I. Bekersky and C.V. Puglisi, *J. Chromatogr. Sci.*, 11 (1973) 547–554.
310 P.J. Howard, J.K. Lilburn, J.W. Dundee, W. Toner, P.D.A. McIlroy, *Anaesthesia*, 32 (1977) 767–770.
311 D.J. Greenblatt, K. Franke and R.I. Shader, *J. Chromatogr.*, 146 (1978) 311–320.
312 S. Higuchi, H. Urabe and Y. Shiobara, *J. Chromatogr.*, 164 (1979) 55–61.
313 J. Vessman, M. Johansson, P. Magnusson and S. Strömberg, *Anal. Chem.*, 49 (1977) 1545–1549.
314 J.A.F. de Silva, I. Bekersky, C.V. Puglisi, M.A. Brooks and R.E. Weinfeld, *Anal. Chem.*, 48 (1976) 10–19.
315 G.M. Pacifici and G.F. Placidi, *J. Chromatogr.*, 135 (1977) 133–139.
316 G. Cuisinaud, E. Peillon, N. Ferry, D. Deruaz and J. Sassard, *J. Chromatogr.*, 178 (1979) 314–319.
317 C.V. Puglisi, J.A.F. de Silva and A.S. Leon, *J. Chromatogr.*, 118 (1976) 371–386.
318 C.V. Puglisi, J.C. Meyer, L. D'Arconte, M.A. Brooks and J.A.F. de Silva, *J. Chromatogr.*, 145 (1978) 81–96.
319 T. Agoh, M. Konishi and Y. Mori, *J. Chromatogr.*, 182 (1980) 171–177.
320 A.W. Missen, *J. Anal. Toxicol.*, 1 (1977) 224–226.
321 S. Caccia, M. Ballabio, G. Guiso and M.G. Zanini, *J. Chromatogr.*, 164 (1979) 100–105.
322 H.P. Gelbke and H.J. Schlicht, *J. Pharm. Sci.*, 67 (1978) 1176–1177.
323 A.F. Williams, W.J. Murray and B.H. Gibb, *Nature (London)*, 210 (1966) 816–817.
324 M.H. Litchfield, *Analyst (London)*, 93 (1968) 653–659.
325 M.T. Rosseel and M.G. Bogaert, *J. Pharm. Sci.*, 62 (1973) 754–758.
326 G.B. Neurath and M. Dünger, *Arzneim.-Forsch.*, 27 (1977) 416–419.
327 Y. Givant and F.G. Sulman, *Experientia*, 34 (1978) 643–644.
328 J.O. Malbica, K. Monson, K. Neilson and R. Sprissler, *J. Pharm. Sci.*, 66 (1977) 384–386.
329 H. Laufen, F. Scharpf and G. Bartsch, *J. Chromatogr.*, 146 (1978) 457–464.
330 E. Doyle, L.F. Chasseaud and T. Taylor, *Biopharm. Drug Dispos.*, 1 (1980) 141–147.
331 L. Richard, G. Klein and J.M. Orr, *Clin. Chem.*, 22 (1976) 2060–2061.
332 D.A. Chin, D.G. Prue, J. Michelucci, B.T. Kho and C.R. Warner, *J. Pharm. Sci.*, 66 (1977) 1143–1145.
333 R.V. Smith and J. Besic, *Microchem. J.*, 23 (1978) 185–196.
334 M.T. Rosseel and M.G. Bogaert, *J. Pharm. Sci.*, 68 (1979) 659–660.
335 R.A. Landowne and S.R. Lipsky, *Nature (London)*, 199 (1963) 141–143.
336 N. Ikekawa, O. Hoshino, R. Watanuki, H. Orimo, T. Fujita and M. Yoshikawa, *Anal. Biochem.*, 17 (1966) 16–23.
337 J.J. Pisano, W.J.A. VandenHeuvel and E.C. Horning, *Biochem. Biophys. Res. Commun.*, 7 (1962) 82–86.
338 R.A. Landowne, *Chim. Anal. (Paris)*, 47 (1965) 589–591.
339 R.W. Zumwalt, K. Kuo and C.W. Gehrke, *J. Chromatogr.*, 57 (1971) 193–208.
340 G. Bengtsson and G. Odham, *Anal. Biochem.*, 92 (1979) 426–443.
341 G. Bengtsson, G. Odham and G. Westerdahl, *Anal. Biochem.*, 111 (1981) 163–175.
342 B.H. Robertson, R.M. Wolcott and J.C. Bennett, *Proc. Soc. Exp. Biol. Med.*, 155 (1977) 287–289.
343 J.E. Stouffer, P.I. Jaakonmäki and T.J. Wenger, *Biochim. Biophys. Acta*, 127 (1966) 261–263.
344 P.I. Jaakonmäki and J.E. Stouffer, *J. Gas Chromatogr.*, 5 (1967) 303–306.
345 J.E. Stouffer, *J. Chromatogr. Sci.*, 7 (1969) 124–126.
346 L.B. Hansen, *Anal. Chem.*, 40 (1968) 1587–1589.
347 E.T. Backer and V.J. Pileggi, *J. Chromatogr.*, 36 (1968) 351–353.
348 N.N. Nihei, M.C. Gershengorn, T. Mitsuma, L.R. Stringham, A. Cordy, B. Kuchmy and C.S. Hollander, *Anal. Biochem.*, 43 (1971) 433–435.
349 B.A. Petersen, R.N. Hanson, R.W. Giese and B.L. Karger, *J. Chromatogr.*, 126 (1976) 503–516.
350 P. Hušek, *J. Chromatogr.*, 91 (1974) 475–481.
351 P. Hušek, *J. Chromatogr.*, 91 (1974) 483–492.
352 L.G. Davis, N.L. Sass, B. Manna and M.L. Nusynowitz, *Clin. Chem.*, 25 (1979) 218–220.
353 K. Imai, N. Arizumi, M.-T. Wang, S. Yoshiue and Z. Tamura, *Chem. Pharm. Bull.*, 20 (1972) 2436–2439.

354 M.H. Joseph, *J. Chromatogr.*, 146 (1978) 33–41.
355 J.D.M. Pearson and D.F. Sharman, *J. Neurochem.*, 24 (1975) 1225–1228.
356 R. Schmid and M. Karobath, *J. Chromatogr.*, 139 (1977) 101–109.
357 P.H. Degen and W. Riess, *J. Chromatogr.*, 117 (1976) 399–405.
358 S. Wilk and M. Orlowski, *Anal. Biochem.*, 69 (1975) 100–113.
359 S.A. Bland, J.W. Blake and R.S. Ray, *J. Chromatogr. Sci.*, 14 (1976) 201–203.
360 J. Vessman and S. Strömberg, *Anal. Chem.*, 49 (1977) 369–373.
361 C. Lindberg, K.-E. Karlsson and P. Hartvig, *Acta Pharm. Suecica*, 15 (1978) 327–336.
362 H. Ehrsson, *Acta Pharm. Suecica*, 8 (1971) 113–118.
363 O. Gyllenhaal, H. Brötell and P. Hartvig, *J. Chromatogr.*, 129 (1976) 295–302.
364 J.E. Greving, J.H.G. Jonkman and R.A. de Zeeuw, *J. Chromatogr.*, 148 (1978) 389–395.
365 B. Davis, *Anal. Chem.*, 49 (1977) 832–834.
366 A. Sioufi and F. Pommier, *J. Chromatogr.*, 181 (1980) 161–168.
367 R.V. Smith and S.L. Tsai, *J. Chromatogr.*, 61 (1971) 29–34.
368 C.A. Alley, J.B. Brooks and G. Choudhary, *Anal. Chem.*, 48 (1976) 387–390.
369 L. Witte, H. Nau, J.-H. Fuhrhop, A. Doenicke and B. Grote, *J. Chromatogr.*, 143 (1977) 329–334.
370 S.C. Chan, *Clin. Chem.*, 26 (1980) 1528–1530.
371 J. Vessman and G. Rietz, *Vox Sang.*, 35 (1978) 75–80.
372 A. Arbin and J. Östelius, *J. Chromatogr.*, 193 (1980) 405–412.
373 J. Vessman and G. Rietz, *J. Chromatogr.*, 100 (1974) 153–163.
374 C.F. Poole and D.G. Wibberley, *J. Chromatogr.*, 132 (1977) 511–518.
375 R. Takeshita, E. Takabatake, K. Minagawa and Y. Takizawa, *J. Chromatogr.*, 133 (1977) 303–310.
376 J. Belisle and D.F. Hagen, *Anal. Biochem.*, 101 (1980) 369–376.
377 L.T. Sennello, *Biochem. Med.*, 8 (1973) 345–351.
378 G. Lukas, K.H. Vyas and S.D. Brindle, *Anal. Lett.*, 11 (1978) 953–966.
379 R. van der Hoeven, R.H. Drost, R.A.A. Maes, F. Dost, T.A. Plomp and G.J.J. Plomp, *J. Chromatogr.*, 164 (1979) 106–108.
380 M. Gilbert, A. Penel, F.V. Kosikowski, J.D. Henion, G.A. Maylin and D.J. Lisk, *J. Food Sci.*, 42 (1977) 1650–1653.
381 J.I. Brook, R.H. Biggs, P.A. St. John and D.S. Anthony, *Anal. Biochem.*, 18 (1967) 453–458.
382 S. Bittner and Z. Even-Chen, *Phytochemistry*, 14 (1975) 2455–2457.
383 J.B. Brooks, C.C. Alley and J.A. Liddle, *Anal. Chem.*, 46 (1974) 1930–1934.
384 D.E. Easterling, T. Walle, E.C. Conradi and T.E. Gaffney, *J. Chromatogr.*, 162 (1979) 439–445.
385 S. Mierzwa and S. Witek, *J. Chromatogr.*, 136 (1977) 105–111.
386 R. Gugler and C. Jensen, *J. Chromatogr.*, 117 (1976) 175–179.
387 A. Karmen and H. Haut, *Biochem. Med.*, 12 (1975) 154–161.
388 E.D. Pellizzari and T.A. Saltzman, *Anal. Lett.*, 11 (1978) 975–989.
389 C.R. Nony, M.C. Bowman, C.L. Holder, J.F. Young and W.L. Oller, *J. Pharm. Sci.*, 65 (1976) 1810–1816.
390 P.G. Simmonds, B.C. Pettitt and A. Zlatkis, *Anal. Chem.*, 39 (1967) 163–167.
391 G.A.F. Harrison and E. Collins, *Amer. Soc. Brew. Chem., Proc.*, (1968) 101–105.
392 A. Frigerio, P. Martelli, K.M. Baker and P.A. Biondi, *J. Chromatogr.*, 81 (1973) 139–141.
393 H. Kallio and R.R. Linko, *J. Chromatogr.*, 76 (1973) 229–232.
394 R.G. Bjork, *Anal. Biochem.*, 63 (1975) 80–86.
395 D.J. Tocco, A.E.W. Duncan, R.M. Noll and D.E. Duggan, *Anal. Biochem.*, 94 (1979) 470–476.
396 G.H. Jouvenaz, D.H. Nugteren, R.K. Beerthuis and D.A. van Dorp, *Biochim. Biophys. Acta*, 202 (1970) 231–234.
397 M. Korteweg, G. Verdonk, P. Sandra and M. Verzele, *Prostaglandins*, 13 (1977) 1221–1223.
398 M.J. Levitt, J.B. Josimovich and K.D. Broskin, *Prostaglandins*, 1 (1972) 121–130.
399 A.J.F. Wickramasinghe and R.S. Shaw, *Biochem. J.*, 141 (1974) 179–187.
400 V.A. Skrinska and A. Butkus, *Prostaglandins*, 16 (1978) 571–582.
401 F.A. Fitzpatrick, R.R. Gorman and M.A. Wynalda, *Prostaglandins*, 13 (1977) 201–208.

402 F.A. Fitzpatrick, M.A. Wynalda and D.G. Kaiser, *Anal. Chem.*, 49 (1977) 1032–1035.
403 E.A.M. de Deckere, D.H. Nugteren and F. ten Hoor, *Nature (London)*, 268 (1977) 160–163.
404 F.A. Fitzpatrick, D.A. Stringfellow, J. Maclouf and M. Rigaud, in J.R. Vane and S. Bergström (Editors), *Prostacyclin*, Raven Press, New York, 1979, pp. 55–64.
405 B.H. Min, J. Pao, W.A. Garland, J.A.F. de Silva and M. Parsonnet, *J. Chromatogr.*, 183 (1980) 411–419.
406 D.G. Ferry, D.M. Ferry, P.W. Moller and E.G. McQueen, *J. Chromatogr.*, 89 (1974) 110–112.
407 L. Helleberg, *J. Chromatogr.*, 117 (1976) 167–173.
408 A. Arbin, *J. Chromatogr.*, 144 (1977) 85–92.
409 A. Arbin, H. Brink and J. Vessman, *J. Chromatogr.*, 170 (1979) 25–33.
410 K. Møller Jensen, *J. Chromatogr.*, 153 (1978) 195–202.
411 R.G. Sibeon, J.D. Baty, N. Baber, K. Chan and M. L'E. Orme, *J. Chromatogr.*, 153 (1978) 189–194.
412 M.A. Evans, *J. Pharm. Sci.*, 69 (1980) 219–220.
413 J.A.F. de Silva, N. Strojny and M.A. Brooks, *Anal. Chim. Acta*, 73 (1972) 283–292.
414 L.T. Sennello, S.-Y. Chu and J.W. Borcherding, *J. Chromatogr.*, 147 (1978) 485–490.
415 D.G. Kaiser, S.R. Shaw and G.J. VanGiessen, *J. Pharm. Sci.*, 63 (1974) 567–570.
416 D.G. Kaiser and R.S. Martin, *J. Pharm. Sci.*, 67 (1978) 627–630.
417 A.F. de Long, A. Polk, G. Martin, R.D. Smyth and N.H. Reavy-Cantwell, *J. Pharm. Sci.*, 67 (1978) 1171–1173.
418 D.H. Chatfield and T.J. Woodage, *J. Chromatogr.*, 153 (1978) 101–106.
419 K.-T. Ng, *J. Chromatogr.*, 166 (1978) 527–535.
420 P. Stenberg, T.-E. Jönsson, B. Nilsson and F. Wollheim, *J. Chromatogr.*, 177 (1979) 145–148.
421 B. Hwang, G. Konicki, R. Dewey and C. Miao, *J. Pharm. Sci.*, 67 (1978) 1095–1098.
422 A.G. Zacchei and T. Wishousky, *J. Pharm. Sci.*, 65 (1976) 1770–1773.
423 S.W. Dziedzic, L.M. Bertani, D.D. Clarke and S.E. Gitlow, *Anal. Biochem.*, 47 (1972) 592–600.
424 B. Sjöquist and E. Änggård, *Anal. Chem.*, 44 (1972) 2297–2301.
425 S.W. Dziedzic, L. Bertani-Dziedzic and S.E. Gitlow, *J. Lab. Clin. Med.*, 82 (1973) 829–835.
426 S.W. Dizedzic and S.E. Gitlow, *J. Neurochem.*, 22 (1974) 333–335.
427 E. Watson and S. Wilk, *Psychopharmacologia*, 42 (1975) 57–62.
428 M.W. Weg, C.R.J. Ruthven, B.L. Goodwin and M. Sandler, *Clin. Chim. Acta*, 59 (1975) 249–251.
429 E. Watson, B. Travis and S. Wilk, *Life Sci.*, 15 (1974) 2167–2178.
430 J.D.M. Pearson and D.F. Sharman, *Br. J. Pharmacol.*, 53 (1975) 143–148.
431 E. Markianos and E. Rüther, *J. Clin. Chem. Clin. Biochem.*, 14 (1976) 437–441.
432 D.S. Walter, P.W. Dettmar, K. Taylor, G.M. Shilcock and A. Cowan, *J. Neurochem.*, 30 (1978) 929–931.
433 J. Chauhan and A. Darbre, *J. Chromatogr.*, 183 (1980) 391–401.
434 S. Wilk, S.E. Gitlow, M. Mendlowitz, M.J. Franklin, H.E. Carr and D.D. Clarke, *Anal. Biochem.*, 13 (1965) 544–551.
435 S. Yoshiue, M. Yoshioka and Z. Tamura, *Chem. Pharm. Bull.*, 23 (1975) 3040–3041.
436 M. Makita, S. Yamamoto, K. Ono and T. Katagi, *J. Chromatogr.*, 166 (1978) 294–298.
437 B.A. Davis, D.A. Durden, P. Pun-Li and A.A. Boulton, *J. Chromatogr.*, 142 (1977) 517–522.
438 H. Morita and W.G. Montgomery, *J. Chromatogr.*, 123 (1976) 454–459.
439 H. Morimoto, I. Imada, T. Amano, M. Toyoda and Y. Ashida, *Biochem. Med.*, 7 (1973) 169–177.
440 J.E. Sinsheimer and R.V. Smith, *J. Pharm. Sci.*, 56 (1967) 1280–1284.
441 F.K. Kawahara, *Anal. Chem.*, 40 (1968) 1009–1010.
442 H. Ehrsson, T. Walle and H. Brötell, *Acta Pharm. Suecica*, 8 (1971) 319–328.
443 N.K. McCallum and R.J. Armstrong, *J. Chromatogr.*, 78 (1973) 303–307.
444 J.R. King, C.R. Nony and M.C. Bowman, *J. Chromatogr. Sci.*, 15 (1977) 14–21.
445 J.C. Ryan and J.F. Lawrence, *J. Chromatogr.*, 135 (1977) 117–122.
446 J.F. Lawrence, J.J. Ryan and R. Leduc, *J. Chromatogr.*, 147 (1978) 398–400.
447 L. Laitem, P. Gaspar and I. Bello, *J. Chromatogr.*, 156 (1978) 267–273.
448 K. Chan and J.F. McCann, *J. Chromatogr.*, 164 (1979) 394–398.

449 D.C. Fenimore, R.R. Freeman and R.R. Loy, *Anal. Chem.*, 45 (1973) 2331–2335.
450 E.R. Garrett and C.A. Hunt, *J. Pharm. Sci.*, 66 (1977) 20–26.
451 A. Sioufi, F. Caudal and D. Grimault, *J. Pharm. Sci.*, 66 (1977) 1166–1168.
452 R.C. Bachmann and M.R. Shetlar, *Biochem. Med.*, 2 (1969) 313–324.
453 P.J. Porcaro and P. Shubiak, *Anal. Chem.*, 40 (1968) 1232–1237.
454 R.S. Browing, Jr., J. Grego and H.P. Warrington, Jr., *J. Pharm. Sci.*, 57 (1968) 2165–2166.
455 P.J. Porcaro, P. Shubiak and M. Manowitz, *J. Pharm. Sci.*, 58 (1969) 251–252.
456 D.G. Ferry and E.G. McQueen, *J. Chromatogr.*, 76 (1973) 233–235.
457 S. Wilk, S.E. Gitlow, D.D. Clarke and D.H. Paley, *Clin. Chim. Acta*, 16 (1967) 403–408.
458 D.F. Sharman, *Br. J. Pharmacol.*, 36 (1969) 523–534.
459 F. Karoum, C.R.J. Ruthven and M. Sandler, *Biochem. Med.*, 5 (1971) 505–514.
460 P.A. Bond, *Biochem. Med.*, 6 (1972) 36–45.
461 R. O'Keefe and B.W.L. Brooksbank, *Clin. Chem.*, 19 (1973) 1031–1035.
462 L. Fellows, P. Riederer and M. Sandler, *Clin. Chim. Acta*, 59 (1975) 255–257.
463 H. Dekirmenjian and J. Maas, *Clin. Chim. Acta*, 52 (1974) 203–210.
464 A.E. Halaris, E.M. Demet and M.E. Halari, *Clin. Chim. Acta*, 78 (1977) 285–294.
465 L. Lombrozo, T.J. Anderson, K. Kanaske and L.E. Hollister, *J. Chromatogr.*, 181 (1980) 1–10.
466 C.M. Davis and D.C. Fenimore, *Anal. Biochem.*, 106 (1980) 517–523.
467 L.M. Cummins, Y.C. Martin and E.E. Scherfling, *J. Pharm. Sci.*, 60 (1971) 261–263.
468 H. Smith and J.W. Thorpe, *J. Chromatogr.*, 134 (1977) 178–181.
469 E.R. Garrett and H.J. Lambert, *J. Pharm. Sci.*, 55 (1966) 812–817.
470 D.D. Breimer, H.C.J. Ketelaars and J.M. van Rossum, *J. Chromatogr.*, 88 (1974) 55–63.
471 D.J. Berry, *J. Chromatogr.*, 107 (1975) 107–114.
472 W.J. Cole, R.G. Mitchell and R.F. Salamonsen, *J. Pharm. Pharmacol.*, 27 (1975) 167–171.
473 G.L. Resnick, D. Corbin and D.H. Sandberg, *Anal. Chem.*, 38 (1966) 582–585.
474 M. Yamamoto, S. Iguchi and T. Aoyama, *Chem. Pharm. Bull.*, 15 (1967) 123–127.
475 J.M. Wal, J.C. Peleran and G. Bories, *J. Chromatogr.*, 168 (1979) 179–185.
476 T. Aoyama and S. Iguchi, *J. Chromatogr.*, 43 (1969) 253–256.
477 T.A. Plomp, R.A.A. Maes and M. Thiery, *J. Chromatogr.*, 121 (1976) 243–250.
478 D.G. Pritchard and W. Niedermeier, *J. Chromatogr.*, 152 (1978) 487–494.
479 A. Sioufi and F. Pommier, *J. Pharm. Sci.*, 69 (1980) 167–169.
480 W.R. Maynard, Jr. and R.B. Bruce, *J. Pharm. Sci.*, 59 (1970) 1346–1348.
481 S. Singhawangcha, C.F. Poole and A. Zlatkis, *J. Chromatogr.*, 183 (1980) 433–439.
482 C.E. Morreal, T.L. Dao and A.J. Spiess, *Anal. Biochem.*, 71 (1976) 125–132.
483 R.M. Smith and M. Dawson, *Analyst (London)*, 105 (1980) 85–89.
484 J.E. Lovelock, P.G. Simmonds and W.J.A. VandenHeuvel, *Nature (London)*, 197 (1963) 249–251.
485 F.H. de Jong and H.J. van der Molen, *J. Endocrinol.*, 53 (1972) 461–474.
486 S.J. Clark and H.H. Wotiz, *Steroids*, 2 (1963) 535–540.
487 R.A. Landowne and S.R. Lipsky, *Anal. Chem.*, 35 (1963) 532–535.
488 H.J. van der Molen, *Ann. Clin. Res.*, 2 (1970) 279–288.
489 C.F. Poole, A. Zlatkis, W.-F. Sye, S. Singhawangcha and E.D. Morgan, *Lipids*, 15 (1980) 734–744.
490 K.T. Koshy, *J. Chromatogr.*, 126 (1976) 641–650.
491 C.F. Poole and E.D. Morgan, *J. Chromatogr.*, 90 (1974) 380–381.
492 R.J. Bopp, H.W. Murphy, J.F. Nash and C.R. Novotny, *J. Pharm. Sci.*, 61 (1972) 1441–1444.
493 W.H. Leitch, L.P. Stuart and E. Forchielli, *Anal. Biochem.*, 56 (1973) 580–583.
494 E. Rossi, A. de Pascale, P. Negrini, A. Frigerio and E. Castegnaro, *J. Chromatogr.*, 169 (1979) 416–421.
495 D.G. Kaiser, R.G. Carlson and K.T. Kirton, *J. Pharm. Sci.*, 63 (1974) 420–424.
496 M. Kirschner and J.P. Taylor, *Anal. Biochem.*, 30 (1969) 346–357.
497 P.G. Devaux and E.C. Horning, *Anal. Lett.*, 2 (1969) 637–648.
498 L. Dehennin, A. Reiffsteck and R. Scholler, *J. Chromatogr. Sci.*, 10 (1972) 224–230.
499 A. Wilson and R. Fraser, *J. Endocrinol.*, 51 (1971) 557–567.

500 A. Wilson, P.A. Mason and R. Fraser, *J. Steroid Biochem.*, 7 (1976) 611–613.
501 B.C. Musial and C.N. Williams, *J. Lipid Res.*, 20 (1979) 78–85.
502 W.P. Pulsinelli, R.K. Tcholakian and K.B. Eik-Nes, *Anal. Biochem.*, 49 (1972) 559–568.
503 L.F. Fabre, Jr., D.C. Fenimore, R.W. Farmer, H.W. Davis and G. Farrell, *J. Chromatogr. Sci.*, 7 (1969) 632–638.
504 G.L. Nicolis, H.H. Wotiz and J.L. Gabrilove, *J. Clin. Endocrinol. Metab.*, 28 (1968) 547–557.
505 D. Sklan, P. Budowski and M. Katz, *Anal. Biochem.*, 56 (1973) 606–609.
506 K.T. Koshy, D.G. Kaiser and A.L. van der Slik, *J. Chromatogr. Sci.*, 13 (1975) 97–104.
507 T. Nambara, K. Kigasawa, T. Iwata and M. Ibuki, *J. Chromatogr.*, 114 (1975) 81–86.
508 R. Wehner and A. Handke, *J. Chromatogr.*, 177 (1979) 237–244.
509 R.A. Mead, G.C. Haltmeyer and K.B. Eik-Nes, *J. Chromatogr. Sci.*, 7 (1969) 554–560.
510 I.A. Muni, C.H. Altshuler, J.C. Neicheril and J.T. Voegeli, *Anal. Lett.*, 5 (1972) 785–799.
511 P. Dugan and D. Lundgren, *Anal. Biochem.*, 8 (1964) 312–318.
512 F.W. Crow, Jr., *Applications of Negative Ion Chemical Ionization Mass Spectrometry, Thesis,* University of Virginia, Charlottesville, VA, 1978, University Microfilms, Ann Arbor, MI, 79 16 226.
513 H. Bechtold and E. Jähnchen, *J. Chromatogr.*, 164 (1979) 85–90.
514 P. Maini, *J. Chromatogr.*, 128 (1976) 174–177.
515 J.L. Love and R.G.C. Donnelly, *J. Ass. Offic. Anal. Chem.*, 59 (1976) 1140–1141.
516 W. Postel and U. Güvenc, *Z. Lebensm.-Unters.-Forsch.*, 161 (1976) 35–44.
517 R.A. Scanlan and R.C. Lindsay, *J. Food Sci.*, 33 (1968) 440–441.
518 P.H. Degen, S. Brechbühler, J. Schäublin and W. Riess, *J. Chromatogr.*, 118 (1976) 363–370.
519 E. Malvoisin, B. Rollmann, G. Lhoest, M. Roberford and M. Mercier, *J. Chromatogr.*, 150 (1978) 345–354.
520 Y. Hoshika and G. Muto, *J. Chromatogr.*, 152 (1978) 224–227.
521 K. Kobayashi, M. Tanaka, S. Kawai and T. Ohno, *J. Chromatogr.*, 176 (1979) 118–122.
522 K. Kobayashi, M. Tanaka and S. Kawai, *J. Chromatogr.*, 187 (1980) 413–417.
523 A. Beckett and R. Achari, *J. Pharm. Pharmacol.*, 29 (1977) 212–216.
524 J. Vessman and G. Rietz, unpublished results.
525 C. Razzouk, E. Evrard, G. Lhoest, M. Roberfroid and M. Mercier, *J. Chromatogr.*, 161 (1978) 103–109.
526 J.B. Brooks, C.C. Alley and R. Jones, *Anal. Chem.*, 44 (1972) 1881–1884.
527 J.B. Brooks, W.B. Cherry, L. Thacker and C.C. Alley, *J. Infec. Dis.*, 126 (1972) 143–153.
528 T.G. Alliston, G.B. Cox and R.S. Kirk, *Analyst (London)*, 97 (1972) 915–920.
529 G.M. Telling, *J. Chromatogr.*, 73 (1972) 79–87.
530 E.A. Walker, M. Castegnaro and B. Pignatelli, in Environ. Carcinog. Sel. Methods Anal., IARC Sci. Publ. 1978, (18) pp. 175–187.
531 R.A. Vukovich, A.J. Triolo and J.M. Coon, *J. Agr. Food Chem.*, 17 (1969) 1190–1191.
532 H.P. Burchfield, E.E. Storrs, R.J. Wheeler, V.K. Bhat and L.L. Green, *Anal. Chem.*, 45 (1973) 916–920.
533 C. Pantarotto, A. Bossi, G. Belvedere, A. Martini, M.G. Donelli and A. Frigerio, *J. Pharm. Sci.*, 63 (1974) 1554–1558.
534 C.R. Vogt, J.C. Liao and A.Y. Sun, *Clin. Chem.*, 26 (1980) 66–68.
535 K. Funazo, M. Tanaka and T. Shono, *Anal. Chem.*, 52 (1980) 1222–1224.
536 J.W. Tesch, W.R. Rehg and R.E. Sievers, *J. Chromatogr.*, 126 (1976) 743–755.
537 R.L. Tanner, R. Fajer and J. Gaffney, *Anal. Chem.*, 51 (1979) 865–870.
538 J.C. Valentour, V. Aggarwal and I. Sunshine, *Anal. Chem.*, 46 (1974) 924–925.
539 H.F. De Brabander and R. Verbeke, *J. Chromatogr.*, 138 (1977) 131–142.
540 J.W. Young and G.D. Christian, *Anal. Chim. Acta*, 65 (1973) 127–138.
541 Y. Shimoishi, *J. Chromatogr.*, 136 (1977) 85–93.
542 C.F. Poole, N.J. Evans and D.G. Wibberley, *J. Chromatogr.*, 136 (1977) 73–83.
543 B.M. Bycroft and D.E. Clegg, *J. Ass. Offic. Anal. Chem.*, 61 (1978) 923–926.
544 K. Kurahashi, S. Inoue, S. Yonekura, Y. Shimoishi and K. Tôei, *Analyst (London)*, 105 (1980) 690–695.
545 M.L. Taylor and E.L. Arnold, *Anal. Chem.*, 43 (1971) 1328–1331.

546 S.W. Gunner and R.C. O'Brien, *J. Ass. Offic. Anal. Chem.*, 52 (1969) 1200–1202.
547 R.E. Weston and B.B. Wheals, *Analyst (London)*, 95 (1970) 680–682.
548 M.D. Solomon, W.E. Pereira and A.M. Duffield, *Anal. Lett.*, 4 (1971) 301–304.
549 J.W. Edmunds, D.F. Lee and C.M.L. Nickels, *Pestic. Sci.*, 4 (1973) 101–105.
550 J.P. Mieure, O. Hicks, R.G. Kaley and P.R. Michael, *J. Chromatogr. Sci.*, 15 (1977) 275–277.
551 J. Vessman, *Applications of Gas Chromatography in the Pharmaceutical Field in Modern Methodology for the Isolation, Identification and Quantification of Drugs and Related Substances; Seminar Uppsala, June 5–9, 1977*, Secretariat to the Convention for the Mutual Recognition of Inspections in Respect of the Manufacture of Pharmaceutical Products, EFTA Secretariat, 9–11 Rue de Varembé, CH-1211 Geneva 20, p. 106.
552 F.F. Farris, E.O. Magarian and F.O. Slininger, *Anal. Biochem.*, 88 (1977) 495–503.
553 H.N. Nigg, I.P. Kapoor, R.L. Metcalfe and J.R. Coats, *J. Agr. Food Chem.*, 20 (1972) 446–448.
554 C.M. Derks, *J. Chromatogr.*, 108 (1975) 222–225.
555 K.W. Kirby, J.E. Keiser, J. Groene and E.F. Slach, *J. Agr. Food Chem.*, 27 (1979) 757–759.
556 G.S. King, B.R. Pettit, K. Blau and P.M. Lax, *Biomed. Mass Spectrom.*, 6 (1979) 482–484.
557 D.F. Hunt and F.W. Crow, *Anal. Chem.*, 50 (1978) 1781–1784.
558 J.R. Shipe, Jr., D.F. Hunt and J. Savory, *Clin. Chem.*, 25 (1979) 1564–1571.
559 W.A. Garland and B.H. Min, *J. Chromatogr.*, 172 (1979) 279–286.
560 W.A. Garland, B.J. Hodshon, G. Chen, G. Weiss, N.R. Felicito and A. MacDonald, *J. Agr. Food Chem.*, 28 (1980) 273–277.
561 J.R. Hass, M.D. Friesen, D.J. Harvan and C.E. Parker, *Anal. Chem.*, 50 (1978) 1474–1479.
562 D.W. Kuehl, M.J. Whitaker and R.C. Dougherty, *Anal. Chem.*, 52 (1980) 935–940.
563 R.A. Landowne and S.R. Lipsky, *Anal. Chem.*, 34 (1962) 726–730.
564 G. Schill, *Separation Methods for Drugs and Related Organic Compounds*, Apotekarsocieteten/The Swedish Academy of Pharmaceutical Sciences, Stockholm, 1978, p. 119.
565 J. Vessman and P. Hartvig, *Acta Pharm. Suecica*, 8 (1971) 235–250.

Chapter 14

Negative ion atmospheric pressure ionization mass spectrometry and the electron-capture detector

E.C. HORNING, D.I. CARROLL, I. DZIDIC and R.N. STILLWELL

CONTENTS

14.1. INTRODUCTION

Electron-capture detection and negative ion atmospheric pressure ionization (API) mass spectrometry are related in that both experimental techniques involve the formation of negative ions from organic compounds at atmospheric pressure. Their paths of development were not related, however, and the technological aspects are different. Negative ion mass spectrometry is much less widely used than positive ion mass spectrometry, but it will very likely be used more frequently in the future in specialized applications. Most manufacturers of mass spectrometers have modified their instruments so that they can be used in either mode, or concurrently in both modes by rapid switching, and numerous older instruments have been modified for negative ion work. The electron-capture detector (ECD) has a relatively long history of use in pesticide analytical studies and in the detection of toxic halogenated compounds, and it is still widely used. The measurement of a change in conductance in the gas phase, however, provides no specific information about the compound or the chemical reactions responsible for the change.

The development of the ECD started as a consequence of observations of the signals given by polychlorinated compounds with argon ionization detectors. One of the earliest studies of comparative responses was that of Lovelock and Gregory [1] in 1962; electron absorption or electron attachment responses relative to chlorobenzene were determined for a number of compounds. Values for polychlorobenzenes (Table 14.1), which have been widely quoted since that time, illustrate a feature of negative ion formation that is important both theoretically and in analytical applications: the observed response (a decrease in conductance due to a decrease in electron density as a consequence of an electron–molecule reaction) may vary by 10^3 to 10^4 for individual compounds. Additional early studies led to the recognition of compound classes generally characterized by negative ion formation through reaction with thermalized electrons. These classes include polyhalides, nitro compounds and quinones. Many pesticides and other toxic compounds show strong ECD responses, but most organic compounds show no response under typical electron-capture conditions.

An ECD response involves the formation at atmospheric pressure of stable negative ions from neutral molecules, through direct electron attachment, or through reactions initiated by electron attachment. Since a relatively large amount of a compound with low electron affinity will show the same response as a small amount of a compound with high electron affinity, and since the response is relatively nonspecific, it was difficult in early work to define the events occurring in an ECD. Two major types of reactions were recognized: (A) the formation of stable radical ions (for example, M^- ions from quinones), and (B) the formation of halide ions, along with a neutral organic radical, by dissociative electron capture. The evidence for these reactions was indirect but strong; semiquinone formation, for example, was recognized many years ago as a step in the reduction of quinones.

Electron–molecule reactions, and ion–molecule reactions leading to negative ions, are best studied by mass spectrometric methods. Conditions involving relatively high energy transfer and short residence times in the source generally lead to much fragmentation. Electron impact ionization (EI) conditions, for example, are not used in negative ion studies because only a few negatively charged low a.m.u. fragment ions are likely to be formed. The low pressure chemical ionization (CI) condition developed by Brandenberger and co-workers [2–5] is highly useful for identification purposes, since numerous classes of compounds can be recognized immediately. This condition leads to reactions with moderately high energy transfer, and the products suggest that both EI and CI processes occur in the source (fragment ions may also be produced by energy transfer involving collisions with activated neutral atoms or ions, as well as by electron–molecule reactions,

TABLE 14.1

RELATIVE ELECTRON-CAPTURE RESPONSES OF CHLOROBENZENES

Compound	Relative response
Chlorobenzene	1
o-Dichlorobenzene	42
m-Dichlorobenzene	30
p-Dichlorobenzene	11
1,2,3-Trichlorobenzene	113
1,2,4-Trichlorobenzene	75
1,2,5-Trichlorobenzene	60
Hexachlorobenzene	1100

and the high energy ion products may be formed in this way). Ordinary CI conditions (0.2–1 Torr source pressure) lead to less fragmentation, and electron attachment reactions which occur in an ECD can usually be duplicated under CI conditions with nitrogen, argon, methane, propane, isobutane, or mixtures such as argon–methane, as source gases. Thermal and chemical equilibration is not attained under these conditions, but fragment ions which are typical of relatively high energy transfer processes are not observed under the usual conditions of operation. The method of operation in API mass spectrometry [6–9] for analytical purposes involves relatively long residence times of reactant and product ions in the source, and this technique provides an ionization condition that closely resembles that existing in an ECD. In a recent study [10], an API ion source was constructed with the physical dimensions and design of an ECD. Schematic drawings of the system, with gas chromatography–mass spectrometry–computer (GC–MS–COM) capabilities, and the API source, are given in Figs. 14.1 and 14.2. The gas stream from the GC unit was split, and was directed both to a commercial ECD and to the API mass spectrometer source. The source was designed so that an ECD response could be obtained, and at the same time mass analysis techniques were used to detect ions in the source. Selected ion detection procedures were used in quantitative studies, and intermittent or repetitive mass scans were used to determine ion products. Examples of electron attachment and related reactions occurring at atmospheric pressure may be studied in this way. Grimsrud and co-workers are currently also using this approach [11,12].

Earlier negative ion studies carried out under CI and API conditions suggested that unsuspected reactions, in addition to those of types (A) and (B), occurred when oxygen was present. An initial electron attachment reaction was followed for some compounds by ion–molecule reactions (O_2^- with M, and O_2 with M^-). The overall result of these

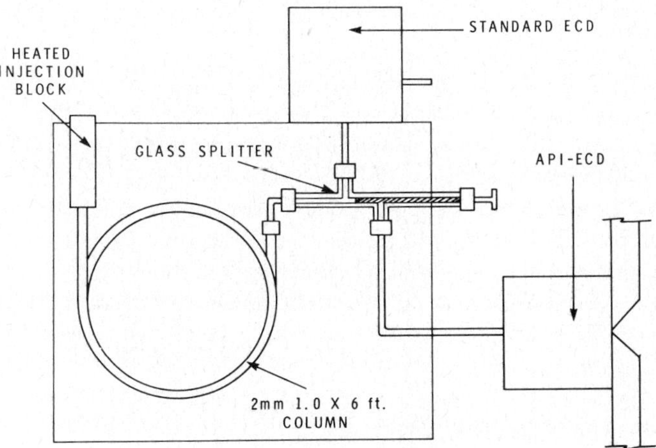

Fig. 14.1. Schematic drawing of part of a modified API GC–MS instrument, with provision for a split effluent stream, for detection by (a) a commercial ECD, (b) an API source constructed with the physical dimensions of an ECD, and which can be operated as an ECD, and (c) an API quadrupole mass spectrometer.

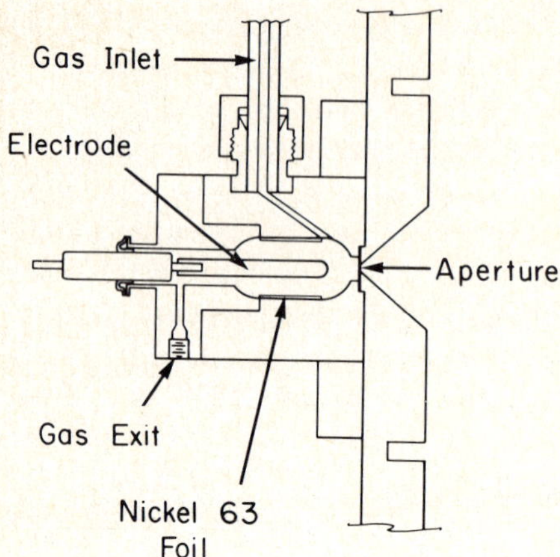

Fig. 14.2. Schematic drawing of an API–ECD source. This may be operated simultaneously as an ECD and as an ion source for API mass spectrometric studies of ions generated in the source.

type (C) reactions is an ECD response which is often very strong, and which was earlier thought to be due to reactions of type (A). Examples have also been found of reactions which involve cleavage with or without rearrangement of radical intermediates, and in these instances the initial step is presumably electron attachment to form an M^- ion which may or may not be observed. Some of these examples may be more properly classified as dissociative electron attachment reactions, but for purposes of discussion ordinary halide elimination reactions are here classified as type (B) reactions, while other types of reactions that involve different or additional steps are discussed as type (D) reactions. The use of special volatile derivatives, leading to a strong ECD response, was originally also based upon the belief that reactions of type (A) or (B) were involved in the observed response. While electron attachment reactions of most electron-capturing derivatives have not been studied in detail, it seems likely that fragmentation occurs to give a stable negative ion and a neutral radical in nearly all instances. When the derivative contains a chlorine or bromine atom, the response is probably due to dissociative electron attachment with chloride or bromide ions as the negative ion products. If the derivatizing group is eliminated as a neutral radical, however, the negative ion product is characteristic of the parent organic molecule. While these gas phase reactions are not different in theoretical sense from direct reactions of underivatized compounds, special reagents are required for derivative formation, and this route to an ECD response is discussed as type (E) reaction sequences.

14.2. TYPES OF ELECTRON–MOLECULE REACTIONS

14.2.1. (A) Formation of stable molecular radical ions

Thermalized electrons will react in the gas phase with some organic compounds to form M^- ions. The corresponding reaction in liquid phase chemistry is one-electron reduction to a relatively stable intermediate which usually is protonated to form a neutral radical ($MH^{\bullet}$). A continuation of the reduction process in solution, through addition of a second electron and proton, leads to MH_2, but the intermediate or one-electron stage is clearly recognizable. Table 14.2 contains a list of compounds, in order of decreasing electron affinity, that form M^- ions and that show ECD responses. The greatest responses are shown by quinones and α-diketones.

Fig. 14.3 shows the observed response for benzil under API–ECD conditions. The records are for an ECD as part of a GC–ECD instrument, the API source functioning only as an ECD, and selected ion detection in the API source for M^- ions. During the period of observation, positive ion mass spectral scans showed that the only compound present in the source was benzil.

All of the compounds and compound types in Table 14.2 show M^- ions under CI conditions. Under API conditions, negative ions are not observed for polycyclic aromatic hydrocarbons (except for azulene) or for benzophenone.

TABLE 14.2

COMPOUNDS, IN ORDER OF DECREASING ELECTRON AFFINITY, THAT FORM M^- IONS

Benzoquinone
1,4-Naphthoquinone
Maleic anhydride
m-Chloronitrobenzene
Phthalic anhydride
Nitrobenzene
Azulene
Polycyclic aromatic hydrocarbons
Benzophenone

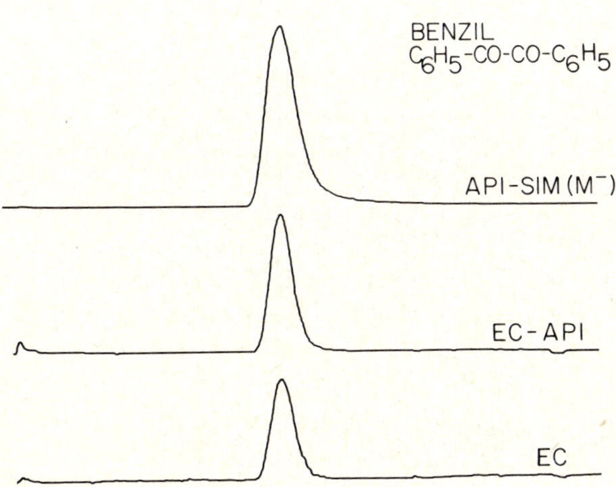

Fig. 14.3. Responses due to negative ion formation from benzil, for: (EC) electron-capture detection; (EC– API) the API ion source in Fig. 14.2 acting as an ECD; [API–SIM(M^-)] the API ion source and mass analyzer operated in selected ion detection mode for M^- ions from benzil. The carrier gas was argon–methane (95:5). The PZ 179 column was operated isothermally at 240°C. Positive ion mass scans showed that benzil was the only organic compound in the source during the response period.

Some compounds show an ECD response because of conversion to negative radical ions and even-electron ions by both electron–molecule and ion–molecule reactions. Phthalate esters, for example, show an ECD response as a consequence of conversion to two negative ions: M^- from phthalic anhydride, and $(M-R)^-$ ions through loss of an alkyl group. The ratio of ions observed under API conditions varies with the temperature and concentration of oxygen in the source. Reactions of this kind are discussed in a later section.

14.2.2. (B) Formation of halide ions by dissociative electron attachment

Many halogenated organic compounds show a strong ECD response. In some instances this is due to reaction with thermalized electrons to form a halide ion and a neutral organic radical. The carbon–halogen bond strength varies for halogens in the order F>Cl>Br>I; bromo and iodo compounds ionize in this way. but fluoro compounds do not. Chloro compounds show great variability in response. Methyl chloride shows almost no response under ordinary electron-capture conditions, and much variability in response is always observed for related polychloroaromatic compounds, as indicated in Table 14.1. The reason for the variability in response is that chloro compounds may ionize by reac-

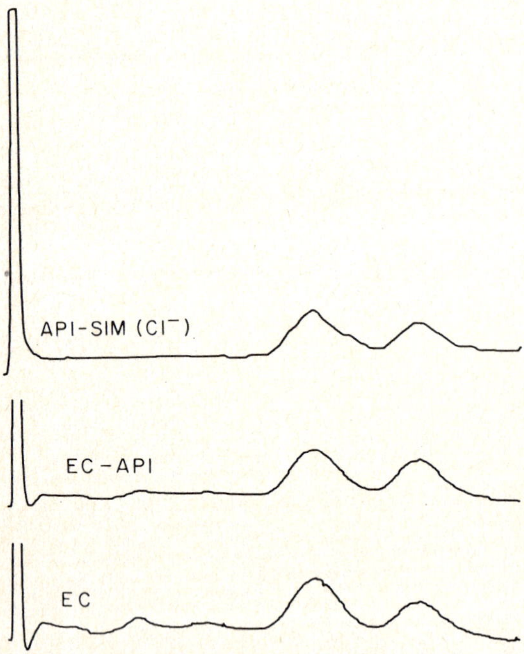

Fig. 14.4. Responses due to negative ion formation from aldrin and dieldrin, for: (EC) electron-capture detection; (EC–API) the API ion source in Fig. 14.2 acting as an ECD; [API–SIM(Cl^-)] the API source and mass analyzer operated in selected ion detection mode for Cl^- ions. The carrier gas was argon–methane (95:5). The PZ 179 column was operated isothermally at 200°C. The samples were 60 pg.

tions of type (A), (B) or (C), and the least well documented reactions, in mass spectral studies, are those of type (B) for chloro compounds.

The present belief is that ECD responses observed for bromo and iodo compounds in the absence of oxygen are always due to dissociative electron attachment, and this view is supported by mass spectral studies indicating that Br^- and I^- are the negative ion products under API and CI conditions. Monochloro compounds show a relatively low ECD response in the absence of oxygen (for example, chloroalkanes), and a strong ECD response in the presence of oxygen. These circumstances are discussed in the next section. Polychloro aliphatic and alicyclic compounds show strong ECD responses, and this (in the absence of oxygen) is due to dissociative electron attachment resulting in chloride ion elimination. An example is shown in Fig. 14.4. The three responses for aldrin and dieldrin, as in Fig. 14.3, are for an ECD, the API source acting as an ECD, and selected ion detection in the API source. In this instance the negative ion reaction product is the chloride ion. Electron–molecule reactions of polyfluoro compounds are discussed as examples of reactions of type (E).

14.2.3. (C) Reactions involving oxygen

In a study of negative ion formation for polychlorobiphenyls, using a Plasma Chromatograph, Karasek [13] found that the presence of a single chlorine atom in one or both rings was not sufficient to form M^- ions, but that the presence of three or more chlorine atoms always led to negative organic ions. The product ions were incorrectly identified as M^- ions through ion mobility determinations and through accord with earlier but wrong views. It is now known that negative ion products from organic halides are dependent upon both structure and conditions of ionization, and that ion–molecule reactions may be involved.

As the degree of chloro-substitution increases for benzene or for biphenyls, reactive M^- ions are formed. In the presence of oxygen the final product is a highly stable phenoxide ion corresponding to $(M-Cl+O)^-$. The reactions which occur, defined by API mass spectrometry [10,14], are:

$$
\begin{array}{ccccc}
O_2 & & & & M \\
(1)\ \downarrow e & & & & e \downarrow\ (2) \\
O_2^- + M & \xrightarrow{(3)} & [MO_2]^- & \underset{(5)}{\overset{(4)}{\rightleftharpoons}} & O_2 + M^- \\
 & & (6)\ \downarrow & & \\
 & & (M-Cl+O)^- + ClO^{\bullet} & &
\end{array}
$$

In the presence of relatively high concentrations of oxygen, reaction 2 is suppressed, and phenoxide ions are formed by reactions 1, 3 and 6. In instances where the proton affinity of M is greater than that of oxygen, it might be expected that the sequence 1, 3, 5 would be followed, but this does not occur because the intermediate ion is immediately converted to more stable products. As the concentration of oxygen is decreased, reaction through the sequence 2, 4 and 6 increases. In the absence of oxygen, reaction 2 would be the sole route of ion formation. This condition is not easily attainable for ordinary mass spectrometers, and in practice the ratio found for $(M-Cl+O)^-$ and M^- in the case of aromatic compounds depends upon the ionization conditions. Both ions are usually observed under CI conditions, and $(M-19)^-$ ions were identified by Dougherty and coworkers [15,16] as phenoxide ions resulting from reaction with oxygen in a CI mass spectrometer source. Under API or CI conditions both ions may be present, but when oxygen is admitted to the source the molecular negative ion is no longer observed. In a Plasma Chromatograph source, the organic ions are phenoxide ions, since oxygen is available for reaction.

Similar reactions occur for *o*- and *p*-nitrochlorobenzene [14], and the formation of *p*-nitrophenoxide ions from *p*-nitrochlorobenzene in a CI or API source is a good test for the presence of oxygen. The fact that oxygen can affect ECD responses, in addition to contributing to a background conductance change through the formation of O_2^- ions, has been known for a long time, and the usual recommendation in analytical work is to remove all oxygen from carrier gases. The deliberate addition of oxygen [17,18] will enhance ECD responses for aliphatic chloro-substituted compounds, however, and this effect involves a related ion–molecule substitution reaction. The radical mechanism is related to an S_N2 process, and it occurs readily in solution [19] as well as in the gas phase.

There are several interesting aspects of this group of ion–molecule reactions that should be noted. Fig. 14.5 shows results obtained in an early study of reactions of

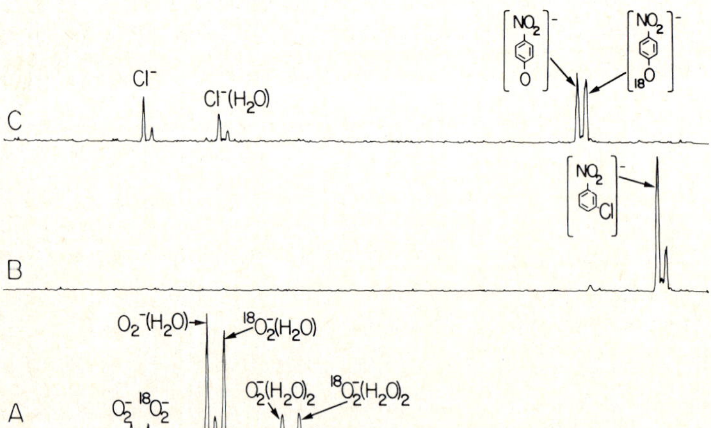

Fig. 14.5. API negative ion mass spectra for *m*-nitrochlorobenzene and for *p*-nitrochlorobenzene observed with a mixture of $^{16}O_2$ and $^{18}O_2$ in the nitrogen source gas. The *meta*-isomer is ionized to M^- ions (B). Oxygen in the source is ionized to O_2^- and hydrated O_2^- ions (A). The products of an *ipso*-substitution reaction for *p*-nitrochlorobenzene are *p*-nitrophenoxide ions and chloride ions (C). The oxygen labels in (C) show that the organic ion products are not due to a gas phase hydrolysis reaction with water in the source, but to a reaction involving oxygen. Neutral radical reaction products which are also present are not observed.

m-chloronitrobenzene and *p*-chloronitrobenzene in an API source. The *m*-isomer reacts through electron attachment [a type (A) reaction], and this leads to a stable M^- radical ion. The reaction sequence for *p*-chloronitrobenzene leads to two negative ions: chloride ion and nitrophenoxide ion. Oxygen in the source was a mixture of $^{16}O_2$ and $^{18}O_2$, and in the presence of excess oxygen the only organic ion product was the phenoxide ion. Since the ratio of labeled oxygen in the product was the same as for the reagent ions, there was no contribution to phenoxide formation by a hydrolysis reaction with water in the source. These facts, together with current knowledge relating to radical substitution reactions, suggests that *ipso*-substitution is involved in the reaction with oxygen, and that the intermediate $[MO_2]^-$ ion has the structure:

$[O_2N\text{-}C_6H_4\text{-}C(Cl)OO]^{-\cdot}$

(6) → (*p*-nitrophenoxide)$^-$ + ClO$^\cdot$

(7) → (*p*-nitrophenyl)$^\cdot$ + O_2 + Cl^-

In our earlier paper [14] it was assumed that chloride ions were derived from the chlorine oxide radical through a secondary reaction. One of the characteristics of *ipso* radical substitution mechanisms, however, is that multiple products are often formed, and in this instance another route of dissociation of the intermediate $[MO_2]^-$ ion is through reaction 7, with the formation of Cl^- and a neutral radical related to a hydroperoxynitrobenzene or to the *p*-nitrophenyl radical and oxygen, as illustrated. This reaction would explain the formation of chloride ions in a direct way, and would not require the assumption that a secondary reaction of the ClO radical also occurs. If this explanation is correct, the relative amounts of negative ion products should indicate the relative rates for reactions 6 and 7. They are approximately equal in this example. This is a general reaction that occurs for tri and tetrachlorobenzenes, for penta- and hexachlorobenzenes, and for polychlorobiphenyls.

Substitution reactions leading to the formation of phenoxide ions from polycyclic aromatic hydrocarbons and oxygen have been observed under CI, but not under API, conditions. This reaction presumably also involves the formation of an intermediate $[MO_2]^-$ ion. While gas phase ion intermediates of this kind have never been observed in this or other instances, they very likely have a structure analogous to the type proposed for chloro compounds:

$[C_{10}H_7\text{-}C(H)OO]^{-\cdot}$

(8) → (phenoxide)$^-$ + HO$^\cdot$

(9) → (phenoxy)$^\cdot$ + HO$^-$

The conversion of the intermediate to phenoxide ions (reaction 8) is well documented; reaction 9 is not found in the literature.

14.2.4. (D) Reactions initiated by electron attachment which involve multiple products and/or bond cleavage, rearrangement or cyclization

Numerous examples are known of reactions which are initiated by electron attachment, and where the products involve elimination, rearrangement or cyclization. In these instances, several different negative ion products may be observed, and the course of the reactions may be affected markedly by source conditions.

A number of organic phosphates and thiophosphates are widely used as pesticides. These compounds show strong ECD responses. Ordinary phosphate esters (tri-*n*-butyl phosphate, triphenyl phosphate) also show ECD responses, although their electron affinity is low when compared to quinones and α-diketones. Fig. 14.6 shows the responses observed for methyl parathion with an ECD, with the API source acting as an ECD, and for $(M-15)^-$ ions in the API source. Fig. 14.7 shows the negative ion products; a negative molecular ion is not observed, but cleavage with or without rearrangement leads to a thiophosphate anion by loss of a methyl radical, to a different thiophosphate anion by loss of a nitrophenyl radical, to the nitrophenoxide ion by cleavage, and to the thionitrophenoxide ion by rearrangement and cleavage. All of these negative ions correspond to anions of strong gas phase acids. Ordinary alkyl or aryl esters of phosphates also do not show M^- ions; the negative ion products are $(M-R)^-$, where R is an alkyl or aryl group, and PO_3^- ions. These are ions of gas phase acids; the metaphosphate anion has also been observed in other studies of phosphates by Harvan et al. [20].

Tetradifon, an acaricide, shows an ECD response. When oxygen is present, the only negative ion product under API conditions corresponds to $(M-Cl+O)^-$. The replacement

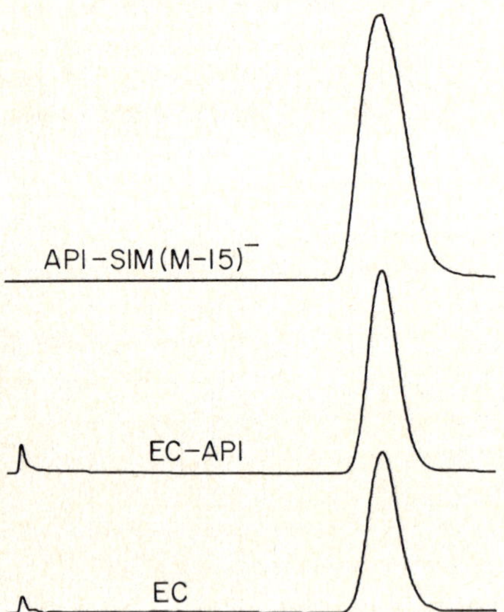

Fig. 14.6. Responses due to negative ion formation from methyl parathion, for: (EC) electron-capture detection; (EC–API) the API source in Fig. 14.2 acting as an ECD; [API–SIM$(M-15)^-$] the API source and mass analyzer operated in selected ion detection mode for $(M-15)^-$ ions. The carrier gas was argon–methane (95:5). The PZ 179 column was operated isothermally at 240°C.

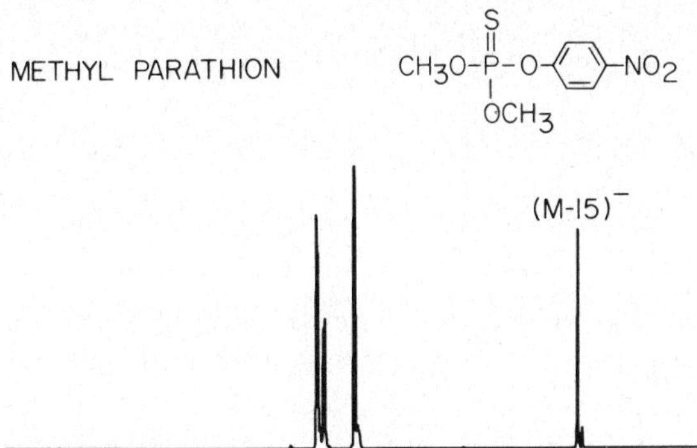

Fig. 14.7. Mass scan for negative ions from methyl parathion. The loss of a methyl radical leads to $(M-15)^-$ ions at m/z 248; *p*-nitrothiophenoxide ions are at m/z 154; loss of a *p*-nitrophenyl radical leads to ions at m/z 141; and *p*-nitrophenoxide ions are at m/z 138. This mass spectrum was taken while methyl parathion was in the source, in the separation shown in Fig. 14.6. The carrier gas was argon–methane (95:5). The PZ 179 column was operated isothermally at 240°C. These are anions of strong gas phase acids. The accompanying neutral radicals are not observed.

of chlorine by oxygen through *ipso* radical substitution probably occurs in the *p*-position of the trichlorophenoxy group, although this is not certain. Even with relatively low concentrations of oxygen, M^- ions are not observed. In the absence of oxygen, the negative ion product corresponds to $(M-HCl)^-$. The reaction route very likely involves an initial electron attachment, followed by a cyclization reaction and elimination of chloride ion and a proton. The product ion is of the M^- type. Radical intermediates are particularly likely to undergo cyclization when this is possible. For analytical purposes, the most satisfactory API condition is that involving oxygen as a reagent.

A third example is the conversion of alkyl phthalates to negative ions under API conditions. The observed negative ion products are M^- ions from phthalic anhydride, and $(M-R)^-$ ions (where R is an alkyl ester group). The conversion of a diester to an anhydride involves the elimination of two groups; in this instance the parent M^- ion is not observed, and R and RO groups are presumably eliminated by successive losses of neutral radicals. The final product of this route, M^- from phthalic anhydride, is a stable negative radical ion. The formation of $(M-R)^-$ may represent the first step in this process, but there are indications that this ion is formed independently, and that it is not a precursor to formation of the M^- radical ion of phthalic anhydride. If this is correct, the elimination of an alkoxy group is the first elimination step in the conversion of these diesters to the phthalic anhydride M^- ion. If an alkoxy group is lost as an RO radical, the last step would be cyclization with loss of an alkyl radical.

2,3,7,8-Tetrachlorodibenzodioxin does not have high electron affinity. Radical molecular anions are observed under CI conditions, but not under API conditions [21], and in this respect the behavior of this compound is similar to benzophenone and to polycyclic aromatic hydrocarbons. In the presence of oxygen, however, under API [21] or CI [22–24] conditions, a highly stable negative ion is observed. The process probably involves reaction with O_2^- to form an *ipso* substitution intermediate which then undergoes bond cleavage to yield a semiquinone anion with an *o*-quinone structure:

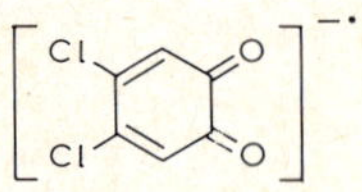

$$\left[\text{4,5-dichloro-}o\text{-benzoquinone}\right]^{-\cdot}$$

In this instance the electron affinity of the chlorinated compound is less than that of oxygen, but in the presence of oxygen a reaction sequence occurs which leads to a stable negative ion. The response observed under ECD conditions varies with the source conditions (temperature and oxygen concentration).

These examples indicate that in some instances an ECD response may be due to more complicated circumstances than to simple reactions of types (A), (B) or (C), and that some compounds react by more than one pathway.

14.2.5. (E) Reactions of derivatives prepared for gas chromatography–electron-capture detection analyses

In order to take advantage of the high sensitivity of detection available through use of ECDs, numerous reactions involving the preparation of electron-capturing derivatives have been studied. Most of these are analogs of widely used derivatization procedures, with reagents containing one or more halogen atoms. The introduction of bromine and iodine atoms in derivatives leads to a considerable increase in retention times, however, and this effect led to early emphasis upon reactions with chloro- or fluoro-substituted reagents. Perfluoro reagents, in particular, have been widely investigated as possible derivatizing agents. These include trifluoroacetic (TFA) anhydride, pentafluoropropionic (PFP) anhydride, heptafluorobutyric (HFB) anhydride and the corresponding imidazoles. Other reagents which have been recommended include pentafluorobenzyl bromide, pentafluorobenzyl chloroformate, pentafluorobenzoyl imidazole, pentafluorobenzaldehyde, pentafluorobenzyl hydroxylamine, and pentafluorophenyldimethylchlorosilane. The introduction of an increasing number of fluorine atoms in a structural unit from a derivatizing reagent has relatively little effect upon retention time; PFP derivatives, for example, show only slightly greater retention times than TFA derivatives, and there is usually very little difference in retention behavior between PFP and the corresponding HFB derivatives. Perfluoroalkyl and perfluorophenyl groups, when included as structural units through derivatization, lead to strong ECD responses. This property is shared by the reagents.

Very few studies have been made of routes of ionization for electron-capturing derivatives. It is generally assumed that bromo-substituted compounds ionize by dissociative electron attachment, with Br^- as the negative ion product, and that chloro-substituted compounds ionize through this route [type (B) reactions] and through type (C) reactions. Most fluorine-containing derivatives probably cleave to stable negative ions and neutral radicals, but few examples have been studied. The PFP derivative of epinephrine, which shows a strong ECD response, ionizes by loss of a C_3F_5CO radical, and one organic negative ion product is a phenoxide ion of the catechol:

$$\left[C_3F_5COO \; O-C_6H_3-\underset{|}{\overset{OCOC_3F_5}{}}CH-CH_2N(CH_3)COC_3F_5 \right]^-$$

A related derivative, with three trimethylsilyl groups in place of the three perfluoroacyl groups attached to oxygen atoms, shows a relatively low ECD response which may be due to M^-, rather than to phenoxide ion formation.

Further studies of derivatives are needed. In mass spectrometric studies, the most valuable kinds of derivatives are those which yield ions that are characteristic of the parent molecule, rather than ions associated with the derivatizing group alone.

14.3. COMPARISON OF ATMOSPHERIC PRESSURE IONIZATION AND CHEMICAL IONIZATION RESPONSES

Negative ions which are observed under the ordinary conditions of operation of an API source are chemically and thermally equilibrated with other ions and molecules in the source. They may be formed under higher energy conditions, but the ions released to the mass analyzer region of the mass spectrometer have the lowest internal energy of all ion samples, when compared to ordinary CI conditions, to the Brandenberger CI condition, and to EI conditions. The closest degree of correspondence, however, is with ordinary CI conditions, and in general API and CI negative ion mass spectra are similar. The differences which have been observed are those which might be expected. Compounds with very low electron affinity, such as polycyclic aromatic hydrocarbons, benzophenone and 2,3,7,8-tetrachlorodibenzodioxin, do not show M^- ions under API conditions. These ions may be formed under higher energy conditions in an API source,

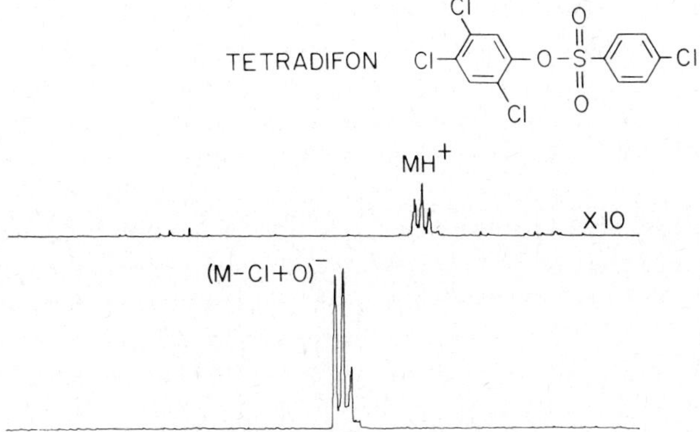

Fig. 14.8. Negative and positive ion API mass spectra observed for tetradifon. The source gas was nitrogen containing a low concentration of oxygen. The positive ion spectrum showed MH^+ ions. The negative ion spectrum showed $(M-Cl+0)$ ions, but no M^- ions. In the absence of oxygen, $(M-HCl)^-$ ions were observed. It is assumed that the substitution reaction occurred for the trichlorophenoxy group. The source temperature was 200°C.

but they are not observed. In a CI source, where electrons of higher energy are available and where the residence time of product ions is shorter than under API conditions, M^- ions are observed for these compounds.

Compounds with low electron affinity normally show weak responses under ECD conditions, but the response may be affected by the presence of oxygen. There is a close correlation between API and CI observations for reactions involving oxygen or O_2^-. The presence of oxygen in a CI or API source can be detected through the direct observation of O_2^- or indirectly through reaction with *p*-nitrochlorobenzene. Ratios of M^- to $(M-Cl+O)^-$ ions, for compounds where chlorine replacement occurs, are generally greater under CI conditions with low oxygen concentration than under comparable API conditions, and this is because equilibration is not complete under CI conditions. Very little oxygen is required for the complete elimination of M^- ions from API mass spectra of these compounds. The ECD response in these instances is enhanced when oxygen doping is employed.

14.4. QUANTITATIVE ANALYTICAL STUDIES BY ATMOSPHERIC PRESSURE IONIZATION NEGATIVE ION MASS SPECTROMETRY

Relatively few studies have been published which describe procedures based on API negative ion mass spectrometry, in contrast to the many reports of quantitative applications of gas chromatography with ECDs. In theory, API negative ion mass spectrometry should show at least as great sensitivity of detection as that of an ECD, and a higher degree of specificity is present because specific ions can be monitored. An example of an analytical procedure is illustrated in Fig. 14.9. The method was developed in order to determine if the flame-proofing agent tris-dibromopropyl phosphate ("tris"), used in children's sleepwear, was being absorbed through the skin. The chart records are for urinary analyses for 2,3-dibromopropanol, the expected urinary metabolic hydrolysis

TABLE 14.3

ANALYSES OF URINE SAMPLES FOR 2,3-DIBROMOPROPANOL

Day	New treated pajamas	2,3-Dibromopropanol (ng/ml)
1	No	0.4
2	No	0.4
3	Yes	11.0
4	Yes	29.0
5	Yes	(sample lost)
6	Yes	21.0
7	Yes	18.0
8	No	9.0
9	No	14.0
10	No	6.0
11	No	6.0
12	No	8.0

product of the phosphate ester. The bioanalytical system included an API mass spectrometer, with a GC unit for sample introduction, and computer-based data acquisition capabilities for selected ion detection and for mass scans. The responses (Fig. 14.9) are for bromide ion isotopes, for 2,3-dibromopropanol (A) and for an internal standard (B). Table 14.3 contains the results of analyses of urine samples before and after new sleepwear was used. This analytical application [25] was based on a type (B) reaction.

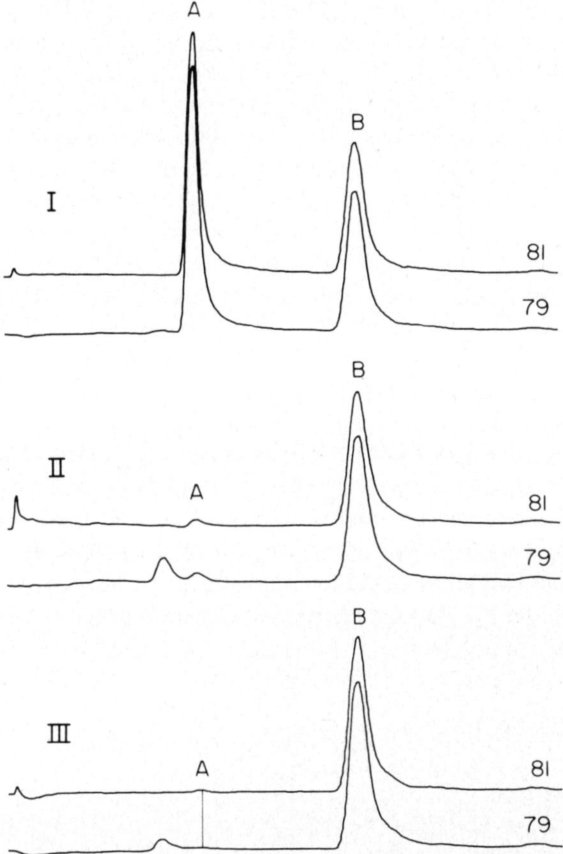

Fig. 14.9. Urinary analyses for 2,3-dibromopropanol for (I) a child wearing new "tris"-treated sleepwear; (II) the same child wearing old "tris"-treated sleepwear before use of the new sleepwear; and (III) a child who had never worn "tris"-treated sleepwear.

Tissue analyses for 2,3,7,8-tetrachlorodibenzo-*p*-dioxin were described by Mitchum et al. [21]. An API GC–MS–COM bioanalytical system was used, with an open tubular capillary column, and with provision for the addition of oxygen–nitrogen (10:90) to the source. A type (D) reaction with O_2^- was employed; the negative ion product was a semiquinone anion corresponding to a dichloro-*o*-benzoquinone. An extensive clean-up procedure was necessary. The method was effective for ppt analyses.

Direct air analyses may be carried out with API instruments when provision is made for the removal of cluster ions. Analytical methods for air are important, but this aspect of API mass spectrometry is not included in this discussion.

14.5. TOXICITY AND ELECTRON-CAPTURE OR ATMOSPHERIC PRESSURE IONIZATION DETECTION

Many organic compounds with broad toxicity to life forms show an ECD response, and this accounts for the wide use of GC–ECD analytical methods for pesticides, fungicides, herbicides and other toxic compounds including many chlorinated and brominated substances used for other purposes. An examination of negative ion products from representative compounds suggests that high electron affinity, with formation of stable M^- ions, is not a property that is generally associated with broad spectrum toxicity, but that reaction with oxygen, leading to a neutral radical and a stable negative ion, is characteristic of compounds of this kind. In both liquid phase chemistry [19] and in gas phase reactions, the transition intermediates are *ipso*-substitution intermediates for aromatic compounds, and radical transition states related to S_N2 reactions for aliphatic and alicyclic compounds. The products which are observed by mass spectrometric methods are stable negative ions, and their formation is responsible for ECD responses. Toxicity affecting cellular processes may be due to the neutral radicals which are formed at the same time, but which have not been observed by direct experiment. These radical entities, as illustrated for *p*-nitrochlorobenzene, may be an inorganic radical, an organic dioxy radical, or an organic radical corresponding to loss of oxygen from a dioxy radical. The initiating step may be reaction with enzymes of the P-450 class, and if oxygen is returned at the end of the sequence, the effect would be that of a toxic chain reaction characterized by the generation of neutral organic radicals through oxygen-containing intermediates, but without overall consumption of oxygen. Further study of this hypothesis will require methods for the detection of radical products.

14.6. ACKNOWLEDGEMENTS

This work was supported by Grant 13901 of the National Institute for General Medical Sciences, by Grant HL-17269 of the National Heart, Lung and Blood Institute, and by Grant Q-125 of the Robert A. Welch Foundation.

REFERENCES

1 J.E. Lovelock and N.L. Gregory, in N. Brenner (Editor), *Gas Chromatography*, Academic Press, New York, 1962, p. 219.

2 D. Frangi-Schnyder and H. Brandenberger, *Z. Anal. Chem.*, 290 (1978) 153.

3 R. Ryhage and H. Brandenberger, *Biomed. Mass Spectrom.*, 5 (1978) 615.

4 H. Brandenberger and R. Ryhage, in A. Frigerio (Editor), *Recent Developments in Mass Spectrometry in Biochemistry and Medicine*, Vol. I, Plenum, New York, 1978, p. 327.

5 H. Brandenberger, in M. Klein, A.V. Kruegel and S.P. Sobol (Editors), *Instrumental Applications in Forensic Drug Chemistry*, U.S. Government Printing Office, Washington, DC, 1979, p. 4147.
6 E.C. Horning, M.G. Horning, D.I. Carroll, I. Dzidic and R.N. Stillwell, *Anal. Chem.*, 45 (1973) 936.
7 D.I. Carroll, I. Dzidic, R.N. Stillwell, M.G. Horning and E.C. Horning, *Anal. Chem.*, 46 (1974) 706.
8 E.C. Horning, D.I. Carroll, I. Dzidic, K.D. Haegele, M.G. Horning and R.N. Stillwell, *J. Chromatogr. Sci.*, 12 (1974) 725.
9 E.C. Horning, D.I. Carroll, I. Dzidic, K.D. Haegele, M.G. Horning and R.N. Stillwell, *J. Chromatogr.*, 99 (1974) 13.
10 E.C. Horning, D.I. Carroll, I. Dzidic, S.-N. Lin, R.N. Stillwell and J.-P. Thenot, *J. Chromatogr.*, 142 (1977) 481.
11 P.L. Gobby, E.P. Grimsrud and S.W. Warden, *Anal. Chem.*, 52 (1980) 473.
12 E.P. Grimsrud, S.H. Kim and P.L. Gobby, *Anal. Chem.*, 51 (1979) 223.
13 F.W. Karasek, *Anal. Chem.*, 43 (1971) 1982.
14 I. Dzidic, D.I. Carroll, R.N. Stillwell and E.C. Horning, *Anal. Chem.*, 47 (1975) 1308.
15 H.P. Tannenbaum and R.C. Dougherty, *21st Annual Conference on Mass Spectrometry and Applied Topics, San Francisco, May 20–25, 1973*, p. 243.
16 P.F. Levonovich, H.P. Tannenbaum and R.C. Dougherty, *J. Chem. Soc., Chem. Commun.*, (1975) 597.
17 E.O. Grimsrud and D.A. Miller, *Anal. Chem.*, 50 (1978) 1140.
18 D.A. Miller and E.P. Grimsrud, *Anal. Chem.*, 51 (1979) 851.
19 W.C. Danen, R.J. Warner and R.L. Arudi, in W.A. Pryor (Editor), *Organic Free Radicals*, A.C.S. Symposium Series No. 69, American Chemical Society, Washington, DC, 1978, p. 244.
20 D.J. Harvan, J.R. Hass, K.L. Busch, M.M. Bursey, F. Ramirez and S. Meyerson, *J. Amer. Chem. Soc.*, 101 (1979) 7409.
21 R.K. Mitchum, G.F. Moler and W.A. Korfmacher, *Anal. Chem.*, 52 (1980) 2278.
22 D.F. Hunt, T.M. Harvey and J.W. Russell, *J. Chem. Soc., Chem. Commun.*, (1975) 151.
23 J.R. Hass, M.D. Friesen, D.J. Harvan and C.E. Parker, *Anal. Chem.*, 50 (1978) 1474.
24 J.R. Hass, M.D. Friesen and M.K. Hoffman, *Org. Mass Spectrom.*, 14 (1979) 9.
25 A. Blum, M.D. Gold, B.N. Ames, C. Kenyon, F.R. Jones, E.A. Hett, R.C. Dougherty, E.C. Horning, I. Dzidic, D.I. Carroll, R.N. Stillwell and J.-P. Thenot, *Science*, 201 (1979) 1020.

Chapter 15

Electron-capture process and ion mobility spectra in plasma chromatography

F.W. KARASEK and G.E. SPANGLER

CONTENTS

15.1. INTRODUCTION

15.1.1. Background

During the late 1960's, plasma chromatography (PC) was developed as a method for ultratrace analysis of organic compounds. By 1969, the basic instrumentation had reached its present configuration. Most of the activity in PC has centered around exploratory studies to demonstrate and understand the scope of the technique. The continued efforts through the 1970's are described in several reviews [1,2]. Quite early, it became clear that a better understanding of PC could be accomplished with mass spectrometric (MS) identification of the ion species observed in the ion mobility spectra of different compounds. Using PC–MS systems, various publications appeared describing in greater detail the data generated and the operation of the instrument [3–5]. The close relationship between PC and atmospheric pressure ionization (API) and chemical ionization mass spectrometry (CIMS) also helped the interpretation of ion mobility spectra. By 1979, sufficient interest had developed in PC that additional efforts were initiated to further develop the instrumentation. Development efforts and exploratory studies are continuing in both fundamental areas of ion–molecule reactions and instrumental areas of creating monitoring devices for trace components in air and universal gas chromatographic (GC) detectors.

PC addresses the same problems as the electron-capture detector (ECD) except that it provides an additional capability to monitor positive ions as well as negative ions. Although its flexibility may be enhanced when used in conjunction with GC separation techniques, PC is also capable of achieving specific results independently of such techniques. This performance is achieved by coupling an ion generator with an ion–molecule reactor and an ion mobility drift tube spectrometer. The technique is based on performing ion–molecule reactions with sample molecules under atmospheric pressure conditions and observing drift time differences of ions which traverse specified drift lengths under the influence of an applied electrostatic drift field. Analogous to the ECD, the reactor region is a small volume irradiated by short-range β-particles to create in selected reagent carrier gases dense distributions of electrons and positive and negative ions. The name plasma chromatography was originally suggested by the use of ions (plasma) which undergo a molecular separation similar to that available from chromatography. However, due to the close relationship of PC to ion mobility spectrometry (IMS) developed by physicists, the terms PC and IMS are used interchangeably in referring to the technique. Earlier publications referred to the spectra as "plasmagrams", a term which has been replaced by the more scientifically correct designation of ion mobility spectra.

15.1.2. Basic principles of plasma chromatography

Mechanically, the plasma chromatograph consists of two electrode regions: one region contains the ^{63}Ni radioactive source and functions as an ion–molecule reactor: the other region contains a pulsed ion injection grid and ion collector which functions as a drift spectrometer for ion mobility separation. Trace sample molecules introduced into the plasma chromatograph are ionized in the reaction region and are separated according to drift times in the drift region. A schematic representation of the essential components and electronics needed to produce a working system is shown in Fig. 15.1.

Sample vapor is transported to the reaction region by means of a carrier gas. As noted previously, the carrier gas may be purified nitrogen or air containing a trace of water vapor (10–100 ppm) or these gases mixed with reagent vapors carefully selected to tailor the ion–molecule reactions. A drift gas flows in the drift region counter to the carrier gas in order to exclude sample vapor and thus to quench continuing ion–molecule reactions in the drift tube. Under these conditions, PC primarily responds to long lived stable ions formed from the sample vapor. As illustrated in Fig. 15.1, both the carrier and drift gases meet in the vicinity of the ion shutter grid to exit from the PC cell at this location. Gas velocities are less than 1% of ion velocities under typical operating conditions and flow in proportion to the cross-sectional areas of the respective regions under optimum conditions.

The 67 keV β-particles from the ^{63}Ni radioactive source ionize the carrier gas to form reactant ions. These ions in turn react with the sample molecules to produce product ions. This series of reactions can be symbolized by

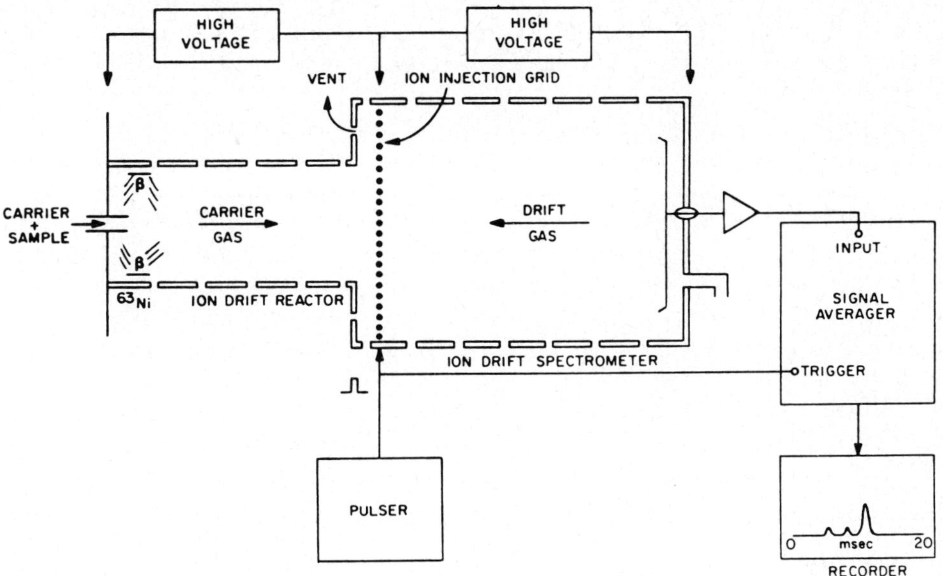

Fig. 15.1. Schematic diagram of the plasma chromatographic instrumental system.

$$\beta \xrightarrow[\text{chemical reagents}]{1} R^{\pm} + M \xrightarrow{2} R + M^{\pm} \tag{15.1}$$

where R is the reactant ion with positive or negative charge and M is the sample molecule. In nitrogen, the primary reactant ions are $(H_2O)_n(N_2)_mH^+$, $(H_2O)_n(N_2)_mH^+$, $(H_2O)_nNH_4^+$ and electrons. When air is used as a carrier gas, $(H_2O)_nO_2^-$ ions are added to this group. Chemical reagents can be added to the front end of these reactions to alter the identity of the reactant ions and thus control the ion–molecule reactions. Step 2 of the reactions can be quite complicated depending on the identity and energetics of the reactant ion.

After the product ions are formed, they are accelerated, along with the remainder of the reactant ions, by a potential gradient toward the ion injection grid. The ions are then pulsed intermittently into the drift region. The pulse of mixed ions is then moved through the drift region by means of a strong electric field. During their transit time, the ions collide with the neutral gas molecules and drift with rates characterized by their collision cross section and mobility. Heavier, larger ions move more slowly than lighter, smaller ions and arrive at the ion collector with longer drift times. The arrival time spectrum derived from PC is called an ion mobility spectrum (Fig. 15.2) with ions of mass 400 a.m.u. arriving in the order of 20–40 ms depending on drift temperature. Positive or negative ions can be observed depending on the polarity of the electrostatic drift field.

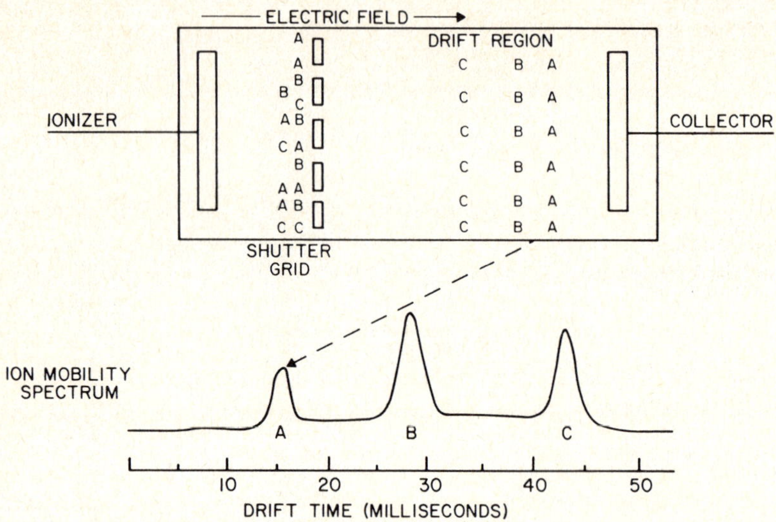

Fig. 15.2. Mobility spectra produced by the plasma chromatographic instrumentation for ions produced by air.

15.1.3. Ion mobility relationships

Theoretically, the velocity of the center of mass of the ion cloud traveling through the drift region is called the drift velocity. This drift velocity, v_d, displays a linear relationship with the electrostatic drift field E applied across the drift region:

$$v_d = \mu E \tag{15.2}$$

where the proportionality constant μ is the (scalar) mobility of the ions. If v_d is replaced by the drift length, d, divided by drift time, t_d, between the ion injection grid and collector, the mobility becomes

$$\mu = \frac{d}{t_d E} \tag{15.3}$$

The mobility is a joint property of the ions and the gas through which the motion occurs. Mason and Schamp [6] have shown that the mobility satisfies

$$\mu = \frac{3}{16}\frac{e}{N}\left[\frac{1}{m}+\frac{1}{M}\right]^{1/2} \times \left[\frac{2\pi}{kT}\right]^{1/2} \frac{1+\Delta}{\pi r_m{}^2 \Omega^{(1,1)*}} \tag{15.4}$$

where N is the molecular number density, k is the Boltzmann constant, e is the electronic charge, T is the absolute temperature in K, Δ is a small correction term for high approximations, r_m is the position of minimum potential for interaction between the ion and molecule, and $\Omega^{(1,1)*}$ is the first order collision integral. The mobility is thus inversely proportional to the collision cross section and the gas density. To remove gas density effects from the results, the mobility is adjusted to standard conditions (T = 237 K, p = 760 mmHg) by exercising the following relationship:

$$K_0 = \frac{d}{t_d E} \; \frac{P}{760} \; \frac{273}{T} \tag{15.5}$$

The parameter K_0 is referred to as the reduced mobility. From the kinetic theory of gases, the reduced mobility depends on the ionic charge, size, and mass; the molecular size and mass; and the drift gas composition and polarizability.

15.2. RELATIONSHIPS OF PLASMA CHROMATOGRAPHY TO ELECTRON-CAPTURE DETECTOR MECHANISMS

15.2.1. Background

From its conception by Lovelock, the ECD has developed into one of the most sensitive and selective devices available for use in gas chromatography. The importance of this detector for analysis of trace compounds of biomedical and environmental significance has led to many studies of its anomalous responses and linearity [7,8], range of relative response factors for different types of compounds [10–12], and the effect of temperature on sensitivity [13–15]. In 1963 Lovelock described the plane parallel detector design and the pulse sampling technique of operation and its effect on anomalous responses, errors, and linearity range. This work, together with that reported by Wentworth [14,16–21] on the mechanism of the electron attachment phenomena, has done much to expand our understanding of the fundamental physical processes involved in the functions of the ECD.

These studies and their conclusions were necessarily made from indirect experimental data derived from observations of total currents under different detector parameters of voltage, temperature, and geometry. While these studies provided improved understanding of the ECD, a greater advance in knowledge required more detailed observations on the behavior of the charged particles present and the kinetics of the reactions involved in the ECD. With the development of PC, this was made possible by creating conditions similar to those found in the ECD, studying both the positive and negative charged species formed, and measuring their response and mobility under changing parameters. Operating at atmospheric pressure with a carrier gas of purified nitrogen, conditions were established in the plasma chromatograph which approximate closely those established in the ECD for the performance of the detector. The identity of the major reactant ions has been established using negative ion PC under these conditions. These ions arise from electron–molecule reactions which are near thermal energy. Direct experimental observation of the reactions which occur between these negative ions and electrons with sample molecules has allowed studies to be performed on many fundamental aspects of electron attachment phenomena and ECD characteristics.

15.2.2. Electron energy considerations

In order to properly interpret PC data relative to ECD data, careful consideration must be given to the electron energies utilized by the two techniques. In PC, the electrons are continuously under the influence of an electrostatic drift field which raises the electron energy above thermal. This is unlike ECD where either methane is added to the carrier gas or a pulsed mode is utilized to assure thermal energy for the electrons. The mean energy of an electron is given by the Townsend energy factor [22]:

$$K = \frac{\text{mean energy of electron}}{\text{mean energy of molecular translation}} = \frac{\frac{1}{2}m\bar{c}^2}{\frac{3}{2}kT} \tag{15.6}$$

where m = electron mass and $\bar{c}$ = mean velocity of electron.
Townsend's energy factor K is an experimental quantity which has been found, at a fixed temperature T (15°C), to be a function of E/p where E is the electric field strength and p is the pressure. From statistical mechanics, the correct expression for the kinetic energy is:

$$\tfrac{1}{2}m\bar{c}^2 = \frac{e}{F}\,\frac{D}{\mu} \tag{15.7}$$

where e is the electron charge, D is the diffusion coefficient, μ is the electron mobility, and F is a factor determined by the nature of the distribution function $f_0(c)$. When the distribution function is that of Maxwell, $F = 2/3$ and Townsend's energy factor K becomes

$$K = \frac{D}{\mu}\,\frac{e}{kT} \tag{15.8}$$

In general, the distribution function is not that of Maxwell and better approximations to F can be made. However, the values of K with $F = 2/3$ provide reasonable estimates for the energy factor.

Combining eqns. 15.1–15.3, the energy of the electron is given by

$$\text{Electron energy} = 3/2\,\frac{D}{\mu} \tag{15.9}$$

Summarized in Table 15.1 are values of D/μ in nitrogen at 293 K as derived from swarm experiments. Values for D/μ are dependent on experimental parameters through E/N where E is the electric field strength and N is the number density of molecules through which the electron must pass. Since typical operating parameters for PC are $E = 214$ V cm^{-1}, $T = 473$ K, and $p = 760$ mmHg, E/N values of 1.33 Townsends (a unit of 10^{-17} when E is expressed in V cm^{-1} and N in cm^{-3}) describe the performance of the instrument. From Table 15.1, this corresponds to an electron energy of 0.53 eV.

The pulsed ECD provides a means whereby electron-capture reactions can be performed with truly thermal electrons. At 200°C, this energy (3/2 kT) is 0.061 eV. Consequently, the electron energy used to study electron-capture reactions in PC is an order of magnitude greater than those used in the ECD. As will be discussed later, care must be taken in comparing the two sets of data for this reason.

TABLE 15.1

TRANSPORT COEFFICIENTS FOR ELECTRONS IN NITROGEN

E/N (T_d)	Drift velocity (10^5 cm s^{-1})	D/μ (10^{-2} V)
1.0	4.43	27.7
1.2	4.74	32.3
1.4	5.03	36.8
1.7	5.49	42.8
2.0	5.98	48.0
2.5	6.79	55.0
3.0	7.67	60.9
3.5	8.51	65.3
4.0	9.33	69.1
5.0	10.95	74.4
6.0	12.60	79.7
7.0	14.09	83.5
8.0	15.52	87.3
10	18.38	93.2
12	20.9	97.9
14	23.5	101.9
17	27.3	
20	30.9	
25	36.5	

15.2.3. Electron-capture reactions

Like the ECD, PC is a member of a much broader class of techniques using principles of ionization for detection purposes. When used with inert carrier gases, free electrons are generated which react with sample molecules to form molecular negative ions. The negative ion may be produced by (a) associative resonance capture (thermal electrons), (b) dissociative resonance capture (0–15 eV electrons), and (c) pair production (>10 eV). All processes may produce vibrationally excited products, and reactions (b) and (c) can also produce translationally excited ones. Since electron energies utilized in ECD and negative ion PC are near thermal, the first two ionization processes are of primary concern and are generally referred to as electron-capture reactions.

15.2.3.1. Negative ion lifetimes

Negative ions of atoms and molecules are generally considered to have been formed when the incoming electron is associated with the atom or the molecule for a time longer than the transit time of the electron. This time may be as short as 10^{-15} s, depending on the velocity of the electron and the size of the target. Thus, negative ion formation can be viewed as proceeding via a negative ion intermediate that can itself be formed either by electron capture in the field of the ground electronic state or by electron capture in the field of an excited electronic state:

$$AB + e^- \underset{}{\overset{\sigma_0}{\rightleftarrows}} AB^- \text{ (or } AB^{-*}) \tag{15.10}$$

The time the electron is retained by the molecule is referred to as the negative ion (autodetachment) lifetime τ_a. Depending on the magnitude of τ_a, three classes of negative ions have been distinguished [23].

(1) Extremely short lived (10^{-15} s $\leqslant \tau_a \leqslant 10^{-12}$ s). These are observed as resonances in electron scattering and/or dissociative attachment studies.

(2) Moderately short lived (10^{-12} s $\leqslant \tau_a \leqslant 10^{-6}$ s). These can be stabilized in high (1 to ca. 1000 Torr) and very high (>1000 Torr) pressure swarm experiments.

(3) Long lived ($\tau_a > 10^{-6}$ s). These can be conveniently studied with conventional time-of-flight mass spectrometers.

The various channels through which a transient molecular negative ion AB^{-*} decays are:

$$AB + e^- \underset{}{\overset{\sigma_0}{\rightleftarrows}} AB^{-*} \begin{cases} \xrightarrow{\text{psc}} AB \text{ (or } AB^*) + e^- \\ \xrightarrow{\text{pda}} A \text{ (or } A^*) + B^- \\ \xrightarrow{\text{pst}} AB^- + \text{energy} \end{cases} \tag{15.11}$$

where σ_0 is the cross section for the formation of AB^{-*} and psc = σ_{sc}/σ_0, pda = σ_{da}/σ_0, pst = σ_{st}/σ_0 are, respectively, the probability that AB^{-*} will decay via autoionization (indirect elastic or inelastic electron scattering), will undergo dissociative attachment, or will be stabilized radiatively or collisionally. All of the above three decay processes can be in competition. A semitheoretical model was presented by Compton et al. [24] using the principle of detailed balancing to derive an expression which relates the average autodetachment lifetime τ_a, the associative attachment cross section σ_a, and the densities of state for the ion ρ^- and its fragments ρ^0 of relative velocity ν through:

$$\bar{\tau}_a = \rho^-/\rho^0 \, (1/\nu\sigma_a) \tag{15.12}$$

Measurement of τ_a and σ_a and a knowledge of the vibrational frequencies can then be used to calculate the electron affinity, which is the dominant parameter in the term ρ^-/ρ^0.

Under atmospheric pressure conditions normally utilized in ECD and negative-ion PC, the transient negative-ion AB^{-*} can be collisionally stabilized through

$$AB^{-*} + M \rightarrow AB^- + M + \text{energy} \tag{15.13}$$

provided collision times of M with AB^{-*} are less than τ_a. Since collision times between molecules is of the order of $6 \cdot 10^{-10}$ s at 760 Torr pressure and 300 K, only negative ions with autodetachment lifetimes greater than 10^{-10} s can be studied by these techniques. Thus only moderately short lived and long lived negative ion species can be studied with ECD and PC.

15.2.3.2. Electron affinity

One of the fundamental properties of a molecular ion is the lowest energy required to remove an electron. This energy is defined as the electron affinity (EA). That is for the associative capture reaction:

$$AB + e^- \rightarrow AB^- \qquad (15.14)$$

the enthalpy ΔH_{min} for the reaction is:

$$\Delta H_{min} = -EA \text{ (of AB)} \qquad (15.15)$$

However, when the reaction is dissociative in nature, the enthalpy includes heats of formation for radicals, ions, molecules, such that

$$AB + e^- \rightarrow A^- + B \qquad (15.16)$$

$$\Delta H_{min} = A_p - E_e = \Delta H_f(A^-) + \Delta H_f(B) - \Delta H_f(AB) \qquad (15.17)$$

where ΔH_{min} is the enthalpy of reaction, A_p is the appearance potential:

$$A_p = D(A\text{–}B) - EA(A) + E_e \qquad (15.18)$$

D(A–B) is the energy of dissociation for A and B, and E_e, the excess energy, is the sum of the electronic, vibrational, rotational, and translational energies. Partitioning of excess energy among the various degrees of freedom in polyatomic molecular ion intermediates is the subject of much discussion.

Electron affinities have been determined experimentally for a large class of compounds. The most recent summary of these data has been provided by Janousek and Brauman [25]. Closely related to the electron affinity is the capture coefficient derived from ECD data. Pellizzari [26] has provided a recent review of these data. Summarized in Fig. 15.3 is a condensed compilation of both sets of data combined with estimated values presented in a format suitable for use as a guide to PC. The oxygen and chloride anions are the major reactant ions of PC which quench or interfere with the electron-capture process in the presence of air. Electron-capture processes can only be studied using dry nitrogen as drift and carrier gases in PC. Due to its high electron affinity, nitrogen dioxide is a major interferent which quenches the electron-capture process.

With the increase in the number of experimentally determined electron affinities, it is now possible to explore the relationship between structure and stability in negative ions. For instance, from the point of view of the electron attracting power of atoms, elements to the upper right in the periodic table have an increasing tendency to bind electrons (increasing ionization potential). Due to instability produced by electron pairing configurations of atoms, the group VA atoms (N, P and As) have low electron affinities while the group IVA atoms (C, Si and Ge) have relatively high electron affinities. Thus, the oxygen and nitrogen components of ambient air are characterized by different electron affinities, −0.07 eV and 1.462 eV, respectively. For this reason, purified carrier gases such as nitrogen and the noble gases have long been considered necessary to allow proper performance of the ECD. Additives such as methane are sometimes added to the carrier gas to provide large collision cross sections for thermalization of the electrons without itself capturing electrons.

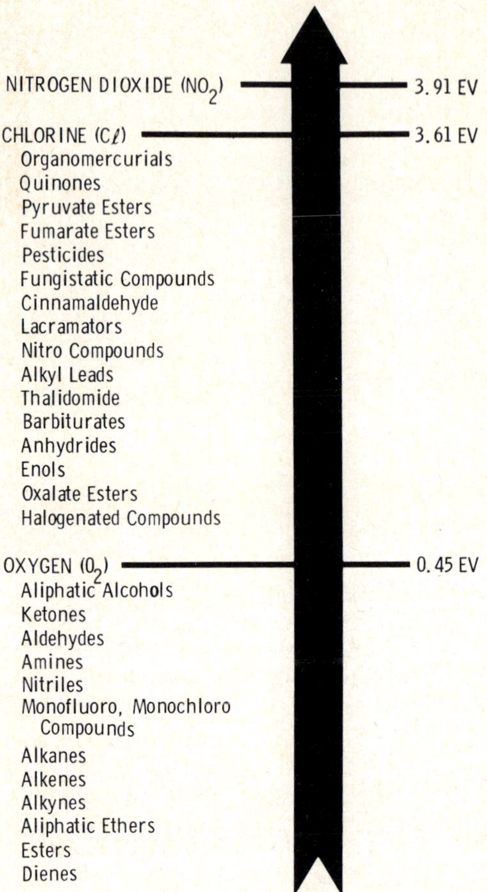

Fig. 15.3. Electron affinity scale for compound classes.

15.3. EXPERIMENTAL STUDIES BY PLASMA CHROMATOGRAPHY

15.3.1. Plasma chromatographic determination on effect of oxygen in electron-capture detection

While these observations on the effect of impurities in the carrier gas used for the ECD were necessarily made from indirect experimental data, further insight was provided by plasma chromatography. Karasek and Kane [27] noted that PC permits direct experimental data to be obtained on the charged species formed in the ECD and their response and mobility under changing parameters. Data collected from the plasma chromatograph indicate that the reason for the restrictions on carrier gas composition is that electrons react with oxygen in air to form ion–molecule complexes of the form $(H_2O)_nO_2^-$ [28–30]. This removes a significant portion of the electrons and prevents them from performing their function of reacting with the trace constituents of sample vapor in the cell (see Fig. 15.4). These constituents must then undergo competing reactions of electron

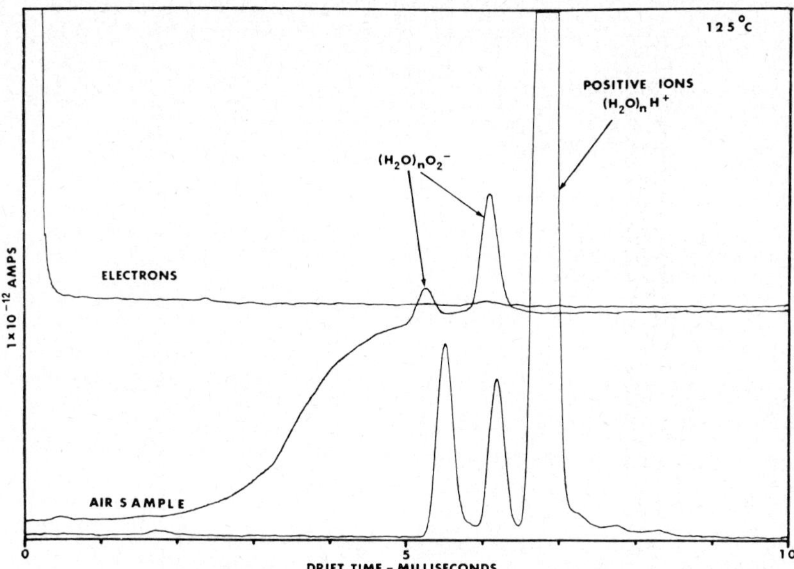

Fig. 15.4. Ion mobility spectra of the charged species present in the plasma chromatograph with nitrogen carrier. The change in electron current and formation of oxygen ions is seen upon injection of an air sample. (Reproduced from *Anal. Chem.*)

capture and ion–molecule reaction with the $(H_2O)_nO_2^-$ species for ionization. Since the ECD detects primarily electrons, this effect will considerably reduce sensitivity and linearity of the ECD, is greater at lower cell temperatures and is somewhat related to water concentration.

The ion mobility spectra of Fig. 15.4 show the charged species present when nitrogen is used as the carrier gas to simulate operating conditions in an ECD. The negative particles are thermal electrons. These electrons appear as a continuum across the ion mobility spectrum because the grids do not close perfectly for such small, high-speed particles. If the grids could exclude the electrons, the ion mobility spectrum would show only a single peak with a drift time of 0.01 ms, the magnitude of this peak being about 15 times greater than that of the continuum. This event is too fast for the response of the recording system. The initial electron peak shown in Fig. 15.4 appears because of the gating technique employed; during that initial interval the injection and scan gates are open simultaneously and the full electron current reaches the detector. The positive particles are predominantly of the form $(H_2O)_nH^+$. When a 10-ml sample of cylinder air (Linde, very dry, purified grade) is injected into the carrier gas of the plasma chromatograph, the charged species present become those shown in Fig. 15.4. One can note the initial disappearance of the electrons and the appearance of the $(H_2O)_nO_2^-$ ions, where n varies from 0 to 3 and is a function of water concentration and temperature. The identity of these species formed with air has been established previously by use of a mass spectrometer directly coupled to the plasma chromatograph. The drift time scan was recorded over a 2-min scan time, so that the rise of the curve toward the original electron level represents depletion of the air sample in the plasma chromatograph by the moving nitrogen carrier gas.

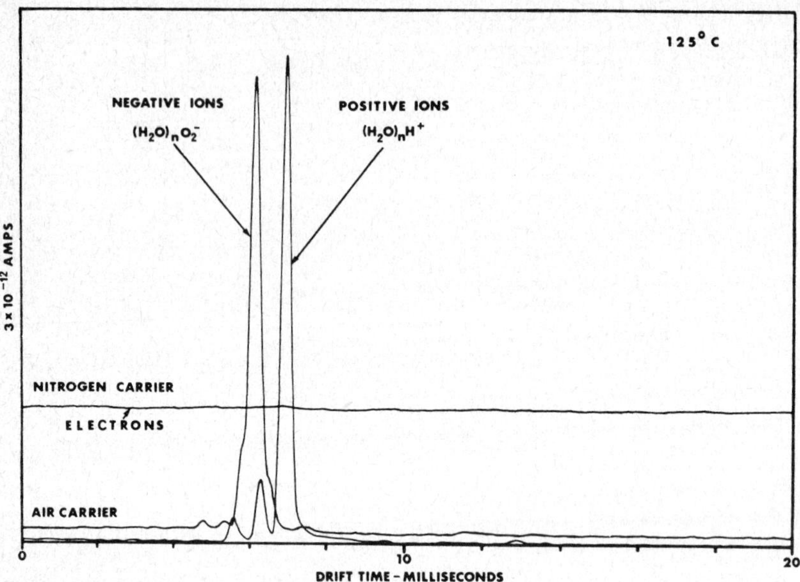

Fig. 15.5. Composite ion mobility spectra obtained at 125°C with a nitrogen carrier and an air carrier gas. The positive ions are the same in each case; the negative species differ. (Reproduced from *Anal. Chem.*)

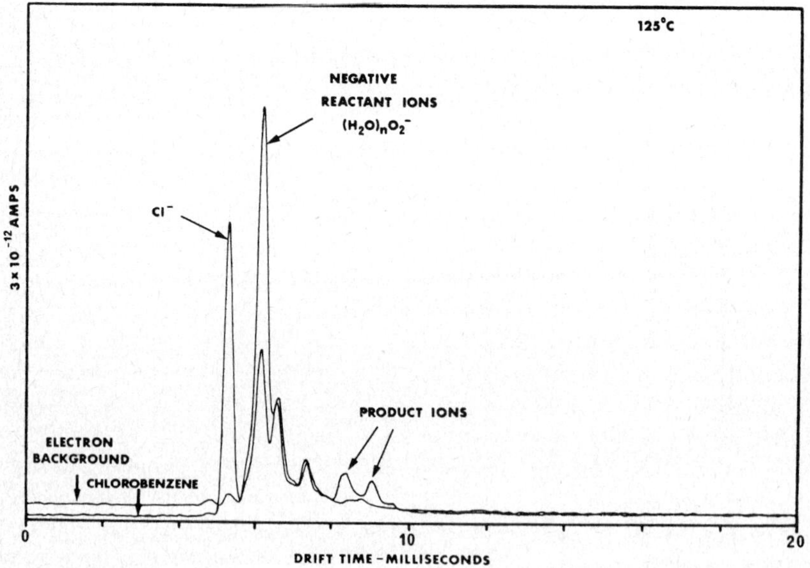

Fig. 15.6. Ion mobility spectra obtained with an air carrier before and after injection of 10^{-8} g of chlorobenzene. (Reproduced from *Anal. Chem.*)

By changing to an air carrier gas, the effect of greater amounts of oxygen can be observed. This results in the appearance of the charged species shown in Fig. 15.5. A greater portion of the electrons, about 90%, is converted to the $(H_2O)_nO_2^-$ particles. No significant change is observed in the positive reactant ions. We would expect to see the initial electron peak, but it is not observed in the 20-ms ion mobility spectra; because of the relative gate widths, the peak is too narrow for the recording system to follow.

The high sensitivity of detection of trace constituents in a sample by the plasma chromatograph depends upon reaction between reactant ions and the trace compounds. Depending upon the amount of air present in the carrier of an ECD there will be a mixture of electrons and the oxygen ions. The trace constituent in a GC peak will undergo reaction with both the electrons and the $(H_2O)_nO_2^-$ species depending upon their relative numbers and reaction rates. Since the ECD produces a GC peak by measuring a decrease in the standing electron current, this dual reactivity leads to anomalous sensitivity and linearity effects when oxygen is present. This is clearly illustrated in Fig. 15.6, where an air carrier gas is used. Along with a low level of electron current, one can observe essentially the same pattern of negative reactant ions as appears in Fig. 15.5. Upon injection of a trace sample of 10^{-9} of chlorobenzene, formation of the chloride anion is seen from reactivity with both the electrons and the main peak of the negative reactant ions. The peaks indicated as product ions are most probably due to an ion–molecule reaction of the chlorobenzene with reactant ions.

Temperature has an effect on the relative distribution of the $(H_2O)_nO_2^-$ species and also on the relative amounts of the electrons and the total of these ions. This is shown by comparison of the data in Fig. 15.7 with those in Fig. 15.5. Ion mobility spectra for

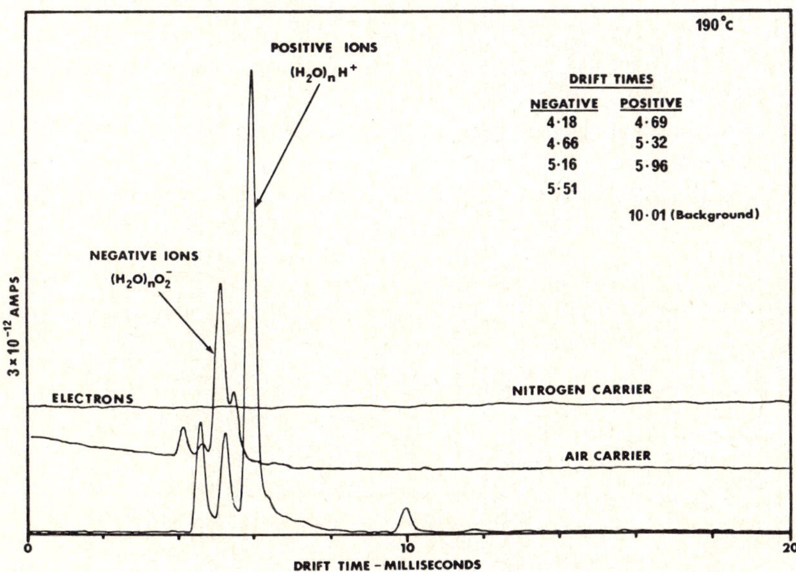

Fig. 15.7. Composite ion mobility spectra obtained with a nitrogen carrier and an air carrier at 190°C. This temperature reduces conversion of electrons to negative ions. (Reproduced from *Anal. Chem.*)

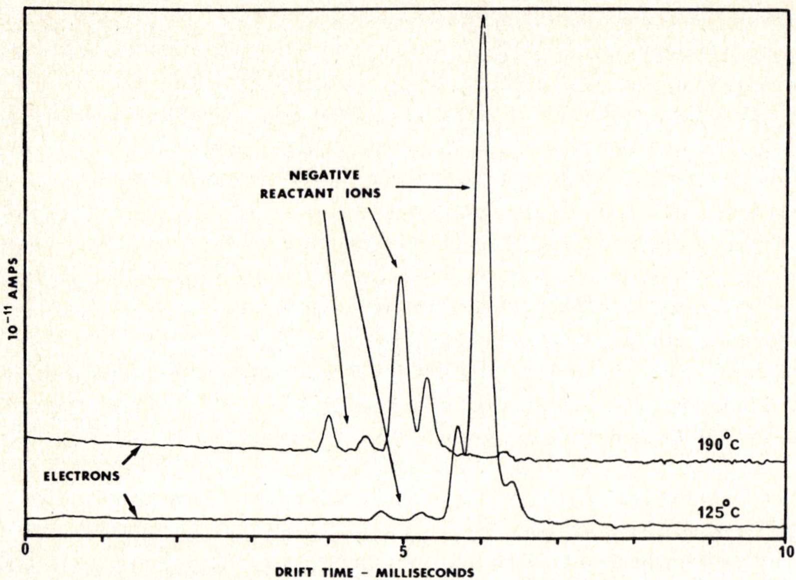

Fig. 15.8. Composite ion mobility spectra of negative ions obtained with an air carrier at different temperatures under high resolution conditions. (Reproduced from *Anal. Chem.*)

Fig. 15.7 were taken under the same instrumental conditions as for Fig. 15.5 except that a temperature of 190°C was used instead of 125°C. At the higher temperature, only about 20% of the electrons are converted to the $(H_2O)_nO_2^-$ ions. These negative ions also show a different relative composition. The identity of equivalent ions at the two temperatures can be established by correspondence of their calculated reduced mobility. Changing to higher resolution conditions by reducing gate widths to 0.1 ms gives the data shown in Fig. 15.8 and Table 15.2 revealing essentially the same reduced mobility for equivalent ions. The presence of both reactive ions and electrons would lead to a different sensitivity and linearity in an ECD at these two temperatures. The dependence of sensitivity on

TABLE 15.2

DRIFT TIMES AND CALCULATED REDUCED MOBILITIES (K_0) FOR NEGATIVE IONS AT TWO TEMPERATURES

Temperature (°C)	Drift time (ms)	K_0 (cm^2 V^{-1} s^{-1})
125	4.90	3.33
	5.43	3.01
	5.87	2.78
	6.18	2.64
190	4.18	3.39
	4.66	3.02
	5.16	2.75
	5.51	2.57

temperature that has been observed for the ECD could result, in part, from a similar situation where traces of electron-capturing compounds are present, whether from air, column bleed, or detector contamination.

These data apply most directly to the d.c. mode of operating the ECD using a nitrogen carrier gas. The nitrogen carrier used was a highly purified grade and was brought to a uniform water concentration, estimated to be 10 ppm or less, by passing through a large molecular sieve dryer. This still provides sufficient water to produce the positive and negative water clustered species observed.

When air is present in the carrier gas, and the negative species are a mixture of electrons and water-clustered negative ions (Fig. 15.7), the addition of water results in a decrease of electrons with the formation of larger amounts of the negative ionic species. These observations tend to suggest that removal of water from the carrier gas of an ECD is more important when air and other electron-capturing contaminants are present.

While more complete and definitive work is needed to determine the exact effect these observations would have on the ECD in its pulsed mode of operation with an argon–methane mixture, some data have been obtained. The addition of a 10-ml sample of argon–methane (90:10) to the nitrogen carrier gas results in some formation of methane ion-molecules of lower mobilities by reaction with the water-clustered protons which remain the dominant species. In the negative mode, the presence of the argon–methane mixture gives about the same results as does water; in the presence of air, the relative electron concentration decreases with formation of greater amounts of ion-molecules; when only electrons are present as in pure nitrogen, the argon–methane has little effect on the electron level. These effects are all a function of temperature.

15.3.2. Plasma chromatographic studies of negative species from dissociative electron capture

15.3.2.1. Halogenated compounds

In general, conjugation helps to stabilize negative ions by charge dispersal. The most obvious kind of conjugation in negative ions is π-delocalization, of which the phenoxide anion provides a good example. The electron affinity of the phenoxyl radical (ca. 2.36 eV [31]) is considerably greater than the electron affinities of alkoxyl radicals (ca. 1.6–1.9 eV [32])

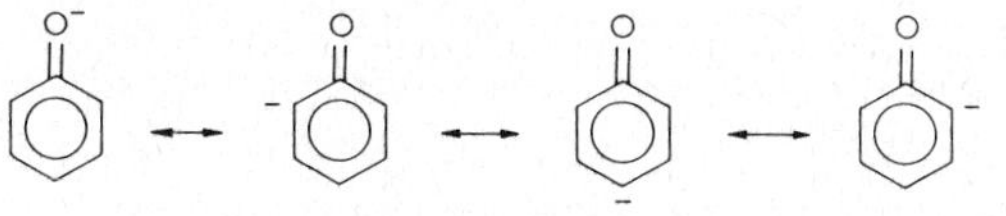

(15.19)

which have no π-delocalization. This suggests that, as expected, the delocalized negative ion is highly stabilized.

An exception to this generalization is the case where the following condition is satisfied:

$$D(\mathrm{A{-}X}) - EA(\mathrm{X}) < 0 \qquad (15.20)$$

TABLE 15.3

IONIZATION AND REDUCED MOBILITIES (K_0) FOR HALOGENATED AND NITROAROMATICS

Compound	Positive ions	K_0 ($cm^2\ V^{-1}\ s^{-1}$)	Negative ions	K_0 ($cm^2\ V^{-1}\ s^{-1}$)
Fluorobenzene	$(C_6H_5F)H^+$	2.11	–	–
	$(C_6H_5F)_2H^+$	–		
Chlorobenzene	$(C_6H_5Cl)H^+$	1.99	Cl^-	2.92
	$(C_6H_5Cl)_2H^+$	–		
Bromobenzene	$(C_6H_5Br)H^+$	1.91	Br^-	2.61
	$(C_6H_5Br)_2H^+$	–		
Iodobenzene	$(C_6H_5I)H^+$	1.81	I^-	2.51
	$(C_6H_5I)_2H^+$	–		
Nitrobenzene	$(C_6H_5NO_2)H^+$	1.88	$C_6H_5NO_2^-$	1.87
o-Chloronitrobenzene	$(C_6H_4ClNO_2)H^+$	1.70	Cl^-	2.92
			$C_6H_4NO_2^-$	1.86
o-,*m*-,*p*-Nitrotoluene	Unknown		$C_6H_4ClNO_2^-$	1.71
			NO_2^-	2.70
2,4-Dinitrotoluene	$(C_6H_3CH_3N_2O_4)H^+$	1.62	$(C_6H_4CH_3NO_2)^-$	1.74
	$(C_6H_3CH_3N_2O_4)NO^+$	1.55	$(C_6H_3CH_3N_2O_4)^-$	1.61
			$(C_6H_3CH_2N_2O_4)^-$	1.67
2,4,6-Trinitrotoluene	$(C_6H_2CH_3N_3O_6)H^+$	1.48	$(C_6H_2CH_3N_3O_6)^-$	1.49
	$(C_6H_2CH_3N_3O_6)NO^+$	1.41	$(C_6H_2CH_2N_3O_6)^-$	1.54
Octachlorobiphenyl	$(C_{12}H_2Cl_8)H^+$	1.16	Cl^-	2.92
			$C_{12}H_2Cl_8^-$	1.16
Decachlorobiphenyl	$(C_{12}Cl_{10})H^+$	1.08	Cl^-	2.92
			$C_{12}Cl_8^-$	1.16
			$C_{12}Cl_{10}^-$	1.08
Carbon tetrachloride	–		Cl^-	2.92
n-Butyl chloride	–		Cl	2.92
			$(C_4H_9Cl)^-$	

such that D (A–X) is the dissociation energy for the substituent and EA (X) is the electron affinity of the substituent X. The most notable examples of dissociative capture arising from condition 15.20 are the halogenated benzene derivatives which undergo dissociative electron-capture reactions even with thermal electrons. Using swarm beam techniques, Christophorou et al. [33,34] showed that the benzene derivatives C_6H_5Cl, o-$C_6H_4Cl_2$, o-$C_6H_4CH_3Cl$, C_6H_5Br, and o-$C_6H_4CH_3Br$ capture low energy (including thermal) electrons dissociatively in a resonance capture process which peaks in the electron energy (E_{max}) range between thermal ($3/2\ kT$) and 1 eV producing only Cl^- or Br^- anions. The magnitude of the cross section at E_{max} was found to increase with decreasing E_{max} (i.e., in the order of Cl, Br, I) suggesting that the formation of AB^{-*} of reactions 15.10 and 15.11 may be visualized as proceeding via a Frank–Condon transition from the initial state – neutral molecule and electron separated at infinite distance – to the final state representing the compound negative ion [35]. It is observed that the electron energy E_{max} lies well below the energy of the lowest known excited electronic states of these molecules. When the negative ion state lies at or above known electronic states of the neutral molecule, the magnitude of the cross section for dissociative attachment decreases because the negative ion can now decay to an electronically excited neutral molecule in addition to a rotationally and/or vibrationally excited molecule.

These findings have been reaffirmed by PC [36,37]. A summary of data compiled from the PC literature is provided in Table 15.3 where data for both the positive and negative ion spectra are presented whenever possible. Ion identification is generally accomplished through mass–drift time correlations [38,39] except for mass spectrometric identification in some cases (nitrotoluene negative ion data [40]). Taken together, the positive and negative ion mobility patterns are quite different and characteristic for each compound. For the negative data, the fluoride compound exhibits little detectable response; all other halogenated compounds show a strong peak for the halogen ion (see Fig. 15.9). These results are consistent with those obtained by Melton using negative ion mass spectrometry which give the appearance potential of F^- from C_6F_6 as 3.9 eV. This indicates the electron energy in PC does not exceed 3.9 eV. However, the appearance potentials of Cl^-, Br^-, and I^- from a large number of halogenated compounds are reported to be either 0 or 0.1 eV at the most, so the appearance of these ions from thermal electrons is to be expected. Since, the electron energy typically employed in PC is 0.50 eV, it becomes clear that PC data can provide a method whereby the details of ECD characteristics can be studied as a function of temperature, pressure and reaction time.

Where two different halogen atoms are present in the molecule, the halogen that dissociates is determined by the general order of reactivity for electron attachment shown by halogens: I > Br > Cl. This leads to PC results as recorded in Table 15.4.

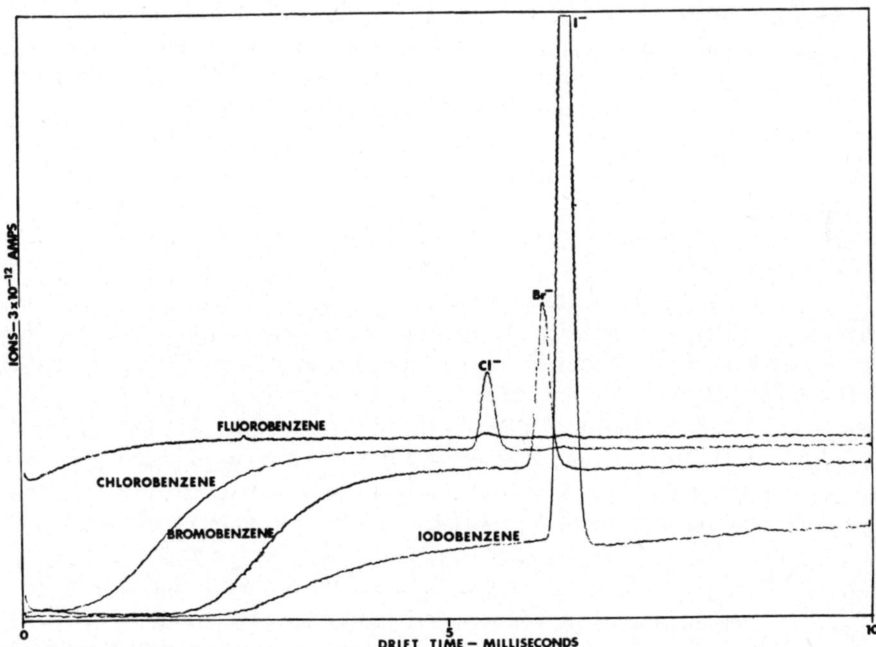

Fig. 15.9. Composite negative ion mobility spectra of the mono-halogenated benzenes. (Reproduced from *Anal. Chem.*)

TABLE 15.4

NEGATIVE SPECIES FORMED BY DISSOCIATIVE ELECTRON CAPTURE

Compound	Negative species from dissociative capture
Fluorobenzene	–
Chlorobenzene	Cl^-
Bromobenzene	Br^-
Iodobenzene	I^-
m-Dichlorobenzene	Cl^-
o-Dibromobenzene	Br^-
o-Fluorochlorobenzene	Cl^-
m-Fluorochlorobenzene	Cl^-
o-Chlorobromobenzene	Br^-
o-Chloroiodobenzene	I^-
m-Chloroiodobenzene	I^-
o-Chlorotoluene	Cl^-
m-Chlorotoluene	Cl^-
4-Chlorobiphenyl	Cl^-
2,2′,4,4′-Tetrachlorobiphenyl	Cl^-

15.3.2.2. Chloro and nitro compounds

Swarm beam techniques in MS indicate that both nitrobenzene and nitrotoluene capture electrons nondissociatively at electron energies below 0.03 eV, only undergoing dissociative capture to form NO_2^- anions for electron energies above 0.43–0.50 eV [23]. Similarly, Karasek et al. [37] reported observing only associative electron capture with formation of a negative molecular ion from nitrobenzene in PC at 125°C with E/p equal to 0.33 V cm^{-1} $Torr^{-1}$ (1.3 Townsends). Spangler and Lawless [40] reported observing only associative electron capture with formation of a negative molecular ion from nitrotoluene in PC at 166°C with E/p equal to 0.28 V cm^{-1} $Torr^{-1}$ (1.2 Townsends). Since the operating conditions selected by both Karasek and Spangler correspond to electron energies near 0.51 eV, the results obtained from PC correspond favorably to those obtained from swarm beam techniques.

Spangler and Lawless, however, pursued further the ionization of chlorobenzene and the nitrotoluene homologous series in PC to determine reaction rates and relative capture coefficients [41–43]. Using a dilution flask technique for delivery of sample to the PC, they found the results for negative ion plasma chromatography (NIPC) recorded in Table 15.5. In addition, Table 15.5 presents data derived from mass spectrometry swarm beam data and electron-capture data for comparison purposes. After acknowledging several difficulties to interpretation, not the least of which is a lack of knowledge of temperature relationships, it was possible to note several trends in the data.

First, the reaction rate for NIPC is related to the attachment rate for swarm beam mass spectrometry by a factor of kT ($2 \cdot 10^{16}$ molecules cm^{-3} $Torr^{-1}$ for Table 15.5). Under reasonable assumptions about the reaction volume in NIPC at similar E/p conditions, reaction rate data for chlorobenzene in NIPC compare favorably with similar swarm beam attachment rate data.

TABLE 15.5

SUMMARY OF PUBLISHED RATE CONSTANT AND ELECTRON CAPTURE DATA

Compound	Mass spectrometry, swarm beam★ (s^{-1} $Torr^{-1}$)	Plasma chromatography★		Electron capture	
		Reaction rate (cm^3 $molecule^{-1}$ s^{-1})	Capture coefficient ($l\ mol^{-1}$)	Chloro-benzene = 1	Capture coefficient ($l\ mol^{-1}$)
Chlorobenzene	$1.4 \cdot 10^6$	$7.1 \cdot 10^{-11}$	$1.2 \cdot 10^8$	1.0	
Bromobenzene	$9.5 \cdot 10^6$			6.0	$8.4 \cdot 10^6$
Nitrobenzene	$1.3 \cdot 10^7$★			39.0	$2.8 \cdot 10^9$
o-Mononitrotoluene		$3.3 \cdot 10^{-11}$	$3.6 \cdot 10^7$		
m-Mononitrotoluene		$8.3 \cdot 10^{-11}$	$9.1 \cdot 10^7$		
p-Mononitrotoluene		$2.1 \cdot 10^{-11}$	$2.3 \cdot 10^7$		
2,4-Dinitrotoluene		$2.3 \cdot 10^{-10}$	$2.4 \cdot 10^8$		
2,4,6-Trinitrotoluene		$8.6 \cdot 10^{-10}$	$8.2 \cdot 10^8$		$8.1 \cdot 10^9$
SF_6	$7.8 \cdot 10^8$★			10^4	
CCl_4				10^4	$3.9 \cdot 10^{10}$

★Except for data marked with an asterisk, all values are for an E/p of 0.282 V cm^{-1} $Torr^{-1}$ and are adjusted to a temperature of 200°C. For data marked with an asterisk, $E/p \leqslant 0.1$ V cm^{-1} $Torr^{-1}$ with the pressure normalized to a temperature of 200°C.

Second, capture coefficient data collected with energetic electrons ($E/p = 0.28$ V cm^{-1} $Torr^{-1}$) in NIPC are high for chlorobenzene and low for trinitrotoluene when compared with thermal electron ECD data. The absolute electron-capture data give trinitrotoluene a relative capture coefficient of about 1000–6000 with respect to chlorobenzene, while the NIPC capture coefficients are in the ratio of 6.8.

Third, assuming SF_6 and CCl_4 have similar thermal ($E/p \leqslant 0.1$ V cm^{-1} $Torr^{-1}$) swarm beam attachment rates, attachment rates to electron-capture coefficient ratios are about $1.3 \cdot 10^{-5}$ s Torr for the electron-capture data referenced to chlorobenzene and 50 s Torr l mol^{-1} for the absolute electron-capture data. Similar ratios are also observed for nitrobenzene: $3 \cdot 10^{-5}$ s Torr for electron-capture data referenced to chlorobenzene and 215 s Torr l mol^{-1} for absolute electron-capture data. On the other hand, the ratios of the energetic (E/p = 0.29 V cm^{-1} $Torr^{-1}$) swarm beam attachment rate for chlorobenzene and bromobenzene to the corresponding electron-capture coefficient are $6.7 \cdot 10^{-7}$ Torr and $6.3 \cdot 10^{-7}$ Torr, respectively. The ratio for chlorobenzene is two orders of magnitude less than the ratios for SF_6, CCl_4, and nitrobenzene and the ratio difference appears to be significant.

Fourth, the ratio of swarm beam attachment rates for bromobenzene relative to chlorobenzene is 6.8. This result for energetic ($E/p = 0.282$ V cm^{-1} $Torr^{-1}$) electron data is similar for thermal electron-capture data collected with the ECD. Hence, for these two compounds, consistent results are obtained for relative data in two electron energy ranges.

Fifth, the reaction rates for the nitrotoluenes in NIPC increase for the higher members of the homologous series and show variation with isomeric substitution. Christophorou and co-workers [35,44] attribute the first characteristic to the ability of the larger molecules to absorb extra energy (comprised primarily of the electron affinity of the molecule and the kinetic energy of the incident electron) due to their larger number of vibrational degrees of freedom. Christophorou et al. [34] attribute the second characteristic to a decrease in the dissociation energy due to the proximity of the CH_3 and NO_2 groups in the *ortho* and *para* positions.

Sixth, the average reaction rate observed under energetic ($E/p = 0.28$ V cm^{-1} $Torr^{-1}$) conditions in NIPC for nitrotoluene is $4.6 \cdot 10^{-11}$ cm^3 $molecule^{-1}$ s^{-1}. This reaction rate corresponds to a swarm beam attachment rate of $9.4 \cdot 10^5$ s^{-1} $Torr^{-1}$ which is low when compared to the thermal ($E/p \leqslant 0.1$ V cm^{-1} $Torr^{-1}$) attachment rate of $1.3 \cdot 10^7$ s^{-1} $Torr^{-1}$ observed for nitrobenzene. Since nitrotoluene is expected to have a faster attachment rate than nitrobenzene because of its ability to more effectively share energy among its larger number of degrees of freedom for vibrational energy, experimental evidence is in apparent contradiction to theory unless the energy of the electron (as indicated by E/p) influences the results.

Seventh, the average capture coefficient of $4.6 \cdot 10^7$ obtained for mononitrotoluene under the energetic ($E/p = 0.28$ V cm^{-1} $Torr^{-1}$) conditions of NIPC is 6% of the NIPC

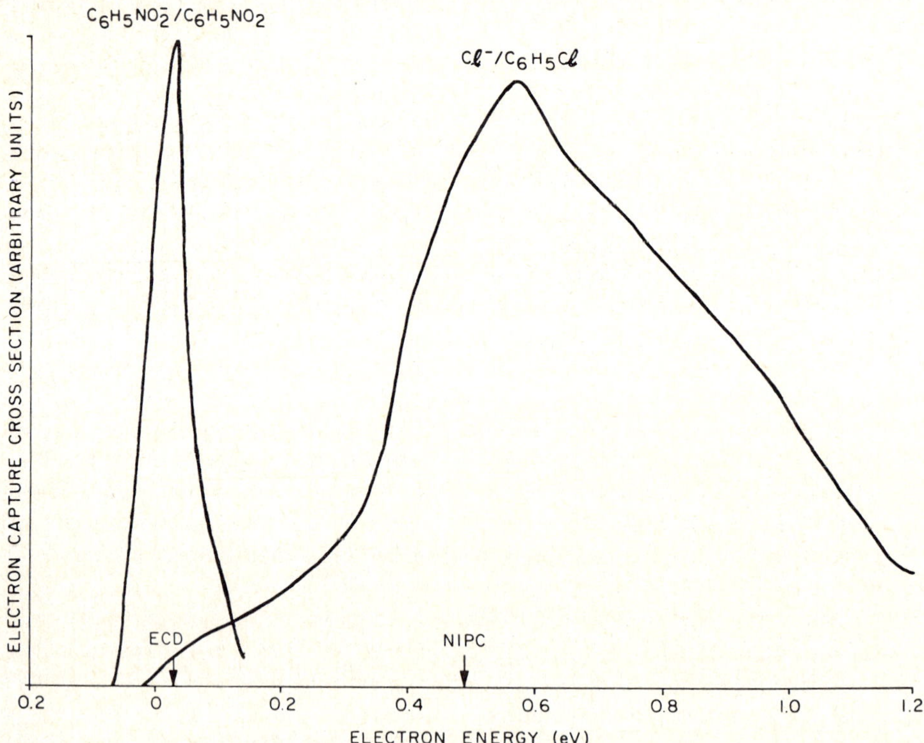

Fig. 15.10. Comparison of electron-capture cross section as a function of electron energy for nitrobenzene and chlorobenzene for electron energies typically employed in NIPC and the ECD. Energy scale calibration is based on SF_6^- and is not corrected for the swarm beam combination. The peak width for nitrobenzene is instrumental [34].

capture coefficient for trinitrotoluene. Assuming that this relative percentage is significant and can be applied equally as well to thermal ECD data, nitrotoluene should have an ECD capture coefficient of $4.9 \cdot 10^8$ l mol^{-1} relative to trinitrotoluene. This proposed value for the ECD capture coefficient for nitrotoluene is an order of magnitude less than that observed for nitrobenzene, and is contrary to considerations based on the effects which vibrational energy has on ion stability as described above. Hence, conclusions derived from relative data support the conclusions derived from absolute data.

Spangler and Lawless noted that these trends are consistent with the resonance nature of the electron-capture process which requires the formation of an intermediate before the formation of an associated or dissociated, yet stabilized, negative product ion. As demonstrated in Fig. 15.10, the electron energy of 0.49 eV utilized in PC lies closer to the resonance energy required to dissociate chlorobenzene than to the thermal energy required to ionize nitrobenzene. By contrast, the thermal electron energy of 0.03 eV utilized in a pulsed ECD lies closer to the energy required to ionize nitrobenzene than to the resonance energy required to dissociate chlorobenzene. From this observation, it was concluded that comparison of capture coefficient data between PC and ECD are complicated by effects relating to differences in electron energies.

Fig. 15.11 shows the ionic species observed from electron capture of 1-chloro-2-nitrobenzene. This compound exhibits only dissociative electron capture. No negative molecular ion appears, but one does observe both the $C_6H_4NO_2^-$ and Cl^- dissociative product ions. Identity of these ions is confirmed by superposition of subsequent ion mobility spectra of nitrobenzene and chlorobenzene on the same ion mobility spectrum of chloronitrobenzene and observation of respective peak coincidence. The relative amounts of the $ClC_6H_4NO_2^-$, $C_6H_4NO_2^-$, and Cl^- ions formed by electron capture with the *ortho*, *meta*, and *para* isomers of chloronitrobenzene are shown in Fig. 15.12. These data indicate that the 1-chloro-3-nitrobenzene isomer forms a stable molecular ion by associative electron attachment, with only a slight amount of dissociation occurring to

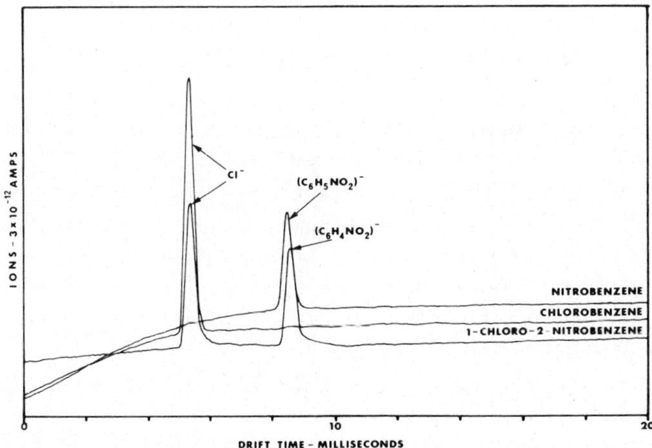

Fig. 15.11. Composite negative ion mobility spectra of 1-chloro-2-nitrobenzene with nitrobenzene and chlorobenzene. (Reproduced from *Anal. Chem.*)

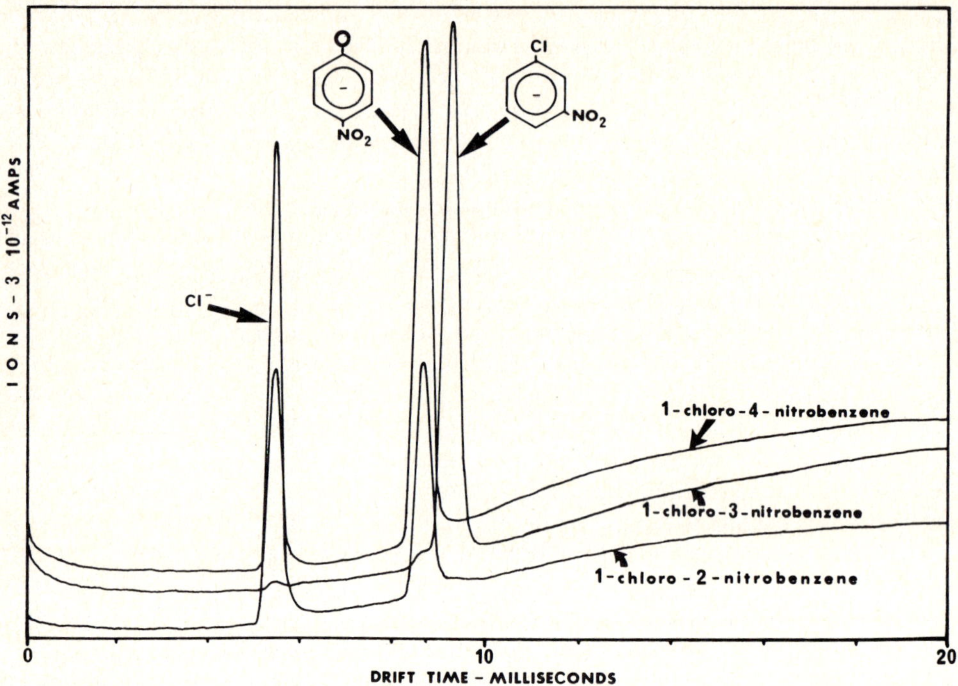

Fig. 15.12. Composite ion mobility spectra obtained for *ortho*, *meta*, and *para* isomers of chloronitrobenzene. (Reproduced from *Anal. Chem.*)

give the Cl^-. The other two isomers dissociate completely to the Cl^- and the $C_6H_4NO_2^-$. The identity of the molecular ion is confirmed by correspondence of its reduced mobility with that of this same ion formed from electron attachment of dichloro-substituted nitrobenzene compounds. The existence of all three ions is predicted by the mechanism previously reported by Wentworth and co-workers [45,46].

Further evidence of multiple ion formation from electron capture is shown in Fig. 15.13, where decachlorobiphenyl is seen to form the three ions identified as Cl^-, $C_{12}Cl_8^-$, and $C_{12}Cl_{10}^-$. Identity of these ions was established by superposition of the ion mobility spectrum of octachlorobiphenyl and from previous work in the plasma chromatography of polychlorobiphenyl compounds [47]. In the previous work, an air carrier gas was used so the reactant negative species were $(H_2O)_nO_2^-$ ions instead of electrons seen in Fig. 15.13. In that case formation of decachlorobiphenyl ions was by an ion–molecule reaction with the $(H_2O)_nO_2^-$ ions. While the negative molecular ion of $C_{12}Cl_{10}^-$ was clearly present with a reduced mobility of 1.09 (compared to a value of 1.08 for data shown in Fig. 15.13), it was not possible to detect dissociative formation of Cl^- because the $(H_2O)_nO_2^-$ reactant and Cl^- ion peaks have almost identical mobilities. For decachlorobiphenyl the negative molecular ion peak predominates at all the different concentration conditions observed while the dissociated Cl^- always remains a very minor one.

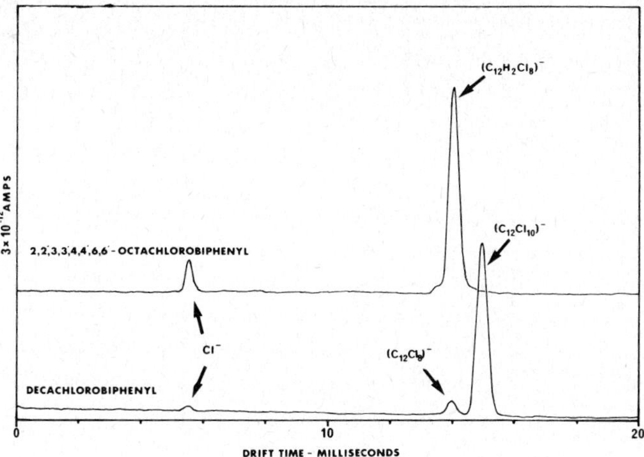

Fig. 15.13. Composite negative ion mobility spectra for octachlorobiphenyl and decachlorobiphenyl. (Reproduced from *Anal. Chem.*)

15.3.3. Ion–molecule reactions in the electron-capture detector

It has long been recognized that impurities in the carrier gas can strongly influence the response of the ECD. For instance, the effect of oxygen contamination in the carrier gas has been described several times and has been universally considered as an undesirable complication in analysis procedures. More recently, it has been noted by various research groups that the presence of impurities in the carrier gas can in some cases enhance the sensitivity of the ECD to specific compounds [48–51]. For instance, Van de Wiel and Tommassen [48] and Grimsrud and co-workers [49,50] noted an increased sensitivity to certain halogenated organic compounds with the addition of oxygen to a nitrogen carrier. Phillips et al. [51] noted an increased sensitivity to methane, carbon dioxide and hydrogen with the addition of nitrous oxide to a nitrogen carrier. In separate experiments, Simmonds [52] also observed an increase in the sensitivity of the ECD to carbon dioxide when oxygen is added to a nitrogen carrier.

The effect of impurities on the response of the ECD has been attributed to ion–molecule reactions augmenting the direct electron attachment mechanism leading to formation of a stable negative ion. In the case of the addition of oxygen to the carrier, the collisional detachment of electrons from O_2^- competes with the attachment process so that O_2^- and electrons are in thermodynamic equilibrium. For this situation, a reaction which converts O_2^- to a stable negative ion produces electron removal from the ECD plasma and generates an ECD response to contain compounds even in the absence of a direct electron attachment mechanism.

For the case of nitrous oxide added to a nitrogen carrier, a reaction sequence

$$e + N_2O \rightarrow O^- + N_2 \qquad (15.21)$$

$$O^- + NO \rightarrow NO^- + NO \qquad (15.22)$$

$$NO^- + N_2 \rightarrow NO + N_2 + e \qquad (15.23)$$

is set up in which electrons are attached (reaction 15.21) and released (reaction 15.23). Ion–molecule reactions involving the O^- anion interrupt the sequence given by reactions 15.21–15.23 resulting in electron loss. This causes an ECD response even in the absence of a direct electron attachment mechanism.

PC is a technique particularly suited to study the effects of impurities on the ECD. The chemistry of the O^- and NO^- ions can be investigated by adding N_2O to a nitrogen carrier and the chemistry of the O_2^- ion can be investigated by either using or adding zero air (<ppm levels of hydrocarbons, water and ammonia) to nitrogen as carrier [53–55].

In PC, when sample molecules are introduced into the carrier, reactions occur between the reactant ions and the sample molecules to produce product ions. Besides the direct electron attachment reactions involving free electrons, other ion–molecule reactions are known to occur. These include charge transfer reactions (associatively or dissociatively)

$$R^- + A \rightarrow R + A^- \qquad (15.24)$$

where R^- is the reactant ion, A is the sample molecule, and A^- is the product ion; proton abstraction reactions

$$R^- + AH \rightarrow RH + A^- \qquad (15.25)$$

where AH is the sample molecule; electrophilic attachment or cluster

$$R^- + A \rightarrow RA^- \qquad (15.26)$$

where RA^- is the product ion; and complex rearrangements. Of these reactions, the three most studied reactions are the charge transfer reactions, the proton abstraction reactions, and the cluster reactions [53,56–59].

Charge transfer reactions are controlled by the same electron affinity scale which controls the electron-capture reactions. That is, an attached electron has a tendency to transfer from a molecule of lower electron affinity to a molecule of higher electron affinity. For endothermic charge transfer:

$$A^- + B \rightarrow A + B^- \qquad (15.27)$$

A^- must be provided translational energy to initiate the reaction. Under these conditions, the electron affinity of B is the electron affinity of A plus the threshold energy.

Charge transfer reactions involving the halogenated aromatic compounds [42,47], the nitrotoluenes [40] and selected diatomic and triatomic molecules [53] have been studied with PC. From these studies, the halogenated aromatics are known to be susceptible to dissociative charge transfer, the nitrotoluenes are known to undergo primarily associative charge transfer reactions and that NO_2^- present in the ambient atmosphere is an important component which quenches these reactions. As noted previously, decachlorobiphenyl reacts with free electrons to produce three ions identified as Cl^-, $C_{12}Cl_8^-$ and $C_{12}Cl_{10}^-$ when nitrogen gas is used as carrier. Similar ions are also observed from ion–molecule charge transfer reactions involving this compound except that the rate for dissociative formation of Cl^- is greatly reduced.

Proton abstraction reactions are the result of one of the oldest concepts of chemistry: the concept of acids and bases. The present definition of acids and bases dates from 1923 when Johannes N. Bronsted and Thomas M. Lowry described an acid as any substance that gives up a proton and a base as any substance that accepts a proton. For the reaction

$$AH \rightarrow A^- + H^+ \qquad (15.28)$$

the enthalpy change ΔH°_{acid} (or proton affinity of A^-) is given by

$$\Delta H^\circ_{acid} \approx DH^\circ\,(A\text{–}H) + IP\,(H\cdot) - EA\,(A) \qquad (15.29)$$

where DH° (A–H) is the bond strength of AH, IP (H•) is the ionization potential of hydrogen, and EA (A) is the electron affinity of A. Since the proton abstraction reaction of eqn. 15.25 is the sum of two reactions

$$\begin{array}{l} R^- + H^+ \rightarrow RH \\ \quad AH \rightarrow A^- + H^+ \\ \hline R^- + AH \rightleftarrows RH + A^- \end{array} \qquad (15.30)$$

involving heterolytic cleavage of a neutral molecule into a proton and negative ion, the enthalpy change ΔH_{25} for eqn. 15.25 can be related to bond dissociation energies and electron affinities through:

$$\Delta H_{25} = D\,(A\text{–}H) - EA\,(A) - D\,(R\text{–}H) + EA\,(R) \qquad (15.31)$$

Since ΔH°_{acid} and ΔH_{25} are related through the ionization potential of hydrogen IP (H•) = 313.6 kcal/mole, gas phase acidity scales describing proton abstraction can either be in terms of ΔH°_{acid} or ΔH_{25}. To experimentally determine absolute data from ΔH_{25}, the standard acid HCl is selected as a reference with established values for D (H–Cl) = 103 kcal/mole and EA (Cl) = 83 kcal/mole.

Refs. 56 and 59 provide excellent reviews of the gas phase acidity scale as it describes proton abstraction reactions. These scales are summarized in Fig. 15.14 where the compounds are organized around the reactant ions (CN^-, Cl^-, O_2^-) for PC. As compounds approach the top of this scale (i.e., low ΔH°_{acid}), a tendency develops for the transfer of a proton (H^+) from that compound to an ionized form of a compound lying further down the scale. Since ΔH°_{acid} for the hydroxyl radical (OH•) is 382.1 kcal/mole, O^- as might be derived from the introduction of N_2O into a PC or ECD would be very active as a proton abstractor.

The proton abstraction of nitrotoluenes has been studied in PC. Spangler and Lawless [40] observed that all the members of the homologous series undergo associative capture with free electrons but also observed that in the presence of $(H_2O)_nO_2^-$ reactant ions, the higher members of the homologous series undergo proton abstraction reactions. This is illustrated in Fig. 15.15 for 2,4,6-trinitrotoluene where the 1.49 $cm^2\,V^{-1}\,s^{-1}$ ion mobility peak, identified as the molecular anion with mass spectrometry, disappeared in favor of the 1.54 $cm^2\,V^{-1}\,s^{-1}$ ion mobility peak, identified as the proton abstracted anion with mass spectrometry, as laboratory air is added to a nitrogen carrier. Since laboratory air also contains NO_2 to form NO_2^- anions as major reactant ions the acidity of 2,4,6-trinitrotoluene lies well above that for HONO or ΔH°_{acid} = 338.3 kcal $mole^{-1}$.

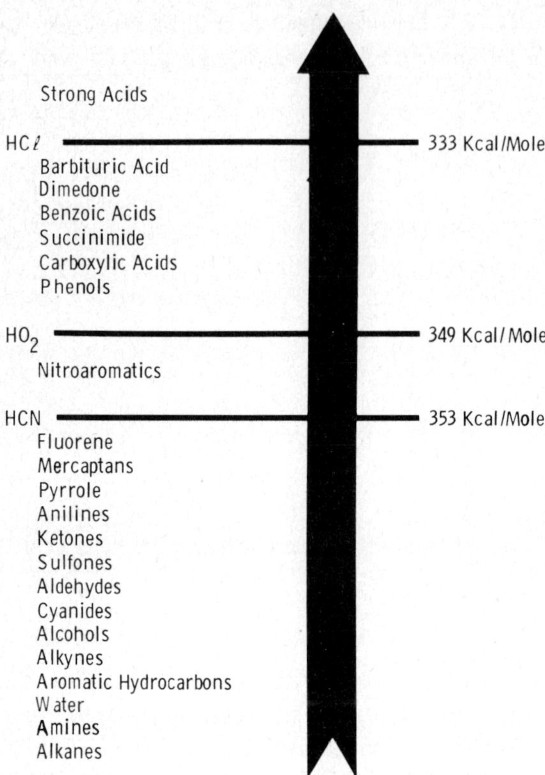

Fig. 15.14. Gas phase acidity scale for compound classes.

A similar phenomenon was observed for 2,4-dinitrotoluene with rate constant data indicating that dinitrotoluene gives up a proton faster than trinitrotoluene. For dinitrotoluene, relative amplitudes for the molecular and proton abstracted anions have also been found to be dependent on the isomer [60]. A similar phenomenon also seems to occur for the mononitrotoluenes where a 1.81 $cm^2\ V^{-1}\ s^{-1}$ ion mobility peak which might be correlated with a proton abstracted anion appears prominently only for the *ortho*- and *para*-substituents (see Fig. 15.16). This result suggests that proton abstraction from aromatics is dependent on electron withdrawing effects of substituents on the π-cloud thus freeing a proton elsewhere in the ring (methyl group for the nitrotoluenes).

The temperature dependence of the dissociative capture reaction is not new to ECD. A similar relationship with temperature exists for the dissociative charge transfer reactions of ion–molecule reaction chemistry with the dissociative charge transfer reaction and the proton abstraction being complementary to each other. This relationship between the two reactions suggests that an electron is attracted to a molecule by the most electronegative moiety attached to that molecule. If the energy associated with the transfer of the electron to the molecule is great enough to exceed an activation energy to allow dissociative capture, dissociative capture will occur. However,

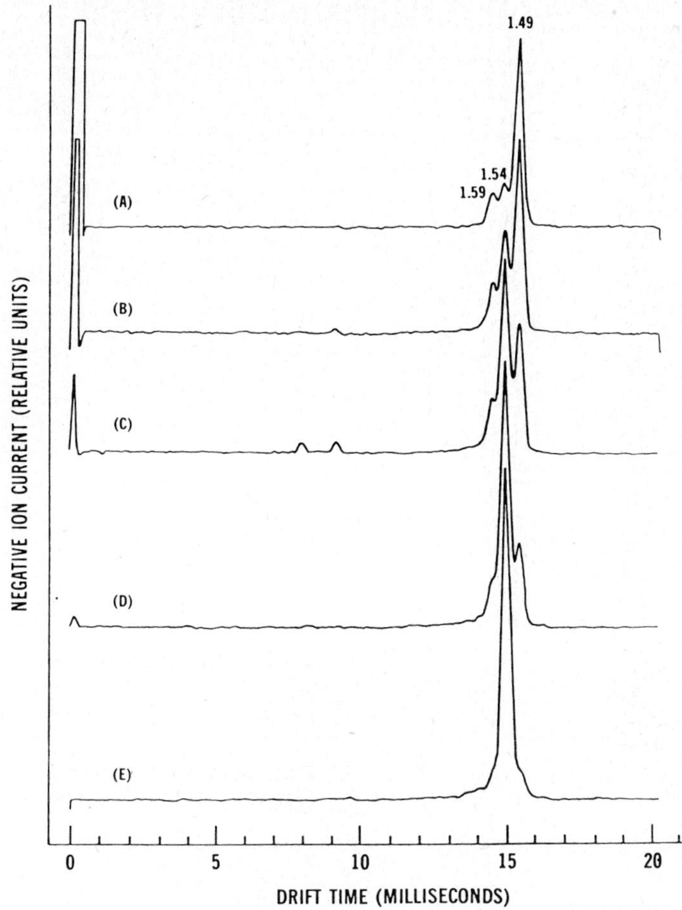

Fig. 15.15. Ion mobility spectra for 2,4,6-trinitrotoluene in air–nitrogen mixtures. Approximate (4% accuracy) mixture percentages are: (A) <2% laboratory air; (B) 8% laboratory air; (C) 10% laboratory air; (D) 15% laboratory air; (E) >21% laboratory air. (Reproduced from *Anal. Chem.*)

if the energy involved in the transfer is insufficient for dissociative capture, the electron then repositions itself within the molecule in accordance with quantum-mechanical positional uncertainty to create acidic protons. These protons can then be removed from the molecule via proton abstraction reactions.

Cluster reactions involve the tendency of a sample to enter into a solvation relationship with the reactant ions. Examples of these type reactions in PC include clustering of the water molecule and CO_2 with the O_2^- reactant ion [53]. The reaction sequence

$$(H_2O)_nO_2^- + H_2O + O_2 \text{ (or } N_2) \rightleftarrows (H_2O)_{n+1}O_2^- + O_2^- + O_2 \text{ (or } N_2) \qquad (15.32)$$

is an equilibrium relationship satisfying the Van't Hoff relation such that the most likely

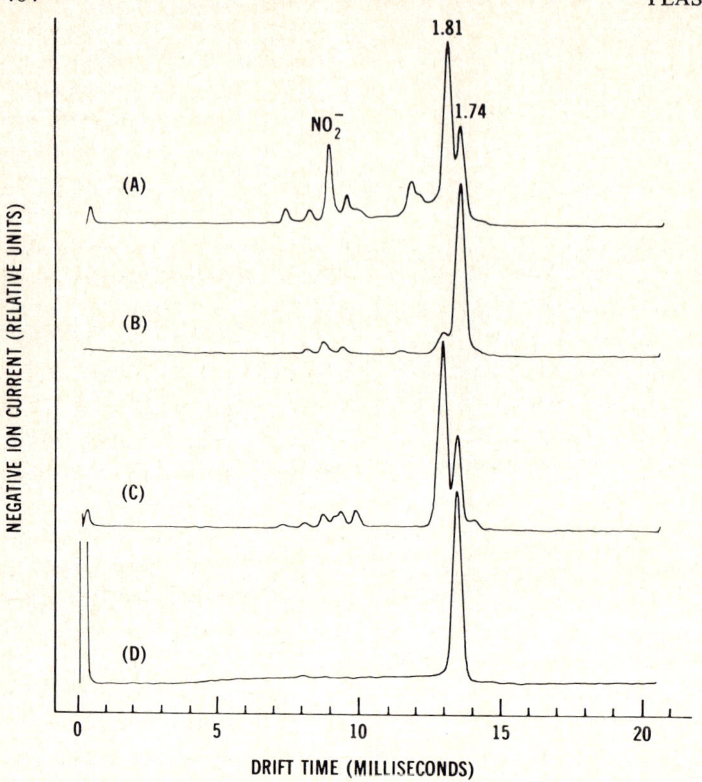

Fig. 15.16. Ion mobility spectra for mononitrotoluenes (MNT). (A) *o*-MNT in zero air; (B) *m*-MNT in zero air; (C) *p*-MNT in zero air; (D) any mononitrotoluene in nitrogen. (Reproduced from *Anal. Chem.*)

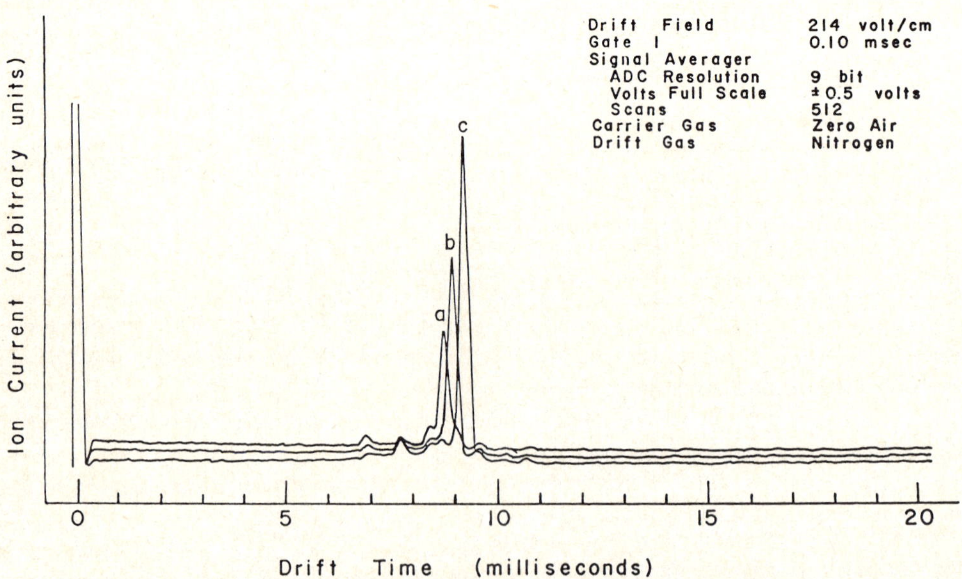

Fig. 15.17. Effect of carbon dioxide on the reactant ion distribution. Spectrum a is the background spectrum, while b and c are for increasing levels of CO_2. (Reproduced from *Anal. Chem.*)

reactant ion formed in PC is $(H_2O)_nO_2^-$. In the presence of CO_2, these reactions are augmented by:

$$O_2^- + CO_2 + M \rightleftarrows O_2^- \cdot CO_2 + M \tag{15.33}$$

$$O_2^- \cdot H_2O + CO_2 \rightleftarrows O_2^- \cdot CO_2 + H_2O \tag{15.34}$$

to form CO_4^-. Displayed in Fig. 15.17 are data collected with PC to demonstrate how cluster reactions with CO_2 do indeed alter reactant ion distributions. In the presence of increased levels of CO_2, the reactant ion normally attributed to $(H_2O)_nO_2^-$ increases in amplitude and shifts to increased drift times. These data are a clear indication of the effect of impurities on the performance of the ECD to CO_2. Similar phenomena might also be expected from other diatomic or triatomic molecules.

In conclusion, these results demonstrate that impurities (e.g. oxygen, nitrous oxide) added intentionally or unintentionally to the carrier gas of an ECD leads to anomalous but controllable results. In the presence of a pure carrier gas, direct electron attachment reactions are responsible for the ionization of sample. As impurities are added to the carrier, this direct electron attachment reaction is replaced by charge transfer, proton abstraction, and/or cluster reactions. Under the conditions of impurities, expressions can be developed for the capture coefficients K_a and K_b but the complicated nature of the expressions provides little basic information to the researcher. Rather, it is better to make the qualitative observation that impurities can enhance the sensitivity of the ECD to certain compounds by activating reactions other than electron attachment for ionization purposes.

REFERENCES

1 F.W. Karasek, *Anal. Chem.*, 46 (1974) 710A.
2 S.H. Kim, *Fundamental and Analytical Aspects of Plasma Chromatography, Ph.D. Thesis*, University of Waterloo, 1978.
3 F.W. Karasek and D.M. Kane, *Anal. Chem.*, 46 (1974) 633.
4 F.W. Karasek, D.W. Denney and E.H. DeDecker, *Anal. Chem.*, 46 (1974) 970.
5 D.I. Carroll, I. Dzidic, R.N. Stillwell, K.D. Haegele and E.C. Horning, *Anal. Chem.*, 47 (1975) 2369.
6 E.A. Mason and H.W. Schamp, *Ann. Phys. (New York)*, 4 (1958) 233.
7 J.E. Lovelock, *Anal. Chem.*, 35 (1963) 474.
8 D.C. Fenimore and C.M. Davis, *J. Chromatogr. Sci.*, 8 (1970) 130.
9 D.C. Fenimore, A. Zlatkis and W.E. Wentworth, *Anal. Chem.*, 40 (1968) 1594.
10 J.E. Lovelock, *Nature (London)*, 189 (1961) 729.
11 J.E. Lovelock, P.G. Simmonds and W.J.A. VandenHeuvel, *Nature (London)*, 197 (1963) 249.
12 L. Dehennin, A. Reiffstech and R. Scholler, *J. Chromatogr. Sci.*, 10 (1972) 224.
13 B.C. Pettitt, P.G. Simmonds and A. Zlatkis, *J. Chromatogr. Sci.*, 6 (1968) 85.
14 W.E. Wentworth and E. Chen, *J. Gas Chromatogr.*, 5 (1967) 170.
15 P.G. Simmonds, D.C. Fenimore, B.C. Pettitt, J.E. Lovelock and A. Zlatkis, *Anal. Chem.*, 39 (1967) 1428.
16 W.E. Wentworth, E. Chen and J.E. Lovelock, *J. Phys. Chem.*, 70 (1966) 445.
17 W.E. Wentworth, R.S. Becker and R. Tung, *J. Phys. Chem.*, 71 (1967) 1652.
18 W.E. Wentworth and E. Chen, *J. Phys. Chem.*, 71 (1967) 1929.
19 J.C. Steelhammer and W.E. Wentworth, *J. Chem. Phys.*, 51 (1969) 1802.

20 W.E. Wentworth, R. George and H. Keith, *J. Chem. Phys.*, 51 (1969) 1791.
21 W.E. Wentworth and J.C. Steelhammer, in E.J. Hart (Editor), *Radiation Chemistry*, American Chemical Society, Washington, DC, 1968.
22 L.G.H. Huxley and R.W. Crompton, *The Diffusion and Drift of Electrons in Gases*, Wiley, New York, 1974.
23 L.G. Christophorou and M.W. Grant, *Advan. Chem. Phys.*, 36 (1977) 413.
24 R.N. Compton, L.G. Christophorou, G.S. Hurst and P.W. Reinhardt, *J. Chem. Phys.*, 45 (1966) 4634.
25 B.K. Janousek and J.I. Brauman, in M.T. Bowers (Editors), *Gas Phase Ion Chemistry*, Vol. 2, Academic Press, New York, 1979, pp. 53–86.
26 E.D. Pellizzari, *J. Chromatogr.*, 98 (1974) 323.
27 F.W. Karasek and D.M. Kane, *Anal. Chem.*, 45 (1973) 579.
28 G.E. Spangler and C.I. Collins, *Anal. Chem.*, 47 (1975) 403.
29 T.W. Carr, *Anal. Chem.*, 49 (1977) 828.
30 M.W. Siegel and M.C. McKeown, *J. Chromatogr.*, 122 (1976) 397.
31 J.H. Richardson, L.M. Stephenson and J.I. Brauman, *J. Chem. Phys.*, 62 (1975) 1580.
32 B.K. Janousek, A.H. Zimmerman, K.J. Reed and J.I. Brauman, *J. Amer. Chem. Soc.*, 100 (1978) 6142.
33 L.G. Christophorou, R.N. Compton, G.S. Hurst and P.W. Reinhardt, *J. Chem. Phys.*, 43 (1965) 4273.
34 L.G. Christophorou, R.N. Compton, G.S. Hurst and P.W. Reinhardt, *J. Chem. Phys.*, 45 (1966) 536.
35 L.G. Christophorou, M.W. Grant and D.L. McCorkle, *Advan. Chem. Phys.*, 36 (1977) 413.
36 F.W. Karasek and O.S. Tatone, *Anal. Chem.*, 44 (1972) 1978.
37 F.W. Karasek, O.S. Tatone and D.M. Kane, *Anal. Chem.*, 45 (1973) 1210.
38 W.D. Kilpatrick, *An Experimental Mass–Mobility Relationship for Ions in Air at Atmospheric Pressure, Proceedings of the 19th Annual Conference on Mass Spectrometry*, Atlanta, GA, May 1971.
39 D.I. Carroll and E.A. Mason, *The Theoretical Relationship Between Ion Mobility and Mass, Proceedings of the 19th Annual Conference on Mass Spectrometry*, Atlanta, GA, May 1971.
40 G.E. Spangler and P.A. Lawless, *Anal. Chem.*, 50 (1978) 884.
41 P.A. Lawless and G.E. Spangler, *Rev. Sci. Instrum.*, 48 (1977) 240.
42 G.E. Spangler and P.A. Lawless, *Anal. Chem.*, 50 (1978) 290.
43 G.E. Spangler and P.A. Lawless, *Anal. Chem.*, 52 (1980) 193.
44 L.G. Christophorou, *Atomic and Molecular Physics*, Wiley, New York, 1971, Ch. 6.
45 W.E. Wentworth, R.S. Becker and R. Tung, *J. Phys. Chem.*, 71 (1967) 1652.
46 W.E. Wentworth, R. George and H. Keith, *J. Chem. Phys.*, 51 (1969) 1791.
47 F.W. Karasek, *Anal. Chem.*, 43 (1971) 1982.
48 H.J. van de Wiel and P. Tommassen, *J. Chromatogr.*, 71 (1972) 1.
49 E.P. Grimsrud and D.A. Miller, *Anal. Chem.*, 50 (1978) 1141; 51 (1979) 851.
50 E.P. Grimsrud and R.G. Stebbins, *J. Chromatogr.*, 155 (1978) 19.
51 M.P. Phillips, R.E. Sievers, P.D. Golden, W.C. Kuster and F.C. Fehsenfeld, *Anal. Chem.*, 51 (1979) 1819.
52 P.G. Simmonds, *J. Chromatogr.*, 166 (1978) 593.
53 G.E. Spangler and C.I. Collins, *Anal. Chem.*, 47 (1975) 393.
54 T.W. Carr, *Anal. Chem.*, 49 (1977) 828.
55 T.W. Carr, *Anal. Chem.*, 51 (1979) 705.
56 P. Kebarle, *Ann. Rev. Phys. Chem.*, 28 (1977) 445.
57 G.W.A. Milne and M.J. Lacey, *CRC Critical Reviews in Analytical Chemistry*, CRC Press, Cleveland, OH, 1974, pp. 45–104.
58 B.K. Janousek and J.I. Brauman, in M.T. Bowers (Editor), *Gas Phase Ion Chemistry*, Vol. 2, Academic Press, New York, 1979, pp. 53–86.
59 J.E. Bartmess and R.T. McIver, Jr., in M.T. Bowers (Editor), *Gas Phase Ion Chemistry*, Vol. 2, Academic Press, New York, 1979, pp. 87–121.
60 M. Asselin, DREV, Valcartier, personal communication.

Chapter 16

The electron-capture detector as a detector in liquid chromatography

U.A.Th. BRINKMAN

CONTENTS

16.1. INTRODUCTION

In recent years, high-performance column liquid chromatography (LC) has emerged as an analytical technique that is equivalent in performance to gas chromatography (GC). To-day, LC and GC are seen as supplementary rather than competitive, each having specific advantages and drawbacks, and each having preferred field(s) of application. A particular drawback in LC is the lack of detectors that can match the sensitivity of GC detectors. Currently, the UV–visible absorption spectrophotometer is the most useful of the LC detectors, on account of its versatility and reliability, and its relatively high sensitivity for a broad range of compounds. Next to absorption spectrophotometers, fluorescence and electrochemical detectors are increasingly used, as they can detect less than 1 pg of a favourable solute, whereas the absorption detector is limited to about 1 ng in a favourable situation.

A variety of other detectors has been developed for use in LC (for overviews see, e.g., refs. 1 and 2). Many of these devices do not provide the level of sensitivity that is generally required for use with modern LC. Some of them, however, have important applications in specific instances. Amongst these can be classified the electron-capture detector (ECD), which displays excellent sensitivity and selectivity towards electron-capturing species such as chlorinated aromatic hydrocarbons and nitroaromatic compounds. Unfortunately, however, the detector has severe limitations as regards its use in practice. It is the purpose of this chapter to give a brief discussion of the design of the ECD for use in LC and to outline the potential and inherent problems of the LC–ECD system.

16.2. THE LIQUID CHROMATOGRAPH–ELECTRON-CAPTURE DETECTOR SYSTEM

In the literature, two different approaches have been propagated for the construction of a LC–ECD system.

In a single early study [3], a ^{63}Ni ECD with a geometry identical with that of the Pye 104 ECD for GC was simply mounted on a solvent transport system, viz., the then commercially available Pye Unicam (Cambridge, Great Britain) moving-wire system. This system was used to detect, amongst other compounds, lindane and dieldrin after LC on alumina with *n*-hexane–ethanol (19:1) as the mobile phase. The minimum detectable concentrations (MDCs)* were typically in the range 10^{-8} g/ml. On account of both this lack of sensitivity, which was no doubt largely due to the low efficiency (0.1–1%) of the wire-transport system, and the fact that a detailed discussion of transport detectors can be found in many textbooks (e.g., ref. 1), we shall omit any further description of the transport LC–ECD system.

The alternative approach is based on the consideration that a significant proportion of the solvents used as mobile phases in LC do not give a response with the ECD. Such solvents can therefore be volatilized together with any solute that is present and passed through the detector, which will then respond to the solute only, provided that the latter has electron-capturing properties. This principle was first developed by Nota and Palombari [4]. In their set-up, the eluent from an LC column passes directly into a nebulizer, the outlet tube from the column terminating at the nozzle of the nebulizer. Part of the nebulized eluent is admitted into the ECD, which is kept at an appropriate temperature in order to prevent condensation and adsorption phenomena. Benzene, *n*-hexane, cyclohexane, pyridine, lower alcohols and even diethyl ether, acetone and water–ethanol (1:19) can be used as eluents, and under the conditions used (nitrogen, 500 ml/min; eluent, 10 μl/min; detector temperature, 150°C) the ECD had only a slightly higher noise level than when it is used in GC. The detection of volatile substances such as chloroform, lindane and Cr(III) tristrifluoroacetylacetonate did not present any problems. With thermolabile and low-volatility substances some tailing was observed. As has been pointed out by Scott [1], the disadvantages of the detector are the necessity to use large volumes of nitrogen and the introduction of significant extra-column band broadening due to the LC–ECD interface.

In 1974, Willmott and Dolphin [5] reported the use of an improved version of the LC–ECD system (Fig. 16.1). In their design, the total eluent from the LC column is vaporized and is passed directly into the ECD. Vaporization is achieved by passing the mobile phase into a stainless-steel transfer tube (400 or 1000 × 0.25 mm I.D.) mounted in an oven which is maintained at a sufficiently high temperature. The volume increase involved in vaporization forces the vapour into the ECD, which is enclosed in a separate oven. A nitrogen purge of about 30 ml/min sweeps the vapour through the detector into a stainless-steel condenser (2000 × 4.5 mm I.D.), thus preventing back-diffusion of eluent into the purge conduit. From the condenser, which acts as a trap for radioactive material, the eluent is collected as a liquid. The ECD signal is amplified using either a

---

*Unless otherwise stated, MDC values are calculated relative to the mobile phase flow-rate.

pulse-mode or a constant-current amplifier (see below); the latter operates with 60-V pulses with a 1-μsec nominal width over a 0–130-kHz frequency range. The standing current can be varied between 0 and $5 \cdot 10^{-9}$ A.

The above detector has been manufactured for several years by Pye Unicam [6] and sold by Philips (Eindhoven, The Netherlands). As will be discussed in more detail below, the reduction in sensitivity of the LC–ECD system compared with standard GC–ECD operation is modest (10- to 100-fold), and MDC values of favourable compounds are of the order of 10^{-10} g/ml. The linear dynamic range of the detector is, however, rather low (100–500) and so is its dynamic range (about 1000).

Recently, Krull and Bushee [7] have reported on the direct interfacing of LC with GC–ECD, using the latter only as a vaporization oven, heated-transfer interface and detector-electronics arrangement. There is no GC column present within the interface, for the LC eluent is passed directly into the GC oven and vaporized, then split with a variable-ratio splitter valve, and a fraction of the vaporized eluent is passed directly into the ECD. The detector response is stated to be linear over four orders of magnitude, which is about the same as that when the ECD is operated in the GC mode. The overall reproducibility, stability and sensitivity of the directly interfaced LC/GC–ECD appear to be as good as or better than those reported for the commercially available Pye Unicam LC–ECD system.

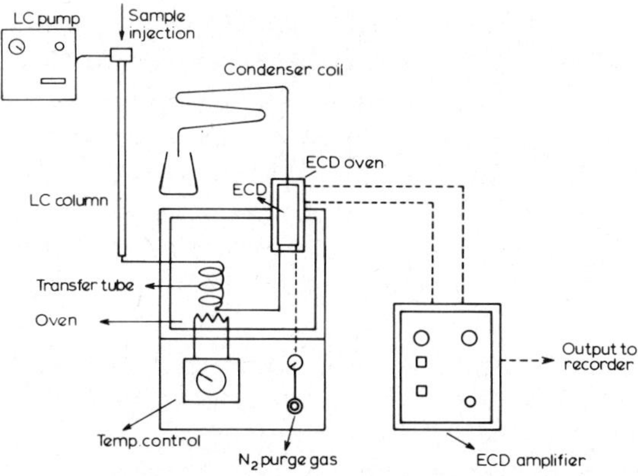

Fig. 16.1. Schematic diagram of the Pye Unicam LC–ECD. (After ref. 6.)

16.3. EVALUATION OF THE LIQUID CHROMATOGRAPH–ELECTRON-CAPTURE DETECTOR SYSTEM

In their first major paper, Willmott and Dolphin [5] compared pulse-sampling and constant-current operation of their detection system and noted that the solvent, in this instance *n*-hexane, fulfills the dual role of providing ionizable species (ionisation potential of *n*-hexane, 10.2 eV) and polyatomic molecules with which the initially high-energy free electrons can undergo non-elastic collisions until their energy is reduced to thermal equilibrium with the solvent molecules. In theory, therefore, a purge gas is unnecessary for the reliable operation of the LC–ECD. However, as the authors used a standard commercial ^{63}Ni ECD for GC which is constructed with a purge line, they advised the use of a purge flow to prevent diffusion of the eluent into the tube, which may lead to high noise levels. Willmott and Dolphin routinely used nitrogen as the purge gas; however, argon, helium and argon plus 10% of methane showed very similar characteristics, the large volume of solvent, i.e., *n*-hexane, playing a major role in determining the detector characteristics.

With aldrin as test compound, plots of log (peak height) vs. log (weight of test solute injected) for the pulse-sampling and constant-current modes showed the upper region of the linear range to extend further for the latter than for the former mode. As the MDC values are comparable, the constant-current method gives the better linear range (about 500) and is the preferred technique. Operation at a low standing current is recommended, because it extends the upper limit of linearity; in general, this is advantageous when working with a very sensitive detector. In the constant-current mode, the detector is virtually insensitive to purge gas flow-rate at low standing currents, but behaves as a concentration-sensitive device at higher currents.

The temperature of the transfer tube must be high enough to vaporize the solvent and solute completely before they enter the detector. With *n*-hexane as solvent and aldrin as test compound, Willmott and Dolphin found 150–250°C to be a suitable temperature range for solvent flow-rates up to 3.5 ml/min. Other workers [8,9] routinely use a higher temperature of 300°C, and Brinkman et al. [10] even stated that they normally kept the oven temperature at its maximum value of 350°C. The latter workers also reported that with decachlorobiphenyl (DCB), increasing the temperature from 320 to 350°C doubled its peak height, and that very high-boiling compounds such as the tetradecachloroterphenyls yield very broad and tailing peaks even at 350°C and therefore cannot be subjected to LC–ECD analysis.

Few relevant data have been published on the contribution of the ECD to extra-column band broadening. Several groups of workers [8,9] suggest that band-broadening effects will be large. Brinkman et al. [10] report $\sigma_t = 4$ s for the ECD at a mobile phase flow-rate of 1.4 ml/min, which does not appear to be an exceptionally high value.

Linearity and sensitivity of the ECD self-evidently are highly dependent on the electron-capturing properties and purity of the solvent used as the mobile phase (see below). Likewise, increasing the solvent flow-rate will lead to a greater flow of electron-capturing molecules through the ECD and therefore to a reduction in the detector current and a consequent loss of response. Experiments show [5] that low flow-rates are indeed advantageous, but that losses are acceptable (40%) up to at least 3.5 ml/min compared with 1.0 ml/min. Similar results were obtained by Demeter and Heyndrickx [9], who reported a 2- to 4-fold decrease in detector sensitivity on increasing the flow-rate from 0.82 to 2.75 ml/min, the actual decrease being dependent on the ECD current used. With their directly interfaced LC/GC–ECD, Krull and Bushee [7] also noted that low flow-rates (0.01–0.08 ml/min) lead to lower detection limits and an overall beter sensitivity.

Data on the sensitivity of the LC–ECD compared with the LC–UV system were reported by Brinkman et al. [10] and Willmott and Dolphin [5]. Table 16.1 summarizes detection limits, expressed as minimum detectable amounts (MDA), for several organo-

TABLE 16.1

MDA VALUES AND RELATED CHARACTERISTICS OF ORGANOCHLORINE PESTICIDES AND DECACHLOROBIPHENYL FOR ELECTRON-CAPTURE AND UV DETECTION

$k' = (t_R - t_0)/t_0$; N_{column} = ca. 3000; LC system silica/dry *n*-hexane.

Compound	k'_{LC}	MDA_{ECD} (g)	MDA_{UV} (g)	log $\epsilon_{max.}$	$\lambda_{max.}$ (nm)	Ref.
Aldrin	1.0	$1 \cdot 10^{-11}$		3.75	210	10
		$4 \cdot 10^{-11}$	$1 \cdot 10^{-8}$	3.40	230	5
DCB	0.5	$1 \cdot 10^{-11}$	$2 \cdot 10^{-10}$	5.05	215	10
p,p'-DDD	17.5	$1 \cdot 10^{-10}$		4,50	202	10
o,p'-DDE	4.5	$5 \cdot 10^{-11}$		4.65	197	10
o,p'-DDT	4.5	$7 \cdot 10^{-11}$	$1 \cdot 10^{-9}$	4.65	201	10
p,p'-DDT	6.0	$5 \cdot 10^{-11}$	$1 \cdot 10^{-9}$	4.65	201	10
Heptachlor	2.5	$3 \cdot 10^{-11}$		4.00	196	10

chlorine pesticides and decachlorobiphenyl. The MDA_{UV} data are of the correct order of magnitude, varying from about 10^{-8} to 10^{-10} g for log $\epsilon_{max.}$ increasing from 3 to 5. With LC–ECD, the compounds considered all display MDA_{ECD} values in the range 10^{-10}–10^{-11} g. Naturally, the superiority of ECD over UV detection is at least partly determined by the log $\epsilon_{max.}$ values of the selected test solutes. In other words, the weakly UV-absorbing (especially at 230 nm) aldrin used in ref. 5 is an attractive model substance from the point of view of propagation of the ECD principle. Similar significant increases in sensitivity upon replacement of UV by electron-capture detection can be read, or at least estimated, from data on the analysis of several pesticides [5], and griseofulvin and fluocinolone acetonide [8], respectively. One should take into account, however, that UV detection was not always performed at the optimal wavelength. Several further examples which are recorded in ref. 10 will be discussed in Section 16.5.

The magnitude of the reduction of sensitivity of LC–ECD compared with GC–ECD is evident when the MDC of aldrin of $1.6 \cdot 10^{-15}$ g/ml in GC is compared [5] with the value of $3.2 \cdot 10^{-13}$ g/ml (calculated with respect to vapour-phase flow through the detector) in LC. This reduction is mainly due to the presence of electron-capturing species in the solvent vapour.

16.4. QUALITY AND SELECTION OF SOLVENTS

Solvent systems which so far have been employed in LC–ECD almost invariably are based on saturated hydrocarbons, and it has repeatedly been stated that, e.g., *n*-hexane, *n*-pentane and 2,2,4-trimethylpentane (isooctane) can be used as mobile phases in LC–ECD at column flow-rates of up to at least 10 ml/min. All workers stress the need to use high-quality solvents, that is, solvents that are dry and contain an extremely low level of electron-capturing impurities. In addition, oxygen must be scrupulously removed from the solvents (and the purge gas), as its presence causes an increase in the noise level and a reduction in the standing current. One should note that the solubility of oxygen increases with solvent polarity, which is of interest if polar modifiers have to be added to the hydrocarbon used as the mobile phase. Some workers [8,11] recommend removal of oxygen from solvents by passing oxygen-free nitrogen or argon at a rate of about 30 ml/min through the solvent by means of a capillary tube; others [9] prefer degassing and increasing the solvent temperature.

As for the quality of the saturated hydrocarbons, Willmott and Dolphin [5] and Chamberlain and Marlow [8] used a commercial special-grade *n*-hexane with a specified low concentration of electron-capturing impurities. Brinkman et al. [10] stated, however, that neither this product nor any of the other brands and types of *n*-hexane and isooctane tested, if used as received from the suppliers, is sufficiently pure for uninterrupted use of the LC–ECD system for a prolonged period of time. Even when using a number of bottles from the same batch, consistent results were not obtained. Two clean-up procedures were therefore studied, viz., treatment with a 10% sodium-in-lead alloy and propan-2-ol or, alternatively, with a 45% dispersion of finely divided sodium metal in solid paraffin. With the former procedure, variable results were obtained and the solvent loss often amounted to 30%. The latter treatment was highly successful with all types of *n*-hexane and isooctane tested; the solvent recovery was typically over 90%. The dramatic improvement in solvent quality when purifying, as a severe test, waste *n*-hexane from clean-up and extraction of polychlorobiphenyl-containing environmental samples is evident from Fig. 16.2. Typical LC–ECD conditions that could be maintained over a prolonged period of time after the treatment with sodium in paraffin were flow-rate of *n*-hexane 1.4 ml/min, d.c. $4 \cdot 10^{-9}$–$4.5 \cdot 10^{-9}$ A, attenuator $\times 64$ or $\times 128$ and peak-to-peak noise 3–4%.

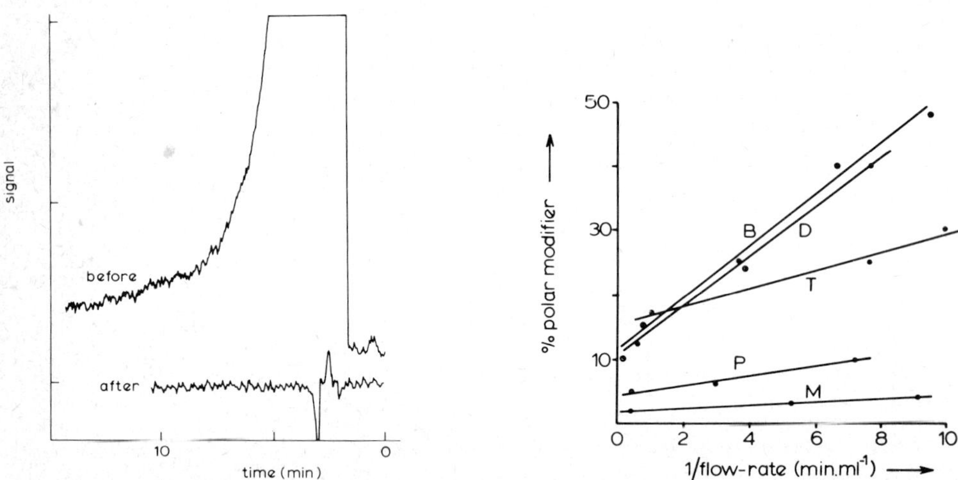

Fig. 16.2. LC–ECD chromatograms of waste *n*-hexane from clean-up and extraction of environmental samples before and after treatment with a dispersion of finely divided sodium metal in solid paraffin. System: silica gel/dry *n*-hexane. (Reproduced from ref. 10.)

Fig. 16.3. Variation of the concentration of polar modifier in *n*-hexane at which amplifier saturation occurs with reciprocal of flow-rate. B = benzene; D = dioxane; T = tetrahydrofuran; P = propan-2-ol; M = methanol. (After ref. 8.)

It should be added that the sodium-in-lead alloy treatment has been used successfully by other workers [9], who combined it with a clean-up step over an activated silica column, using nitrogen as purge gas, for the further removal of polar impurities.

In general, solvent mixtures more polar than *n*-hexane or isooctane are required for efficient separations in adsorption LC. Chamberlain and Marlow [8] therefore studied the possibility of using polar solvent mixtures in LC—ECD. It was found that solvents such as benzene, dioxane and tetrahydrofuran require dilution in saturated hydrocarbons to typically 10—20 vol.% to prevent saturation of the constant-current amplifier electronics; propan-2-ol and methanol should be diluted to a level of 2—5 vol.%. In addition, not unnaturally the concentration of a given polar solvent in *n*-hexane at which saturation occurs was found to vary inversely with the solvent flow-rate (Fig. 16.3). With acetonitrile, a 1% addition caused saturation even at a flow-rate of less than 0.4 ml/min. The authors rightly stated that the suitability of relatively polar solvent systems for use with LC—ECD greatly extends the applicability of this technique.

16.5. APPLICATIONS

Probably the most interesting application of LC—ECD was reported by Dolphin et al. [12], who developed a laboratory prototype of an automatic instrument for the analysis of organochlorine pesticides in milk. Limiting ourselves to the chromatographic part of the so-called milk pesticide monitor, we can summarize the procedure as follows (cf., Fig. 16.4).

A 100-μl volume of a milk extract is injected on to a 50 × 2.1 mm I.D. pre-column filled with 5-μm silica. During T_1 (80 s), the early eluting pesticides hexachlorobenzene, *p,p'*-DDE, *p,p'*-DDT, *p,p'*-TDE and α-hexachlorocyclohexane pass into the 150 × 3.1 mm I.D. analytical column, which is packed with 10-μm silica. In period T_2 (680 s) the late eluting γ- and β-hexachlorocyclohexane, heptachlorepoxide and dieldrin, which have been separated on the pre-column, are led straight to the ECD. Next, the flow is switched again and in period T_3 (225 s) the early eluting compounds are separated on the

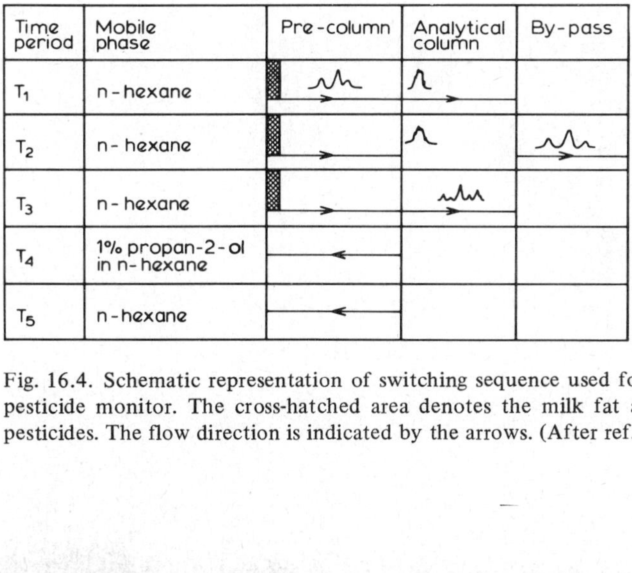

Time period	Mobile phase	Pre-column	Analytical column	By-pass
T_1	n-hexane			
T_2	n-hexane			
T_3	n-hexane			
T_4	1% propan-2-ol in n-hexane			
T_5	n-hexane			

Fig. 16.4. Schematic representation of switching sequence used for analysis of pesticides in the milk pesticide monitor. The cross-hatched area denotes the milk fat and the peak patterns indicate the pesticides. The flow direction is indicated by the arrows. (After ref. 12.)

analytical column and detected. The primary pump is now stopped, the backflush pump is started and *n*-hexane–propan-2-ol (99:1) is pumped through the pre-column in the backflush mode to remove adsorbed fat, which is determined with a refractive index detector (T_4, 180 s). Finally, the pre-column is regenerated with *n*-hexane (T_5, 600 s). MDC values of the pesticides were calculated to be about $2 \cdot 10^{-9}$–$5 \cdot 10^{-9}$ g/ml (see also ref. 13). This can be shown to correspond to MDC values in the milk fat of less than 0.1 mg/kg, which is below the tolerance levels suggested by the U.S. Food and Drug Administration for pesticide residues in dairy products.

Dolphin and Willmott [14] also used LC–ECD to compare the retention of α- and β-hexachlorocyclohexane, which have negligible UV responses, with that of polychlorinated dibenzo-*p*-dioxins. On a 250 × 4.6 mm I.D. column packed with 10-μm alumina, and with dry *n*-hexane as the mobile phase, both isomers were found to be clearly separated from each other and from all dibenzodioxins, which eluted much more rapidly.

The applicability of LC–ECD to environmental analysis has been illustrated [5] by using an extract of the final outfall from a plant treating domestic sewage. The peaks were not identified, but the contamination level of the major components in the original sample was calculated to be ca. 30 ppt, based on a calibration graph for aldrin. This is a realistic value for organochlorine pesticides in sewage effluents.

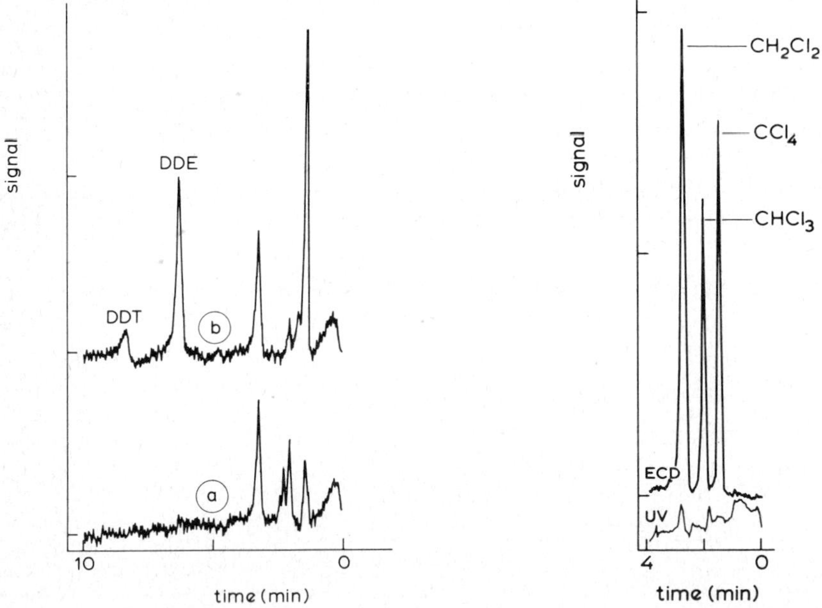

Fig. 16.5. LC–ECD chromatograms of *n*-hexane extract of (a) non-spiked river-water sample and (b) sample spiked with 0.03 ppb of *p,p'*-DDT and 0.10 ppb of *o,p'*-DDE. System: silica gel/dry *n*-hexane; flow-rate, 1.4 ml/min. ECD: d.c., $4 \cdot 10^{-9}$ A; attenuation, ×64. (Reproduced from ref. 10.)

Fig. 16.6. Comparison of ECD and UV recordings of a liquid chromatogram of di-, tri- and tetrachloromethane run in the system silica gel/dry *n*-hexane (flow-rate 1.4 ml/min). ECD: d.c., $3.5 \cdot 10^{-9}$ A; attenuation, ×128; UV, detection at 230 nm; 0.02 a.u.f.s. (Reproduced from ref. 10.)

Brinkman et al. [10] extracted a 500-ml sample of river water spiked with low levels of p,p'-DDT (0.03 ppb) and o,p'-DDE (0.10 ppb). Results obtained after a single extraction with 25 ml of n-hexane, which was evaporated to 5 ml, and a 50-μl injection on to a silica/dry n-hexane LC system are shown in Fig. 16.5. Obviously, with a slightly more efficient experimental procedure, detection limits of 1–10 ppt can be achieved. As another application, the authors separated the virtually non-UV-absorbing di-, tri- and tetrachloromethanes in the same LC system (Fig. 16.6); MDA_{ECD} values were about 10^{-10} g for all three compounds.

Brinkman et al. also ran chromatograms of a large series of polychlorobiphenyl (PCB; Aroclor 1221–1268) and polychloronaphthalene (PCN; Halowax 1031–1051) mixtures in the system silica/dry n-hexane, and compared the peak profiles found with electron-capture and UV detection. The electron-capture and UV detector patterns are similar for all PCN mixtures and also for the more highly chlorinated PCBs (Aroclor 1242–1268); such similarity is, however, completely lost with the low-chlorinated Aroclor 1221. Using the prominent peaks present in the various patterns, the gain in sensitivity that occurs on substituting the electron-capture for the UV detector was shown to increase from 5–10-fold for Aroclor 1221 to 30-fold for Aroclor 1268; in the Halowax series the results were between 0.5 for Halowax 1031 and about 15 for Halowax 1051.

Demeter and Heyndrickx [9] have reported on the use of LC–ECD in the determination of α- and β-endosulfan in serum, urine, liver and kidney samples. After repeated extraction with benzene and, if required, an additional clean-up in an n-hexane–acetonitrile partitioning system, LC is carried out with silica as stationary phase and n-hexane containing 0.05% of propan-2-ol as mobile phase, using a pre-column/analytical column set-up with a backflush step. Although the authors did not report detection limits, from their experimental data these may be estimated to be about $2 \cdot 10^{-10}$ g for α-endosulfan and 10^{-9} g for β-endosulfan.

A variety of possible applications of LC–ECD is presented in refs. 8 and 11 and relevant details are summarized in Table 16.2. The detection limits of griseofulvin and

TABLE 16.2

SUMMARY OF LC–ECD APPLICATIONS REPORTED IN REFS. 8 AND 11

LC system: silica/n-hexane or isooctane with added polar modifier if required.

Sample	Polar modifier	Remarks
Milk extract	–	
Chlorinated pesticides and PCBs	–	DDTs, DDE, aldrin, hexachlorobenzene, heptachlor, lindane, dieldrin, endrin, Aroclor 1260
Organophosphorus pesticides	5% tetrahydrofuran	Parathion, malathion
Substituted carbamates	1% propan-2-ol	CIPC, Swep
Sewage outfall, water and effluent extracts	–	
Antibiotics	5% propan-2-ol	Griseofulvin
Steroids	5% propan-2-ol	Fluocinolone acetonide
Nitro compounds	1% propan-2-ol	2,4- and 2,6-dinitrotoluene

TABLE 16.3

MDC VALUES FOR NITROAROMATICS

Compound	MDC (g/ml)
Nitrobenzene	$8 \cdot 10^{-10}$
2-Nitrotoluene	$3 \cdot 10^{-9}$
3-Nitrotoluene	$3 \cdot 10^{-9}$
4-Nitrotoluene	$5 \cdot 10^{-9}$
2,4-Dinitrotoluene	$9 \cdot 10^{-10}$
2,6-Dinitrotoluene	$8 \cdot 10^{-10}$

fluocinolone acetonide are better than $3 \cdot 10^{-10}$ and $1.5 \cdot 10^{-10}$ g/ml, respectively. Data on the MDC values of a number of nitroaromatics are recorded in Table 16.3; in ref. 7, MDA values for compounds such as nitrobenzene and 2,6-dinitrotoluene are stated to be about 1 ng. In a recent review on the analysis of explosives by LC, Krull and Camp [15] have emphasized that ECD is uniquely suited for the trace analysis of organic nitro compounds and, thus, in the detection of many commonly used explosives.

The wire-transport LC–ECD system has been applied [3] to the analysis of the iodomethyldimethylsilyl ethers of androsterone, etiocholanolone, dehydroepiandrosterone and testosterone using an alumina/cyclohexane–acetone (99:1) separation system. Detection limits were estimated to be about 10^{-8} g/ml for all compounds.

16.6. CONCLUSION

Two main reasons may be quoted to explain why ECD as yet has not emerged as a popular mode of detection in column LC. First, most of the LC studies made to-day are carried out using reversed-phase systems, i.e., partly aqueous, rather polar mobile phases which are particularly ill-suited for (direct) monitoring in an ECD. Secondly, LC–ECD – even if used for adsorption LC, i.e., with relatively apolar solvent mixtures – is a much more difficult detection mode than LC–UV. Here the main problem apparently is that the overall performance of LC–ECD is highly dependent on the nature and quality of the solvents used as the mobile phase. No doubt these factors are responsible for the fact that the only commercially availablc LC–ECD system has recently been removed from the market [16].

In spite of the above serious problems and disadvantages of LC–ECD, some positive aspects shold also be noted. The construction of an LC–ECD system, either along the lines of the Pye Unicam apparatus, or along those recently published by Krull and Bushee [7], is relatively simple. Also, in adsorption LC, solvent purification and storage no longer constitute a major problem with saturated hydrocarbons. It has also been shown that the addition of fairly large proportions (10–20%) of several polar modifiers to those solvents is allowable, and that even a few per cent of lower alcohols can be tolerated. Further, the reduction in sensitivity of LC–ECD compared with GC–ECD has been shown to be modest and has been estimated to be from 10- to 100-fold. The drawback of higher MDA values in LC is, moreover, often largely offset in practice owing to the larger injection volumes in LC than in GC (typically 100 vs. 1 μl). Lastly, one should take into account that with miniaturization, and therefore lower flow-rates, the full advantage of higher sensitivity in LC–ECD due to lower flow-rates can be enjoyed.

In summary, at present LC–ECD cannot be used for chromatographic systems employing polar mobile phases. Here, a breakthrough can be expected only if, for example, an efficiently working wire- or belt-transport system is constructed or, alternatively, if an extraction reactor is inserted between the outlet of the LC column and the ECD inlet in order to extract the trace constituents of interest from the polar LC effluent into an apolar solvent compatible with the ECD. Such reactors are currently being used for, e.g., post-column derivatization [17] and in LC–mass spectrometry work [18]. In adsorption LC, the ECD can be recommended as a detector that can be used with a wide range of mobile phase compositions and that is suited to the detection of, e.g., organochlorine pesticides, PCBs, nitroaromatics and chlorinated aliphatic hydrocarbons. The sensitivity of the ECD is excellent, MDA values for favourable compounds being in the 10^{-10}–10^{-11} g range; the selectivity of the detection principle will often enable one to simplify clean-up procedures and thereby reduce the time of analysis. However, the small number of workers who have so far explored LC–ECD certainly have not been able to optimize the system itself or to investigate its full potential. In other words, the field remains one which calls for continued active research and improvements.

REFERENCES

1 R.P.W. Scott, *Liquid Chromatography Detectors*, Elsevier, Amsterdam, 1977.
2 L.R. Snyder and J.J. Kirkland, *Introduction to Modern Liquid Chromatography*, 2nd ed., Wiley-Interscience, New York, 1979, Ch. 4.
3 R.J. Maggs, *Column*, 2(4) (1968) 5–7.
4 G. Nota and R. Palombari, *J. Chromatogr.*, 62 (1971) 153–155.
5 F.W. Willmott and R.J. Dolphin, *J. Chromatogr. Sci.*, 12 (1974) 695–700.
6 *LC–EC Chromatograph*, Pye Unicam, Cambridge, 1976.
7 I.S. Krull and D. Bushee, *Anal. Lett.*, 13 (A14) (1980) 1277–1282.
8 A.T. Chamberlain and J.S. Marlow, *J. Chromatogr. Sci.*, 15 (1977) 29–31.
9 J. Demeter and A. Heyndrickx, *10th Int. Symp. Chromatography and Electrophoresis, Venice, 19–20 June 1979.*
10 U.A.Th. Brinkman, P.M. Onel and G. de Vries, *J. Chromatogr.*, 171 (1979) 424–430.
11 *Liquid Chromatography Application No. 13*, Pye Unicam, Cambridge, 1976.
12 R.J. Dolphin, F.W. Willmott, A.D. Mills and L.P.J. Hoogeveen, *J. Chromatogr.*, 122 (1976) 259–268.
13 R.J. Dolphin and F.W. Willmott, *J. Chromatogr. Sci.*, 14 (1976) 584–588.
14 R.J. Dolphin and F.W. Willmott, *J. Chromatogr.*, 149 (1978) 161–168.
15 I.S. Krull and M.J. Camp, *Amer. Lab.*, 12 (May 1980) 63–76.
16 Philips/Pye Unicam, personal communication.
17 U.A.Th. Brinkman, J.F. Lawrence, C. van Buuren and R.W. Frei, in A. Frigerio and M. McCamish (Editors), *Recent Developments in Chromatography and Electrophoresis 10*, Elsevier, Amsterdam, 1980, pp. 247–259.
18 B.L. Karger, D.P. Kirby, P. Vousos, R.L. Foltz and B. Hidy, *4th Int. Symp. Column Liquid Chromatography, Boston, MA, 7–10 May 1979.*

SUBJECT INDEX